Advanced Catalysts Based on Metal-organic Frameworks (Part 1)

Edited by

Junkuo Gao
School of Materials Science and Engineering
Zhejiang Sci-Tech University
China

&

Reza Abazari
Department of Chemistry, Faculty of Science
University of Maragheh
Iran

Advanced Catalysts Based on Metal-organic Frameworks (Part 1)

Editors: Junkuo Gao and Reza Abazari

ISBN (Online): 978-981-5079-48-7

ISBN (Print): 978-981-5079-49-4

ISBN (Paperback): 978-981-5079-50-0

need for a court order if at any point you breach any terms of this License Agreement. In no event will any delay or failure by Bentham Science Publishers in enforcing your compliance with this License Agreement constitute a waiver of any of its rights.

3. You acknowledge that you have read this License Agreement, and agree to be bound by its terms and conditions. To the extent that any other terms and conditions presented on any website of Bentham Science Publishers conflict with, or are inconsistent with, the terms and conditions set out in this License Agreement, you acknowledge that the terms and conditions set out in this License Agreement shall prevail.

Bentham Science Publishers Pte. Ltd.
80 Robinson Road #02-00
Singapore 068898
Singapore
Email: subscriptions@benthamscience.net

BENTHAM SCIENCE

CONTENTS

PREFACE

The energy crisis combined with environmental pollution has been recognized as one of the most serious global concerns. Therefore, huge attempts have been devoted in recent decades to resolving these challenges by introducing more advanced materials with higher efficiency. Specially-designed catalysts are among the first candidates proposed to combat environmental contaminations. Metal-organic frameworks (MOFs) are a novel class of crystalline porous substances possessing high surface area, excellent structural properties, chemical adjustability, and stability which can be used in different applications including catalysis. The significant benefits of MOFs in the field of catalysis can be found in their tunable porosity, uniformly-distributed active sites, and excellent porosity offering accessible active sites through their open channels facilitating the mass transport and diffusion, and finally, their robust structure ensuring recyclability. Thus, MOFs can efficiently offer the positive features of both homogeneous and heterogeneous catalysts, at high reaction efficiency and recyclability. Catalytic properties of MOFs can be further enhanced when used as precursors and/or templates. Their combination with other compounds as a hybrid can further improve their catalytic activities due to synergic effects. The huge attempts to reinforce and modify this class of materials raise our hopes in the bright future of MOFs in the field of catalysis.

Due to unique features of inorganic-organic hybrid compositions, MOFs, compared with traditional porous materials, have a variety of advantages: (1) Good crystallinity. MOFs with highly ordered structures, could be precisely and intuitively analyzed by X-ray diffraction technology, which is helpful to determine structure-property relationships; (2) Good designability and facile functionalization. Applying to crystal engineering, MOFs can not only be predesigned with expected structures (topologies) and functions, even the coordination diversity of metal ions and organic ligands, but also easily operated by post synthetic methods; (3) High porosity. MOFs are highly porous materials with a large specific surface area (exceeding to 7000 m2 g-1), and more importantly, the size, shape and composition of pores can be well tuned by a lot of methods, which is beneficial for host-guest studies; (4) Flexibility. Due to the flexibility of coordination bond and organic linkers, most of the MOFs are somewhat flexible, which endows MOFs with peculiar properties like dynamic irritating response to external conditions (temperature, pressure, humidity, *etc.*), and these features make MOFs more intelligent in applications.

The MOF-based materials offer favorable catalytic performance owing to their unique structural attributes and subsequent modulation. Their range of chemical functionalities and porosities facilitate to adsorb/activate other substrates/CO_2 leading to facile CO_2 conversion. The rise in the number of MOF based catalytic materials with improved performance has opened a new avenue for CO_2 capture and conversion. One of the most important attributes an MOF has is its chemical tunability along with its interactions with other substrates. MOFs as photocatalysts are benefited from a large surface area, suitable band-gap, the ideal structure for charge transfer, and high photo-corrosion resistance.

Junkuo Gao
School of Materials Science and Engineering
Sci-Tech University
China

Reza Abazari
Department of Chemistry, Faculty of Science
University of Maragheh
Iran

List of Contributors

Aleksander Ejsmont Department of Chemical Technology, Faculty of Chemistry, Adam Mickiewicz University in Poznań, Uniwersytetu Poznańskiego 8, 61-614 Poznań, Poland

Agata Chełmińska Department of Chemical Technology, Faculty of Chemistry, Adam Mickiewicz University in Poznań, Uniwersytetu Poznańskiego 8, 61-614 Poznań, Poland

Anita Kubiak Department of Chemical Technology, Faculty of Chemistry, Adam Mickiewicz University in Poznań, Uniwersytetu Poznańskiego 8, 61-614 Poznań, Poland

Aleksandra Galarda Department of Chemical Technology, Faculty of Chemistry, Adam Mickiewicz University in Poznań, Uniwersytetu Poznańskiego 8, 61-614 Poznań, Poland

Anna Olejnik Department of Chemical Technology, Faculty of Chemistry, Adam Mickiewicz University in Poznań, Uniwersytetu Poznańskiego 8, 61-614 Poznań, Poland

Chunxian Guo Institute of Materials Science and Devices, School of Material Science and Engineering, Suzhou University of Science and Technology, Suzhou, Jiangsu, 215009, PR China
Jiangsu Laboratory for Biochemical Sensing and Biochip, Suzhou, Jiangsu, 215009, PR China
Collaborative Innovation Center of Water Treatment Technology & Material, Suzhou, Jiangsu, 215009, PR China

Chuan Qin School of Power and Mechanical Engineering, Hubei International Scientific and Technological Cooperation Base of Sustainable Resource and Energy, Hubei Province Key Laboratory of Accoutrement Technique in Fluid Machinery & Power Engineering, Wuhan University, Wuhan University, Wuhan 430072, China

Chunyan Zhou Department of Chemistry, Key Laboratory of the Ministry of Education for Advanced Catalysis Materials Zhejiang Normal University, Jinhua 321004, China

Dinesh De Department of Basic Science, Vishwavidyalaya Engineering College, Ambikapur, CSVTU-Bhilai, Chhatisgarh-497001, India

Fabiola Hernandez-García Universidad Autónoma del Estado de Hidalgo. Área Académica de Química. Carretera Pachuca-Tulancingo Km 4.5. C.P 42184. Mineral de la Reforma, Hidalgo, México

Giaan A. Álvarez-Romero Universidad Autónoma del Estado de Hidalgo. Área Académica de Química. Carretera Pachuca-Tulancingo Km 4.5. C.P 42184. Mineral de la Reforma, Hidalgo, México

Humberto Mendoza-Huizar Universidad Autónoma del Estado de Hidalgo. Área Académica de Química. Carretera Pachuca-Tulancingo Km 4.5. C.P 42184. Mineral de la Reforma, Hidalgo, México

Jingsha Li Institute of Materials Science and Devices, School of Material Science and Engineering, Suzhou University of Science and Technology, Suzhou, Jiangsu, 215009, PR China

Joanna Goscianska Department of Chemical Technology, Faculty of Chemistry, Adam Mickiewicz University in Poznań, Uniwersytetu Poznańskiego 8, 61-614 Poznań, Poland

Junkuo Gao Institute of Functional Porous Materials, School of Materials Science and Engineering, Zhejiang Sci-Tech University, Hangzhou 310018, China

J. Antonio Cruz-Navarro Universidad Autónoma del Estado de Hidalgo. Área Académica de Química. Carretera Pachuca-Tulancingo Km 4.5. C.P 42184. Mineral de la Reforma, Hidalgo, México

J. Ángel Cobos-Murcia Universidad Autónoma del Estado de Hidalgo. Área Académica de Química. Carretera Pachuca-Tulancingo Km 4.5. C.P 42184. Mineral de la Reforma, Hidalgo, México

Kayode Adesina Adegoke Department of Chemical Sciences, University of Johannesburg, South Africa

Martyna Kotula Department of Chemical Technology, Faculty of Chemistry, Adam Mickiewicz University in Poznań, Uniwersytetu Poznańskiego 8, 61-614 Poznań, Poland

Marcelina Kotschmarów Department of Chemical Technology, Faculty of Chemistry, Adam Mickiewicz University in Poznań, Uniwersytetu Poznańskiego 8, 61-614 Poznań, Poland

Mingyue Ding School of Power and Mechanical Engineering, Hubei International Scientific and Technological Cooperation Base of Sustainable Resource and Energy, Hubei Province Key Laboratory of Accoutrement Technique in Fluid Machinery & Power Engineering, Wuhan University, Wuhan University, Wuhan 430072, China

Mayank Guptac Department of Physics, National University of Singapore, 3 Science Drive 3, 117543, Singapore

Nobanathi Wendy Maxakato Department of Chemical Sciences, University of Johannesburg, South Africa

Ning Yuan Department of Chemical Engineering, School of Chemical and Environmental Engineering, China University of Mining and Technology, Beijing 100083, China

Nobanathi Wendy Maxakato Department of Chemical Sciences, University of Johannesburg, South Africa

Olugbenga Solomon Bello Department of Pure and Applied Chemistry, Ladoke Akintola University of Technology, P.M.B 4000, Ogbomoso, Oyo State, Nigeria
Department of Physical Sciences, Industrial Chemistry Programme, Landmark University, Omu-Aran, Nigeria

Olugbenga Solomon Bellob Department of Pure and Applied Chemistry, Ladoke Akintola University of Technology, P.M.B 4000, Ogbomoso, Oyo State, Nigeria
Department of Physical Sciences, Industrial Chemistry Programme, Landmark University, Omu-Aran, Nigeria

Reza Abazari Department of Chemistry, Faculty of Science, University of Maragheh, Iran

Rhoda Oyeladun Adegoke	Department of Pure and Applied Chemistry, Ladoke Akintola University of Technology, P.M.B 4000, Ogbomoso, Oyo State, Nigeria
Soheila Sanati	Institute of Functional Porous Materials, School of Materials Science and Engineering, Zhejiang Sci-Tech University, Hangzhou 310018, China
Simona M. Coman	Department of Organic Chemistry, Faculty of Chemistry, Biochemistry and Catalysis, University of Bucharest, Bd. Regina Elisabeta, 4-12, 030018 Bucharest, Romania
Solomon Oluwaseun Akinnawo	Department of Pure and Applied Chemistry, Ladoke Akintola University of Technology, P.M.B 4000, Ogbomoso, Oyo State, Nigeria Department of Chemical Sciences, Olusegun Agagu University of Science and Technology, P.M.B. 353, Okitipupa, Ondo State, Nigeria
Shuxuan Liu	Department of Chemistry, Key Laboratory of the Ministry of Education for Advanced Catalysis Materials Zhejiang Normal University, Jinhua 321004, China
Tuyuan Zhu	Department of Chemistry, Key Laboratory of the Ministry of Education for Advanced Catalysis Materials Zhejiang Normal University, Jinhua 321004, China
Vivekanand Sharma	Advanced Membranes and Porous Materials Center, Bldg. 4. Level 3, Office 3266-WS10, King Abdullah University of Science and Technology (KAUST), Thuwal 23955-6900, Kingdom of Saudi Arabia
Vera V. Khrizanforova	Arbuzov Institute of Organic and Physical Chemistry, FRC Kazan Scientific Center, Russian Academy of Sciences, Kazan, Russian Federation
Verónica Salazar-Pereda	Universidad Autónoma del Estado de Hidalgo. Área Académica de Química. Carretera Pachuca-Tulancingo Km 4.5. C.P 42184. Mineral de la Reforma, Hidalgo, México
Xinling Zhang	Department of Chemical Engineering, School of Chemical and Environmental Engineering, China University of Mining and Technology, Beijing 100083, China
Xuehui Gao	Department of Chemistry, Key Laboratory of the Ministry of Education for Advanced Catalysis Materials Zhejiang Normal University, Jinhua 321004, China
Yushan Wu	School of Power and Mechanical Engineering, Hubei International Scientific and Technological Cooperation Base of Sustainable Resource and Energy, Hubei Province Key Laboratory of Accoutrement Technique in Fluid Machinery & Power Engineering, Wuhan University, Wuhan University, Wuhan 430072, China
Yanfei Xu	School of Power and Mechanical Engineering, Hubei International Scientific and Technological Cooperation Base of Sustainable Resource and Energy, Hubei Province Key Laboratory of Accoutrement Technique in Fluid Machinery & Power Engineering, Wuhan University, Wuhan University, Wuhan 430072, China
Yingchong Huang	Department of Chemistry, Key Laboratory of the Ministry of Education for Advanced Catalysis Materials Zhejiang Normal University, Jinhua 321004, China

Yulia H. Budnikova Arbuzov Institute of Organic and Physical Chemistry, FRC Kazan Scientific Center, Russian Academy of Sciences, Kazan, Russian Federation

Zuo-Xi Li Institute of Materials Science and Devices, School of Material Science and Engineering, Suzhou University of Science and Technology, Suzhou, Jiangsu, 215009, PR China

Zhanguo Jiang Department of Chemistry, Key Laboratory of the Ministry of Education for Advanced Catalysis Materials Zhejiang Normal University, Jinhua 321004, China

Strategies, Synthesis, and Applications of Metal-Organic Framework Materials

Zuo-Xi Li[1,*] and **Chunxian Guo**[1,2,3,*]

[1] *Institute of Materials Science and Devices, School of Material Science and Engineering, Suzhou University of Science and Technology, Suzhou, Jiangsu, 215009, PR China*

[2] *Jiangsu Laboratory for Biochemical Sensing and Biochip, Suzhou, Jiangsu, 215009, PR China*

[3] *Collaborative Innovation Center of Water Treatment Technology & Material, Suzhou, Jiangsu, 215009, PR China*

Abstract: Metal-Organic Frameworks (MOFs), as one type of famous porous material with many advantages (good crystallinity, design ability, facile modification and flexibility), show a wide range of applications in gas adsorption and separation, ion exchange, fluorescent recognition, nonlinear optics, molecular magnets and ferroelectrics, heterogeneous catalysis, semiconductors, and so on. The research of MOFs span many disciplines, such as inorganic chemistry, organic chemistry, coordination chemistry, supramolecular chemistry, crystal engineering and materials science. The design, synthesis, and applications of MOFs have attracted tremendous attention in broad scientific areas. Therefore, it is worth releasing a professional publication to elucidate so many related issues. In this chapter, we start with the introduction of MOFs, including the definition, classification, concepts, terminologies, and some well-known research. Then we carefully summarize the design and synthesis of MOFs from three aspects of raw materials, synthetic methods, and design strategy, aiming to get the goal of controllable syntheses of MOFs. Following this, we report the developments and applications of MOF materials in adsorption and separation, organic catalysis, luminescence, and drug delivery. Finally, we briefly outline challenges and perspectives of MOF materials, and provide some promising research subjects in this area.

Keywords: Controllable Syntheses, Crystal Engineering, Long-range Ordered Pores, Properties.

* **Corresponding authors Zuo-Xi Li and Chunxian Guo:** Institute of Materials Science and Devices, School of Material Science and Engineering, Suzhou University of Science and Technology, Suzhou, Jiangsu, 215009, PR China, Jiangsu Laboratory for Biochemical Sensing and Biochip, Suzhou, Jiangsu, 215009, PR China & Collaborative Innovation Center of Water Treatment Technology & Material, Suzhou, Jiangsu, 215009, PR China; E-mail: cxguo@usts.edu.cn

Junkuo Gao & Reza Abazari (Eds.)

1. INTRODUCTION OF METAL-ORGANIC FRAMEWORKS

Metal-Organic Frameworks (MOFs), also known as porous coordination polymers (PCPs), are crystalline porous materials with periodic networks formed by the self-assembly of metal ions (or metal clusters) and organic ligands through coordination bonds. The concept of MOFs was proposed and firstly reported by the group of Yaghi in 1995 [1]. In their work, it is proved that MOFs are microporous framework materials, which are adjusted through selecting proper organic ligands and metal ions. Furthermore, MOFs can adsorb guest molecules and remain stable after the guest molecules are removed. With the quick development of MOF materials, other some similar terms were proposed from different perspectives as well, such as MILs (Materials of Institute Lavoisier Frameworks) by Férey's group [2], ZIFs (Zeolitic Imidazolate Frameworks) by Yaghi's group [3], MAF (Metal Azolate Frameworks) by Chen's group [4], PCPs (Porous Coordination Polymers) by Kitagawa's group [5] and PCN (Porous Coordination Networks) by Zhou's group [6]. In recent years, the bonding interactions in MOFs have not only referred to coordination bonds, but also included other interactions, such as hydrogen bonds, van der Waals force, π-π interactions between aromatic rings, *etc.*Due to the abundant interactions, the structures and functionalities of MOFs are becoming more and more diversified. In 2013, to classify coordination polymers (CPs), coordination networks (CNs) and MOFs, International Union of Pure and Applied Chemistry (IUPAC) published a set of terms and definitions [7]. According to the recommendations, MOFs are CNs with potential voids, where CPs refer to coordination compounds that extend through repeating coordination entities in one dimension (1D, including cross-links between two or more individual chains, loops or spiro-links), or coordination compounds that extend through repeating coordination entities in two or three dimensions (2D or 3D). That is to say, MOFs are a subset of CNs, also a branch of CPs.

Due to unique features of inorganic-organic hybrid compositions, MOFs, compared with traditional porous materials, have a variety of advantages: (1) Good crystallinity. MOFs with highly ordered structures, could be precisely and intuitively analyzed by X-ray diffraction technology, which is helpful to determine structure-property relationships; (2) Good designability and facile functionalization. Applying to crystal engineering, MOFs can not only be predesigned with expected structures (topologies) and functions, even the coordination diversity of metal ions and organic ligands, but also easily operated by post synthetic methods; (3) High porosity. MOFs are highly porous materials with a large specific surface area (exceeding to 7000 m^2 g^{-1}), and more importantly, the size, shape and composition of pores can be well tuned by a lot of methods, which is beneficial for host-guest studies; (4) Flexibility. Due to the flexibility of coordination bond and organic linkers, most of the MOFs are somewhat flexible, which endows MOFs with peculiar properties like dynamic irritating response to external conditions (temperature, pressure, humidity, *etc.*), and these features make MOFs more intelligent in applications.

Nowadays, as a new type of functional molecular material, the design and synthesis of MOFs with the desired structure and properties have become one of the frontier fields of coordination chemistry, supramolecular chemistry, crystal engineering and materials science. The research of MOFs span many disciplines and categories, such as inorganic chemistry, organic chemistry, coordination chemistry, material chemistry, and synthetic chemistry, which have shown broad applications in heterogeneous catalysis, molecular recognition, gas adsorption, ion exchange, molecular magnets, ferroelectric materials, fluorescent materials, nonlinear optical materials, and so on. In this chapter, we aim to introduce the synthesis methods, construction strategies and potential applications of MOFs, as well as some recent developments in this area.

2. SYNTHESIS OF MOFS

As a kind of coordination compounds, MOFs are composed of inorganic metal ions, organic ligands and guest molecules inside the frameworks. The synthesis process of MOFs is very similar with that of other coordination compounds, and the key for the synthesis of MOFs is the formation of coordination bonds between metal centers and coordination atoms from organic ligands. Compared with covalent bonds, the bond energy of coordination bonds is much smaller, and so most of the MOFs have a simple and mild synthesis condition. Due to great potential applications of MOFs, some of them have begun to be commercialized. Therefore, to meet requirements of rapid, controllable and large-scale production, new methods including microwave synthesis, ultrasonic synthesis, electrochemical method, mechanochemical method, spray drying and mobile chemical synthesis have been gradually developed, besides traditional methods.

2.1. Raw Materials

2.1.1. Meal Nodes

In the synthesis of MOFs, various central metal nodes provide empty orbitals for the formation of coordination bonds, which can be regarded as binders to anchor organic ligands. Most of the metal nodes have relatively definite coordination numbers and configurations, which are one key factor to determine the structures of final products. It is worth mentioning that metal nodes in MOFs are not only limited to simple mononuclear metal ions, but also polynuclear clusters with different sizes and spatial configurations. With the development of coordination chemistry, the types of metal ions or polynuclear clusters for the synthesis of MOFs have been expanded and updated.

1. Transition metal ions

The most commonly used central metal ions come from the transition elements of periodic table, including *d*, *ds* and *f* regions. Most *d* and *ds*-region metals in the 4th period often exhibit fixed coordination number and configuration, which could guide the arrangement of organic ligands, thus providing opportunities for designing MOFs with specific structures. In addition, several metals from the fifth period, are also commonly used in recent years, for example, the Zr^{IV} ion for the assembly of highly stable MOFs. Meanwhile, Lanthanide metals with large and variable coordination numbers are often applied for the construction of luminescent MOFs, due to their fluorescence properties.

2. Main group metal ions

The often-used main group metals to construct MOFs include alkali and alkaline earth metals of IA and IIA groups, respectively. These metals are easily obtained with non-toxicity and safety for human bodies and environments, which are suitable for applying MOFs into biomedicine areas. However, there is also large variability in coordination numbers, which brings troubles for the prediction of MOF structures. Additionally, the IIIA group metals, such as Al and In, are also widely utilized in the set-up of MOFs.

3. Metal clusters

Metal centers could be simplified as nodes in the topology analysis of MOFs. Except for single metal nodes, metal clusters formed by either *in-situ* synthesis or pre-synthesis could be viewed as special nodes as well, which has a fairly large proportion in MOF structures. Metal clusters, also known as secondary building units (SBUs), often have specific configuration and different number of directional coordination sites, which are easily replaced by organic ligands to make the targeted construction of MOFs much achievable. Fig. (**1**) shows some famous SBUs that are often used in the construction of MOFs [8].

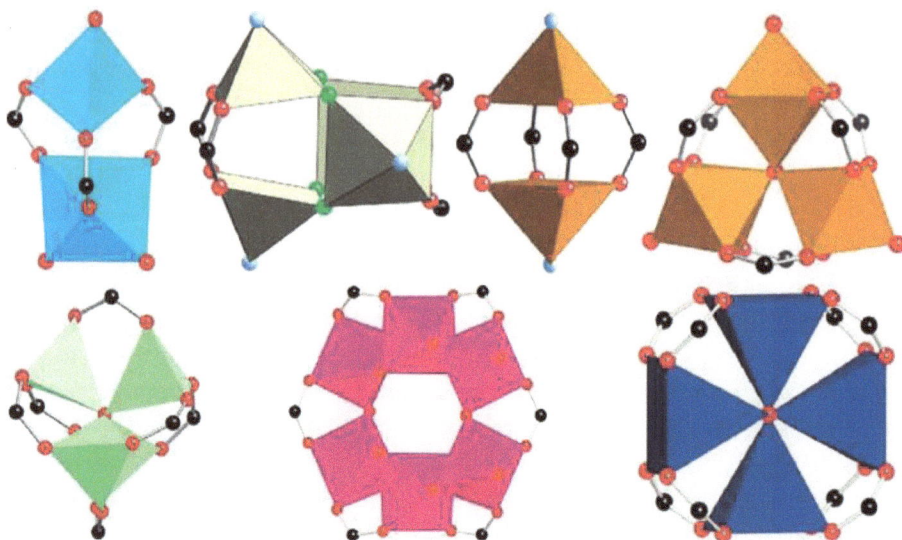

Fig. (1). Some classical SBUs in MOFs [8].

2.1.2. Organic Linkers

Organic ligands play a role of linkers for metal nodes. Theoretically, matters that could coordinate with metal ions are all potential ligands for the construction of MOFs, including organic and inorganic ligands. Among these used ligands, organic ligands with O, N and/or S as coordination atoms have an overall majority, in spite of some inorganic anions (OH⁻, F⁻, Cl⁻, SiF_6^{2-}, *etc.*), they may also act as linkers in the structures of MOFs (Fig. **2**).

1. Ligands containing O/S donors

These ligands are mainly multi-carboxylate ligands, which own plenty of coordination modes and provide various linking manners for the diversity of MOF structures. For example, MOF-5 was synthesized from PTA ligands, and HKUST-1 from TMA ligands. A small part of ligands contain coordination groups like -SH, -OH or mixed groups of -COOH and -SH/-OH, such as HHTP, HTTP and 2,5-dimercaptoterephthalic acid (dmpta).

2. Ligands containing N donors

These kinds of ligands are mainly organic compounds including nitrogenous heterocyclic rings, such as pyridine, azole, or other N-containing groups like -NH$_2$ and -CN. Besides coordination with metals of N-donor groups, the rest non-metal N sites could call for property studies.

3. Ligands containing both O and N donors

Through pre-designation, these ligands can be obtained with both O and N donor groups, which would not only enrich the structural topologies but also endow MOFs with more properties. For example, incorporated with -COOH and triazole, these ligands could provide abundant coordination modes to link metal centers, meanwhile act as guest binding sites inside the resulting architectures.

Fig. (2). Various ligands used for the construction of MOFs.

2.1.3. Other Compositions

Due to high porosity, the as-prepared MOF samples are always obtained with some guest matters in their pores. Firstly, because MOFs are synthesized in some solutions, their pores are usually filled with water or organic solvent molecules. Secondly, if the frameworks are ionic, there must be cations or anions inside the pores as counter charges for maintaining electroneutrality. Additionally, some guest-directed synthesis or post-synthetic methods may introduce some guest molecules as templates.

2.2. Synthesis Methods

2.2.1. Solution Methods (Stirring, Evaporation and Diffusion)

Solution method is the primitive synthesis of MOFs, offering several ways of operation.

(1) Stirring. A common process is as below: directly mix metal salts and organic ligands in specific solvents (such as water or organic solvent), and adjust pH values by deprotonation reagents if necessary; the reaction system is then stirred in an open environment below 100°C, and MOF products will be precipitated with the progress of reaction.

(2) Evaporation. This operation is suitable for organic ligands with high solubility. Similar to the stirring way, the reactants (metals and ligands) are dissolved in certain solvents (maybe heating), and then keep standing for some time. Perfect crystals will grow with supersaturation during the cooling and evaporating process.

(3) Diffusion. In this method, different solvents with less mutual solubility are often used to dissolve the reactants, and single crystals with high quality are easier to produce in the interface between the solvents. One specific operation is: after dissolving metal salts and organic ligands in different solvents, the solution A with smaller density is carefully spread over solution B with larger density; or put the reaction system C into the vapor atmosphere of solvent D that will slowly diffuse to adjust the solubility of reaction solution C. Another operating way is the gel diffusion method by using a U-tube as a reaction container, in which agar or gelatin is firstly added as gels, and then solutions of metal salts or organic ligands is added from different sides, respectively. The solutions in both sides will slowly diffuse to the gels, and single crystals are generated at the junction of two solutions.

Usually, the stirring synthesis has advantages of short reaction time and large yield, but it also faces problems such as low purity and poor crystallinity. While MOF crystals, prepared by the evaporation and diffusion methods, show features of large size and high quality (the reaction rate is always slow enough), but the resultant framework structures are often unstable and not attractive. Therefore, these time-killing methods are not suitable for the study of unknown MOFs.

2.2.2. Hydrothermal Method (Water and Organic Solvents)

Hydrothermal method was first used by geologists for the simulated study on natural mineralization, which was then widely applied for the synthesis of functional materials. For MOF synthesis, hydrothermal methods have been the most popular way in recent two decades. Generally, metal salts and organic ligands are directly mixed in a specific solvent (such as water or organic solvent) or mixed solvents, which were put into a well-closed high-pressure autoclave; then the reaction system is heated by a temperature controlling program, and coordination reactions will happen under the self-generated pressure of the solvent system (Fig. **3**) [9]. The reaction temperature is usually in the range of 80~200 °C, and many MOFs can be synthesized at around 150°C. Because solvents in the closed high-pressure vessel could reach supercritical states when temperatures rise to a certain extent, the insoluble reactants under normal conditions can be dissolved in supercritical liquid and thus the self-assemble reactions occur, which have an overwhelming superiority over many other methods.

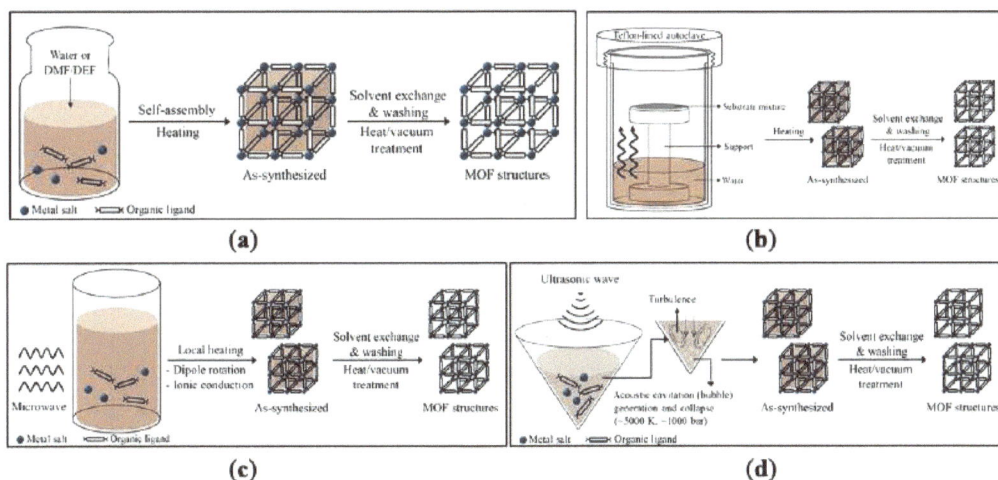

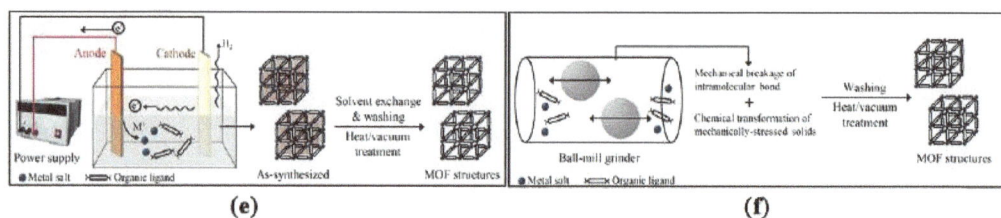

Fig. (3). Synthesis methods of MOFs [9]: **(a)** hydrothermal synthesis; **(b)** ionothermal synthesis **(c)** microwave-assisted synthesis; **(d)** sonochemical synthesis; **(e)** electrochemical synthesis; **(f)** mechanochemical synthesis.

There are also shortcomings of hydrothermal methods. 1) it is very easy to form mechanical mixtures of different compound crystals in the product, which are very difficult to separate. Recently, some strategies, such as solvent-assisted separation of mixed MOFs based on density disparity [10], seed-mediated synthesis of phase-pure MOFs [11] and purification of MOF mixtures through precisely modulating reaction conditions [12], have been proposed and proved to be very useful and effective for solving this problem; 2) Reaction vessels are opaque, so that the reacting progress could not be real-time monitored. To overcome this difficulty, some other reaction containers such as hard glass tubes were used instead of high-pressure autoclave, making it easier for observation. In addition, when the solvent with high boiling point and low reaction temperature are used, the glass bottle with cover can also be used as the reaction container. However, it should be noted that the reaction temperature and time must be well controlled in the experimental process, lest container cracking or boiling dry.

2.2.3. Ionthermal Method

Ionothermal synthesis, conducted in ionic liquids, is a novel method used in the synthesis of MOFs. The synthesis process is similar to hydrothermal methods, except using ionic liquid instead of conventional solvents.

2.2.4. Sublimation Method

When raw materials have high vapor pressure below the decomposition temperature, it is possible to prepare single crystals of MOFs by a sublimation method. Heat the raw materials to sublimate, make the reaction happen in gas phase or gas-solid multiphases, and MOF products will crystallize in the cool part of reactor. MOF crystals obtained by this method are usually of high purity, but there are only a few raw materials meeting the sublimation requirements, so this method has not been popularized at present.

2.2.5. Solid-State Reaction Method

Solvent-free methods, especially high-temperature solid-state reactions, have been widely used in the synthesis of various inorganic materials. For MOF synthesis, it only needs to mix raw materials in a certain proportion within a mortar and the reaction will be finished by grinding the mixture for a while. This reaction process not only needs no solvent, but also has a very flexible scale and is very easy for mass production, which is beneficial to environmental protection and cost reduction. However, these solid-state reactions are always of low completion and there may be some raw materials leaving in the products. However, this method can be improved by heating the grinded mixture at a certain temperature. For example, high purity MAF-4 (ZIF-8) can be produced by heating ZnO and 2-methylimidazole to 180°C for 3h with the yield of almost 100% and no by-product except for water vapor, which can be used in adsorption measurements without any pretreatment and has a better performance than that of samples obtained by solvothermal methods [13].

2.2.6. Microwave-Assisted Synthesis

Microwave-assisted synthesis, which uses microwave as a heating mode, can be considered as an improvement in energy sources of hydrothermal methods. The main advantages of microwave-assisted hydrothermal synthesis of MOFs are quick reaction, energy saving, pure phase, high yield and uniform crystal size. However, this method is usually difficult to grow large-sized single crystals which can be used in single-crystal diffractometers. In some cases, by exploring and optimizing reaction conditions, for example, using continuous and multi-step microwave heating to raise the temperature, it is possible to obtain single crystals with a relatively large size.

2.2.7. Sonochemical Synthesis

Although not often used, the sonochemical synthesis has proved that crystallization time can be much shortened and crystal particle sizes can be controlled by adjusting nucleation rates in the crystallization process. Raw materials are mixed and placed in a trumpet-shaped Pyrex reactor, and treated by ultrasonic waves. During the process of ultrasonic degradation, the so-called cavitation effect occurs that a large number of bubbles will appear in the reaction system. When these bubbles are broken, it could locally generate high temperature (about 5000 K) and high pressure in the reactor within a short time, rapidly accelerating the MOF crystallization during the rapid rise and fall process of temperature. However, pore structures of MOFs are diverse, leading to impurity of as-synthesized products.

2.2.8. Electrochemical Synthesis

The target of electrochemical synthesis is preparing MOFs through the direct reaction of organic ligands in the electrolyte with metal ions continuously supplied by the anodic dissolution in the electrolytic cell. To prevent metal deposition on the cathode, ionic solvents are usually selected. This method can realize the continuous synthesis of MOFs and presents unparalleled advantages over others. 1) The influence of anion on the synthesis system is avoided, due to metal source from anodic dissolution instead of metal salts. 2) The reaction can be controlled autonomously through the turn-on/off of power. 3) The reaction process can be finished in a short time, which can also be real-time recorded by electrochemical workstation so that people can make a better perception of the reaction mechanism. However, it should be noted that as-prepared samples by electrochemical synthesis may have some differences in properties. For example, the electrochemically synthesized MIL-53 and NH_2-MIL-53 samples exhibit suppressed flexibility compared to their solvothermally synthesized samples [14].

3. DESIGN STRATEGIES

3.1. Reticular Chemistry

The reticular chemistry strategy was initially proposed by O'Keeffe and Yaghi groups [15]. In brief, the strategy is based on topological analysis to assemble judiciously selected rigid building blocks into predesigned ordered networks, which are held together by strong bonding (Fig. **4**). Firstly, all networks constructed from fixed nodes and linkers can be defined. Then, the framework structures are simulatively constructed by actual SBU and organic ligands instead of fixed nodes and linkers. Furthermore, the most possible network is determined by calculating the energy of different MOF structures, and based on this, specific experimental routes are later designed. Reticular chemistry has yielded many targeted materials with predetermined structures, compositions and properties, which is not only suitable for the synthesis of MOFs, but also significant for the design and preparation of other long-range ordered structures, such as covalent organic frameworks (COFs) and hydrogen-bonded organic frameworks (HOFs) [16].

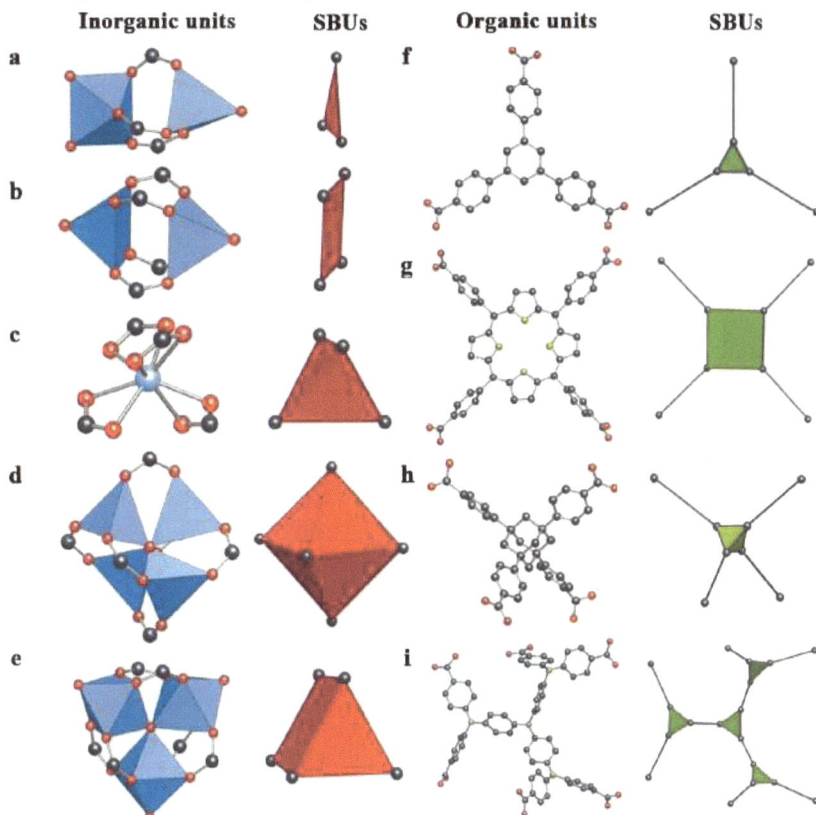

Fig. (4). Examples of reticular chemistry strategy [15].

3.2. Natural Mineral Structure Simulation

Natural minerals with long-range ordered structures are original modes of artificial synthesis. For MOFs, it is of great possibility to obtain coordination frameworks similar to that of natural minerals, by selecting central metals and ligands with specific coordination configuration to simulate structural motifs of minerals. Zeolite molecular sieves are the most imitative minerals, due to their structural unit of TO_4 with regular tetrahedron configuration and the T-O-T angle of 145°. Therefore, through employing metal nodes with four-connected tetrahedron configuration and ligands with certain coordination angles (around 145°), zeolite-like coordination frameworks may be prepared when minimizing the energy (Fig. **5**). For example, using Zn, Co and other metals to combine with imidazole ligands that just have 145° bridging angle, Chen and Yaghi' groups have successively synthesized a series of MOFs with zeolite structures, *i.e.* Zeolitic Imidazolate Frameworks (ZIFs).

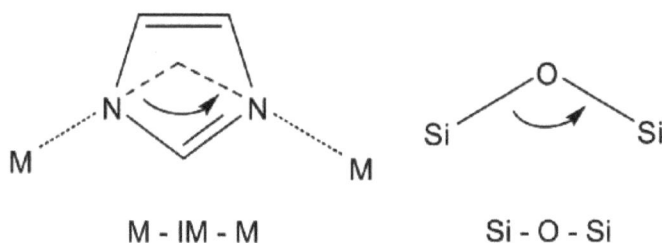

Fig. (5). The bridging angles in ZIFs (**1**) and zeolites (**2**) [3].

3.3. Stepwise Assembly

Generally, this strategy can be described in two steps. Coordination fragments with fixed coordination patterns and intrinsic properties are pre-synthesized aforehand, which are called molecular building motifs. Then, the synthesis of MOFs is further realized by connecting the pre-synthesized coordination motifs with other metals or ligands. There are mainly two advantages for stepwise assembly synthesis. On one hand, it greatly improved the predictability and regulation of final structures. On the other hand, the unique features of molecular building motifs are also brought into final frameworks.

3.4. Synthesis of Homologues with the Same Frameworks

This method is suitable for the cases of MOFs with fixed structures that can be specifically prepared under certain reaction conditions. One or several raw materials in the reaction system can be finely adjusted, so as to accurately obtain related homologues with the same overall structures but different properties. The implementation method includes changing the present metal into other kinds of metal ions with similar coordination modes and charge numbers, and replacing organic ligands with homologous molecules with different functional groups. The most successful examples of such strategies are the preparation of IRMOF-5 and IRMOF-74 series materials. In 1999, the famous MOF-5 was synthesized from terephthalic acid (TPA) by Yaghi and coworkers with pore size of 12.9 Å [17]. Then, in 2002, while they used $Zn_4O(CO_2)_6$ as the fixed SBU and expanded TPA-like organic ligands with functional groups, a series of IRMOF (isoreticular MOF) homologues of good thermal stability and high specific surface area were successfully constructed with pore sizes span from 3.8 Å to 28.8 Å (Fig. **6**) [18]. Moreover, according to the structure of Mg-MOF-74, Yaghi's group have successfully extended the pore size of this series of IRMOFs from 14 Å to 98 Å by gradually lengthening these linear dicarboxylic ligands [19]. Adding different homologues into the original system to prepare solid-solution MOFs can also be regarded as an extension of this strategy. In 2010, Yaghi's team further fabricated a series of multivariate MOF-5 samples by mixing various analogues of PTA ligand with different functional groups into one framework (Fig. **7**) [20].

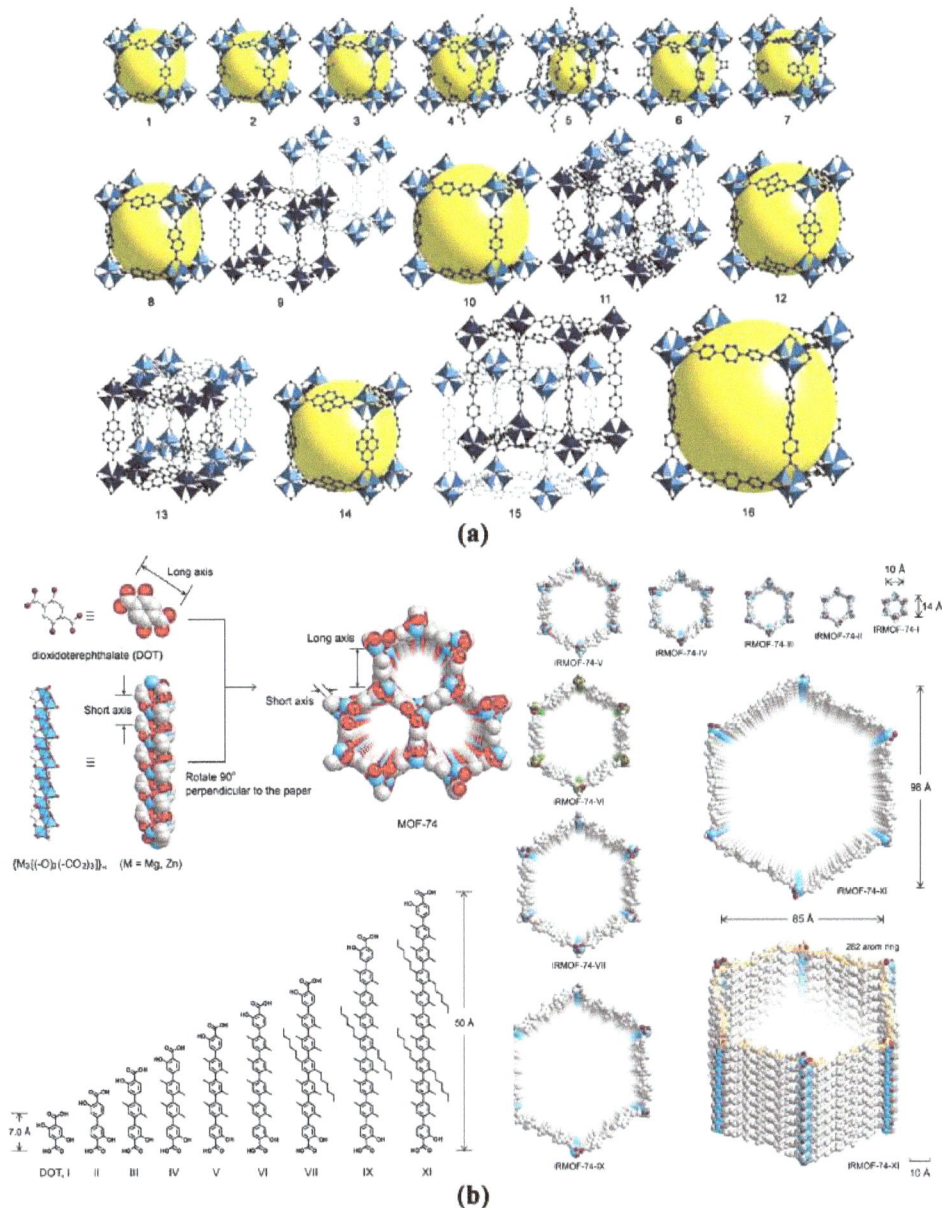

(a)

(b)

Fig. (6). Structures of the IRMOFs [18, 19].

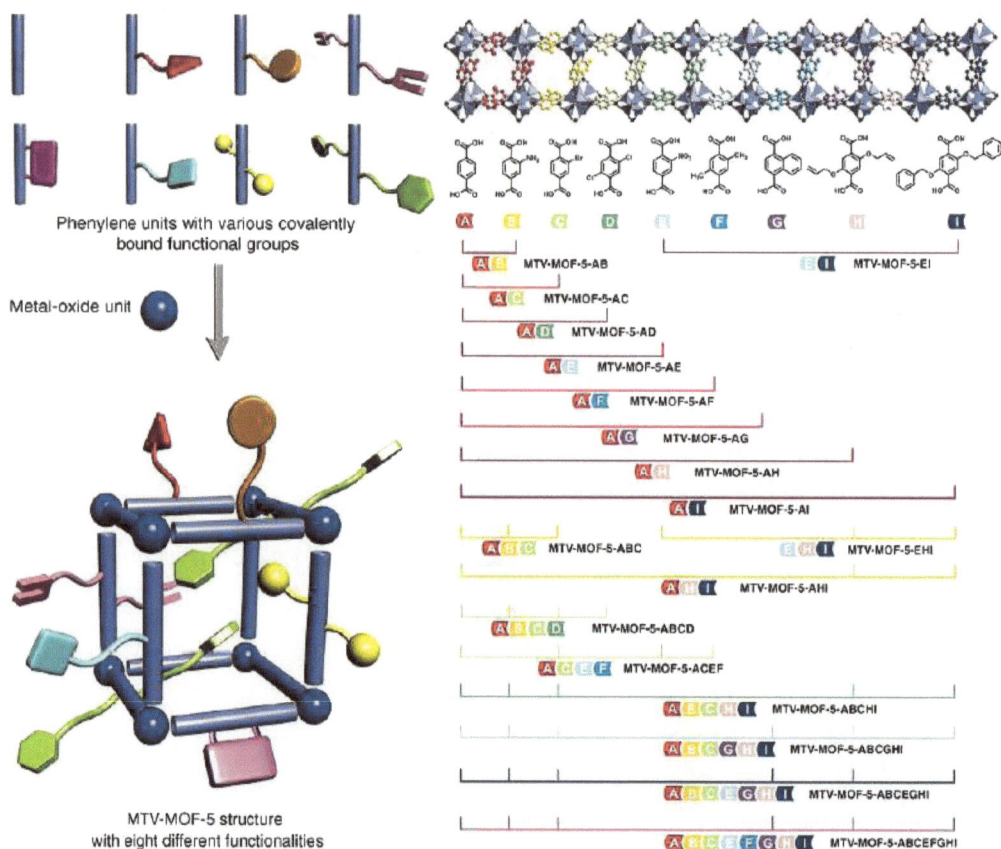

Phenylene units with various covalently bound functional groups

Metal-oxide unit

MTV-MOF-5 structure with eight different functionalities

Fig. (7). Construction of the multivariate MOF-5 samples [20].

The successful synthesis of coordination pillared-layer (CPL) structures by Kitagawa' group is also a classic example of this strategy [21], which provides a very popular method due to good predictability of pillared-layer frameworks and adjustability of pore structures (Fig. **8**) [22]. This kind of structures are generally constructed by mixing carboxylic acids and bipyridyl ligands, in which stably 2D layers could be obtained by the connection of carboxylic acids with metal ions in certain conditions, and bipyridyl ligands as pillars further connect these 2D layers into 3D frameworks by substituting coordinated solvent molecules in these 2D layers. As a result, the pore sizes, shapes and properties of pillared-layer MOFs can be well adjusted by replacing different "pillars".

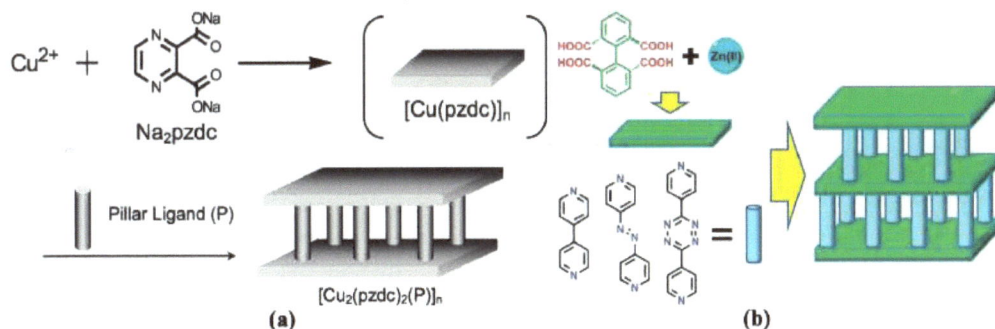

Fig. (8). Examples of CPL structures [21, 22].

3.5. Postsynthetic Modification Strategy

The ultimate goal of MOF research is practical applications. Therefore, it is a priority to achieve excellent performance through structural design and control. However, due to the complexity of assembly, MOFs cannot be designed to reach the level that any necessary functional groups can be directly introduced through the reactions within raw materials. Therefore, the method of postsynthetic modification (PSM) was developed (Fig. **9**) [23]. PSM allows for the introduction of diverse chemical functional groups into pre-existing frameworks [24]. Active sites can be decorated in these frameworks by various possible chemical reactions (electrophilic bromination [25], "Click" chemistry [26], amide couplings [27], imine condensation [28], reduction [29], *etc.*) without changing the original structure of MOFs. A common process of PSM is to introduce active components into original frameworks by the reactions with specific functional groups in MOFs under mild conditions, and directly replacing the central metals or organic ligands by changing related properties of original frameworks through the redox activity of central metals.

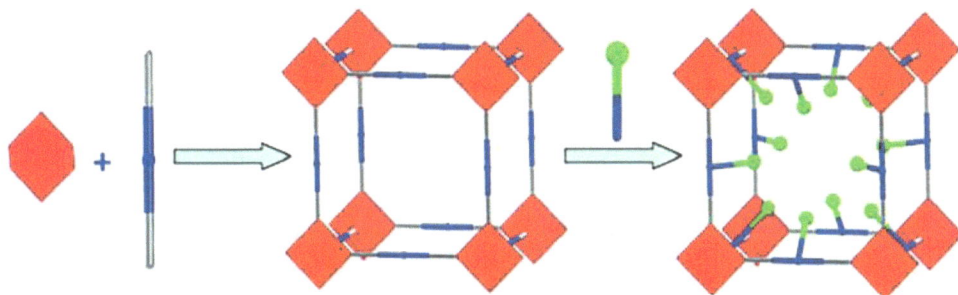

Fig. (9). A general scheme illustrating the PSM concept of porous MOFs [23].

Recently, Mandal *et al.* have presented a summary of PSM, which are divided into four main aspects [30], *i.e.* metal-based, ligand-based, metal & ligand-based and guest-based modification. For example, the structures and properties of MOFs can be well controlled through the single-crystal-to-single-crystal conversion, which are realized by UV irradiation [31], heating [32], ligand exchange [33, 34] or adsorption/desorption of guest species [35]. The surface features of MOFs can also be modified by PSM techniques to increase structural stability as well as inducing desired properties (Fig. **10**).

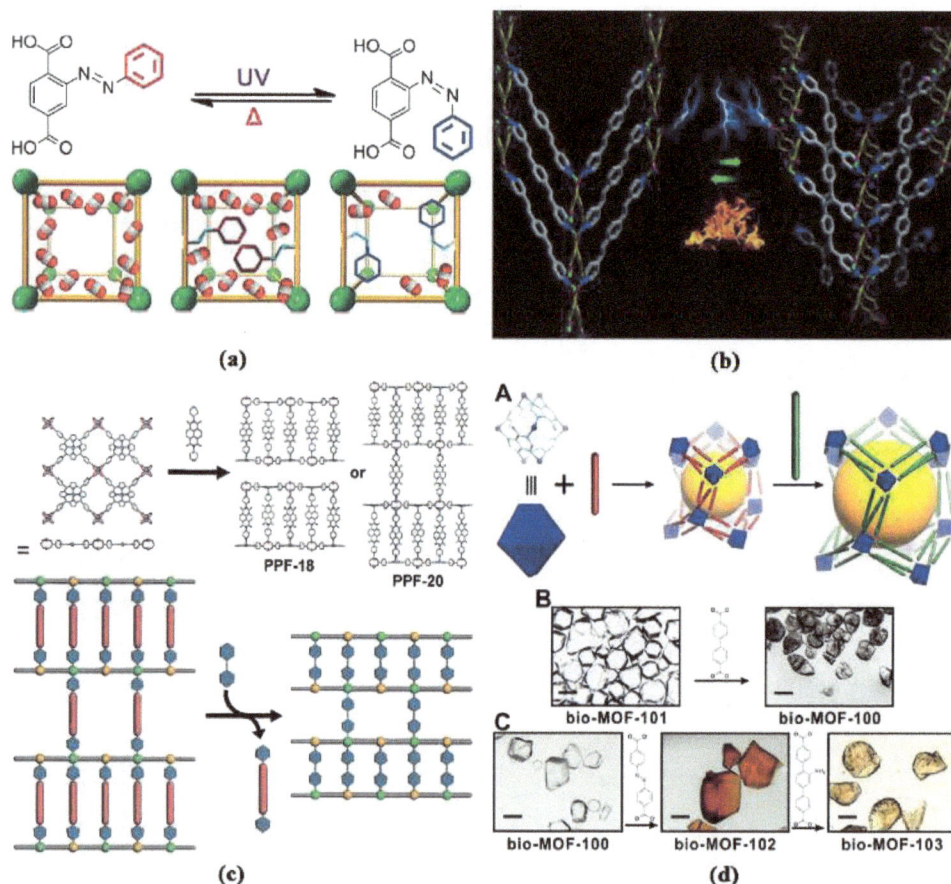

Fig. (10). Single-crystal-to-single-crystal conversion realized by UV irradiation **(a)** [31], heating **(b)** [32] and ligand exchange **(c, d)** [33, 34].

4. APPLICATIONS

4.1. Adsorption and Separation

Due to regular pore structures, adjustable pore sizes, large specific surface area, MOFs were preliminarily explored for potential applications in adsorption and separation. We would describe this section from following three aspects, according to adsorbed guests.

4.1.1. Gas Storage and Separation

Hydrogen storage. In 1999, Yaghi *et al*. reported for the first time hydrogen storage properties of MOF-5 with CaB_6 topology constructed from the reaction of terephthalic acid and zinc nitrate[17]. Because of its structural stability, high porosity and excellent hydrogen adsorption performance under 77 K, it has attracted extensive attention and led to a hot topic of MOF researches. Further studies revealed that MOF-5 can adsorb up to 4.5 wt% of hydrogen at 77 K and 20 atm, and 1.0wt% at 300K and 2MPa [36]. Long *et al*. have improved the synthetic route of MOF-5 and the as-synthesized samples under restrict anhydrous and anaerobic conditions present excess hydrogen uptake of 76 mg g^{-1} at 77 K and 40 bar, and the total adsorption uptake reached to 130 mg g^{-1} at 170 bar [37]. In 2010, Yaghi and co-workers reported MOF-210 with a very large specific surface area (6240 m^2 g^{-1}) and the excess and total hydrogen adsorption capacity of MOF-210 was as high as 86 mg g^{-1} (7.9 wt%) and 176 mg g^{-1} (15 wt%) at 77 K and 60 bar, respectively (Fig. **11**) [38]. MOF-210 has also an uptake of 2.7 wt% for hydrogen at room temperature and 80 bar. Meanwhile, Hupp *et al*. have reported the synthesis of NU-100, which was pre-designed by theoretical simulation [39]. The BET specific surface area of NU-100 reaches 6143 m^2 g^{-1}, which shows a high excess hydrogen adsorption capacity of 99.5 mg g^{-1} at 77 K and 56 bar and total adsorption capacity of 164 mg g^{-1} at 70 bar. Over the past twenty years, hundreds of thousands of MOF materials have been used for the study of hydrogen storage. Although MOFs have good physical adsorption of hydrogen and show fast adsorption-desorption kinetics, the hydrogen storage capacity is relatively low near room temperature, which has seriously hindered the practical application of MOFs in hydrogen storage.

Methane storage. CH_4 is the main component of natural gas. As one kind of cleaner, cheaper and more widely distributed fuel, it has considerable environmental, economic and political advantages compared with oil. However, the bulk density of natural gas is very low under ambient temperature and pressure. Therefore, it is still a great challenge for the storage of natural gas with high density, especially for mobile applications with small space (such as natural gas vehicles). Common storage methods, such as compressed natural gas technology (CNG), have very harsh storage conditions (200-300 bar), resulting in unnecessary energy waste. Absorbed natural gas (ANG) by porous materials has been a promising technology, because it is expected to achieve CH_4 storage capacity equivalent to CNG at room temperature and a relatively low pressure. The key of this field is the development of efficient methane adsorbent. However, the adsorption capacity of traditional porous materials is relatively low, and theoretical simulations show that the transport capacity of almost all traditional adsorbents will not be higher than 196~206 V (STP)/V.

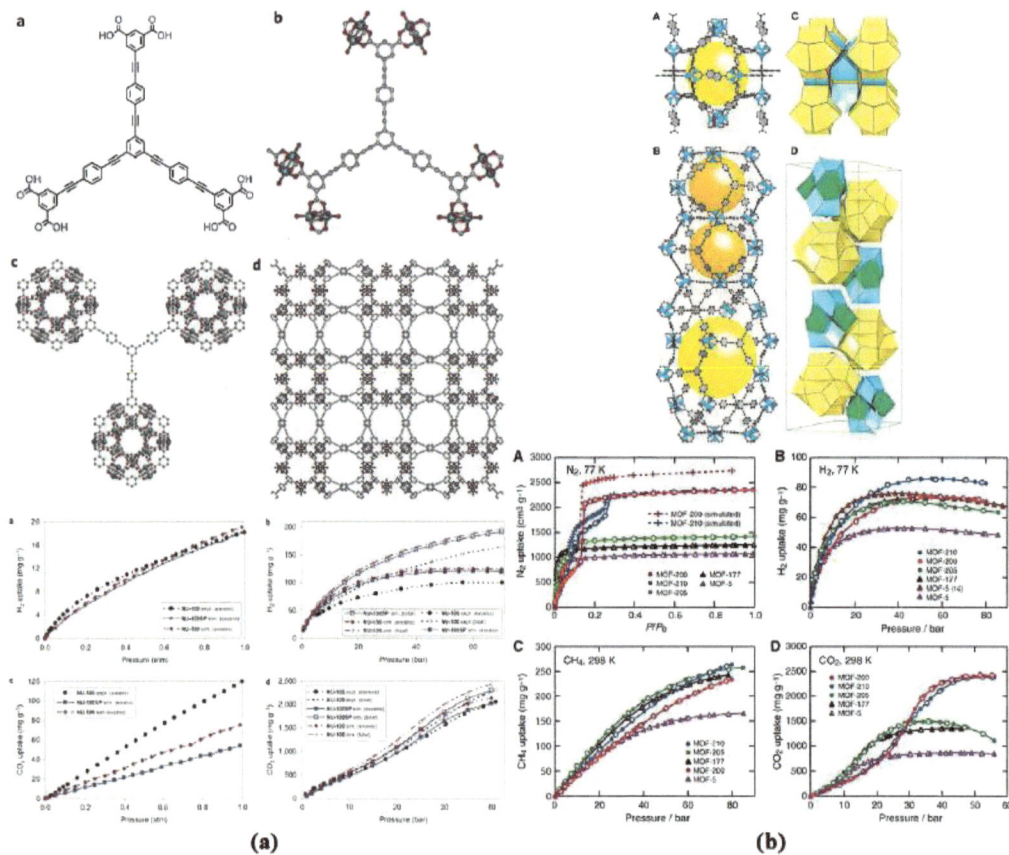

Fig. (11). Structures and hydrogen storage of NU-100 **(a)** and MOF-210 **(b)** [38, 39].

As an excellent absorbent, MOFs have naturally become an ideal candidate for CH_4 storage. Compared with H_2 storage, high capacity of CH_4 adsorption by MOFs under room temperature could be much easier to be realized for practical application, due to moderate interactions between MOFs and CH_4. Recently, the CH_4 uptake capacity of some well-designed MOFs have reached the application level, especially these MOFs developed by Yaghi have been applied in natural gas vehicles produced by BASF and Ford in 2013, which proved the great application prospect of MOFs in the future. New energy vehicles have been one of the most important applications of MOFs that is the closest to the public, just as Yaghi said "to a large extent, the problem of methane storage in motor vehicles has been solved". Depending on actual requirements of practical applications, such adsorbents not only have high adsorption capacity, but also present high available working capacity: it should be able to absorb a large amount of CH_4 at storage pressure (30-60 bar) while release most of the CH_4 at delivery pressure (5.8 bar, working pressure of internal combustion engine using natural gas as fuel).

In 1997, for the first time, Kitagawa *et al.* reported that MOFs can be used as potential CH_4 adsorbents [40]. In 2002, Yaghi *et al.* synthesized a series of IRMOFs with 3D cubic networks and systematically studied their CH_4 adsorption performance [19]. Among them, IRMOF-6 has the largest adsorption capacity, reaching 155 V (STP)/V at 298 K and 36 bar. Since then, a large number of researchers have realized potential value of MOFs in CH_4 storage, and made a breakthrough in 2008-2009. For example, PCN-14 synthesized by Zhou *et al.* and NiMOF-74 synthesized by Wu *et al.*, they both show ultra-high CH_4 capacity [6, 41, 42], reaching 230 V (STP)/V and 200 V (STP)/V at room temperature and 35 bar, respectively, exceeding the target of 180 V (STP)/V set by U.S. Department of Energy (DOE) at that time. Meanwhile, molecular simulation technology has become a powerful tool to study the adsorption behavior of MOFs [43, 44], which provides a reasonable theoretical basis for the systematic design and performance research of MOFs. These results and subsequent studies have opened up a new chapter for MOFs as excellent CH_4 adsorption materials [45 - 48].

In 2012, in order to continue to promote the research in the field of ANG, DOE launched a new program "Methane Opportunities for Vehicular Energy (MOVE) Program", and reset the goal as following: at room temperature and 35 bar, the CH_4 adsorption capacity of adsorbent should reach 0.5 g (CH_4)/g (sorbent) by weight and 263 V (STP)/V by volume. If 25% compression loss is taken into account, the required volume capacity should reach 330 V (STP)/V.

To improve the MOFs gravimetric CH_4 capacity, a viable strategy is seeking or developing MOFs with high porosity. In 2015, Alezi *et al.* reported an aluminum MOF (Al-soc-MOF-1) showing exceptionally high pore volume (2.3 cm^3 g^{-1}) and BET specific surface area (5585 m^2 g^{-1}). Sorption curves revealed that this MOF exhibited the highest total gravimetric CH_4 uptake so far, that is ~580 cm^3 (STP)/g^{-1} (0.42 g/g) at 298 K and 65 bar, achieving 83% of DOE gravimetric target (Fig. **12**) [49].

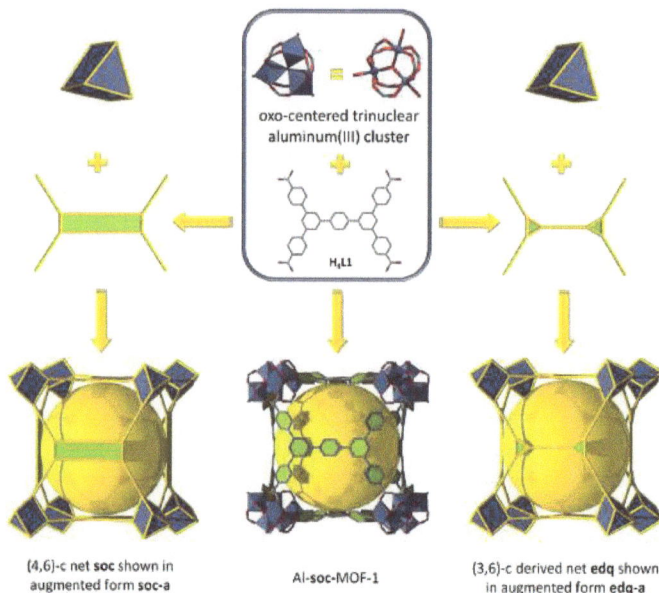

Fig. (12). (a) Crystal structure and topological analysis of Al-soc-MOF-1; **(b)** Single-component CH_4 adsorption isotherms for Al-soc-MOF-1 at different temperatures; **(c)** Comparison of the CH_4 volumetric working capacities (5–80 and 5–65 bar) for Al-soc-MOF-1 with USTA-76, HKUST-1, Ni-MOF-74 and PCN-14 [49] at different temperatures (258, 273, and 298 K).

HKUST-1 is one of the MOF materials that has been widely studied and many groups have investigated its high pressure CH_4 storage. However, the reported adsorption data were not completely consistent, which might be caused by the differences of synthetic and activated methods. In 2013, Peng *et al.* studied the effect of MOF shaping and densification on CH_4 adsorption [46]. Using the theoretical crystal density (0.883 g cm^{-3}), HKUST-1 showed a maximum CH_4 uptake amount up to 270 cm^{-3} (STP) cm^{-3} at 65 bar, so HKUST-1 was identified as the only MOF material that can achieve the volumetric DOE target. According to the linear fitting, they also estimated that to reach the gravimetric target, a hypothetical MOF might have pore volume of 3.2 cm^3 g^{-1} and specific surface area of 7500 m^2 g^{-1}. However, if HKUST-1 was densified experimentally, the volumetric adsorption capacity became 180 cm^{-3} (STP) cm^{-3}, which is 35% lower than that of the theoretical maximum value. This could be attributed to the partial mechanical collapse of internal pore structure. One main challenge for the practical applications of MOFs in industry is indeed the densification and pelletization of MOFs, since the powder MOFs produced by conventional synthesis methods display very low packing density, generally three to four times loss compared to the theoretical crystal density. To solve this problem, Tian *et al.* reported the improved synthesis of $_{mono}$HKUST-1 by using a sol–gel method (Fig. 13) [50]. After densification, the methane storage capacity can reach 259 cm^3/cm^3, which is the first CH_4 adsorbent meeting the goal of US DOE, and is more than 50% higher than any of previously reported materials. Considering the early accurate calculation model of CH_4 storage, the researchers deduced that this material might have reached the physical limit of CH_4 storage capacity at ambient temperature. Nanoindentation test on the $_{mono}$HKUST-1 sample showed strong mechanical properties, and the hardness is at least 130% higher than that of traditional MOFs. This research is an important step for the application of high volume and dense MOFs in energy storage industry.

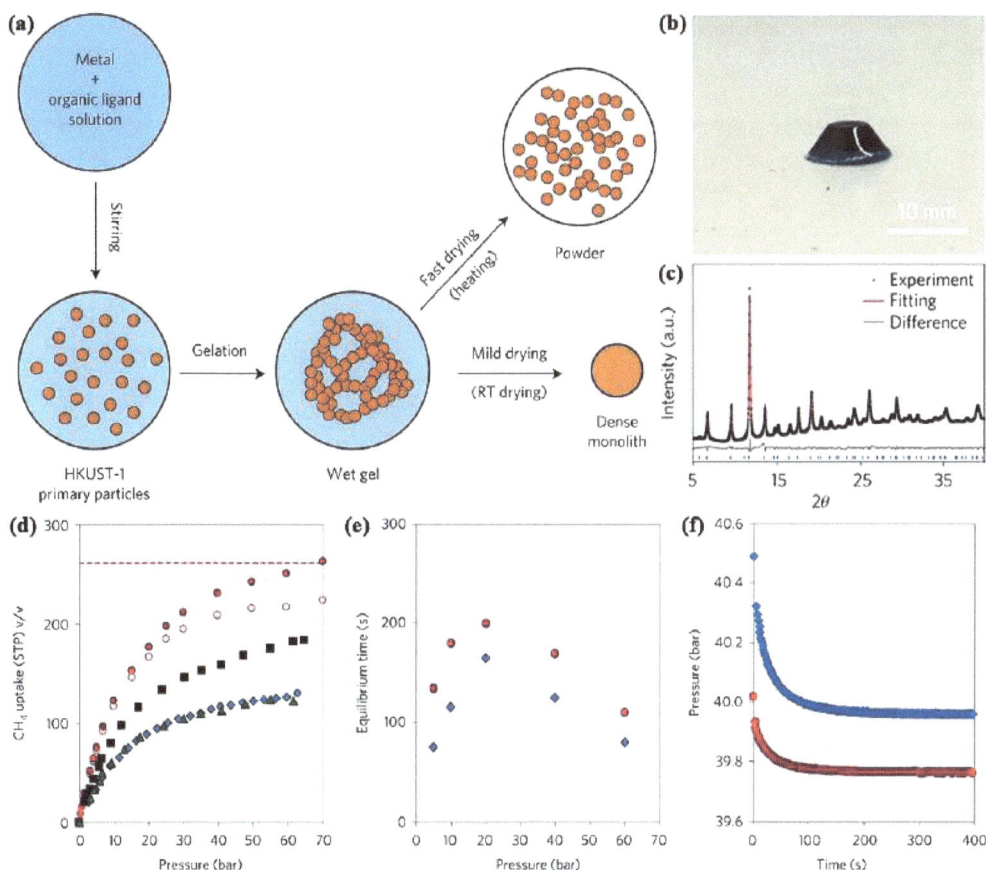

Fig. (13). (a) The synthesis follows a sol–gel process of $_{mono}$HKUST-1; **(b)** Optical image of the $_{mono}$HKUST-1; **(c)** Comparison of PXRD patterns of $_{mono}$HKUST-1; **(d)** Comparison of absolute volumetric CH_4 adsorption isotherms at 298 K on $_{mono}$HKUST-1 (red filled circles), excess volumetric uptake on $_{mono}$HKUST-1 (red open circles), HKUST-1 pellets under hand packing (blue diamonds), HKUST-1 pellets packed under 27.6 MPa (black squares), HKUST-1 pellets under 68.9 MPa (green triangles) and the DOE target of 263 V (STP)/V (red dashed line); **(e)** Equilibrium time of CH_4 adsorption at 298 K as a function of equilibrium pressure; **(f)** Decay of pressure over time for $_{mono}$HKUST-1 (blue diamonds) and $_{powd}$HKUST-1 (red circles) [50] at 40 bar.

Chen's group have found that the introduction of Lewis basic N atoms could greatly enhance the CH_4 storage ability of MOFs (Fig. **14**). In 2016, they synthesized a new MOF (UTSA-76a) using an organic linker with pyrimidine groups, which exhibited very high volumetric CH_4 uptake exceeding to that of NOTT-101 with the same structure, up to ~260 cm^3 (STP) cm^{-3} at 298 K and 65 bar, and record CH_4 capacity of ~200 cm^3 (STP) cm^{-3} (5 ~ 65 bar) at that time [51]. As revealed by computational studies and neutron scattering experiments, the central "dynamic" pyrimidine groups within the framework of UTSA-76a could adjust their orientation to optimize the methane packing at high pressure, which lead to such exceptionally high working capacity of this MOF. Subsequently, through the incorporation of different groups (*e.g.* pyridine, pyridazine and pyrimidine) into the organic linkers, this research group synthesized a series of NOTT-101-like MOFs that contain Lewis basic N sites [52]. Because these immobilized functional groups are not beneficial to the storage capacity at 5 bar but could significantly enhance the methane storage capacities at high pressure of 65 bar, the total volumetric CH_4 storage capacity at 298 K and 65 bar can be promoted from 237 cm^3 (STP) cm^{-3} (NOTT-101a) to ~249–257 cm^3 (STP) cm^{-3} (the functionalized MOFs), which also exhibit impressively high working capacity of ~188–197 cm^3 (STP) cm^{-3}. The authors also mentioned that there might exists an up-limit on the CH_4 storage capacity of MOF materials at room temperature and 65 bar, due to the materials genome studies [53] and their rough empirical formulas.

Although some MOFs illustrate excellent CH_4 uptakes, the enthalpy of CH_4 adsorption is relatively low for MOF materials, due to the weak interaction between adsorbed CH_4 molecules and frameworks. Since MOFs can be easily designed with large specific surface area and pore volume, the researchers have mainly focused on how to improve the binding affinity of CH_4, and the most common method is to introduce open metal sites (OMSs) inside the frameworks. However, it is very hard to improve the concentration of OMSs, even not considering the undesired adsorbent weight increase. Lin *et al.* reported the realization of very high CH_4 uptake in a new OMS-free MOF (MAF-38) with a novel pore structure packed by large quasi-cuboctahedral cages and small octahedral cages [54]. MAF-38 with BET specific surface area of 2022 cm^2 g^{-1} showed a novel adsorption mechanism, which presented exceptional CH_4 uptakes of 263 cm^3 (STP) cm^{-3} at 298 K and 65 bar with adsorption enthalpies of 21.6 kJ mol^{-1} that is as high as OMS-MOFs, even beyond Ni-MOF-74 (21.4 kJ mol^{-1}). The working capacities for MAF-38 in the range of 5 ~ 35, 65 and 80 bar are 150, 187 and 197 cm^3 (STP) cm^{-3}, respectively. Computational simulations revealed that there are suitable sizes/shapes and organic binding sites in the hierarchical pore structure that consist of single-wall nanocages, which could enforce not only host–CH_4 and CH_4–CH_4 interactions but also dense packing of

CH_4 molecules (that is 0.481 g cm^{-3} even exceeding the value of liquid CH_4 (0.423 g cm^{-3})).

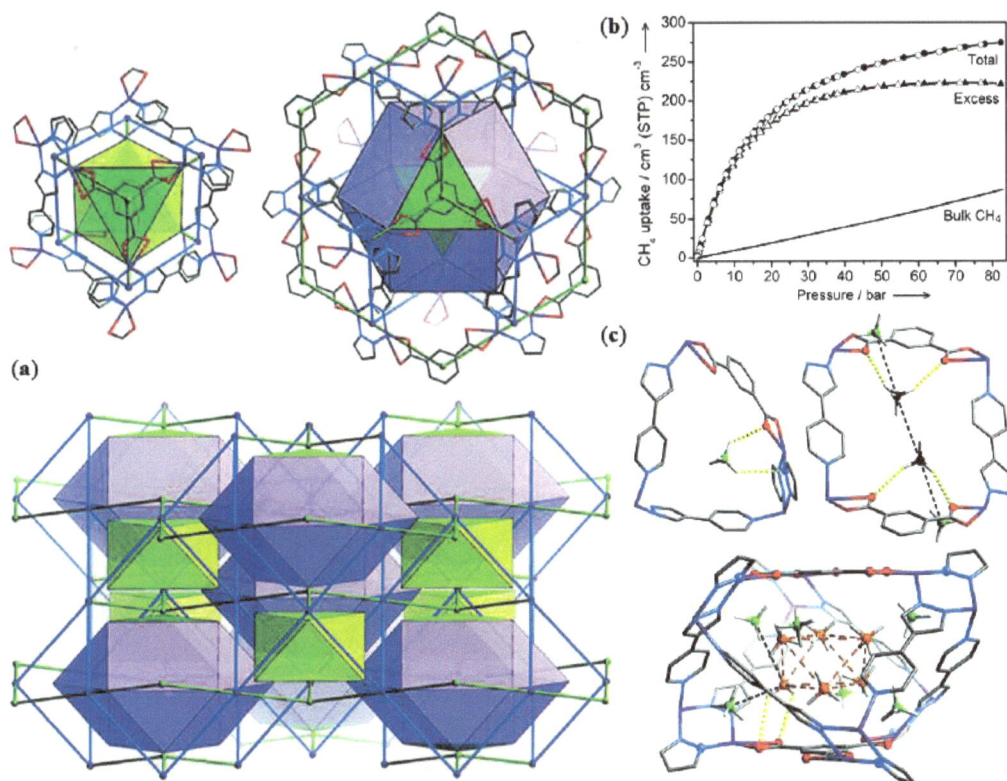

Fig. (14). UTSA-76a: **(a)** structure and topology; **(b)** total and excess CH_4 adsorption (solid) and desorption (open) isotherms at 298 K; **(c)** The primary (green), secondary (black), and ternary (orange) CH_4 adsorption sites, as well as the unique supramolecular CH_4 dimer and hexamer, as revealed by computational simulation [54].

For practical application, it is crucial to improve the working capacity of MOFs. Employing flexible MOFs might be a promising way to realize this target. This is because that flexible MOFs with S-shaped or stepwise adsorption isotherms usually exhibit well-known "gate-opening" behavior making a non-porous structure expand to a porous framework when the gas pressure reaching a certain threshold. If the gate-opening pressure can be well controlled around 5.8 bar, it will obviously maximize the CH_4 working capacity in the range after 5.8 bar, which might realize the lossless delivery of CH_4. In 2015, Mason *et al.* made the attempt in this way [55]. They synthesized a flexible MOF (Co(bdp)) by using 1,4-benzenedipyrazolate ligand (Fig. **15**). *In situ* powder X-ray diffraction experiments demonstrated the reversible structural phase transition of Co(bdp), while in its CH_4 adsorption isotherm at 298 K, a sharp step was observed at 16 bar. At 298 K, the working capacities of Co(bdp) reached 155 cm^3 (STP) cm^{-3} at 35 bar and 197 cm^3 (STP) cm^{-3} at 65 bars, which are among the highest values under these conditions. In 2016, researchers of this group further controlled the pressure of CH_4-induced framework expansion systematically by ligand functionalization, providing tools for the optimization of phase-change MOFs towards better CH_4 storage performance [56].

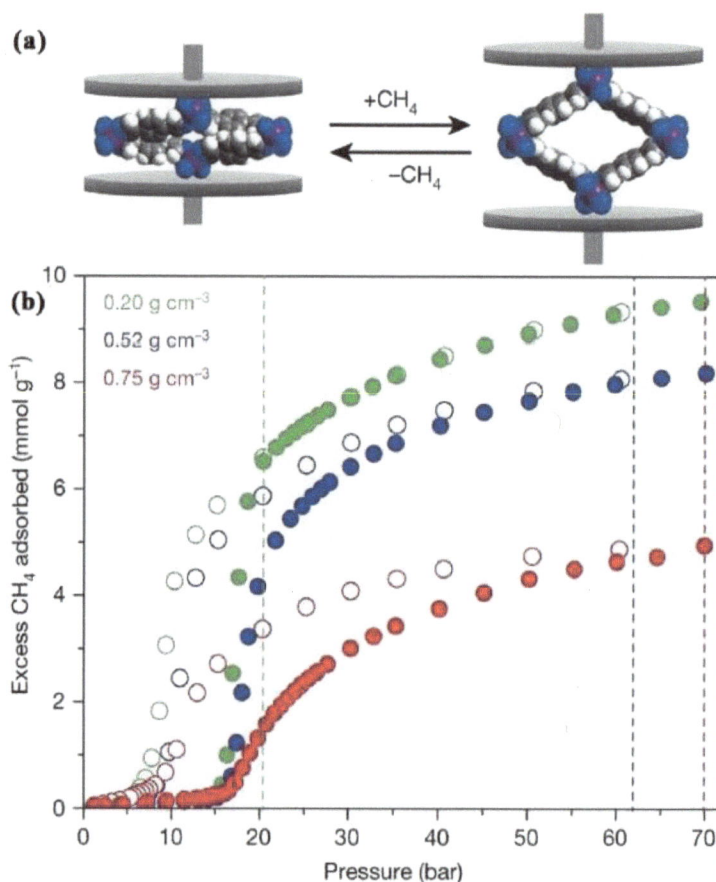

Fig. (15). (a) Effect of mechanical pressure on CH_4 storage in Co(bdp); **(b)** Excess CH_4 adsorption isotherms for Co(bdp) at 25 °C with different external mechanical pressure [55].

In 2018, Yang *et al.* reported another flexible MOF (NiL$_2$, L=4-(4-pyridy-)-biphenyl-4-carboxylic acid) with a diamondoid topology, which switched among non-porous (closed) and several porous (open) phases at specific CH_4 and

CO_2 pressures [57]. The total volumetric and working capacity of CH_4 uptakes for NiL_2 is 189 cm^3 (STP) cm^{-3} and 149 cm^3 (STP) cm^{-3} at 298 K and 65 bar, respectively. NiL_2 showed a potential CH_4 working capacity approaching that of compressed natural gas but at much lower pressures. The detailed structural characterization, including single-crystal XRD, synchrotron powder XRD, *in-situ* variable pressure powder XRD, pressure-gradient differential scanning calorimetry (P-DSC) and molecular modeling, proved that the phase switching were enabled by the ligand linker contortion, interactions between interpenetrated networks and sorbate–sorbent interactions.

Recently, Kundu *et al.* reported the solvent-induced changes in the particle size and stability of different breathing phases of a group of MIL-53 MOF series with flexible frameworks [58]. Researchers found that simple adjustment of synthesis solvent would cause great changes in MOF flexibility. Based on this, a series of MIL-53(Al)-NH_2 materials were synthesized by changing the water/DMF ratio while keeping the same ratio of metal to ligand. Interestingly, when there was 20 vol% DMF in the solvent, the CH_4 adsorption curve of as-synthesized materials significantly changed, while the desorption curve was less affected, which could be attributed to the shape memory effect caused by the decreasing particle size of MOFs. Compared with the very low gravimetric CH_4 working capacity (only 62 mg g^{-1}) for MIL-53(Al)-NH_2 traditionally synthesized in aqueous phase, the CH_4 storage and working capacity of MOF samples synthesized by their method have been increased by 33%. The similar phenomenon is also reflected in another kind of water stable flexible MOF materials MIL-53(Al)-OH. This simple modulation method brought new opportunities for flexible porous materials in challenging applications such as natural gas storage and transportation.

Greenhouse gas capture. CO_2 is the main greenhouse gas, which could lead to the increasing of global temperature. Natural ecosystems will be significantly impacted by the climate change, which could threaten the survival and development of human beings. Although the greenhouse effect of CO_2 is relative weaker compared with other greenhouse gases (CH_4, O_3, N_2O, *etc.*), CO_2 has the largest ratio, which take up 60% of total temperature increasing effect of all greenhouse gases. In recent years, studies of CO_2 capture have received great attention, which is also one of the most researched areas of MOFs. To enhance the CO_2 capture ability of MOFs, several strategies have been developed in the past two decades, such as introducing open metal sites (OMSs) with unsaturated coordination centers, doping metal ions in the frameworks, producing narrow pore-size distribution by controlling the interpenetration of frameworks, and modifying pore channels by ligand size or functionalization.

In 2005, Yaghi's group firstly studied the CO_2 storage capacity of MOFs at room temperature [59]. Gravimetric CO_2 isotherms for MOF-2, MOF-505, $Cu_3(BTC)_2$, MOF-74, IRMOFs-11, IRMOFs-3, IRMOFs-6, IRMOFs-1 and MOF-177 are tested up to the pressure of 42 bar. Type-I isotherms are observed in all situations except for the $Zn_4O(O_2C)_6$ cluster-based MOFs, which show a sigmoidal kind isotherm with one step. The onset pressures of isotherm steps increase as pore sizes, which indicates the potential application of these MOFs in gas separations. The amine-functionalized IRMOF-3 with modified pores shows an increasing affinity for CO_2. Adsorption capacity qualitatively multiplies with surface area, and ranges from 3.2 mmol/g for MOF-2 to 33.5 mmol/g (320 $cm^3(STP)/cm^3$, 147 wt %) for MOF-177, the highest CO_2 capacity up to now.

M-MOF-74 (M = Mg, Mn, Fe, Co, Ni and Zn) series are typical MOFs with open metal sites (OMSs). Due to strong coordination of OMSs inside the frameworks with adsorbed CO_2 molecules, M-MOF-74 generally present high uptake amount at low pressure. Among them, with the high initial CO_2 adsorption enthalpy of 47 kJ mol^{-1}, Mg-MOF-74 exhibit the largest CO_2 capacity at 298 K, and could adsorb 23.6 wt% of CO_2 at 0.1 atm with the value increasing to 35.2 wt% at 1 atm, which was also the highest CO_2 capacity of MOFs under the same conditions [60, 61].

Besides OMSs, the introduction of special group to enhance the framework-CO_2 interaction is another mostly used method towards improving the CO_2 capture ability of MOFs. Zhang *et al.* found that the reaction of Ni(II) ion with co-ligands of 2,4,6-tri(4-pyridinyl)-1,3,5-triazine (tpt) and o-phthalicacid (OPA) could result into highly porous frameworks, $Ni(tpt)(OPA)(H_2O)$ (TKL-101) [62]. Through analyzing the coordination environment of OPA in the framework, there was enough space inside the pores for the ligand functionalization. As a result, functionalized MOFs isostructural with TKL-101 were prepared by applying a series of OPAs with different functional groups including $-NH_2$, $-NO_2$, and -F. Interestingly, when fluorine was introduced into the framework *via* the functionalization, both the stability and adsorption capacity towards CO_2 were enhanced significantly, while other groups (amino and nitro groups) decorated frameworks were not stable after activated. More detailed analysis revealed that the fluorine atoms of 3-position (3-F) in the OPA ligands played a significant role in the stabilization of the frameworks, due to the additional weak interaction between the introduced fluorine and ambient atoms of tpt ligand, which were confirmed by PXRD and theoretical calculation of Mayer bond order. Therefore, these MOFs (3-F, $3,6-F_2$ and $-F_4$) with 3-F decorated pores present more stable frameworks than that of MOFs only with 4-F (and $4-NH_2$, $4-NO_2$) decorated pores. The porosity and adsorption properties could also be fine turned by the number of fluorine atoms, and CO_2 uptakes are remarkably improved with the increasing the number of F atoms. TKL-107 with the largest number of F atoms (four) in the OPA ligand presented the highest CO_2 storage capacity of 150 cm^3 g^{-1} (29.5wt%) at 273K and 1.2 bar, increasing up to 42.5% till 30 bars, although its BET specific surface area is the lowest (1454 m^2 g^{-1}) among 3-F decorated frameworks (3-F, 1509 m^2 g^{-1} and 3,6-F$_2$, 1636 m^2 g^{-1}). Moreover, the authors also used TKL-107 to fabricate mixed matrix membranes (MMMs) to examine the potential application for CO_2 separation. It revealed that at 20% TKL-107 loading, the ideal selectivity and mixed-gas selectivity values both reached a maximum of 64.6 and 50.3, respectively, showing the promising applications of these TKL-MOFs as filter in MMMs.

In addition, it is also an efficient way to enhance the CO_2 capture of MOFs by constructing pores which is suitable for the CO_2 molecule. For example, Zhao *et al.* demonstrated that through a ligand-insertion pore-space partition strategy [63], crystalline porous materials (CPMs) could be created with superior capacity (Fig. **16**). Specifically, they synthesized a MIL-88-type of MOFs without OMSs (CPMs, [M$_3$O(H$_2$O)$_2$X(L)$_3$]·guest, where M = Fe, Cr; X = F, Cl, acetate; L = fdc, bdc, ndc, bpdc) by a series of terpyridine ligands. The 1D hexagonal nano-channels along the c-axis of MIL-88 structure were partitioned into octahedron cages by the terpyridine ligand. Significantly, CPM-33b exhibited CO_2 uptake capacity of 173.9 cm^3 g^{-1} at 273 K and 126.4 cm^3 g^{-1} at 298 K, which are comparable to that of MOF-74 at 1 bar. In another example, Zaworotko and Eddaoudi *et al.* reported SIFSIX series MOFs with no OMSs, which also showed great potential in selective CO_2 uptake [64 - 66]. These MOFs were synthesized by linear dipyridyl ligands with structural characteristics of 'pillared square grids', in which that 2D grids are pillared *via* SiF$_6^{2-}$ anion ('SIFSIX') to form 3D networks with a primitive cubic topology. Interestingly, the CO_2 uptake ability of these MOFs was strongly affected by the pore volume. Although possessing the largest BET specific surface area (3140 m^2 g^{-1}), pore volume (1.15 cm^3 g^{-1}) and pore size (13.05 Å) among these MOF series, SIFSIX-2-Cu presented the lowest CO_2 uptake amount (1.8 mmol g^{-1} at 298 K and 1 bar). Conversely, SIFSIX-2--u-i, which has a similar but two-fold interpenetrated framework of SIFSIX-2-Cu, exhibited a much higher CO_2 uptake of 5.4 mmol g^{-1} at 298 K and 1 bar, despite its much lower BET specific surface area (735 m^2 g^{-1}), pore volume (0.26 cm^3 g^{-1}) and pore size (5.15 Å) than those of non-interpenetrated SIFSIX-2-Cu. SIFSIX--Cu and SIFSIX-3-Zn constructed by the shortest pyrazine ligand presented the lowest pore sizes of 3.5 Å and 3.84 Å, respectively. They almost had no adsorption of N$_2$ even under 77 K due to their pore sizes close to the dynamic diameter of N$_2$ (3.64 Å), but could adsorb moderate amounts of CO_2 (2.5 mmol g^{-1}) at 298 K and 1 bar. Further analysis revealed that although presenting lower CO_2 uptake amount compared with SIFSIX-2-Cu-i, SIFSIX-3-Zn showed a 2.4 mmol g^{-1} of CO_2 uptake at 0.1 bar, which is higher than that of SIFSIX-2-Cu-i (1.7 mmol g^{-1}). The mixed-gas breakthrough experiments revealed that SIFSIX-3-Zn showed much higher selectivity (495 for CO_2/N_2:10/90 and 109 for CO_2/CH_4:50/50, respectively) than SIFSIX-2-Cu-i, and CO_2 was retained for longer time (for example, ~2000 s *versus* 300s for CO_2/N_2).

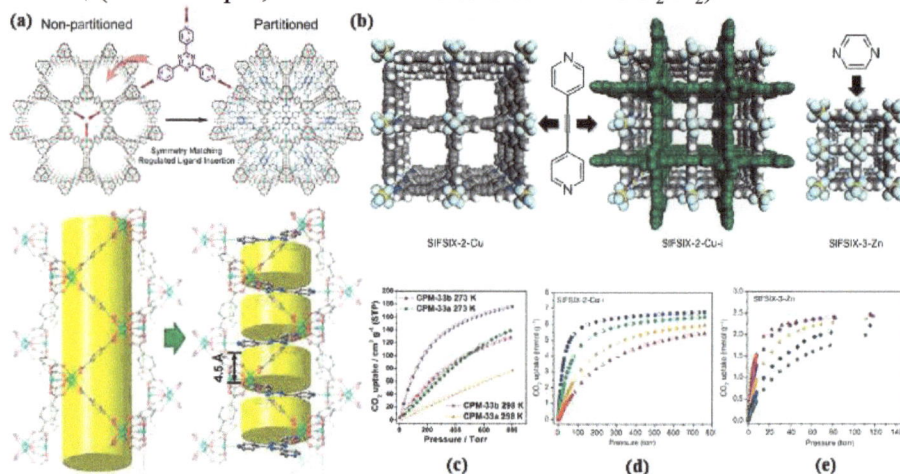

Fig. (16). (a) CPMs with pore-space partition; **(b)** the variable pore-size channels of SIFSIX-2-Cu, SIFSIX--Cu-i and SIFSIX-3-Zn; **(c)**, **(d)**, **(e)** CO_2 adsorption capture properties of CPM-33, SIFSIX-2-Cu-i and SIFSIX-3-Zn, repectively [63, 64].

Light hydrocarbon separation. The light hydrocarbon (LH) mixtures separation and purification have been one of the most significant but energy demanding processes in the petrochemical industry. As an alternative technology to traditional separation methods (distillation and extraction), adsorptive separation which uses porous solid as adsorbents can not only reduce energy cost but also bring higher efficiency. Therefore, the development of solid porous materials for the efficient selective adsorption of LH molecules under mild conditions, is of great significance and urgency. In recent several years, MOFs have got substantial achievement in the field of light hydrocarbon separation, and many well-designed strategies have been proposed. Herein, we mainly discuss the recent studies of C2 (C_2H_6, C_2H_4, C_2H_2) and C3 (C_3H_8, C_3H_6, C_3H_4) separations using MOFs as passivating agents. Ethylene (C_2H_4), which is produced by thermal decomposition of ethane (C_2H_6) or steam cracking, is the largest feedstock in petrochemical industry. It requires an ethylene purity of at least 99.95% for polymer production, and therefore, the separation of residual ethane and ethylene should be considered in this process. However, there are great challenges for efficient separation of C_2H_4/C_2H_6, because of their similar physical and chemical properties (similar sizes and volatilities). In addition, cleavage products both contain ethylene and acetylene (C_2H_2), which could lead to the catalyst poisoning in the ethylene polymerization. Thereby, it is also very important for the efficient C_2H_4/C_2H_6 separation in industry.

In 2015, Yang *et al.* reported the supramolecular binding and separation of hydrocarbons by using NOTT-300 with a formula of $Al_2(OH)_2(L)$ (H_4L = biphenyl-3,3′,5,5′-tetracarboxylic acid) [67]. Comprehensive studies combining synchrotron X-ray and neutron diffraction, neutron scattering and computational modelling, allow defining detailed binding of ethane, ethylene and acetylene at a molecular level within the pores of NOTT-300. At 293 K and 1.0 bar, the total adsorption uptakes for CH_4, C_2H_6, C_2H_4 and C_2H_2 in NOTT-300 were measured as 0.29 0.85, 4.28 and 6.34 mmol g^{-1}, respectively. After analyzing pure-component gas adsorption isotherms at 293 K by the ideal adsorbed solution theory (IAST), the selectivity of C_2H_4/C_2H_6 and C_2H_2/C_2H_4 for NOTT-300 was calculated to be 48.7 and 2.30 for equimolar mixtures at 1.0 bar, respectively. At the same conditions, the C_2H_6/CH_4, C_2H_4/CH_4 and C_2H_2/CH_4 selectivity of NOTT-300 were estimated as 5, ~380 and >1,000 by IAST analysis, respectively.

Lin *et al.* reported an ultramicroporous MOF $Ca(C_4O_4)(H_2O)$ (termed as UTSA-280) possessing rigid 1D cylindrical channels (Fig. **17**), which could be easily and environment-friendly synthesized by calcium nitrate and squaric acid at the kilogram scale [68]. These apertures with slightly different shapes (3.2×4.5 Å^2 and 3.8×3.8 Å^2) displayed cross-sectional areas of about 14.4 Å^2, which are smaller than that of C_2H_6 (15.5 Å^2) but larger than that of C_2H_4 (13.7 Å^2), suggesting potential size/shape sieving for C_2H_4/C_2H_6 separation. Although UTSA-280 showed a low BET specific surface area of 331 m^2 g^{-1}, it could adsorb a high amount of C_2H_4, reaching 2.5 mmol g^{-1} (88.1 cm^3 cm^{-3}) at 298 K and 1 bar, while the C_2H_6 molecule was excluded by the pore apertures with only a negligible C_2H_6 uptake amount of 0.098 mmol g^{-1} under the same condition, and the sieving effect for C_2H_6 did not decline by even deceasing temperature to 195 K. Single-crystal X-ray diffraction demonstrated a weak C–H$\cdots$O hydrogen bond (3.32–3.44 Å), van der Waals (vdW) interaction (shortest C–H$\cdots$π distance of 3.32 Å) and π$\cdots$π stacking (3.31 Å) between aromatic rings of organic ligands and discrete C_2H_4 molecules. Considering the vdW radii, the tilted C_2H_4 molecule was well-fitted with the narrow pore channel. Calculated from the measured isotherms, record-high C_2H_4/C_2H_6 selectivity was obtained exceeding 10,000 that is several orders of magnitude larger than those of Fe-MOF-74 (13.6) [69], NOTT-300 (48.7) [67] and π-complexation sorbents like PAF-1-SO$_3$Ag (27) [70]. Breakthrough experiments at 298 K presented an amount of C_2H_4 enriched up to 1.86 mol kg^{-1} from an equimolar C_2H_4/C_2H_6 mixture. Furthermore, specific C_2H_4 enrichment was realized by UTSA-280 for a quaternary $C_3H_8/C_2H_6/C_2H_4/CH_4$ mixture (5/25/25/45) and even more complicated octonary cracking stream of $C_4H_8/C_3H_8/C_3H_6/C_2H_6/C_2H_4/C_2H_2/CH_4/H_2$ (2/1/2/40/45/1/5/4) on the basis of sieving effect.

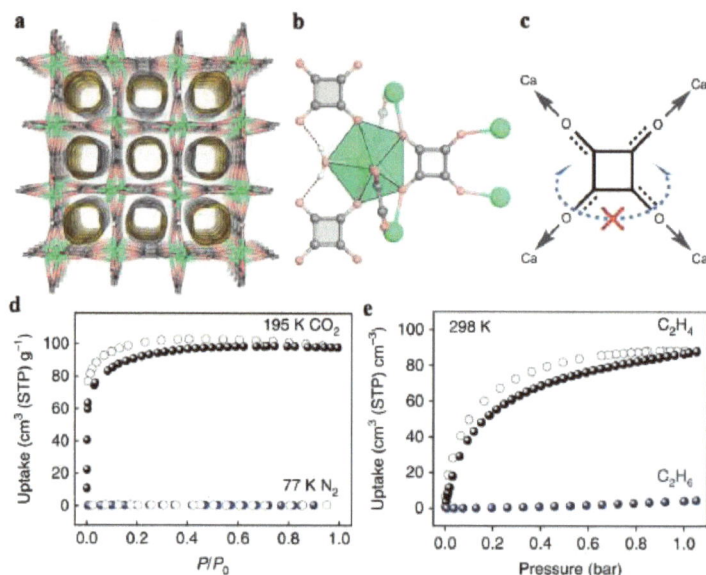

Fig. (17). Structure and gas sorption properties of UTSA-280 [68].

Many absorbents take up lower amounts of C_2H_6 than C_2H_4, mainly because of the weaker interaction of C_2H_6 with immobilized metal sites. If C_2H_6 is preferentially adsorbed, the later C_2H_4 can be directly recovered in the adsorption area. Furthermore, this approach can reduce about 40% of energy consumption (0.4 to 0.6 GJ/ton of ethylene) for the C_2H_4/C_2H_6 separation on pressure swing adsorption (PSA) technology, compared with C_2H_4-selective adsorbents. To date, there are scarce adsorbents exhibiting C_2H_6 adsorption preferred over C_2H_4, and only a few MOFs have been reported for selective C_2H_6/C_2H_4 separation, for example, MAF-49 [71], ZIF-7 [72], Ni(bdc)(ted)$_{0.5}$ [73] and PCN-250 [74]. Although these MOFs showed quite low separation selectivity and productivity, these pioneering works lead to a feasible and reverse way to develop new porous materials with preferential adsorption of C_2H_6. In 2018, Li *et al.* realized highly reversed selective separation of C_2H_6/C_2H_4 in a microporous MOF [75], $Fe_2(O_2)$(dobdc) (dobdc^{4-} = 2,5-dioxido-1,4-benzenedicarboxylate), with Fe(III)–peroxo sites as the preferential binding of C_2H_6 over C_2H_4 (Fig. **18**). $Fe_2(O_2)$(dobdc) with a BET specific surface area of 1073 m^2 g^{-1} showed a C_2H_6 uptake amount of 74.3 cm^3 g^{-1} (about 1.1 C_2H_6 per formula), which is much higher than the C_2H_4 adsorption at 298 K and 1 bar. However, the pristine Fe_2(dobdc) without Fe(III)–peroxo sites could take up more C_2H_4 than C_2H_6, indicating the significance of Fe(III)–peroxo sites in improving the C_2H_6 binding affinity of the framework confirmed by the high Q_{st} value of C_2H_6 adsorption (66.8 kJ mol^{-1}) than that of C_2H_4 adsorption (36.5 kJ mol^{-1}) in $Fe_2(O_2)$(dobdc). IAST calculations revealed that the adsorption selectivity for C_2H_6/C_2H_4 (50/50) at 298 K and 1 bar was estimated to be 4.4, which is higher than previously reported best-performing MOFs (MAF-49: 2.7) and all-silica zeolite structures (highest value: 2.9) [76]. Transient breakthrough simulation validated the feasibility of using $Fe_2(O_2)$(dobdc) toward C_2H_6/C_2H_4 mixtures separation in a fixed bed. Then breakthrough experiments for the C_2H_6/C_2H_4 (50/50) mixture at 298 K revealed a clean and sharp C_2H_6/C_2H_4 separation in $Fe_2(O_2)$(dobdc), which is in accord with the simulation. Meanwhile, $Fe_2(O_2)$(dobdc) showed high recyclability and regeneration capability, with no obvious decrease in the mean residence times within five continuous cycles for both C_2H_6 and C_2H_4. This research group also reported an approach to boost the C_2H_6/C_2H_4 separation by controlling pore structures in two isoreticular ultramicroporous MOFs, namely Cu(ina)$_2$ (Hina = isonicotinic acid) and Cu(Qc)$_2$ (HQc = quinoline-5-carboxylic acid) [77], presenting weakly polar pore surface that strengthen the binding affinity of C_2H_6 over C_2H_4. Cu(ina)$_2$ possessing a larger pore size of 4.1 Å showed a quite small adsorption difference and selectivity for C_2H_6/C_2H_4 under ambient conditions, whereas its isoreticular analogue Cu(Qc)$_2$ with smaller pores (3.3 Å) exhibited a very large adsorption ratio of 237%, that is 60.0/25.3 cm^3 cm^{-3}, significantly improving the selectivity of C_2H_6/C_2H_4. Moreover, Cu(Qc)$_2$ exhibited a calculated C_2H_6/C_2H_4 selectivity up to 3.4 for the corresponding binary equimolar mixture by using IAST at 298 K and 100 kPa, while the value was only 1.3 for Cu(ina)$_2$.

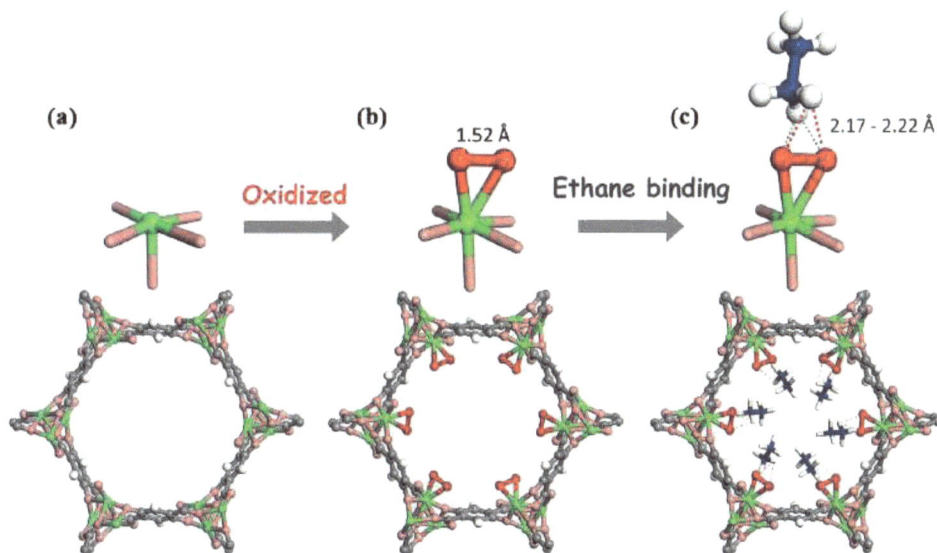

Fig. (18). Structures determined from NPD studies of **(a)** Fe_2(dobdc), **(b)** $Fe_2(O_2)$(dobdc), and **(c)** $Fe_2(O_2)$(dobdc)⊃C_2D_6 at 7 K. Note the change from the open Fe(II) site to Fe(III)-peroxo site for the preferential binding of ethane. Fe, green; C, dark gray; H or D, white; O, pink; O$_2^{2-}$, red; C in C_2D_6, blue [75].

Separating acetylene from ethylene is one of the important procedures in high-purity polymer production. In 2016, Cui *et al.* reported the pore-size control in hybrid porous materials for C_2H_2 capture from C_2H_4 as the first time [78]. They used SIFSIX-MOFs as target adsorbents. Because of the preferential binding of C_2H_2 molecule enabled by the geometric disposition of SiF_6^{2-} moieties, SIFSIX--Cu-i and SIFSIX-1-Cu both exhibit exceptional C_2H_2 capture performance. SIFSIX-2-Cu-i could rapidly adsorb C_2H_2 at very low pressure (≤ 0.05 bar) with an uptake reaching 2.1 mmol g^{-1} at 298 K and 0.025 bar, indicating the promise prospect for capturing minor C_2H_2 in a gas mixture (1/99). The C_2H_2/C_2H_4 selectivity was calculated by IAST to be a record of 39.7~44.8 as the pressure increasing. SIFSIX-1-Cu displayed moderate C_2H_2/C_2H_4 selectivity (7.1~10.6). The amounts of C_2H_2 captured by SIFSIX-1-Cu and SIFSIX-2-Cu-i from the 1/99 mixture is 0.38 and 0.73 mmol g^{-1}, respectively and 6.37 and 2.88 mmol g^{-1}, respectively from the 50/50 mixture during the breakthrough process, which agreed well with simulated results. This unprecedented performance could be attributed to the "sweet spots" in pores that enable the occurring of highly specific recognition and high uptake of C_2H_2 in the same material.

Recently, Xue *et al.* have successfully prepared a series of MOFs with excellent C_2H_2 storage capacity and ultra-high C_2H_2/CO_2 separation by adopting precise pore partition strategy and introducing high-density hydrogen bonding receptors [79]. They selected a typical MOF, MIL-88 as pore-space-partition architecture. The 1D hexagonal channels of MIL-88 was divided into finite segments with well-modulated pore sizes (SNNU-26,4.5 Å; SNNU-27, 6.4 Å; SNNU-28, 7.1 Å; SNNU-29, 8.1 Å). Coupled with 6 or 12 bare N sites of tetrazole within one cage as high-density H-bonding acceptors of C_2H_2, the target MOFs offered good combination of high C_2H_2/CO_2 adsorption selectivity and high C_2H_2 uptake capacity besides the good stability. The optimized SNNU-27-Fe demonstrated a very high C_2H_2 uptake of 182.4 $cm^3 g^{-1}$ and extraordinary breakthrough time of C_2H_2/CO_2 up to 91 min g^{-1} under ambient conditions.

The purity of propylene (C_3H_6) is also very important for the synthesis of polypropylene in industrial production, and the removal of minor amount of propane (C_3H_8) and/or propyne (C_3H_4) is the key step in the purification process. Cadiau *et al.* successfully realized the separation of C_3H_6/C_3H_8 mixture [80], using a chemically stable fluorinated MOF (KAUST-7 with the formula of $NiNbOF_5(pyrazine)_2 \cdot 2H_2O$) synthesized by tuning SIFSIX platform (using Nb^{5+} instead of Si^{4+}) (Fig. **19**). Compared with the parent SIFSIX-3-Ni, the judiciously selected bulkier $(NbOF_5)^{2-}$ in KAUST-7 caused the looked-for hindrance of previously free-rotating pyrazine moieties, delimiting the pore system and dictating the aperture size maximally open. Therefore, the resultant square-shaped channels of KAUST-7 was obtained with aperture size reducing to 3.0471(1) Å compared to SIFSIX-3-Ni [5.032(1) Å]. Due to small pore size, KAUST-7 did not adsorb N_2 (3.64 Å) at 77K, showing an estimated BET specific surface area of 280 m^2 g^{-1} with a pore volume of 0.095 cm^3 g^{-1} by using room temperature CO_2 sorption isotherms. At 298 K and 1 bar, KAUST-7 showed a C_3H_6 uptake of 60 mg g^{-1} while almost no adsorption for C_3H_8. Breakthrough measurements for the mixed-gas of C_3H_6/C_3H_8 at 50/50 revealed that KAUST-7 presented an obvious longer breakthrough time of C_3H_6 than C_3H_8 and excellent repeatability for the C_3H_6/C_3H_8 separation. Under ambient conditions, KAUST-7 could adsorb 0.6 mol kg^{-1} of C_3H_6 every cycle.

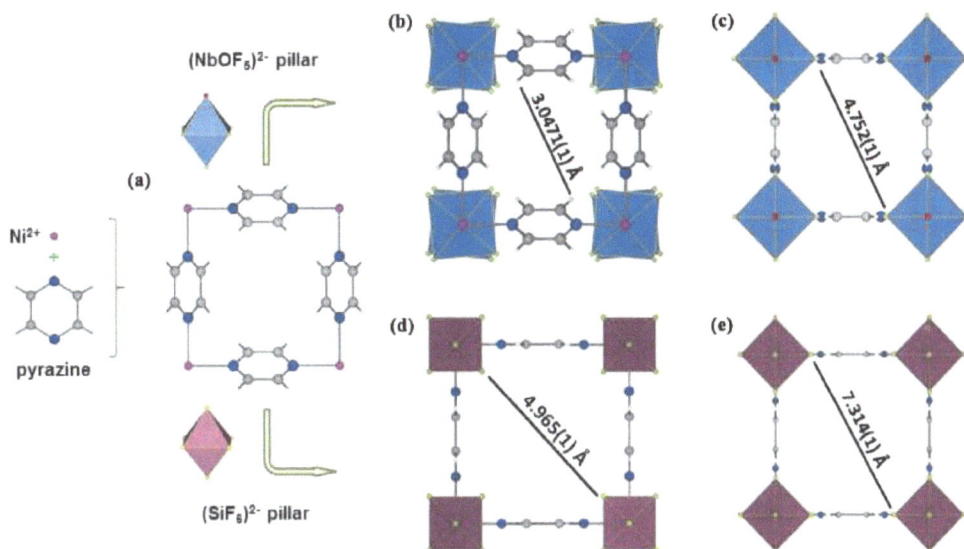

Fig. (19). (a) Illustration of the building blocks arrangement; Structure description of KAUST-7 **(b, c)**, and its comparison with parent SIFSIX-3-Ni **(d, e)** [80].

Li *et al.* showed the first example of MOFs with efficient C_3H_4/C_3H_6 separation [81]. The MOF material selected in this work was a previously reported flexible–robust framework, ELM-12, with the formula $Cu(bpy)_2(OTf)_2$ (bpy = 4,4′-bipyridine, OTf⁻ = trifluoromethanesulfonate) [82, 83]. ELM-12 contained two kinds of cavities, that is, the dumbbell-shaped (type I) with a small pocket of 6.1 Å × 4.3 Å × 4.3 Å at each end which is connected together *via* a small aperture of 3.2 Å × 4.3 Å, and the ellipsoid-shaped (type II) with the size of 6.8 Å × 4.0 Å × 4.2 Å, which is isolated from type I through the dynamic dangling OTf⁻ groups. Considering the shape and size of C_3H_4 (6.2 Å × 3.8 Å × 3.8 Å) and C_3H_6 (6.5 Å × 4.0 Å × 3.8 Å), these cavities obviously matched well with the C_3H_4. The C_3H_4 uptake of ELM-12 with a very high Q_{st} value of 60.6 kJ mol^{-1} was 1.83 mmol g^{-1} at 298 K and 0.01 bar, and increased to 2.55 mmol g^{-1} at 0.1 bar, about 93% of total uptake at 298 K and 1.0 bar, which is critical for C_3H_6 purification. However, for the C_3H_6 uptake, ELM-12 only showed a relatively low adsorption amount (0.67 mmol g^{-1} at 0.1 bar and 298 K) with a far smaller Q_{st} value of 15.8 kJ mol^{-1}, implying the much weaker host–guest interaction than that of C_3H_4. IAST calculations revealed that ELM-12 exhibited excellent C_3H_4/C_3H_6 selectivity, up to 84 for the 1/99 mixture and 279 for the 50/50 mixture, and it could capture 0.881 mmol g^{-1} of C_3H_4 from the C_3H_4/C_3H_6 (1/99) mixture. The purity of obtained C_3H_6 is over 99.9998%, as demonstrated by experimental breakthrough curve for 1/99 C_3H_4/C_3H_6 mixture. The high uptake capacity and selectivity of ELM-12 are attributed to its suitable pore confinement and strong binding affinity for C_3H_4, as evidenced by density functional theory (DFT) calculations and neutron powder diffraction studies.

Other gas adsorptions. Except for the gas mentioned above, MOFs also show wide potential applications in the adsorption and separation of some other harmful gases, such as C_2H_2, Xe, O_2, NO, SO_2, CO and so on.

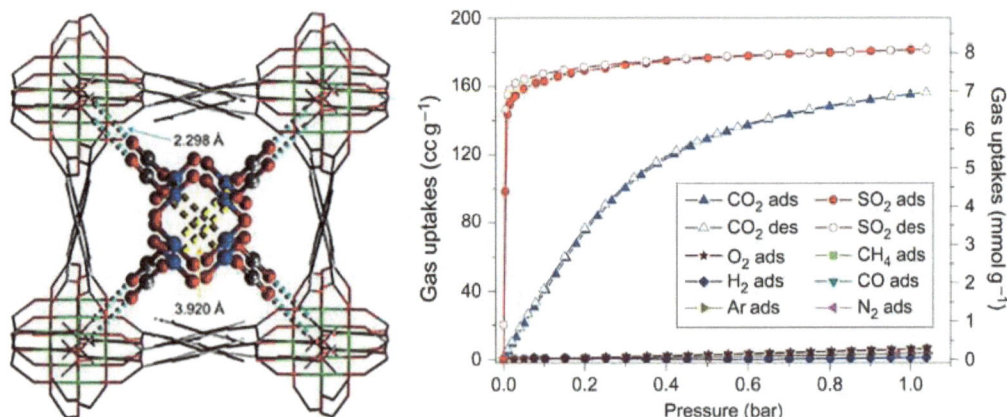

Fig. (20). View of the structure of NOTT-300·3.2CO$_2$ obtained using PXRD analysis and gas adsorption properties of NOTT-300 [84].

In 2012, Yang *et al.* reported a non-amine-containing MOF (NOTT-300) constructed by biphenyl-3,3′,5,5′-tetracarboxylic acid and Al(NO$_3$)$_3$·9H$_2$O (Fig. 20) [84]. Gas adsorption experiments revealed that NOTT-300 with high thermal and chemical stability, presented high uptake capacity and selectivity for SO_2 and CO_2 over a lot of gases (N_2, O_2, H_2, CO, Ar, CH_4). The SO_2 selectivity was 4974, 2518, 105, 6522, 3105 and 3620 for Ar, O_2, H_2, N_2, CO and CH_4, respectively, which were calculated from the initial slope ratio of isotherms. Interestingly, compared with CO_2 (kinetic diameter, 3.30 Å), the adsorption of SO_2 (kinetic diameter, 4.11 Å) exhibits higher uptakes, and the maximum SO_2 capacity is 8.1

mmol g^{-1} at 273 K and 1.0 bar, which is the highest value at that time. *In situ* powder X-ray diffraction and inelastic neutron scattering studies, combined with modelling, revealed that the Al–OH group in the framework could take part in moderate interaction with SO_2 and CO_2, which is supplemented by cooperative interaction with adjacent C–H group of benzene ring. This study offered potential application of new SO_2 'easy-on/easy-off' capture system that brought fewer environmental and economic penalty.

Rare gases such as Xe and Kr are widely used in lighting, medical diagnosis, semiconductor, aerospace military industry, flat panel TV, electronic chip, insulating glass, satellite science and other research fields, known as "gold gas". The Xe/Kr mixture separation is very important to industry, but limited selectivity for the adsorption of Xe and Kr is obtained with available porous materials. Recently, Wang *et al.* reported an anion-pillared ultramicroporous material (ZU-62) with finely tuned pore size and structure flexibility, which as the first time enables an inverse size-sieving effect in separation along with record Xe/Kr selectivity and ultrahigh Xe capacity [85]. Evidenced by single-crystal X-ray diffraction, the rotation of anion and pyridine ring adapts cavities to the shape/size of Xe and allows strong host-Xe interaction, while the smaller-size Kr is excluded. Breakthrough experiments confirmed that ZU-62 has a real practical application for producing high-purity Kr and Xe from air-separation byproducts, showing record Kr productivity (206 mL g^{-1}) and Xe productivity (42 mL g^{-1}, in desorption) as well as good recyclability.

Many industrial processes produce CO, which could be used as a chemical feedstock, but separation of CO from other gases, especially N_2, is too difficult to be economically viable. Sato *et al.* reported a soft nanoporous crystalline MOF with adaptable pores that could selectively adsorb CO, and presented crystallographic evidence that the CO molecule can coordinate with Cu^{2+} ion [86]. Unprecedented high CO selectivity was obtained, due to the synergetic effect of local interaction between the accessible metal sites and CO molecules (CO can be Cu^{2+} ion can selectively bound through serial structural changes). This MOF material with transformable crystalline structure realized the CO separation from mixture with N_2 (the most competitive gas to CO). This efficient and dynamic molecular trapping and releasing system make the CO-N_2 mixture separation achieve a low input energy for the desorption of CO.

The former mentioned Al-soc-MOF-1 reported by Eddaoudi' group also showed excellent O_2 storage property [49]. Al-soc-MOF-1 exhibited a record amount of absolute gravimetric O_2 uptake (29 mmol g^{-1} at 140 bar) that is much higher than those of NU-125 (17.4 mmol g^{-1}) and HKUST-1 (13.2 mmol g^{-1}), and a record deliverable capacity (27.5 mmol g^{-1} between 5 and 140 bar) [87]. Compared to the conventional empty cylinder [88], a 1 L cylinder that is filled with Al-soc-MOF-1 will potentially enhance the volumetric O_2 storage capacity (172 cm^3 cm^{-3}) by 70% at 100 bar, if the effect of void space and packing density are neglected. Notably, a 25% enhancement of the volumetric O_2 storage capacity for Al-so--MOF-1 still could be obtained over an empty cylinder, even if a prospective 25% loss is considered associated with packing density.

In addition, MOFs can also be applied in the adsorption of chemical warfare agents, which is of great importance in military fields. For example, Montoro *et al.* reported the capture of mustard gas analogues and nerve agents by a hydrophobic robust MOF [89], [Zn$_4$(μ_4-O)(μ_4-4-carboxy-3,5-dimethyl-4-c-rboxy-pyrazolato)$_3$], whose crystal structure showing remarkable mechanical, thermal, and chemical stability resembled that of MOF-5. Based on the adsorption heat values and Henry constant for different essayed adsorbates along with the H$_2$O/VOC partition coefficients that is gained from variable-temperature reverse gas chromatography experiments, it revealed that this MOF could selectively capture harmful VOCs (including the chemical warfare agent models of Sarin and mustard gas), even competing with ambient moisture (for example, under conditions that mimick operative ones). Further study showed that the capture ability of this MOF is comparable to carbon molecular sieve Carboxen, and better than HKUST-1. Liu and coworkers reported a sodalite-type porous MOF [90], H$_3$[(Cu$_4$Cl)$_3$(BTC)$_8$]$_2$[PW$_{12}$O$_{40}$]·(C$_4$H$_{12}$N)$_6$·3H$_2$O (NENU-11; BTC = 1,3,5-benzenetricarboxylate), by using polyoxometalate as templates. NENU-11 with a relatively low BET area (572 m^2 g^{-1}) and pore volume (0.39 cm^3 g^{-1}) displayed repaid adsorption behavior for nerve gas dimethyl methylphosphonate (DMMP) with an uptake of 19.2 mmol g^{-1} at 298 K, that is, about 15.5 molecules, exceeding that of HKUST-1 (8.2 per unit) and MOF-5 (6.0 per unit) with larger BET specific surface area under the same conditions. Due to the catalytic activity of polyoxometalate guest, NENU-11 could also efficiently accelerate the nerve gas mimic degradation by a hydrolysis reaction with methyl alcohol, methyl methylphosphonic acid (MMPA) and methylphosphonic acid (MPA), and the conversion ratio gradually increased with temperature, achieving 93% at 50 °C.

4.1.2. Vapor Adsorption

Organic vapor adsorption. In 2004, Lee *et al.* obtained a MOF, {[Ni(cyclam)(bpydc)]·5H$_2$O}$_n$, which was constructed by nickel macrocyclic complex [Ni(cyclam)](ClO$_4$)$_2$ and Na$_2$bpydc ligand [91]. In this MOF, [Ni(cyclam)]$^{2+}$ and bpydc^{2-} were connected together to form 1D chain structures. Neighboring chains were further linked with each other to generate a 3D supramolecular framework with 1D honeycomb channels (about 5.8 Å) by π-π

stacking interaction and hydrogen bonds. The MOF showed high stability and BET specific surface area of 817 m^2 g^{-1}, and could selectively adsorb ethanol, benzyl alcohol, pyridine and benzene, but not toluene.

Galli *et al.* reported the harmful organic vapors adsorption by flexible hydrophobic MOFs [92], Zn(bpb) and Ni(bpb) (H$_2$bpb = 1,4-(--bispyrazolyl)benzene). The obtained Zn(II) and Ni(II) MOFs showed square and rhombic channels, respectively, which account for 65 and 57% of total cell volume. The BET specific surface area and pore volume of two MOFs were 2200 m^2 g^{-1} and 0.71 cm^3 g^{-1} for Zn(bpb), and 1600 m^2 g^{-1} and 0.38 cm^3 g^{-1} for Ni(bpb), respectively. The pore-size, hydrophobic nature and flexibility of their channels made two MOFs adequate to incorporate organic vapors. At 303 K, Zn(bpb) and Ni(bpb) could adsorb a large amount of benzene vapors, that is, 5.8 and 3.8 mmol g^{-1}, respectively. The authors also explored breakthrough measurements, which revealed that the thiophene uptake amount for Zn(bpb) and Ni(bpb) could reach up to 0.34 g per gram of adsorbents from a flow of CH$_4$/CO$_2$ that contain 30 ppm of thiophene. Importantly, the excellent thiophene removal performance of Ni(bpb) could not be significantly affected, even at the presence of relative humidity (RH) 60%, while Zn(bpb) and HKUST-1 became ineffective under moisture conditions.

Yang *et al.* demonstrated the high affinity and capacity to C6–C8 hydrocarbons of oil components of the highly hydrophobic porous fluorous MOFs (FMOFs) materials [93]. In their work, FMOF-1, which was constructed from 3,5-bis(trifluoromethyl)-1,2,4-triazolate and Ag(I) center, exhibited reversible adsorption and high capacity for n-hexane (190 kg m^{-3}), cyclohexane (300 kg m^{-3}), benzene (290 kg m^{-3}), toluene (270 kg m^{-3}), and p-xylene (265 kg m^{-3}), without detectable water adsorption even at near RH 100%, which drastically outperform zeolite and activated carbon porous materials. Obtained from the anneal of FMOF-1, FMOF-2 showed double toluene adsorption *vs* FMOF-1 due to its enlarged channels and cages.

In 2018, Xie *et al.* reported a new type of MOFs, Zr$_6$(μ_3-O)$_4$(μ_3-OH)$_4$(BDB)$_6$ (BUT-66, H$_2$BDB = 4,4'-(benzene-1,3-diyl)dibenzoate) and Zr$_6$ (μ_3-O)$_4$ (μ_3-OH)$_4$(NDB)$_6$ (BUT-67, H$_2$NDB = 4,4'-(naphthalene-2,7-diyl) dibenzoate) as excellent adsorbent towards capturing trace amount of aromatic VOCs in air (Fig. **21**) [94]. BUT-66 and BUT-67 both have two-fold interpenetrated 3D pcu-type frameworks with permanent porosity, small pore size, hydrophobic pore and crystal surface, and high hydrolytic stability. Particularly, BUT-66 represented high volumetric benzene adsorption capacity of 1.75 mmol cm^{-3} at 0.12 kPa and 353 K even at high temperature and low pressure, better than some benchmark adsorbents, such as hydrophobic ZIF-8 or MAF-4 (0.05 mmol cm^{-3}), highly porous MIL-101-Cr (0.19 mmol cm^{-3}), highly porous and hydrophobic PAF-1 (0.08 mmol cm^{-3}), mesoporous MCM-41 (0.01 mmol cm^{-3}), and commercial carbon molecular sieve Carboxen 1000 (0.83 mmol cm^{-3}). Furthermore, BUT-66 was capable of capturing benzene in air of 50% RH with high capacity of 0.28 mmol cm^{-3}. The high VOCs capture performance of BUT-66 resulted from mutually less-interfered hydrophobic, hydrophilic adsorption sites, small hydrophobic pores and rigid framework with local rotatable phenyl rings, suggested by its single-crystal structure of guest-loaded phase.

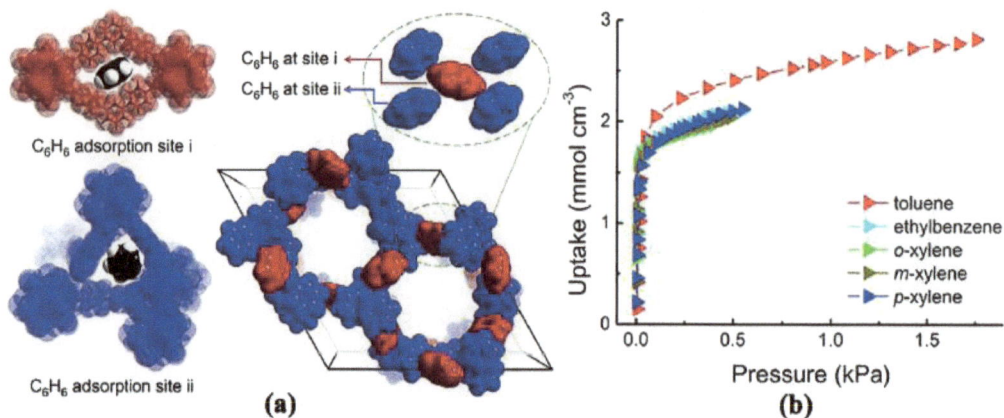

Fig. (21). (a) Single-Crystal X-Ray Diffraction Elucidation of Benzene Adsorption in BUT-66; **(b)** Selected Aromatic VOCs Adsorption in BUT-66 and BUT-67 [94].

Water Adsorption

Water is everywhere in nature. Stable MOFs as water adsorbents could be applied in thermal batteries, dehumidification, and drinking water delivery in remote areas.

Yaghi's group have recently researched the water adsorption performances of a series of water-stable MOFs, which show excellent performance meeting the basic requirements [95 - 102]. They proposed the criteria for water adsorption application of MOFs: water condensation pressure in MOF pores, adsorption capacity, and water stability and recyclability. In 2014, they investigated twenty MOFs and compared their water adsorption properties [95]. Among them, MOF-801-P (microcrystalline powder of MOF-801, Zr$_6$O$_4$(OH)$_4$(fumarate)$_6$) and MOF-841 (Zr$_6$O$_4$(OH)$_4$(MTB)$_2$(HCOO)$_4$(H$_2$O)$_4$, H$_4$MTB = 4,4',4'',4'''-Methanetetr ayltetrabenzoic acid) with water-stable frameworks are the highest performers according to the three criteria mentioned above. The two mentioned MOFs could rapidly adsorb water with high capacities (at 298 K: MOF-801-P, 22.5 wt% at P/P$_0$ = 0.1 and 36 wt% at P/P$_0$ = 0.9; MOF-841, 44 wt% at P/P$_0$ = 0.3 and 51.4

wt% at P/P_0 = 0.9) under well-defined, low relative pressure. Importantly, capacity loss did not happen after five cycles for the two MOFs that are readily regenerative at room temperature.

In 2017, Yaghi *et al.* reported the water harvesting performance of MOFs from air powered *via* natural sunlight [96]. They designed the first-generation device based on MOF-801 and demonstrated that it could capture water from atmosphere under ambient conditions through using low-grade heat from natural sunlight under a flux of less than 1 sun (1 kilowatt per square meter). Water could be adsorbed at night and released the next day with the sun heat, and the water vapor condensed on the inside surface of container. 2.8 liters of water could be daily harvested per kilogram of MOF by this device at RH level as low as 20% with no additional requirement of energy input.

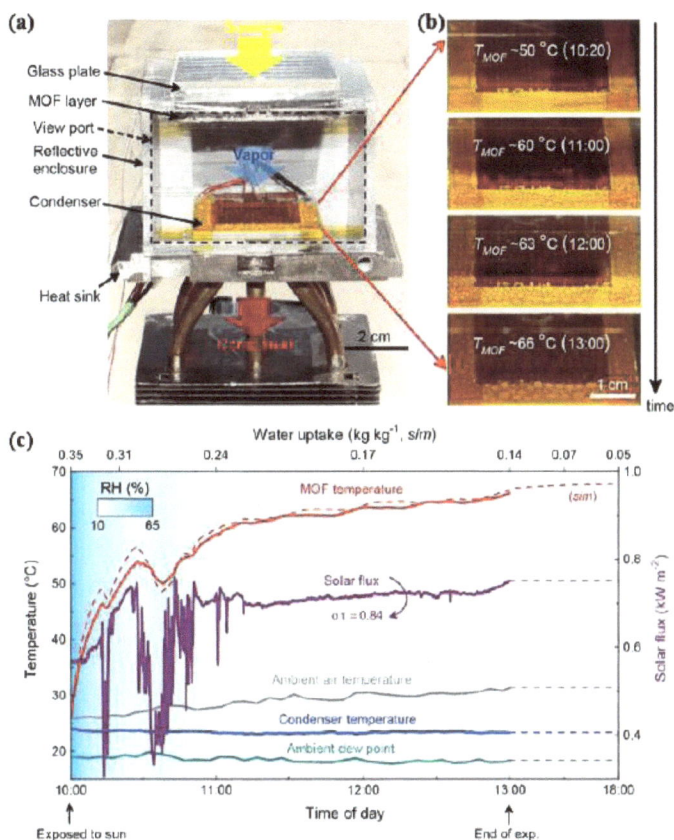

Fig. (22). Proof-of-concept water-harvesting prototype with 1.34 g activated MOF-801, whose packing porosity of **(a)** ~0.85 and outer dimension of $7 \times 7 \times 4.5$ cm³; **(b)** Formation and growth of water droplets as temperatures and local time; **(c)** Representative temperature profiles for MOF-801 layer [96].

In 2018, Yaghi's team further reported the practical production of water from desert air and turned the proof-of-concept device into a second-generation harvester [103]. They did laboratory *vs* desert experiments where a prototype that used up to 1.2 kg of MOF-801 was measured in laboratory and later in desert of Arizona, USA (Fig. **22**). The obtained device could produce 141.8 g and 100 g water per kilogram of MOF-801 in one day-and-night cycle under laboratory condition or only employing natural cooling and ambient sunlight as energy source at 5 to 40% RH. In this work, they also reported a new aluminum-based MOF-303 (Al(OH)(HPDC), HPDC = 1H-pyrazole-3,5-dicarboxylate) by the use of relatively cheap aluminum to replace zirconiuma, with an **xhh** topology containing infinite Al(OH)(COO)$_2$ SBUs and HPDC linkers. MOF-303 featured 1D hydrophilic pores with a diameter of 6 Å and free pore volume of 0.54 cm^3 g^{-1}, which facilitated the water capture capacity with a large maximum of 0.48 g g^{-1}. Up to 233 g of water could be produced by MOF-303 at laboratory conditions, which is more than twice the amount of water for MOF-801, thus bringing the production of water at low RH a step that is closer to practical application.

4.1.3. Hazards and Pollutant Concentration in Solutions

Due to modifiable pores, MOFs can also be used as potential absorbents in liquid adsorption and separation, such as heavy metal ions, anions and organic compounds.

Ion exchanges and separation. MOFs with good ion adsorption properties often have ionic frameworks countered by cations or anions inside their pores, which can be reversibly exchanged by other ions. For example, Dincă *et al.* reported an anionic MOF, Mn$_3$[(Mn$_4$Cl)$_3$(BTT)$_8$(CH$_3$OH)$_{10}$]$_2$ (1-Mn^{2+}; BTT = 1,3,5-benzenetristetrazolate), countered with hydrated Mn^{2+} ions filled in the pores [104]. Cation exchanges led to the generation of a class of compounds 1-M (M = Li$^+$, Cu$^+$, Fe^{2+}, Co^{2+}, Ni^{2+}, Cu^{2+}, Zn^{2+}) with isostructural frameworks. Similar to original 1-Mn^{2+}, obtained 1-M materials were stable to desolvation with microporosity which showed a modulable high H$_2$ storage capacity (2.00 ~ 2.29 wt % under 77 K and 900 torr). The initial isosteric heat of H$_2$ adsorption revealed a 2 kJ/mol difference between the weakest (1-Cu^{2+}, 8.5 kJ mol^{-1}) and strongest H$_2$-binding materials (1-Co^{2+}, 10.5 kJ mol^{-1}), which could be assigned to the variations in the interaction strength between the H$_2$ molecules and the unsaturated metal centers inside its own framework.

Wang' group have reported a series of cationic MOFs showing great promise in anionic pollutant removal. For example, they presented a solution to solve the challenge of selective ^{99}TcO$_4^-$ remediation in the face of a large excess of SO$_4^{2-}$ and NO$_3^-$ from natural waste systems based on a stable cationic MOF,

$Ni_2(tipm)_3(NO_3)_4$ (SCU-102) (Fig. **23**) [105]. ReO_4^- (a surrogate for TcO_4^-) exchange experiment with SCU-102 exhibited fast sorption kinetics, large capacity (291 mg g^{-1}), high distribution coefficient, and record-high uptake selectivity. SCU-102 could almost quantitatively remove TcO_4^- in the presence of a large excess of NO_3^- and SO_4^{2-}. The researcher also practically used this MOF for the decontamination of a simulated waste stream with low activity and contaminated groundwater at the Hanford site, which confirmed the representation of SCU-102 as the optimal Tc scavenger that has the highest clean up efficiency (distribution coefficient (K_d) as high as 5.6 × 10 mL g^{-1}). DFT simulation revealed that the exceptional selectivity was derived from the TcO_4^- recognition *via* the plentiful hydrophobic pockets inside the framework.

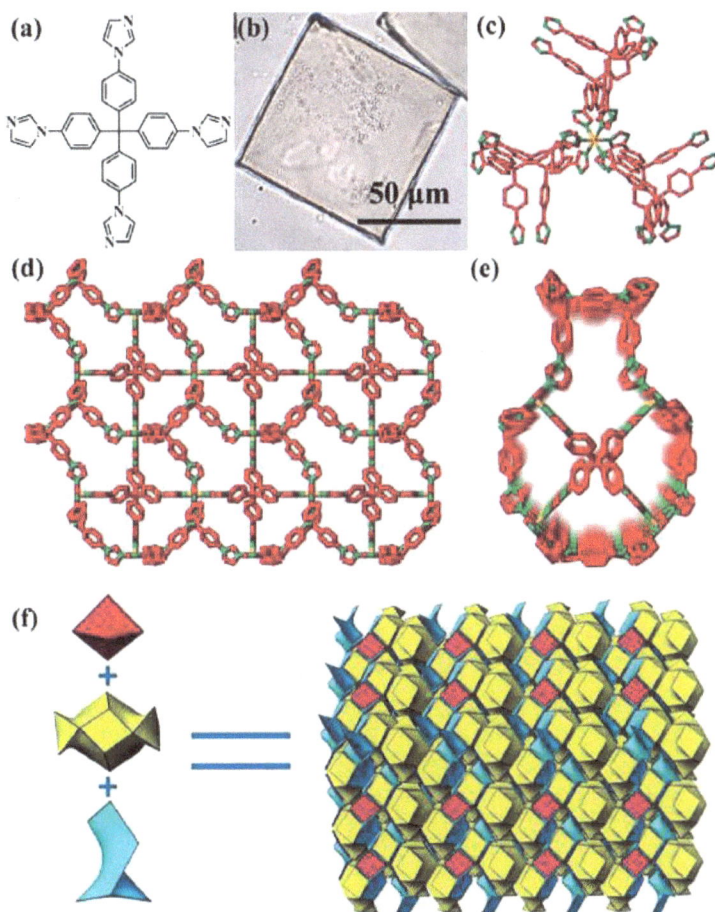

Fig. (23). (a) Chemical structure of tipm ligand; **(b)** optical image of SCU-102 crystals; **(c)** coordination environment of Ni^{2+} with six tipm ligands; **(d, e)** perspective packing structure of SCU-102 viewed along *a* axis; **(f)** three tiles of SCU-102 [105].

Except for inorganic ions, MOFs can also effectively adsorb organic pollutant ions, such as organic dyes. For example, Han *et al.* synthesized a anionic MOF, $[(CH_3)_2NH_2][Co_2NaL_2(CH_3COO)_2]$ (BUT-51, H_2L = 5-(pyridine-4-yl)isophthalic acid) containing two kinds of nanotubular channels, and reported its dye adsorption properties [106]. Through ion exchanges of counter $(CH_3)_2NH_2^+$ inside the pores, BUT-51 could quickly adsorb cationic dyes, such as Methylene Blue (MB), Acriflavine Hydrochloride (AH) and Acridine Red (AR), but Methylene Violet (MV), Methyl Orange (MO) and neutral Solvent Yellow 2 (SY2) with larger size, could hardly be adsorbed. Obviously, the selective cationic dyes adsorption properties of BUT-51 followed not only a charge- but also size- and shape-exclusive effect. Moreover, because of the coordination of AH with Co^{2+} in its framework, BUT-51 also presented preferential adsorption of AH over MB and AR in their acetone solution, respectively. Li *et al.* ionothermally synthesized the first mesoporous cationic MOF [107], $[Th_3(bptc)_3O(H_2O)_{3.78}]Cl\cdot(C_5H_{14}N_3Cl)\cdot 8H_2O$ (SCU-8, H_3bptc = [1,1′-biphenyl]-3,4′,5-tricarboxylicacid) containing channels with large inner diameter of 2.2 nm and possessing specific surface area of 1360 m^2 g^{-1} (Fig. **24**). SCU-8 exhibited superior anion-exchange capability toward various anionic environmental pollutants, containing small oxo-anions ReO_4^- and $Cr_2O_7^{2-}$ as well as organic dyes such as MB, and the persistent organic pollutant perfluorooctane sulfonate (PFOS). Particularly, SCU-8 showed efficient removal of PFOS, which could quickly adsorb 88 and 96% of PFOS anion in 30 s and 2 min (and sequentially reached equilibrium), respectively, as determined by HPLC-MS/MS.

Organic Isomer Separation In natural and crude petrochemical products, isomers of many important chemical matters usually coexist with each other. Therefore, it is of great significance for the separation of isomers. Luckily, porous MOF materials bring a new idea to resolve this issue.

Alaerts *et al.* firstly studied the C8 alkylaromatic isomer separation properties of MOFs (using HKUST-1, MI-53-Al and MIL-47 as adsorbents) in 2007 [108]. They found that MIL-47 with the best performance among three MOFs was an excellent adsorbent for the separation of ethylbenzene, *meta*-xylene, and *para*-xylene. MIL-47 showed both high selectivity of 9.7 for *para*-xylene/ethylbenzene and 2.9 for *para*-xylene/*meta*-xylene. Breakthrough experiment revealed that the separation selectivity for a 1:1 mixture of *para*-xylene/ethylbenzene and *para*-xylene/*meta*-xylene was 7.6 and 2.5, respectively. Obvious separated peaks could be obtained in the chromatographic experiment, and the calculated selectivity from the chromatography reached up to 9.7 and 3.1 for *para*-xylene/ethylbenzene and *para*-xylene/*meta*-xylene, respectively. Rietveld refinements of X-ray powder diffraction patterns of MIL-47 samples saturated with each aromatic compound demonstrated that there existed strong π–π interaction between para-xylene molecules and weak interaction between the protons of CH_3 group and terephthalate ligand.

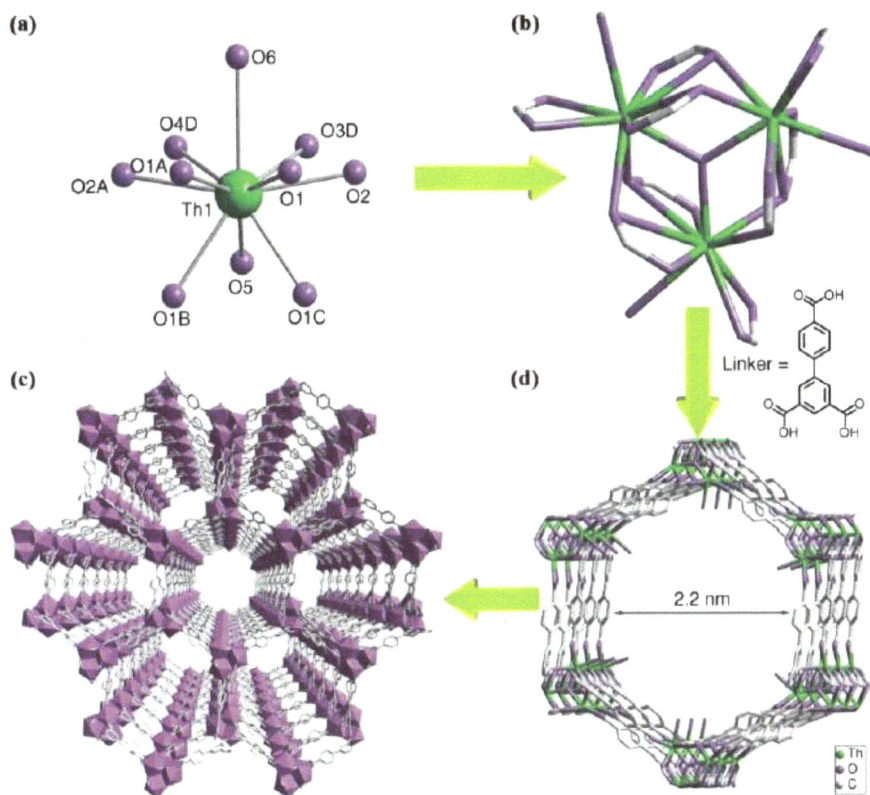

Fig. (24). Crystal structure of SCU-8. **(a)** coordination geometry of Th^{4+}; **(b)** the cationic cluster of $[Th_3(COO)_9O(H_2O)_{3.78}]^+$; **(c)** hexagonal tubular channels; **(d)** the cationic mesoporous framework along c axis [107].

Liu *et al.* reported separation performance of a series of 3D functionalized MOF (CdL_2, L = 4-amino-3,5-bis(4-pyridyl-3-phenyl)-1,2,4-triazole) with 1D channels (pore size: 11×11 Å^2) [109]. Directly performing adsorptions and separations on single crystals, CdL_2 displayed very rigorous selectivity for the aromatic isomers with substituted reactive group and could fully isolate the guest isomers (*i.e.*, 2-thenaldehyde *vs* 3-thenaldehyde, 2-furaldehyde *vs* 3-furaldehyde, and *o*-toluidine *vs* *m*-toluidine *vs* *p*-toluidine) in both vapor and liquid phases under mild conditions (Fig. **25**).

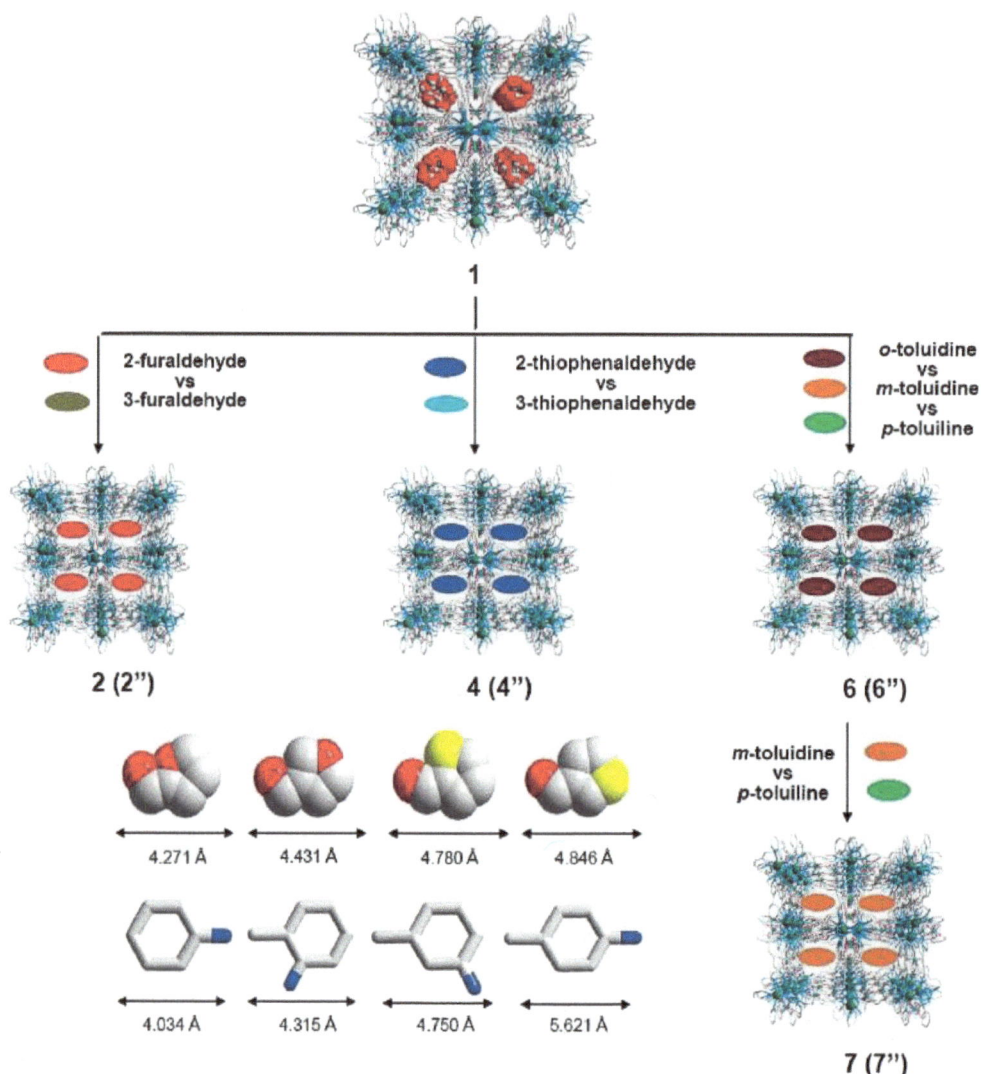

Fig. (25). Isomer separation based on single crystals of CdL_2 MOF [109].

The design of chiral MOFs towards enantiomers resolution is also an important research area. For example, Seo *et al.* firstly reported homochiral MOFs for enantioselective separation [110]. They synthesized a 2D layered MOF, $[Zn_3(\mu_3-O)(1-H)_6]\cdot2H_3O\cdot12H_2O$ (L-POST-1), using a L-tartaric acid ligand. Stacking along *c* axis, the 2D layers showed a mean interlayer isolation of 15.47 Å, which was further evidently stabilized by efficient vdW interaction between 2D layers. Notably, there existed 1D triangle chiral channels along *c* axis, and its side length was 13.4 Å. After the suspension of L-POST-1 into a methanol solution of racemic $[Ru(2,2'\text{-bipy})_3]Cl_2$ (bipy = bipyridine), the white color of L-POST-1 turned to reddish yellow. Nuclear magnetic resonance (NMR), circular dichroism and ultraviolet-visible measurements revealed a 80% of protons exchanging with $[Ru(2,2'-bipy)_3]^{2+}$ (0.8 Ru complex per $[Zn_3(\mu_3-O)(1-H)_6]^{2-}$ unit) with 66% enantiomeric excess in favor of Δ form. Peng *et al.* reported two isostructural robust porous chiral MOFs, $[Mn_2L^1(DMF)_2 (H_2O)_2]\cdot 3DMF\cdot 2H_2O$ (1a) and $[Mn_2L^2(DMF)_2(H_2O)_2]\cdot 3DMF\cdot H_2O$ (1b), decorated with chiral dihydroxy or methoxy auxiliaries from enantiopure tetracarboxylate-bridging ligands of 1,10-biphenol and manganese carboxylate chain, and their potential in enantio-separation of racemic primary and secondary amines through adsorption and liquid chromatography[111]. 1a and 1b both showed permanent porosity with BET specific surface area of 2145 and 1746 m^2 g^{-1}, respectively. The racemic 1-phenylethylamine (1-PEA) adsorption was used as a model substrate for (S)-1a, which allowed for the R-enantiomer in excess with 88.5% ee. Then, the authors investigated a variety of electron-deficient/rich substituents on the aromatic ring. Notably, an ee value up to 98.3% was obtained for 1-(4-fluorophenyl) ethylamine, which is the highest among the analytes. Importantly, the host MOF materials could be easily recycled with no obvious loss of performance. The resolution of racemic 1-PEA in large-scale by 1a-packed column for HPLC further demonstrated the utility of present adsorption separation for this MOF. The chiral recognition and separation could be assigned to the different specific binding energies and orientations of the enantiomers in the framework microenvironment, suggested by molecular simulation and control experiments.

4.2. Catalysis

The development of MOFs-based catalysts is also one of the most popular research directions, which plays an important role in the development of catalytic chemistry. In view of MOFs-based catalysts, investigations mainly contain preparing (1) metal particles embedded in N-doping nanocarbons through high temperature calcination, (2) incorporation noble metal species into MOF frameworks, or (3) applying MOFs themselves directly as catalyst. Compared with two former methods, the last one which are based on the high-density and uniform dispersion of active sites inside MOFs should be more cheering while it also faces big challenges, such as chemical stability and relative low activity. Different heterogeneous and homogeneous catalysts can be synthesized by rational design of MOFs. To reach this purpose, MOFs need be synthesized with unsaturated metal sites in an *in-situ* way, using metal clusters as building blocks, or obtaining active sites by post-synthetic methods [112]. Herein, we mainly introduce the design of MOFs showing catalytic activity in various kinds of reactions, with metal nodes, organic ligands and guest molecules of MOFs as catalytic active centers.

4.2.1. Metal Nodes as Active Sites

Solvent molecules often participate in the coordination of MOFs as terminal ligands. Unsaturated metal centers will be generated after removing coordination solvent which become strong Lewis acid sites and show good catalytic activity for the reactions of carbon dioxide cycloaddition, cyanosilicate, olefin reduction, olefin epoxidation, *etc.* The introduction of unsaturated coordination metal centers into MOFs is an effective way to produce catalytic performance. At the same time, the regular arrangement of unsaturated metal sites, pore size and shape are very favorable for the selectivity of catalytic reactions.

Through the assembly of $Zn(NO_3)_2$ and a tripodal organic ligand, H_3TCPB (1,3,5-tri(4-carboxyphenoxy)benzene), Lin *et al.* synthesized two isoreticular MOFs, $[Zn_3(TCPB)_2 \cdot 2DEF] \cdot 3DEF$ and $[Zn_3(TCPB)_2 \cdot 2H_2O] \cdot 2H_2O \cdot 4DMF$, both possessing a (3,6)-connected 2D network that contain 1D channels with effective size of 3.0×2.9 Å^2 and 4.0×4.0 Å^2, respectively [113]. Variable-temperature PXRD patterns demonstrated that the solid phase of $[Zn_3(TCPB)_2 \cdot 2DEF] \cdot 3DEF$ can be converted into $[Zn_3(TCPB)_2 \cdot 2H_2O] \cdot 2H_2O \cdot 4DMF$ during dynamic structural transformation induced by temperature, thereby suggesting that the most thermally stable polymorph is the latter framework. After removing coordination water molecules, the coordinatively unsaturated Zn^{II} sites in the obtained $Zn_3(TCPB)_2$ framework can serve as Lewis acid catalytic centers for carbonyl cyanosilylation and the Henry reaction. It was found that a loading of 2.5 mol% $Zn_3(TCPB)_2$ could lead to 100% conversion of benzaldehyde after 13 hours. However, it showed clearly decreasing trend of the conversion with the particle size enlarging.

Zhang *et al.* reported the construction and hydrogenation performance enhancement of Cu-based MOFs by integrating open electropositive metal sites [114]. They synthesized two 2D Cu-based MOF, namely $Cu_2(L^1)_2 \cdot (DMF)_2$ and $Cu(L^2) \cdot (dpye)$ (L^1 = 2-nitrobiphenyl-2,4'-dicarboxylicacid; L^2 = 2-nitrobipheny--3,4'-dicarboxylicacid; dpye=1,2-di[Jpyridin-4-yl)ethane). Interestingly, it was found that the presence of paddle-wheel $Cu_2(COO)_4$ SBU as open metal sites endowed $Cu_2(L^1)_2 \cdot (DMF)_2$ with an effectively improvement of catalytic hydrogenation performance, while for $Cu(L^2) \cdot (dpye)$, the binuclear Cu cluster was coordinatively saturated by the dpye ligand. As a benchmark reaction, the hydrogenation of 4-nitrophenol (4-NP) to 4-aminophenol (4-AP) under mild conditions (25°C, 1atm) was carried out employing $NaBH_4$ as a reducing agent, and it revealed that the reaction could quickly finished with a complete conversion within only 5 min, which showed a rate constant (k) of 0.850 min^{-1} that is about 8 and 16 times higher than that of $Cu(L^2) \cdot (dpye)$ and CuO NPs, respectively. In addition, $Cu_2(L^1)_2 \cdot (DMF)_2$ which could be reused at least eight times with on obvious decline in activity, also exhibited high hydrothermal stability, indicating a way to improve future catalysts with structure control. Recently, by using a functional metal-oxo cluster that is embedded into the frameworks, this group reported the structural modulation and catalytic performance of a series of Co-MOFs in aerobic oxidation reaction [115]. Three pillar-layered MOFs (Co$_6$-MOF-1–3) that is isoreticular based on 4,4′,4″-tricarboxyltriphenylamine (H$_3$TCA) and a unique $Co_6(\mu_3\text{-OH})_6$ cluster were designed, synthesized, and structurally characterized (Fig. **26**). Tuning the second bis(pyridines) ligands, the framework backbone of Co$_6$-MOFs could be well modulated, which result in adjustable apertures of 8.8-13.4 Å and high BET surfaces of 1896-2401 m^2 g^{-1}. Because of the variable valences of $Co_6(\mu_3\text{–OH})_6$ cluster, these MOFs were then used as possible heterogeneous catalysts with high conversion (>90%) for highly selective styrene and benzyl alcohol oxidation. The selectivity was over 80% for styrene to styrene oxide and up to 98% for benzyl alcohol to benzaldehyde. The TOF values showed the positive correlation between the reaction rate increasement and the pore sizes enlargement in these MOFs, which also had good stability and recyclability.

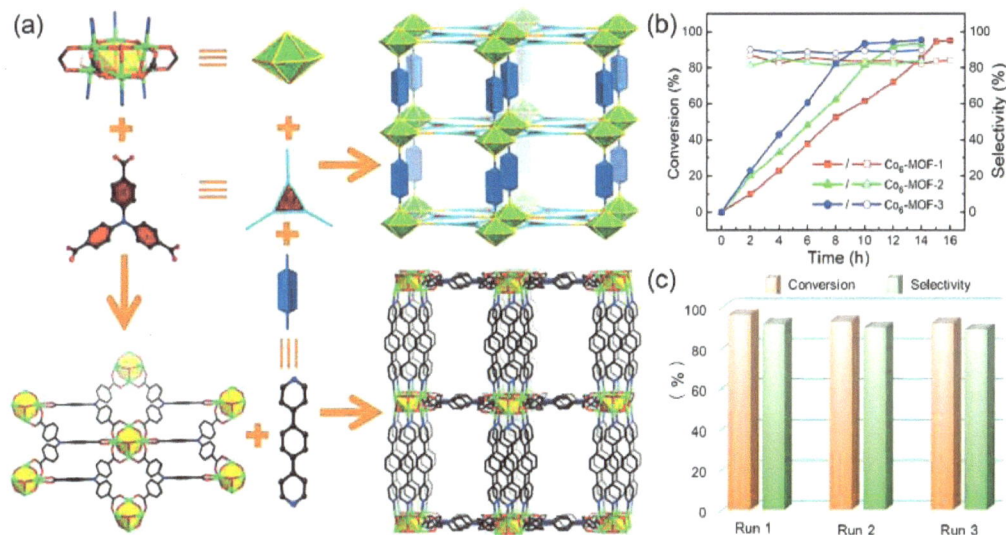

Fig. (26). (a) Crystal structures of Co$_6$-MOFs based on a unique $Co_6(\mu_3\text{-OH})_6$ cluster and H$_3$TCA; (b) the epoxidation of styrene catalyzed by Co$_6$-MOFs using air as the oxidant source; (c) conversion and selectivity in three cycles of styrene epoxidation catalyzed by Co$_6$-MOF-3 [115].

Leus *et al.* reported catalytic properties of V-MIL-47 with coordinatively saturated V^{+IV} sites connected by terephthalic linkers [116]. They evaluated V-MIL-47 for the catalysis of the cyclohexene epoxidation with tert-butylhydroperoxide (TBHP) as the oxidant. It was found that if dissolving the oxidant TBHP in water, a significant V-species leaching into the solution could be discovered, and the products were mainly cyclohexene oxide and tert-butyl-2-cyclohexenyl-1-peroxide, with cyclohexene oxide subsequently conversing to cyclohexane-1,2-diol as time goes. In contrast, if TBHP was dissolved in decane, V-MIL-47 could maintain its structural integrity during successive runs with negligible leaching, and the selectivity toward the cyclohexene oxide was very high in these circumstances. NMR and EPR measurements confirmed the co-existence of at least two parallel catalytic cycles: one with pre-oxidized V^{+V} sites and one with V^{+IV} sites.

Although very scarce, pristine MOFs could also be used in electrocatalysis. For example, Mao *et al.* firstly reported the usage of copper-based MOF (Cu-bip--BTC) as the electrocatalyst for ORR in 2012 [117]. The structure of Cu-bip--BTC was stable in neutral solution due to the Cu^{II} centers coordination mode. Cu-bipy-BTC showed a pair of clear redox peaks at *ca.* -0.15 V in a phosphate buffer with pH of 6.0. The presence of O_2 into the buffer could obviously increase reduction peak current, while decrease reversed oxidation peak current in redox wave.

4.2.2. Organic Ligands as Active Sites

Functional groups or small molecules with catalytic activity can also be introduced into MOFs in the form of ligands and participate the construction of frameworks, thus providing active sites for catalysis of MOFs. For example, the integration of pyridine, amino, aldehyde, urea and sulfonic groups into organic ligands can effectively catalyze transesterification, Knoevenagel condensation, Friedel Crafts and alkyd esterification reactions, while porphyrin, Schiff base, bipyridine, chiral bisphenol and catechol incorporated into MOFs can realize catalytic activity for the reactions of styrene oxidation, cyclohexane hydroxylation, carbon dioxide/epoxy cycloaddition, olefin oxidation, asymmetric synthesis, *etc.*This provides rich sources for the design of MOFs with high catalytic activity.

Hasegawa *et al.* created an amide functionalized MOF, [Cd(4-btapa)$_2$ (NO$_3$)$_2$]·6H$_2$O·2DMF (4-btapa = 1,3,5-benzene tricarboxylic acid tris[*N*-(--pyridyl)amide]) [118]. The tridentate amide ligand could effectively restrain the formation of hydrogen-bond in these amide groups, and produce attractive interactions with guest molecules. The amide groups, functioning as sites of guest interaction, occurred on the channel (4.7 × 7.3 Å^2) surfaces. They triggered a Knoevenagel condensation reaction of benzaldehyde with each of active methylene compounds (malononitrile, ethyl cyanoacetate, and cyano-acetic acid

tert-butyl ester) catalyzed by this Cd-MOF, which showed reactant-siz-
-dependent selective heterogeneous base catalytic properties. Therefore, the
malononitrile was proven to be a good substrate, with adduct conversion of 98%.
The guest-selective catalytic reactions suggest that the channels, not the surface,
supply place for the reaction. Moreover, the solid catalyst that could be easily
recycled maintained its framework structure after the reaction.

Commonly, ureas present aggregation and self-recognition behavior, leading to
the loss of their catalytic activities. With the mind to achieve catalyst spatially
isolated in a MOF environment, Roberts *et al.* synthesized a new urea-containing
MOF acting as a heterogeneous catalyst [119]. The combination of
$Zn(NO_3)_2·6H_2O$, and a symmetrical urea-tetracarboxylate strut (1-4H), 4,4′-
bipyridine (BIPY) afforded a microporous MOF, $Zn_2(BIPY)_2(1-4H)$ (NU-601)
under solvothermal conditions. This material was found to be an effective catalyst
with hydrogen-bond-donor for Friedel–Crafts reactions between nitroalkenes and
pyrroles, in spite of the much less competency of a homogeneous urea. For the
reaction of N-methylpyrrole and (*E*)-1-nitroprop-1-ene with concentration of 1.0
M, 98% consumption of (*E*)-1-nitroprop-1-ene was observed by the catalysis with
10 mol % NU-601 at 333 K after 36 h. A higher reaction rate of smaller substrates
versus larger ones strongly indicated the catalysis primarily occurred within the
pores of NU-601 rather than on its exterior. This work is the first example that
hydrogen-bonding catalysis engineered into MOFs.

Feng *et al.* reported a series of highly stable MOFs (PCN-224) with 3-D
nanochannels, which are assembled by tetrakis(4-carboxyphenyl)-porphyrin
(MTCPP) and six-connected Zr_6 cluster through a linker-elimination strategy
[120]. The PCN-224 series MOFs not only exhibited high BET specific surface
area of 2600 m^2 g^{-1} but also remained intact in pH = 0~11 aqueous solution. The
metalloporphyrins in the frameworks could be employed as active centers for
catalytic CO_2/propylene oxide coupling reaction. In a representative reaction,
tetrabutylammonium chloride (71.6 μmol), propylene oxide (35.7 mmol), and
PCN-224(Co) (32.1 μmol) were added into an autoclave reactor, which was then
pressurized to 2 MPa with CO_2 under 100 °C for 4 h. The results revealed that
PCN-224(Co) showed a TON of > 461 and TOF of 115 h^{-1}, which is comparable
to homogeneous cobalt porphyrin catalysts. Thanks to 3-D nanochannels of PCN-
224, the reactants could contact all the cobalt porphyrin centers and at the same
time the products could also easily come out from the pores, thus causing
excellent catalytic activity of PCN-224(Co). Importantly, the catalyst was highly
crystalline even after three consecutive catalysis runs confirmed by PXRD,
suggesting that PCN-224(Co) possess the reusability as a good heterogeneous
catalyst.

Catalytic cycloaddition of CO_2 with epoxides by using MOFs-based catalysts is another popular research hotspot for CO_2 conversion. However, the addition of molecular co-catalysts, *e.g.* tetrabutylammonium bromide (TBAB), is usually essential for the proceeding of CO_2 conversion in this kind of reactions, which complicates the separation process due to the homogeneous nature of co-catalysts. Liang *et al* [121]. firstly synthesized a bifunctional imidazolium functionalized zirconium MOF, (I^-)Meim-UiO-66, by isoreticular synthesis and a post-synthetic medication (PSM) method from the Zr-MOF (Im-UiO-66) containing imidazole (Fig. **27**). Im-UiO-66 was efficiently constructed using a delicate linear linker 2-(imidazol-1-yl)-terephthalicacid (Im-H_2BDC), and (I^-)Meim-UiO-66 was post synthesized by the reaction of Im-UiO-66 with CH_3I in CH_3CN solution under 353 K for 48 h. Pore features and crystal sizes of Im-UiO-66 was found to be tuned at the nanoscale. The (I^-)Meim-UiO-66 with bifunctions, containing both iodide ions and Brønsted acid sites, was a recyclable heterogeneous catalyst that is efficient for the cycloaddition of CO_2 with epoxides, without using any co-catalyst at ambient pressure. High conversion of 92% could be obtained for the epoxide allyl glycidyl ether with satisfactory carbonate selectivity of >81%. However, phenylglycidyl ether afforded moderate conversion of 76% and high selectivity of >92%. In contrast, moderate or low epoxide conversion and selectivity were obtained for the chain epoxides (propyleneoxide,1,2-epoxyhexane and 1,2-epoxyoctane). The carbonate selectivity of aromatic epoxides is obviously higher than those of chain epoxides, which could be possibly attributed to the hydrogen bonding, π...π and cationic...π interactions between frameworks and aromatic substrates.

(a)

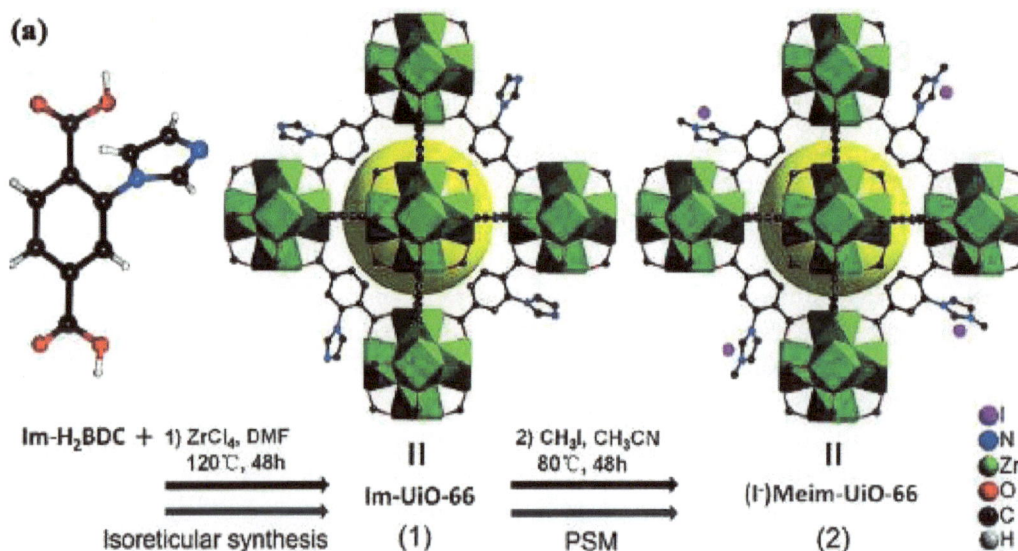

Im-H_2BDC + 1) ZrCl$_4$, DMF   2) CH$_3$I, CH$_3$CN
              120℃, 48h   80℃, 48h

                            II   II
                       Im-UiO-66   (I⁻)Meim-UiO-66

Isoreticular synthesis   (1)   PSM   (2)

● I
● N
● Zr
● O
● C
● H

(Fig. 27) contd.....

(b)

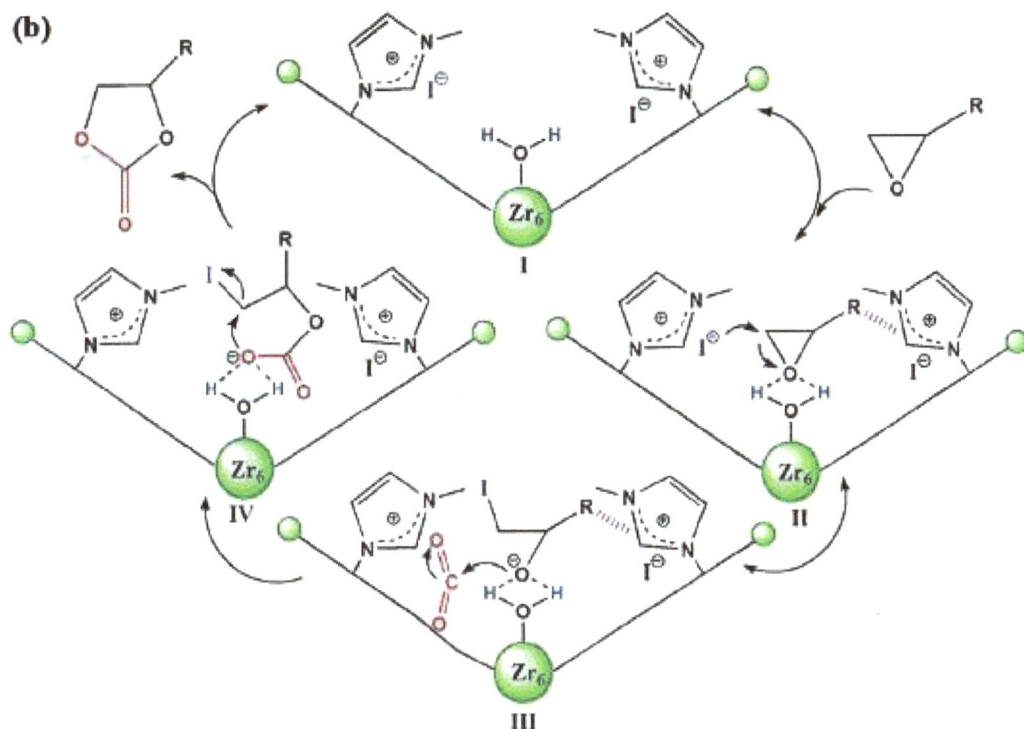

Fig. (27). (a) Syntheses of Im-UiO-66 and (I⁻)Meim-UiO-66 *via* reticular chemistry and PSM method, respectively. **(b)** Plausible mechanism of (I⁻)Meim-UiO-66 catalyzing cycloaddition reaction of CO_2 with epoxides (hashed bonds represent π---π, cationic---π or hydrogen bonding interactions) [121].

In addition, using privileged chiral small molecules or chiral organometallic compounds as active sites to take part in the construction of MOFs [122], Lin's group, Jeong's group and Cui's group have successfully developed a series of privileged chiral ligands, such as chiral metal binaphthol (BINOL)/biphenylphosphine (BINAP)/biphenol and chiral metal schiff base (salen) ligands, and synthesized a lot of chiral MOF-based heterogeneous asymmetric catalysts with excellent performances. For example, guided by the isoreticular expansion strategy and modulated synthesis, Cui's group recently synthesized five chiral Zr-MOFs that are chemically stabile, with **flu** or **ith** topologies through the assembly of Zr_6, Zr_9 or Zr_{12} clusters and enantiopure 1,1'-biphenol-derived tetracarboxylate linkers (Fig. **28**) [123]. The two **flu** MOFs which feature dihydroxyl groups of biphenol in large and open cages, could be used as recyclable heterogeneous catalysts that are highly efficient for α-dehydroamino acid esters hydrogenation with high ee value up to 98%, after sequential postsynthetic modification with P(NMe₂)₃ and [Ir(COD)Cl]₂. However, the three **ith** MOFs which feature dihydroxyl groups in small cages could not be installed with P(NMe₂)₃ for the supporting of Ir complexes. Moreover, the durability, chemical stability, and even stereoselectivity of these Zr-MOFs could be greatly enhanced, after the incorporation of Ir-phosphorus catalysts.

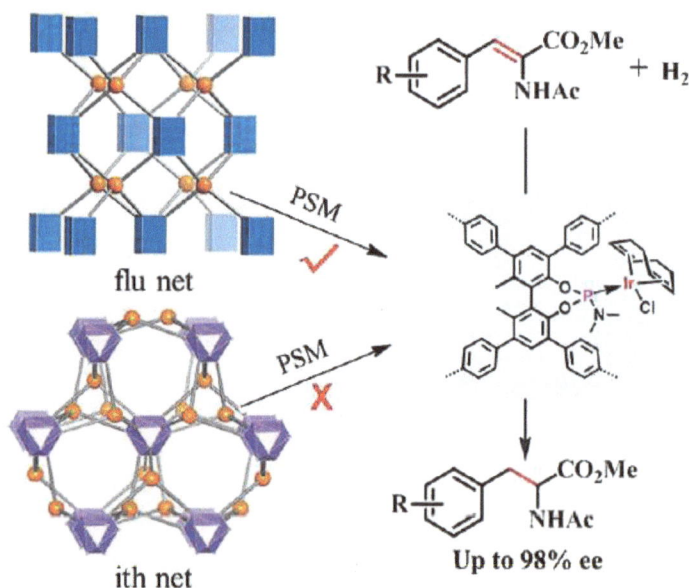

Fig. (28). Chiral Zr-MOFs with **flu** or **ith** topologies for asymmetric catalysis [123].

4.2.3. Guests as Active Sites

The small guest molecules inside MOFs pores can also provide catalytic centers to play a synergistic role towards excellent catalytic performance. Lions *et al.* studied the electrocatalytic ORR performance of Co-Al-PMOF, namely $Al_2(OH)_2(Co(tcpp)$ (tccp = tetrakis(4-carboxyphenyl)porphyrin) [124]. Shown from the electrochemical characterization for Co-Al-PMOF, it was found that in an O_2-saturated 0.1 M H_2SO_4, the onset potential (E_{onset}) is 0.75 V *versus* reversible hydrogen electrode (*vs* RHE) and the diffusion limiting current density (j_L) is -0.6 mA cm^{-2}. This investigation indicated the feasibility to catalyze ORR reaction by grafting homogenous guest molecular catalysts with ORR activity into MOFs as active linkers.

Recently, Sun and co-authors reported an easy way to modify MOFs with ionic liquids (ILs), which is useful in heterogeneous catalysis of CO_2 chemical fixation [125]. The authors firmly grafted an amino-functionalized imidazolium-based ionic liquid into MIL-101−SO_3H through the acid−base attraction between negatively charged sulfonate groups from MIL-101−SO_3H and positively charged ammonium groups on the IL. Analyzing from X-ray photoelectron spectroscopy, ^{1}H NMR, Fourier transform infrared spectroscopy, and N_2 sorption experiments, it revealed the obtained IL@MIL-101−SO_3H material was still intact while acting as a recyclable heterogeneous catalyst capable of efficiently converting epichlorohydrin and CO_2 into chloropropene carbonate with no cocatalyst addition. Compared with the reactions with no catalyst or MIL-101−SO_3Na, IL@MIL-101−SO_3H showed great activity enhancement under the same conditions. Further studies revealed that optimizing the reaction conditions, it could reach a high yield of 98% with CO_2 of 1 atm at 90 °C after 24 h.

By encapsulating Fe(III)-protoporphyrin IX (Fe(III)PPIX) into a Zinc MOFs ([$H_2N(CH_3)_2$][$Zn_3(TATB)_2(HCOO)$]·$HN(CH_3)_2$·DMF·$6H_2O$, TATB = 4,4′,4″---triazine-2,4,6-triyl-tribenzoic acid), Dare *et al.* prepared a Fe(III)PPIX@Zn-MOF host-guest system [126], which can efficiently catalyze oxidation of various substrates (including hydroquinone, thymol, benzyl alcohol and phenyl ethanol) and showed dramatic improvement compared to Fe(III)PPIX.

4.3. Luminescence

Luminescent MOFs combining two characteristics of porosity and luminescence have shown great potential applications in light-emitting diode (LED), chemical sensing and bioimaging. The luminescence of MOF can be divided into two mainstreams. One is the luminescence of framework itself which can be realized by using π-conjugated ligands or metal centers (lanthanide, metal clusters) as luminescent sources. In this case, the stability and rigidity of organic ligands can be reinforced by the metal-ligand coordination, which will decrease the non-radiation energy loss, thus improving ligand-based luminescent properties. For the metal-based luminescence, for example, lanthanide ions could be sensitized to produce luminescence by the antenna effect of organic ligands through resonant coupling or d-block metals through d→f energy transfer. The other is the luminescence induced by host-guest interaction between the frameworks and internal guests. At this situation, utilizing controllable porosity of MOFs, guest molecules or ions with special structural features and functions can generate host-guest interactions, thus realizing the modulation of luminescence or luminescence-based chemical sensing of specific matters. In this section, we mainly discuss recent development of luminescent MOFs towards some promising applications.

4.3.1. Chemical Sensors

The luminescent properties of MOFs are closely related to structural topology, coordination environment and pore surface, and very sensitive to weak interactions between the frameworks and guests, such as hydrogen bonds, Van der Waals force, π-π interaction. Therefore, it has been a promising research area using luminescent MOFs as fluorescent probe for small molecules and ions.

The most common MOF-based chemical sensors are for the detection of metal ions with a mechanism of fluorescence quenching or enhancement. For example, Liu *et al.* reported a Cd^{II}-based MOF [127], $[Cd_2(bbib)_2(ndc)_2] \cdot 2DMF$ (JXUST-1; bbib=1,3-bis(benzimidazolyl)benzene, H_2ndc=1,4-naphthalenedicarboxylic acid) with pcu topology as turn-on fluorescent sensor for Al^{3+} with detection limit of 0.048 ppm. Fluorescence studies showed that the naked-eye recognition of suspension JXUST-1 with 10^{-3} M Al^{3+} ion was detected by λ_{ex} =365 nm with a UV lamp with color changing from light blue to blue, while other common metal ions (K^+, Na^+, Li^+, Ag^+, Ca^{2+}, Mg^{2+}, Co^{2+}, Ni^{2+}, Mn^{2+}, Cu^{2+}, Cd^{2+}, Zn^{2+}, Hg^{2+}, Sm^{3+}, Gd^{3+} and Cr^{3+}) did not present this behavior. Clearly, the sensitive and selective recognition for Al^{3+} was attributed to a fluorescence enhancement effect. Furthermore, JXUST-1 displayed good thermal and chemical stabilities as well as reusability.

Qian's team reported a series of Ln-MOF (Ln = Eu^{3+} and Tb^{3+}) based fluorescence sensors towards the recognition of small organic molecules, anions and metal ions [128 - 130]. Eu(BTC) (BTC = benzene-1,3,5-tricarboxylate) showed a porous framework with a pore size of 6.6 × 6.6 Å^2, which presented different fluorescence response (enhancing with DMF and quenching with acetone) to different solvents. Tb(BTC) exhibited high-sensitivity sensing with respect to F$^-$ ion from methanol solutions of different anions (F$^-$, Cl$^-$, Br$^-$, NO_3^-, CO_3^{2-} and SO_4^{2-}) based on fluorescence enhancing effect derived from F$^-$$\cdots$H-OCH$_3$ hydrogen bond, which prevented quenching effect for the bond stretching of O-H. For $[Eu(pdc)_{1.5}(DMF)] \cdot (DMF)_{0.5}(H_2O)_{0.5}$ (pdc = pyridine-3,5-dicarboxylate) with a 1D hexagonal channel of about 6.3 × 8.5 Å^2, the free Lewis basic pyridyl sites inside the framework could coordinated with incorporated Cu^{2+} ion, which restricted the antenna effect. So, this MOF exhibited sensor capability toward Cu^{2+} ion based on strong quenching effects.

Detecting high explosives has attracted increasing attention owing to earth security, humanitarian implication and environmental protection. Li's group has developed high explosive detection functionality of luminescent MOFs [131 - 135]. In 2009, they reported the first example of high explosive sensing based on a luminescent MOF [131], $Zn_2(bpdc)_2(bpee)$ (bpdc = 4,4'-biphenyldicarboxylate; bpee = 1,2-bipyridylethene), which can rapidly and reversibly detect 2,4-dinitrotoluene (DNT) and plastic explosive taggant 2,3-dimethyl-2-3-dinitrobutane (DMNB) vapors with unprecedented sensitivity, through redox fluorescence quenching. At room temperature, this MOF was found to show a fluorescence intensity quenching over 80% when exposed to equilibrated DNT (ca. 0.18 ppm at 298 K) and DMNB (ca. 2.7 ppm at 298 K) vapors within 10 s. Additionally, both DNT and DMNB detections were completely reversible. The photoluminescence could be recovered after quenching by simply heating at 423 K for about 1 min. Interestingly, red-shifts of fluorescence peaks were obviously recorded upon exposure to DNT and DMNB, indicating the guest-dependent interaction between analytes and host framework.

Usually, chemosensors can only detect single targeted molecule from several kinds of molecules. Takashima *et al.* reported a molecular decoding strategy, and found that single host domain can accommodate a class of molecules and differentiate them with corresponding readout [136]. The researchers synthesized a two-fold interpenetrated MOF, $Zn_2(bdc)_2(dpNDI)$ (bdc = bdc, 1,4-benzenedicarboxylate; dpNDI = N,N'-di(4-pyridyl)-1,4,5,8-naphthalenediimide), undergoing dynamic structure transformation due to the dislocation of two non-interconnected frameworks (Fig. **29**). It confined a class of tropospheric air pollutants, or aromatic VOCs, and the chemical substitution information of aromatic species were decoded into recognizable photoluminescence signals in visible light region. The fluorescence of the as-synthesized and dried samples was very weak with short average lifetime ($\tau \sim 150$ ps) and very low quantum yield ($\Phi_f < 0.01$), which was recognized as a luminescence-off state. However, after incorporating a series of aromatic compounds, an intensive turn-on emission was observed, and different luminescent colors dependent on chemical substituent of aromatic guest were obtained. Red shift gradually happened with the increasing of VOCs electron-donating capability after excitation at 370 nm, as exhibited in the normalized emission spectra. The encapsulation of aromatic species benzene, toluene, xylene, anisole and iodobenzene into MOFs were indicated as blue (λ_{em} = 439 nm), cyan (λ_{em} = 476 nm), green (λ_{em} = 496~518 nm), yellow (λ_{em} = 592 nm) and red (λ_{em} = 640 nm), respectively. Furthermore, the iodobenzene-encapsulated MOF has the highest average lifetime of 139 μs and the xylene-encapsulated MOF has the highest Φ_f of 22%. This chemoresponsive and multicolor luminescence was derived from host-guest intensive interaction caused by the structure transformation.

Besides the detection mentioned above, MOFs can also be used as excellent sensors for gases and biomolecules [137 - 145]. For example, for the first time, Cui *et al.* reported the incorporation of an organic dye perylene into a lanthanide MOF, $[Eu_2(QPTCA)(NO_3)_2(DMF)_4] \cdot (CH_3CH_2OH)_3$ (ZJU-88; H_4QPTCA = 1,1':4',1":4",1'''-quaterphenyl-3,3''',5,5'''-tetracarboxylic acid) to form ZJU-88⊃perylene composite[146], which featured a red-emitting at 615 nm and appended blue emitting around 473 nm, arising from Eu^{3+} ion and perylene dyes, respectively. ZJU-88⊃perylene was able to be highly temperature-sensitive over the physiological temperature range with a maximum sensitivity of 1.28% $°C^{-1}$ at 20 °C, due to the unique energy transfer between the perylene molecule and Eu^{3+} ion, and the luminescent intensity ratios between Eu^{3+} ion and perylene dye were linearly dependent very well on the temperature from 20 to 80 °C.

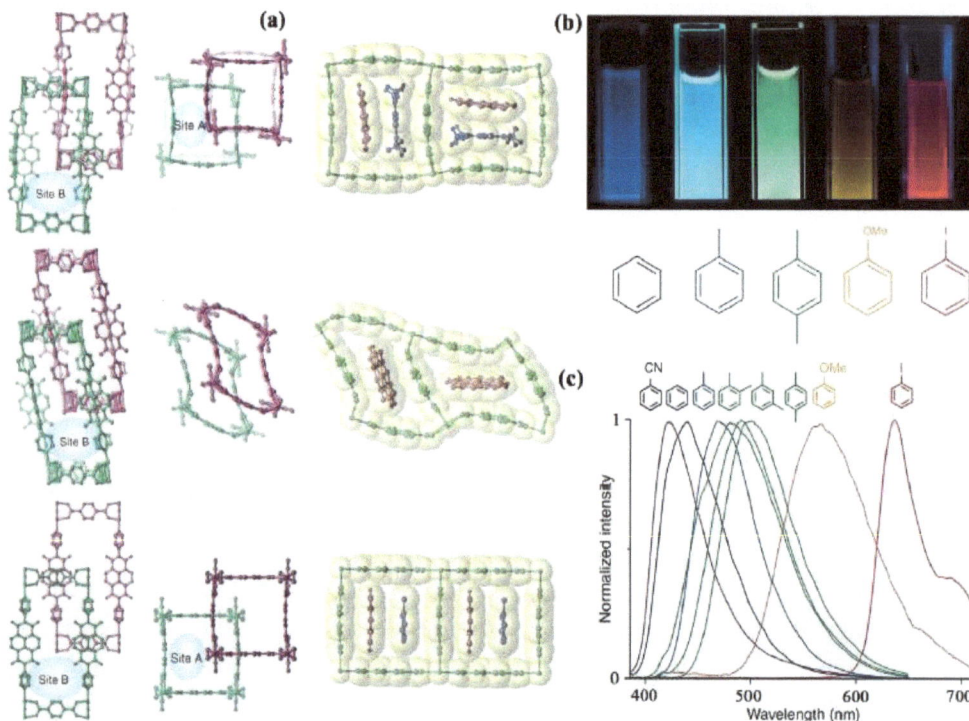

Fig. (29). (a) Structural dynamics of $Zn_2(bdc)_2(dpNDI)$ during the removal and incorporation of guest molecules; **(b)** the resulting luminescence when suspended in VOC liquids after excitation at 365 nm; **(c)** height-normalized luminescent spectra of $Zn_2(bdc)_2(dpNDI)⊃VOCs$ after excitation at 370 nm [136].

Through careful design, some MOFs have been proved to be very sensitive to external condition changes, and recent studies on luminescent sensing of MOFs are expanding from small molecule or ion detection to other detection stimulus such as temperature or pressure. Yao *et al.* reported a new dual-stimuli-responsive MOF, $[Cd_3(TPPA)_2(NDA)_3]·(DMF)_{10}·(H_2O)_6$ (NKU-121, TPPA = tri(4-(pyridin--4-yl)phenyl)a-mine, NDA = 2,6-naphthalenedicarboxylic acid), which deformed in a reversible and smooth way under either pressure or thermal stimulation and featured a reversible luminescence-tuning behavior in a wide range (Fig. **30**) [147]. In this system, organic fluorophores (TPPA molecules) were bound to the channel by coordination in a face-to-face arrangement, responding to the structure deformation through fine adaption of their arrangement and conformation.

Therefore, NKU-121 exhibited a remarkable luminescence shift in a dual-stimul--responsive manner across almost the whole visible region. The color of emission for NKU-121 crystals gradually altered from cyan to green upon heating, which then turned to red upon compression of pressure. Moreover, each stage was fully reversible, and both emission intensity and maximum exhibited linear dependence on the stimulus.

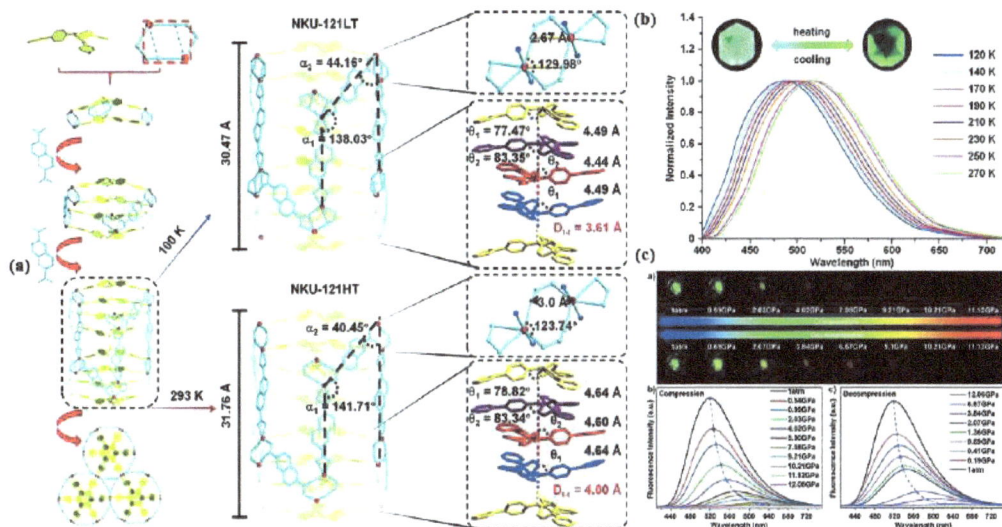

Fig. (30). **(a)** Dynamic structures of hinged-fence-like scaffold of NKU-121 at 100 and 293 K; **(b)** Normalized fluorescence spectra of NKU-121 (ex = 365 nm) in response to temperature; **(c)** Photographs of NKU-121 during a compression–decompression cycle under UV illumination at 365 nm and *in-situ* emission spectra [147].

4.3.2. MOF-based LED Materials

White light LEDs (WLEDs) are important in daily life, and one of the white-light sources that is most widely used is mercury-containing fluorescent lighting. White light is composite light, and commonly, the methods generating white light include: (i) monochromatic emitters emitting in the whole visible spectrum; (ii) dichromatic emitters blending yellow and blue light; (iii) trichromatic emitters combining blue, green and red lights. Among these approaches, the dichromatic and trichromatic emitters are mainly focused due to their higher luminescent efficiency and finer color-rendering properties. Spectrum of good WLED materials should meet the criteria of Commission International de I'Eclairage (CIE) coordinates close to (0.33, 0.33), correlated color temperature (CCT) between 2500~6500 K and color-rendering index (CRI) higher than 80.

Recent studies on luminescent MOFs have demonstrated that it is very promising to design MOF-based WLED materials. For example, Wang *et al.* reported a direct white-light MOF, [Ag(4-cyanobenzoate)]$_n$·nH$_2$O, obtained by the reaction of AgNO$_3$ and deprotonated 4-cyanobenzoic acid in water [148]. Interestingly, this MOF showed tunable photoluminescence of yellow-to-white by use of different excitation light with two main emission peaks at 427 nm and 566 nm, which could be attributed to intraligand $\pi \rightarrow \pi^*$ electron transition and metal to ligand charge transfer (MLCT), respectively. Moreover, this MOF presented close pure white emission with CIE coordinates of (0.33, 0.34) if excited by UV light of 349 nm, which present compatibility to light output of deep UV LED. Also, some main-group-metal based MOF (for example, bismuth and lead) have been reported with direct white light emission [149, 150].

Lanthanide ions have obvious advantages to obtain white light emission, due to their colorful emissions, good monochromaticity and luminescent stability. WLED materials can be conveniently modulated through synthesizing rare earth MOFs or doping. For example, Sava *et al.* presented a indium MOF, In(BTB)$_{2/3}$(OA)(DEF)$_{3/2}$ (SMOF-1, BTB = 1,3,5-Tris(4-carboxyphenyl)benzene, OA = oxalic acid) featuring inherent broad-band direct white emission, which can be improved by incorporating red-emitting Eu^{3+} ion into the framework [151]. After Eu^{3+} introducing, Eu-doped SMOF-1 with 2.4%, 4.8% and 9.0% Eu^{3+} contents were successfully obtained, and the 9.0% Eu-doped SMOF-1 showed the CRI of 90 and CCT of 3200 K closer to the target. In this system, the structural-property relationship is driven by two unique structural features that is complementary with each other: interpenetration and corrugation.

In addition, combining yellow, green or red emitted organic luminescent dyes or other luminescent species into MOFs with blue emission features can also obtain white light. Xia *et al.* reported a Zn(II) based MOF, Zn$_4$OL$_2$·xDMF, showing a highly porous framework structure and intense blue-light emission based on carbazole, by using 4,4′,4″-(9H-carbazole-3,6,9-triyl)-tribenzoicacid (H$_3$L) [152]. These characteristics make the MOF an ideal platform of energy transfer to accommodate and sensitize green and red fluorescent dyes for the modulation of white light emission. By the simultaneous introduction of 4-(dicyanomethylene--2-methyl-6-(p-dimethylaminostyryl)-4H-pyran (DCM) (green fluorescer) and coumarin 6 (C6) (red fluorescer) into the host MOF framework, and careful control of their contents, the MOF emission could be easily modulated based on the effective framework to dye energy transfer. Finally, a luminophore (Zn$_4$OL$_2 \supset$ DCM/C6, 0.04 wt% DCM, 0.03 wt% C6) with high-performance white-ligh--emitting was obtained and showed a wide-range emission band from 350 to 675 nm. The luminophor revealed CIE coordinates of (0.32, 0.31), a CRI of 91, a CCT of 6186 K, and a high QY up to 39.4%.

Sun *et al*. realized efficient white-light emission through the encapsulation of an iridium complex into the cavity of a MOF [153]. They prepared a blue-emitting anionic MOF (λ_{em} = 425 nm) with mesoporosity, $[(CH_3)_2NH_2]_{15}[(Cd_2Cl)_3$ $(TATPT)_4]\cdot12DMF\cdot18H_2O$ (Cd-MOF, H_6TATPT = 2,4,6-tris(2,5-dicarboxyl phenyl-amino)-1,3,5-triazine) as host to encapsulate a yellow emitting iridium complex, $[Ir(ppy)_2(bpy)]^+$ (λ_{em} = 570 nm) (Fig. **31**). The resultant composites showed dual wavelength emission at 425 and 570 nm, and could emit good-colo- -quality bright white light with 3.5 wt% $[Ir(ppy)_2(bpy)]^+$ concentration. The CIE coordinates, CCT and CRI were calculated to be (0.31, 0.33), 5409 K and 84.5, respectively, and when excited at 377 nm, high quantum yield (Φ_f) was also obtained up to 20.4%.

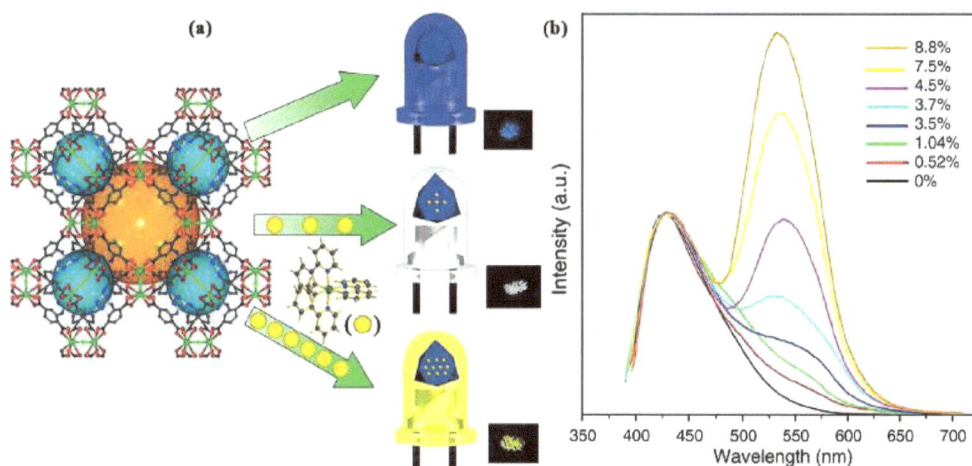

Fig. (31). (a) The encapsulation of $[Ir(ppy)_2(bpy)]^+$ into Cd-MOF; **(b)** room temperature emission spectra of Cd-MOF and $[Ir(ppy)_2(bpy)]^+$@Cd-MOF with different $[Ir(ppy)_2(bpy)]^+$ concentrations [153].

Much attention has been attracted by organic donor-acceptor systems because of their plenty of potential applications. However, it has been a challenge to rationally construct and modulate highly ordered donor-acceptor systems because the self-assembly process between donor and acceptor species is very complicated. Bu's group reported the potential application of crystalline host–guest systems as an ideal platform for systematic investigations of donor–acceptor materials [154]. In this work, as a typical example, the researchers selected a well-established MOF $[Cd_3(tpt)_2(PTA)_3(H_2O)_3]\cdot2H_2O$ (NKU-111, tpt = 2,4,6-tri(4-pyridinyl)-1,3,5-triazine, PTA = p-phthalic acid) featuring electron-deficient triazine motif as a host framework, and through rationally introducing the electron-rich aromatic guests by a one-pot method, a series of host-guest donor-acceptor MOFs, NKU-111⊃guest, were obtained (Fig. **32**). Due to the confined cage-based porous framework of host NKU-111, the guest molecules could be arranged in a highly ordered manner, ensuring the proper (guest molecules) donor-acceptor (host framework) interaction. Therefore, NKU-111⊃ guests revealed highly tunable charge-transfer based emissions (green to red, with high Φ_f up to 28.34% for NKU-111⊃dibenzothiophene) and conductivities $(10^{-10} \sim 10^{-7}$ S cm$^{-1})$, which are guest-dependent. It should be noted that the successful synthesis of highly tunable NKU-111⊃guests should be attributed to special framework structure of NKU-111: (1) although NKU-111 showed a three-fold interpenetrated **tfz** framework, it is still high porosity of 42.2%; (2) interestingly, a triangular prism-shaped cage was defined by the stacked tpt ligand pairs in the individual framework and PTA ligands. Then, one cage with a hexagonal prism-shaped interspace was generated by the interlocking of two cages from distinct networks, due to framework interpenetration; (3) the guest accessible pore volume is about 4 Å (height) × 12 Å (side) × 14 Å (corner), presenting only narrow apertures allowing for the encapsulation of electron-rich guests with proper dimensions in a monodisperse manner and promoting donor-acceptor interaction with electron-deficient tpt motif. Later, this group further used NKU-111 as a host bind to various electron-rich guests (donors) and reported the engineering donor–acceptor heterostructure MOF crystals for photonic logic computation [155]. NKU-111 heterocrystals exhibited spatially segregated multi-color emission that is resulted from guest-dependent charge-transfer emission, and spatially effective mono-directional energy transfer could be obtained through selecting donor guest molecules to tune the energy gradient between adjacent domains.

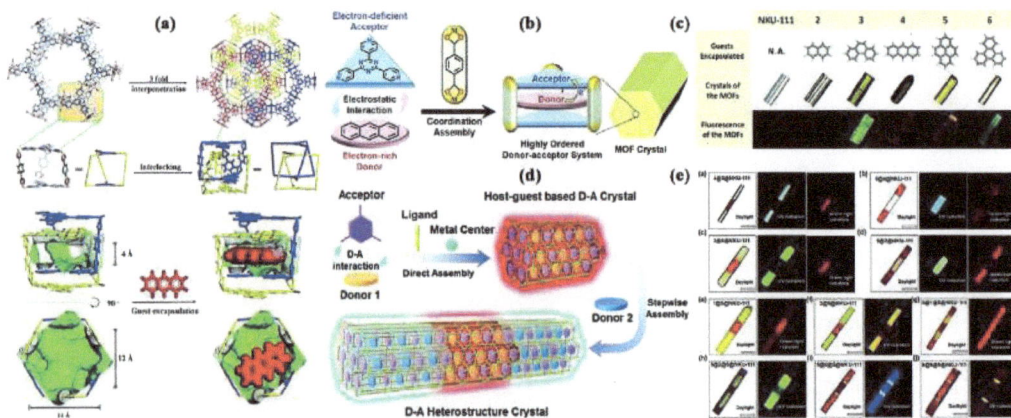

Fig. (32). (a) Structures of NKU-111; **(b, c)** Schematic illustration for the construction of donor–acceptor systems utilizing host–guest MOFs, and the guest-dependent optical (middle) and luminescent (bottom) images of NKU-111⊃guest crystals; **(d, e)** the targeted design and construction of D–A heterostructure crystals, and three domains and five domains heterostructure crystals under daylight, UV radiation (340–380 nm) and green light radiation (515–560 nm) [154, 155].

4.3.3. Bioimaging

Research and development of new nano drugs for disease diagnosis and treatment have become an important goal in the field of nano medicine. Due to highly modulable structures of MOFs as well as their easy surface modification, nanoscale MOFs (NMOF) have shown great potential applications in biomedical science such as bioimaging, biomarkers and drug delivery.

Lin's group have done many researches in this field [156 - 160]. In 2006, they reported NMOF as multimodal contrast enhancing agents for imaging [156]. $Gd(BDC)_{1.5}(H_2O)_2$ nanorods (BDC = 1,4-benzenedicarboxylate) with a length of ~100 nm and ~40 nm in diameter was synthesized by a microemulsion method (Fig. **33**). This NMOF exhibited extraordinarily large longitudinal ($R1$ = 35.8 s^{-1} per mM of Gd^{3+}) and transverse ($R2$ = 55.6 s^{-1} per mM Gd^{3+}) relaxivity, which was higher than commercial Gd-DTPA and made it efficient T1 and T2 contrast agents for MRI, since up to tens of millions of Gd^{3+} centers present in each nanoparticle. This NMOF could also show high luminescence through doping with Eu^{3+} or Tb^{3+} centers, suggesting the potential application in contrast agents for multimodal imaging by using NMOFs. Anther NMOF [158], $Gd_2(bhc)(H_2O)_6$ (bhc = benzenehexacarboxylate) was also prepared by Lin's group, which showed similar functionalities with relaxivity of $R1$ =1.5 mL $(mmol·s)^{-1}$ and $R2$ = 122.6 mL $(mmol·s)^{-1}$. They also synthesized Mn-based NMOFs[159], $Mn(BDC)(H_2O)_2$ and $Mn_3(BTC)_2(H_2O)_6$ (BTC = trimesic acid), showing excellent MR contrast enhancement with $R1$ values of 5.5 and 7.8 mL $(mmol·s)^{-1}$, and $R2$ values of 80 and 78.8 mL $(mmol·s)^{-1}$, respectively. Further surface modification of these Mn NMOFs with a silica shell allowed the covalent attachment of organic fluorophore and a cyclic RGD peptide. The delivery to cancer cells was enhanced by the cell-targeting molecules on Mn NMOFs, allowing for *in vitro* target-specific MR imaging. The enhancement of MR contrast was also proven *in vivo* using a mouse model, and T1-weighted contrast enhancement was observed in the mouse liver, kidneys, and aorta after tail vein injection of $Mn_3(BTC)_2(H_2O)_6$ at a 10 μmol/kg Mn dose (~1 h). This work provides an ideal platform for targeted delivery of other therapeutic and imaging agents to diseased tissues by core-shell hybrid nanostructures.

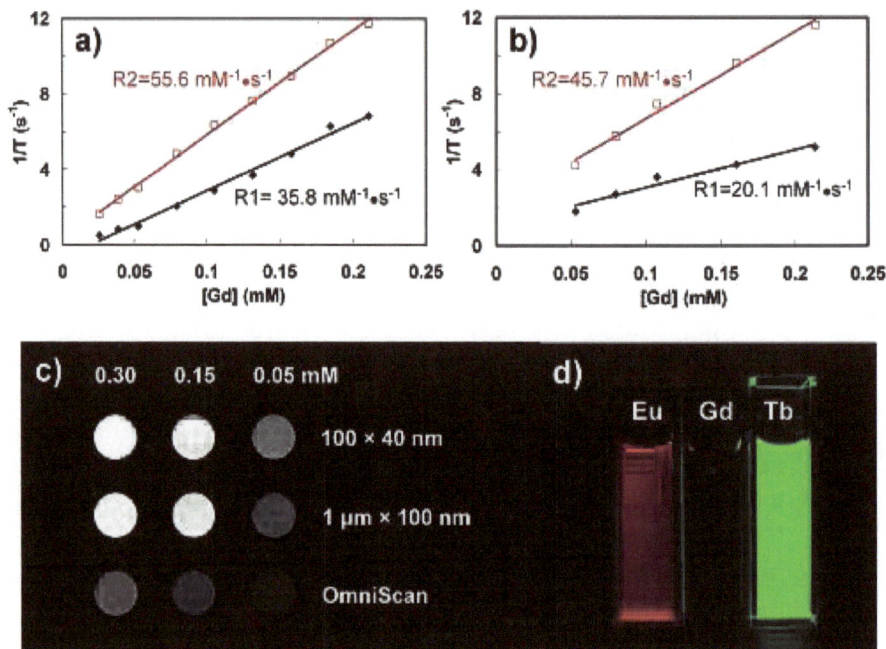

Fig. (33). (a, b) R1 and R2 relaxivity curves of Gd(BDC)$_{1.5}$(H$_2$O)$_2$ and Gd$_2$(bhc)(H$_2$O)$_6$, respectively. **(c)** T1-weighted MR images of Gd-NMOF in water containing 0.1% xanthan gum. **(d)** Luminescence images of ethanolic suspension of Gd-NMOF, Gd-NMOF doped with 5 mol % of Eu^{3+} and 5 mol % of Tb^{3+} ions [156].

Rowe *et al.* demonstrated incorporation of MRI contrast agent (Gd(BDC)$_{1.5}$(H$_2$O)$_2$ nanoparticle), a cellular level imaging agent (fluorescein O-methacrylate), a targeting moiety (H-glycine-arginine-glycine-aspartate-serine-NH$_2$), and an antineoplastic drug into a single nanoscale theragnostic device, which preserved all properties of individual components. These versatile and biocompatible nanoscale scaffolds had bimodal imaging, cancer cell targeting, and disease treatment capabilities.

Luminescent reporter that emits in near-infrared (NIR) region is very beneficial to biological imaging applications, mainly because that 1) in the NIR window, biological materials present low autofluorescence, allowing facile discrimination between the background and desired signal of reporter, which lead to the enhancement of the signal-to-noise ratio and improvement of the detection sensitivity; 2) there are less scatters of NIR light than visible light, therefore resulting into the increased resolution of the optical imaging; 3) compared with visible photons, there are less interaction between biological materials and NIR photons, thus decreasing the disturbing or damaging risk of biological systems. MOFs with NIR properties have been reported recently by researchers and obviously they might have the possible applications for NIR bioimaging. For example, Foucault-Collet *et al.* created a unique NIR–emitting NMOF (Nano-Y--PVDC-3) incorporating Yb^{3+} cation and phenylenevinylenedicarboxylate (PVDC) sensitizers derived from phenylene [161], which could be injected into living cells for NIR imaging. Nano-Yb-PVDC-3 was stable in certain biological media, did not photobleached, and had an IC_{50} of 100 µg/mL, which ass sufficient to allow live cell imaging. Therefore, by the usage of Nano-Yb-PVDC-3, an increased number of emitted NIR photons per unit volume, NIR microscopy images were able to be observed in living cells based on the signal arising from Yb^{3+} ion, which was sensitized *via* the antenna effect of PVDC.

4.4. Drug Delivery

In 2006, Férey's group exposed remarkable capacity for drug hosting and controlled delivery in MOFs, opening the way to design and construct MOFs-based platform for drug delivery [162]. When using MOFs as drug delivery hosts, some important issues must be concerned: 1) the tolerance of frameworks under human body environment; 2) drug loading capacity; 3) biocompatibility to human body; 4) controllable release of drug. The advantages of large surface area, versatile structures and weak coordination bond make MOFs easily meet the need of first three issues. However, construction of controlled MOFs-based drug release systems is still full of challenge. Luckily, MOFs can be introduced with various host-guest interaction between frameworks and drugs. Recent studies have proposed a variety of MOFs-based delivery systems showing high drug loading capacity and controllable release [163].

NU-1000 is a MOF containing mesopores of 3.1 nm and micropores of 1.2 nm in diameters, which is derived from $Zr_6(\mu_3\text{-O})_4(\mu_3\text{-OH})_4(H_2O)_4(OH)_4$ nodes and tetratopic 1,3,6,8-tetrakis(*p*-benzoate)pyrene (TBAPy) linkers [164], and has recently been applied as a promising drug delivery system. Pander *et al.* achieved the incorporation of drug molecules inside MOFs by using a solvent-assisted ligand incorporation (SALI) technique [165]. Three drug molecules containing carboxylate groups: ketoprofen, nalidixic acid and levofloxacin are successfully anchored in NU-1000 with drug loading of 30 wt%, finally yielding guest@NU-1000 materials. The drug release kinetics strongly depends on the size of guest molecules, and drug molecules can be released from guest@NU-1000 by the phosphate concentration in blood plasma without decomposition of MOF frameworks.

Oral insulin delivery is very important for the treatment of diabetes. Chen *et al.* [166] have found that NU-1000 could sever as a potential carrier for the immobilization of insulin, which shows high loading capacity of ~40 wt% only in

30 min. The nanopores of NU-1000 are size suitable for insulin (13 Å × 13 Å) diffusion through the framework to facilitate the encapsulation, and forbid the entrance of pepsin (48 Å × 64 Å) to suppress its proteolysis. Moreover, NU-1000 is acid-stable and can be disassemble in the presence of phosphate ion in the blood stream to initiate insulin release. Therefore, the NU-1000 capsules are found to effectively prevent insulin from degrading in the presence of stomach acid (pH = 1.29) with the digestive enzyme, pepsin, and only 10% of insulin are released after 60 min. However, the encapsulated insulin can be gradually released from NU-1000 under simulated physiological conditions (pH = 7.0), and most of encapsulated insulin (91%) is released from insulin@NU-1000 after 1 h. There results suggest that insulin@NU-1000 could withstand low pH environment of stomach and release insulin in desired environment.

Teplensky *et al.* engineered an optimal MOF nanosystem, which can provide long-term controlled release and super-resolution for verifying *in vitro* phenomena [167]. They presented an improved method to develop a mild temperature treatment protocol, which could cause partial pore decomposing of NU-901 and NU-1000 so that to delay the release of a model compound calcein and anticancer therapeutic α-Cyano-4-hydroxycinnamic acid (α-CHC). Importantly, a super-resolution microscopy technique, namely structured illumination microscopy (SIM), was applied for the first time in this work to visualize the process of MOF taken into HeLa cells. When using temperature-treated NU-901 and NU-1000 to capture α-CHC (denote α-CHC@t.t.MOFs), the loading capacity increased from 54.6 and 68.5 wt% to 79.0 and 81.0 wt%, respectively. With the increment of α-CHC loading, the therapeutic efficacy is improved and far beyond free drugs.

Besides NU-1000, the MIL-, UiO, ZIF-series or other kinds of MOFs are also used as carrier for the studies of drug delivery [168 - 178]. Dong *et al.* reported the guest release control profiles by dialing-in desirable interaction between MOF pores and guest molecules [179]. In this work, by the multivariate strategy [20], $-H$, $-NH_2$, and $-C_4H_4$, as different functional groups, were introduced into MIL-101(Fe) with three molecular guests (guests@MIL-101(Fe), rhodamine B (RhB), ibuprofen (Ibu), and doxorubicin (DOX)), which were encapsulated into the pores for continuous drug release (Fig. **34**). Interestingly, it is found that host-guest interaction can be obtained by the rate constants (k) that are quantitatively correlated with the functional group type and proportion in the MOF structure, which thus lead to predicted and programmed release of guest molecules. The probe molecules release rate (r) can be well modulated in the range of $0.005 \sim 0.16$ d^{-1} (RhB), $0.30 \sim 1.14$ d^{-1} (Ibu), $0.008 \sim 0.022$ d^{-1} (DOX), respectively, and the maximum guests release time can be programmed in the range of $17 \sim 29$ days for DOX. Additionally, the corelease of drug molecules can also be accomplished between Ibu and DOX in the guest@MIL-101(Fe) system.

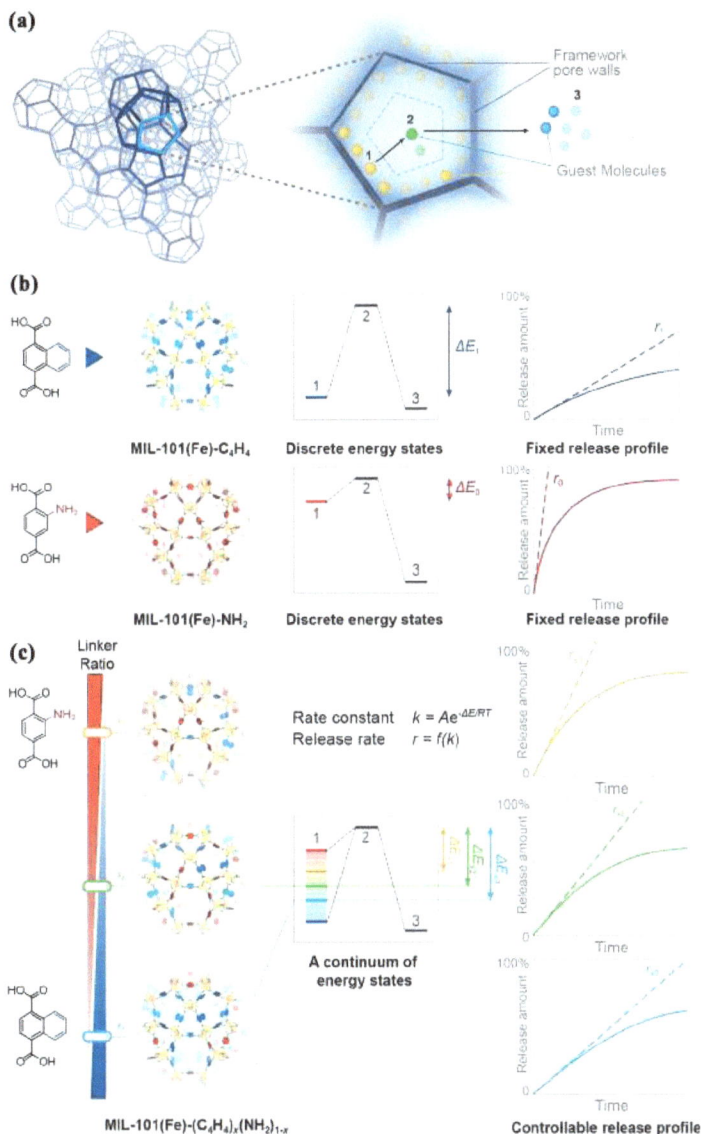

Fig. (34). Comparison of multivariate MOFs with their single-component counterparts in the binding energy states and guest molecules release profiles. **(A)** the release process of guest molecules from porous frameworks through three energy states; **(B)** single-component MOFs provide discrete energy states representing specific interaction with guest molecules; **(C)** multivariate MOFs with various linker ratios present a series of continuous energy states from which guest release kinetics can be dialed-in over a wide range [179].

Prior to the modulation of MOFs guest-encapsulation properties, a critical problem is essential be solved: how to prevent or delay the encapsulated molecules leaching from open pores, considering that in such porous hosts the guests encapsulation/release is naturally a reversible process. Wang *et al.* developed a novel MOF-based crystalline capsules by locking the channels of MOFs using size-matching ligand bolts, which might be potentially used for drug delivery [180]. The researchers synthesized a Na-Zn-MOF with a NbO topology by a solvothermal reaction of Zn^{2+}, 2,5-furandicarboxylic acid, and NaOH (Fig. 35). Interestingly, each Na^+ ion coordinated with two water molecules could provide two OMSs upon activation, which act as anchor sites for the size-matching auxiliary ligands to graft on. The ligand 2,4,6-tris(4-pyridyl)pyridine with a matching size was selected and by a milling process the coordination N atoms could conveniently graft onto the OMSs of Na-Zn-MOF. Thus, a two-step "loading-locking" process has been successfully provided for the implementation of guest encapsulation.

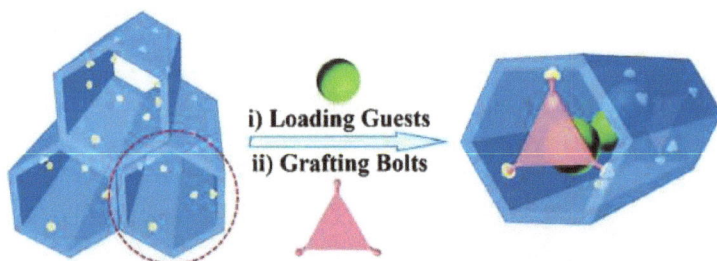

Fig. (35). Schematic representation of proposed two-step "loading-locking" guest-encapsulation process [180].

CHALLENGES AND PERSPECTIVES/CONCLUSION

Despite significant progress achieved in the area of MOFs, their practical applications in daily activities of people still have a long way to go, and the spectacle blueprint is full of great challenges. Firstly, chemical, thermal and mechanical stability are the main weakness of MOFs, and they are necessary to be further promoted to conquer the hindrance of actual utilization of MOF materials. Secondly, the precise prediction of MOF architectures is a challenging scientific endeavor, because the assembly of MOFs is accompanied with so many uncertain factors, and the outcome is usually occasional. The last but not the least, the structure analysis and refinement of MOFs are highly dependent on single-crystal X-ray diffraction, but sometimes, the growth of millimeter-scale MOF crystals is very difficult, which will certainly restrict the exploration of novel MOFs. We need develop more structure characterizations, especially powder diffraction refinement methods.

Although with such great challenges, MOFs show great values in the commercial market. Besides above stated potential applications, MOFs also present noticeable perspective in supercapacitors and secondary batteries, because of their long-range ordered structures, hierarchical pores, and abundant of redox active metal sites, which meet basic requirements for energy storage studies. However, there are very rare MOFs applied in the energy storage system, and MOF-based supercapacitors or cells are promising projects in the future research area of MOFs.

Overall, the exploration of MOF materials has achieved large progress, albeit there are some unresolved issues about the design, assembly and properties of MOFs. We expect that future researches could focus on unprecedented ligands, newly synthesis strategies and interdisciplinary properties, which will present novel high-performance MOFs with fantastic structures. Opportunities come with challenges, and we must be confident to overcome these challenges. we believe that MOF materials will soon arrive at practical applications, allowing for on-going worldwide interest in this research area.

CONSENT FOR PUBLICATION

Not applicable.

CONFLICT OF INTEREST

The authors declare no conflict of interest, financial or otherwise.

ACKNOWLEDGEMENTS

The authors are grateful for financial support from the National Natural Science Foundation of China (21972102, 21673177, 22171205), Jiangsu Key Laboratory for Micro and Nano Heat Fluid Flow Technology and Energy Application.

REFERENCES

[1]   Yaghi, O.M.; Li, G.; Li, H. Selective binding and removal of guests in a microporous metal–organic framework. *Nature,* **1995,** *378*(6558), 703-706.
      [http://dx.doi.org/10.1038/378703a0]

[2]   Férey, G.; Serre, C.; Mellot-Draznieks, C.; Millange, F.; Surblé, S.; Dutour, J.; Margiolaki, I. A hybrid solid with giant pores prepared by a combination of targeted chemistry, simulation, and powder diffraction. *Angew. Chem. Int. Ed.,* **2004,** *43*(46), 6296-6301.
      [http://dx.doi.org/10.1002/anie.200460592] [PMID: 15372634]

[3]   Park, K.S.; Ni, Z.; Côté, A.P.; Choi, J.Y.; Huang, R.; Uribe-Romo, F.J.; Chae, H.K.; O'Keeffe, M.; Yaghi, O.M. Exceptional chemical and thermal stability of zeolitic imidazolate frameworks. *Proc.*

Natl. Acad. Sci. USA, **2006,** *103*(27), 10186-10191.
[http://dx.doi.org/10.1073/pnas.0602439103] [PMID: 16798880]

[4]   Huang, X.-C.; Zhang, J.-P.; Lin, Y.-Y.; Yu, X.-L.; Chen, X.-M. Two mixed-valence copper(I,II) imidazolate coordination polymers: metal-valence tuning approach for new topological structures. *Chem. Commun. (Camb.),* **2004,** *9*(9), 1100-1101.
[http://dx.doi.org/10.1039/b401691b] [PMID: 15116204]

[5]   Uemura, T.; Hoshino, Y.; Kitagawa, S.; Yoshida, K.; Isoda, S. Effect of Organic Polymer Additive on Crystallization of Porous Coordination Polymer. *Chem. Mater.,* **2006,** *18*(4), 992-995.
[http://dx.doi.org/10.1021/cm052427g]

[6]   Ma, S.; Sun, D.; Simmons, J.M.; Collier, C.D.; Yuan, D.; Zhou, H.C. Metal-organic framework from an anthracene derivative containing nanoscopic cages exhibiting high methane uptake. *J. Am. Chem. Soc.,* **2008,** *130*(3), 1012-1016.
[http://dx.doi.org/10.1021/ja0771639] [PMID: 18163628]

[7]   Batten, S.R.; Champness, N.R.; Chen, X.M.; Garcia-Martinez, J.; Kitagawa, S.; Öhrström, L.; O'Keeffe, M.; Paik Suh, M.; Reedijk, J. Terminology of metal–organic frameworks and coordination polymers (IUPAC Recommendations 2013). *Pure Appl. Chem.,* **2013,** *85*(8), 1715-1724.
[http://dx.doi.org/10.1351/PAC-REC-12-11-20]

[8]   Tranchemontagne, D.J.; Mendoza-Cortés, J.L.; O'Keeffe, M.; Yaghi, O.M. Secondary building units, nets and bonding in the chemistry of metal–organic frameworks. *Chem. Soc. Rev.,* **2009,** *38*(5), 1257-1283.
[http://dx.doi.org/10.1039/b817735j] [PMID: 19384437]

[9]   Lee, Y.R.; Kim, J.; Ahn, W.S. Synthesis of metal-organic frameworks: A mini review. *Korean J. Chem. Eng.,* **2013,** *30*(9), 1667-1680.
[http://dx.doi.org/10.1007/s11814-013-0140-6]

[10]   Farha, O.K.; Mulfort, K.L.; Thorsness, A.M.; Hupp, J.T. Separating solids: purification of metal-organic framework materials. *J. Am. Chem. Soc.,* **2008,** *130*(27), 8598-8599.
[http://dx.doi.org/10.1021/ja803097e] [PMID: 18549210]

[11]   Xu, H.-Q.; Wang, K.; Ding, M.; Feng, D.; Jiang, H.-L.; Zhou, H-C. Alleviating Luminescence Concentration Quenching in Upconversion Nanoparticles through Organic Dye Sensitization. *J. Am. Chem. Soc.,* **2016,** *138*, 5316-5320.
[http://dx.doi.org/10.1021/jacs.6b01414] [PMID: 27016046]

[12]   Zhang, X.; Liu, P.; Wang, N.; Zhang, D.S. Controllable assembly of three copper-organic frameworks: Structure transformation and gas adsorption properties. *Polyhedron,* **2017,** *126*, 83-91.
[http://dx.doi.org/10.1016/j.poly.2017.01.005]

[13]   Lin, J.B.; Lin, R.B.; Cheng, X.N.; Zhang, J.P.; Chen, X.M. Solvent/additive-free synthesis of porous/zeolitic metal azolate frameworks from metal oxide/hydroxide. *Chem. Commun. (Camb.),* **2011,** *47*(32), 9185-9187.
[http://dx.doi.org/10.1039/c1cc12763b] [PMID: 21755072]

[14]   Martinez Joaristi, A.; Juan-Alcañiz, J.; Serra-Crespo, P.; Kapteijn, F.; Gascon, J. Electrochemical synthesis of some archetypical Zn^{2+}, Cu^{2+}, and Al^{3+} metal organic frameworks. *Cryst. Growth Des.,* **2012,** *12*(7), 3489-3498.
[http://dx.doi.org/10.1021/cg300552w]

[15]   Yaghi, O-M.; O'Keeffe, M.; Ockwig, N-W.; Chae, H-K.; Eddaoudi, M.; Kim, J. ChemInform Abstract: High-Spin Molecules: A Novel Cyano-Bridged Mn 9 II Mo 6 V Molecular Cluster with a S = 51/2 Ground State and Ferromagnetic Intercluster Ordering at Low Temperatures. *Nature,* **2003,** *423*, 705-714.
[http://dx.doi.org/10.1038/nature01650] [PMID: 12802325]

[16]   Wang, B.; Lin, R.B.; Zhang, Z.; Xiang, S.; Chen, B. Hydrogen-bonded organic frameworks as a tunable platform for functional materials. *J. Am. Chem. Soc.,* **2020,** *142*(34), 14399-14416.

[http://dx.doi.org/10.1021/jacs.0c06473] [PMID: 32786796]

[17]   Li, H.; Eddaoudi, M.; O'Keeffe, M.; Yaghi, O.M. Design and synthesis of an exceptionally stable and highly porous metal-organic framework. *Nature,* **1999**, *402*(6759), 276-279.
[http://dx.doi.org/10.1038/46248]

[18]   Rowsell, J.L.C.; Yaghi, O.M. Metal–organic frameworks: a new class of porous materials. *Microporous Mesoporous Mater.,* **2004**, *73*(1-2), 3-14.
[http://dx.doi.org/10.1016/j.micromeso.2004.03.034]

[19]   Eddaoudi, M.; Kim, J.; Rosi, N.; Vodak, D.; Wachter, J.; O'Keeffe, M.; Yaghi, O.M. Systematic design of pore size and functionality in isoreticular MOFs and their application in methane storage. *Science,* **2002**, *295*(5554), 469-472.
[http://dx.doi.org/10.1126/science.1067208] [PMID: 11799235]

[20]   Deng, H.; Doonan, C.J.; Furukawa, H.; Ferreira, R.B.; Towne, J.; Knobler, C.B.; Wang, B.; Yaghi, O.M. Multiple functional groups of varying ratios in metal-organic frameworks. *Science,* **2010**, *327*(5967), 846-850.
[http://dx.doi.org/10.1126/science.1181761] [PMID: 20150497]

[21]   Kitagawa, S.; Matsuda, R. Chemistry of coordination space of porous coordination polymers. *Coord. Chem. Rev.,* **2007**, *251*(21-24), 2490-2509.
[http://dx.doi.org/10.1016/j.ccr.2007.07.009]

[22]   Chang, Z.; Zhang, D.S.; Chen, Q.; Li, R.F.; Hu, T.L.; Bu, X.H. Rational construction of 3D pillared metal-organic frameworks: synthesis, structures, and hydrogen adsorption properties. *Inorg. Chem.,* **2011**, *50*(16), 7555-7562.
[http://dx.doi.org/10.1021/ic2004485] [PMID: 21776953]

[23]   Wang, Z.; Cohen, S.M. Postsynthetic modification of metal–organic frameworks. *Chem. Soc. Rev.,* **2009**, *38*(5), 1315-1329.
[http://dx.doi.org/10.1039/b802258p] [PMID: 19384440]

[24]   Meek, S.T.; Greathouse, J.A.; Allendorf, M.D. Metal-organic frameworks: a rapidly growing class of versatile nanoporous materials. *Adv. Mater.,* **2011**, *23*(2), 249-267.
[http://dx.doi.org/10.1002/adma.201002854] [PMID: 20972981]

[25]   Jones, S.C.; Bauer, C.A. Diastereoselective heterogeneous bromination of stilbene in a porous metal-organic framework. *J. Am. Chem. Soc.,* **2009**, *131*(35), 12516-12517.
[http://dx.doi.org/10.1021/ja900893p] [PMID: 19678638]

[26]   Goto, Y.; Sato, H.; Shinkai, S.; Sada, K. "Clickable" metal-organic framework. *J. Am. Chem. Soc.,* **2008**, *130*(44), 14354-14355.
[http://dx.doi.org/10.1021/ja7114053] [PMID: 18839949]

[27]   Wang, Z.; Cohen, S.M. Postsynthetic covalent modification of a neutral metal-organic framework. *J. Am. Chem. Soc.,* **2007**, *129*(41), 12368-12369.
[http://dx.doi.org/10.1021/ja074366o] [PMID: 17880219]

[28]   Haneda, T.; Kawano, M.; Kawamichi, T.; Fujita, M. Direct observation of the labile imine formation through single-crystal-to-single-crystal reactions in the pores of a porous coordination network. *J. Am. Chem. Soc.,* **2008**, *130*(5), 1578-1579.
[http://dx.doi.org/10.1021/ja7111564] [PMID: 18189406]

[29]   Morris, W.; Doonan, C.J.; Furukawa, H.; Banerjee, R.; Yaghi, O.M. Crystals as molecules: postsynthesis covalent functionalization of zeolitic imidazolate frameworks. *J. Am. Chem. Soc.,* **2008**, *130*(38), 12626-12627.
[http://dx.doi.org/10.1021/ja805222x] [PMID: 18754585]

[30]   Mandal, S.; Natarajan, S.; Mani, P.; Pankajakshan, A. Post-Synthetic Modification of Metal–Organic Frameworks Toward Applications. *Adv. Funct. Mater.,* **2021**, *31*(4), 2006291.
[http://dx.doi.org/10.1002/adfm.202006291]

[31] Park, J.; Yuan, D.; Pham, K.T.; Li, J.R.; Yakovenko, A.; Zhou, H.C. Reversible alteration of CO_2 adsorption upon photochemical or thermal treatment in a metal-organic framework. *J. Am. Chem. Soc.,* **2012**, *134*(1), 99-102.
[http://dx.doi.org/10.1021/ja209197f] [PMID: 22148550]

[32] Park, I.H.; Chanthapally, A.; Zhang, Z.; Lee, S.S.; Zaworotko, M.J.; Vittal, J.J. Metal-organic organopolymeric hybrid framework by reversible [2+2] cycloaddition reaction. *Angew. Chem. Int. Ed.,* **2014**, *53*(2), 414-419.
[http://dx.doi.org/10.1002/anie.201308606] [PMID: 24338857]

[33] Li, T.; Kozlowski, M.T.; Doud, E.A.; Blakely, M.N.; Rosi, N.L. Stepwise ligand exchange for the preparation of a family of mesoporous MOFs. *J. Am. Chem. Soc.,* **2013**, *135*(32), 11688-11691.
[http://dx.doi.org/10.1021/ja403810k] [PMID: 23688075]

[34] Burnett, B.J.; Barron, P.M.; Hu, C.; Choe, W. Stepwise synthesis of metal-organic frameworks: replacement of structural organic linkers. *J. Am. Chem. Soc.,* **2011**, *133*(26), 9984-9987.
[http://dx.doi.org/10.1021/ja201911v] [PMID: 21671680]

[35] Zhou, Y.; Yu, F.; Su, J.; Kurmoo, M.; Zuo, J.L. Tuning Electrical- and Photo-Conductivity by Cation Exchange within a Redox-Active Tetrathiafulvalene-Based Metal–Organic Framework. *Angew. Chem. Int. Ed.,* **2020**, *59*(42), 18763-18767.
[http://dx.doi.org/10.1002/anie.202008941] [PMID: 32652797]

[36] Rosi, N.L.; Eckert, J.; Eddaoudi, M.; Vodak, D.T.; Kim, J.; O'Keeffe, M.; Yaghi, O.M. Hydrogen storage in microporous metal-organic frameworks. *Science,* **2003**, *300*(5622), 1127-1129.
[http://dx.doi.org/10.1126/science.1083440] [PMID: 12750515]

[37] Kaye, S.S.; Dailly, A.; Yaghi, O.M.; Long, J.R. Impact of preparation and handling on the hydrogen storage properties of Zn4O(1,4-benzenedicarboxylate)3 (MOF-5). *J. Am. Chem. Soc.,* **2007**, *129*(46), 14176-14177.
[http://dx.doi.org/10.1021/ja076877g] [PMID: 17967030]

[38] Furukawa, H.; Ko, N.; Go, Y.B.; Aratani, N.; Choi, S.B.; Choi, E.; Yazaydin, A.Ö.; Snurr, R.Q.; O'Keeffe, M.; Kim, J.; Yaghi, O.M. Ultrahigh porosity in metal-organic frameworks. *Science,* **2010**, *329*(5990), 424-428.
[http://dx.doi.org/10.1126/science.1192160] [PMID: 20595583]

[39] Farha, O.K.; Özgür Yazaydın, A.; Eryazici, I.; Malliakas, C.D.; Hauser, B.G.; Kanatzidis, M.G.; Nguyen, S.T.; Snurr, R.Q.; Hupp, J.T. De novo synthesis of a metal–organic framework material featuring ultrahigh surface area and gas storage capacities. *Nat. Chem.,* **2010**, *2*(11), 944-948.
[http://dx.doi.org/10.1038/nchem.834] [PMID: 20966950]

[40] Kondo, M.; Yoshitomi, T.; Matsuzaka, H.; Kitagawa, S.; Seki, K. Three-Dimensional Framework with Channeling Cavities for Small Molecules:[M2(4, 4′-bpy)3(NO3)4]·xH2On(M = Co, Ni, Zn). *Angew. Chem. Int. Ed. Engl.,* **1997**, *36*(16), 1725-1727.
[http://dx.doi.org/10.1002/anie.199717251]

[41] Hausdorf, S.; Baitalow, F.; Böhle, T.; Rafaja, D.; Mertens, F.O.R.L. Main-group and transition-element IRMOF homologues. *J. Am. Chem. Soc.,* **2010**, *132*(32), 10978-10981.
[http://dx.doi.org/10.1021/ja1028777] [PMID: 20698648]

[42] Wu, H.; Zhou, W.; Yildirim, T. High-capacity methane storage in metal-organic frameworks M2(dhtp): the important role of open metal sites. *J. Am. Chem. Soc.,* **2009**, *131*(13), 4995-5000.
[http://dx.doi.org/10.1021/ja900258t] [PMID: 19275154]

[43] Martín-Calvo, A.; García-Pérez, E.; Manuel Castillo, J.; Calero, S. Molecular simulations for adsorption and separation of natural gas in IRMOF-1 and Cu-BTC metal-organic frameworks. *Phys. Chem. Chem. Phys.,* **2008**, *10*(47), 7085-7091.
[http://dx.doi.org/10.1039/b807470d] [PMID: 19039342]

[44] Kannappan, K.; Werblowsky, T.L.; Rim, K.T.; Berne, B.J.; Flynn, G.W. An experimental and

theoretical study of the formation of nanostructures of self-assembled cyanuric acid through hydrogen bond networks on graphite. *J. Phys. Chem. B,* **2007**, *111*(24), 6634-6642.
[http://dx.doi.org/10.1021/jp0706984] [PMID: 17455975]

[45]   Barin, G.; Krungleviciute, V.; Gomez-Gualdron, D-A.; Sarjeant, A-A.; Snurr, R-Q.; Hupp, T. Yildirim J.-T.; Farha, O.-K. Trap-limited exciton diffusion in organic semiconductors. *Chem. Mater.,* **2014**, *26*, 1912-1917.
[http://dx.doi.org/10.1021/cm404155s]

[46]   Peng, Y.; Krungleviciute, V.; Eryazici, I.; Hupp, J.T.; Farha, O.K.; Yildirim, T. Methane storage in metal-organic frameworks: current records, surprise findings, and challenges. *J. Am. Chem. Soc.,* **2013**, *135*(32), 11887-11894.
[http://dx.doi.org/10.1021/ja4045289] [PMID: 23841800]

[47]   Gándara, F.; Furukawa, H.; Lee, S.; Yaghi, O-M. Methane storage in metal-organic frameworks: current records, surprise findings, and challenges. *J. Am. Chem. Soc.,* **2014**, *136*, 5271-5274.
[http://dx.doi.org/10.1021/ja501606h] [PMID: 24661065]

[48]   He, Y.; Zhou, W.; Yildirim, T.; Chen, B. A series of metal–organic frameworks with high methane uptake and an empirical equation for predicting methane storage capacity. *Energy Environ. Sci.,* **2013**, *6*(9), 2735-2744.
[http://dx.doi.org/10.1039/c3ee41166d]

[49]   Alezi, D.; Belmabkhout, Y.; Suyetin, M.; Bhatt, P.M.; Weseliński, Ł.J.; Solovyeva, V.; Adil, K.; Spanopoulos, I.; Trikalitis, P.N.; Emwas, A.H.; Eddaoudi, M. MOF Crystal Chemistry Paving the Way to Gas Storage Needs: Aluminum-Based **soc** -MOF for CH $_4$, O $_2$, and CO $_2$ Storage. *J. Am. Chem. Soc.,* **2015**, *137*(41), 13308-13318.
[http://dx.doi.org/10.1021/jacs.5b07053] [PMID: 26364990]

[50]   Tian, T.; Zeng, Z.; Vulpe, D.; Casco, M.E.; Divitini, G.; Midgley, P.A.; Silvestre-Albero, J.; Tan, J.C.; Moghadam, P.Z.; Fairen-Jimenez, D. A sol–gel monolithic metal–organic framework with enhanced methane uptake. *Nat. Mater.,* **2018**, *17*(2), 174-179.
[http://dx.doi.org/10.1038/nmat5050] [PMID: 29251723]

[51]   Li, B.; Wen, H.M.; Wang, H.; Wu, H.; Tyagi, M.; Yildirim, T.; Zhou, W.; Chen, B. A porous metal-organic framework with dynamic pyrimidine groups exhibiting record high methane storage working capacity. *J. Am. Chem. Soc.,* **2014**, *136*(17), 6207-6210.
[http://dx.doi.org/10.1021/ja501810r] [PMID: 24730649]

[52]   Li, B.; Wen, H.M.; Wang, H.; Wu, H.; Yildirim, T.; Zhou, W.; Chen, B. Porous metal–organic frameworks with Lewis basic nitrogen sites for high-capacity methane storage. *Energy Environ. Sci.,* **2015**, *8*(8), 2504-2511.
[http://dx.doi.org/10.1039/C5EE01531F]

[53]   Simon, C.M.; Kim, J.; Gomez-Gualdron, D.A.; Camp, J.S.; Chung, Y.G.; Martin, R.L.; Mercado, R.; Deem, M.W.; Gunter, D.; Haranczyk, M.; Sholl, D.S.; Snurr, R.Q.; Smit, B. The materials genome in action: identifying the performance limits for methane storage. *Energy Environ. Sci.,* **2015**, *8*(4), 1190-1199.
[http://dx.doi.org/10.1039/C4EE03515A]

[54]   Lin, J.M.; He, C.T.; Liu, Y.; Liao, P.Q.; Zhou, D.D.; Zhang, J.P.; Chen, X.M. A Metal-Organic Framework with a Pore Size/Shape Suitable for Strong Binding and Close Packing of Methane. *Angew. Chem. Int. Ed.,* **2016**, *55*(15), 4674-4678.
[http://dx.doi.org/10.1002/anie.201511006] [PMID: 26948156]

[55]   Mason, J.A.; Oktawiec, J.; Taylor, M.K.; Hudson, M.R.; Rodriguez, J.; Bachman, J.E.; Gonzalez, M.I.; Cervellino, A.; Guagliardi, A.; Brown, C.M.; Llewellyn, P.L.; Masciocchi, N.; Long, J.R. Methane storage in flexible metal–organic frameworks with intrinsic thermal management. *Nature,* **2015**, *527*(7578), 357-361.
[http://dx.doi.org/10.1038/nature15732] [PMID: 26503057]

[56] Taylor, M.K.; Runčevski, T.; Oktawiec, J.; Gonzalez, M.I.; Siegelman, R.L.; Mason, J.A.; Ye, J.; Brown, C.M.; Long, J.R. Tuning the Adsorption-Induced Phase Change in the Flexible Metal–Organic Framework Co(bdp). *J. Am. Chem. Soc.,* **2016**, *138*(45), 15019-15026.
[http://dx.doi.org/10.1021/jacs.6b09155] [PMID: 27804295]

[57] Yang, Q.Y.; Lama, P.; Sen, S.; Lusi, M.; Chen, K.J.; Gao, W.Y.; Shivanna, M.; Pham, T.; Hosono, N.; Kusaka, S.; Perry, J.J., IV; Ma, S.; Space, B.; Barbour, L.J.; Kitagawa, S.; Zaworotko, M.J. Reversible switching between highly porous and nonporous phases of an interpenetrated diamondoid coordination network that exhibits gate-opening at methane storage pressures. *Angew. Chem. Int. Ed.,* **2018**, *57*(20), 5684-5689.
[http://dx.doi.org/10.1002/anie.201800820] [PMID: 29575465]

[58] Kundu, T.; Wahiduzzaman, M.; Shah, B.B.; Maurin, G.; Zhao, D. Solvent-Induced Control over Breathing Behavior in Flexible Metal–Organic Frameworks for Natural-Gas Delivery. *Angew. Chem. Int. Ed.,* **2019**, *58*(24), 8073-8077.
[http://dx.doi.org/10.1002/anie.201902738] [PMID: 30913352]

[59] Millward, A.R.; Yaghi, O.M. Metal-organic frameworks with exceptionally high capacity for storage of carbon dioxide at room temperature. *J. Am. Chem. Soc.,* **2005**, *127*(51), 17998-17999.
[http://dx.doi.org/10.1021/ja0570032] [PMID: 16366539]

[60] Caskey, S.R.; Wong-Foy, A.G.; Matzger, A.J. Dramatic tuning of carbon dioxide uptake *via* metal substitution in a coordination polymer with cylindrical pores. *J. Am. Chem. Soc.,* **2008**, *130*(33), 10870-10871.
[http://dx.doi.org/10.1021/ja8036096] [PMID: 18661979]

[61] Mason, J.A.; Sumida, K.; Herm, Z.R.; Krishna, R.; Long, J.R. Evaluating metal–organic frameworks for post-combustion carbon dioxide capture *via* temperature swing adsorption. *Energy Environ. Sci.,* **2011**, *4*(8), 3030-3040.
[http://dx.doi.org/10.1039/c1ee01720a]

[62] Zhang, D.S.; Chang, Z.; Li, Y.F.; Jiang, Z.Y.; Xuan, Z.H.; Zhang, Y.H.; Li, J.R.; Chen, Q.; Hu, T.L.; Bu, X.H. Fluorous metal-organic frameworks with enhanced stability and high H_2/CO_2 storage capacities. *Sci. Rep.,* **2013**, *3*(1), 3312.
[http://dx.doi.org/10.1038/srep03312] [PMID: 24264725]

[63] Zhao, X.; Bu, X.; Zhai, Q.G.; Tran, H.; Feng, P. Pore space partition by symmetry-matching regulated ligand insertion and dramatic tuning on carbon dioxide uptake. *J. Am. Chem. Soc.,* **2015**, *137*(4), 1396-1399.
[http://dx.doi.org/10.1021/ja512137t] [PMID: 25621414]

[64] Nugent, P.; Belmabkhout, Y.; Burd, S.D.; Cairns, A.J.; Luebke, R.; Forrest, K.; Pham, T.; Ma, S.; Space, B.; Wojtas, L.; Eddaoudi, M.; Zaworotko, M.J. Porous materials with optimal adsorption thermodynamics and kinetics for CO2 separation. *Nature,* **2013**, *495*(7439), 80-84.
[http://dx.doi.org/10.1038/nature11893] [PMID: 23446349]

[65] Shekhah, O.; Belmabkhout, Y.; Chen, Z.; Guillerm, V.; Cairns, A.; Adil, K.; Eddaoudi, M. Made-t--order metal-organic frameworks for trace carbon dioxide removal and air capture. *Nat. Commun.,* **2014**, *5*(1), 4228.
[http://dx.doi.org/10.1038/ncomms5228] [PMID: 24964404]

[66] Burd, S.D.; Ma, S.; Perman, J.A.; Sikora, B.J.; Snurr, R.Q.; Thallapally, P.K.; Tian, J.; Wojtas, L.; Zaworotko, M.J. Highly selective carbon dioxide uptake by [Cu(bpy-n)$_2$(SiF6)] (bpy-1 = 4,4′-bipyridine; bpy-2 = 1,2-bis(4-pyridyl)ethene). *J. Am. Chem. Soc.,* **2012**, *134*(8), 3663-3666.
[http://dx.doi.org/10.1021/ja211340t] [PMID: 22316279]

[67] Yang, S.; Ramirez-Cuesta, A.J.; Newby, R.; Garcia-Sakai, V.; Manuel, P.; Callear, S.K.; Campbell, S.I.; Tang, C.C.; Schröder, M. Supramolecular binding and separation of hydrocarbons within a functionalized porous metal–organic framework. *Nat. Chem.,* **2015**, *7*(2), 121-129.
[http://dx.doi.org/10.1038/nchem.2114] [PMID: 25615665]

[68]   Lin, R.B.; Li, L.; Zhou, H.L.; Wu, H.; He, C.; Li, S.; Krishna, R.; Li, J.; Zhou, W.; Chen, B. Molecular sieving of ethylene from ethane using a rigid metal–organic framework. *Nat. Mater.,* **2018**, *17*(12), 1128-1133.
[http://dx.doi.org/10.1038/s41563-018-0206-2] [PMID: 30397312]

[69]   Bloch, E.D.; Queen, W.L.; Krishna, R.; Zadrozny, J.M.; Brown, C.M.; Long, J.R. Hydrocarbon separations in a metal-organic framework with open iron(II) coordination sites. *Science,* **2012**, *335*(6076), 1606-1610.
[http://dx.doi.org/10.1126/science.1217544] [PMID: 22461607]

[70]   Li, B.; Zhang, Y.; Krishna, R.; Yao, K.; Han, Y.; Wu, Z.; Ma, D.; Shi, Z.; Pham, T.; Space, B.; Liu, J.; Thallapally, P.K.; Liu, J.; Chrzanowski, M.; Ma, S. Introduction of π-complexation into porous aromatic framework for highly selective adsorption of ethylene over ethane. *J. Am. Chem. Soc.,* **2014**, *136*(24), 8654-8660.
[http://dx.doi.org/10.1021/ja502119z] [PMID: 24901372]

[71]   Liao, P.Q.; Zhang, W.X.; Zhang, J.P.; Chen, X.M. Efficient purification of ethene by an ethane-trapping metal-organic framework. *Nat. Commun.,* **2015**, *6*(1), 8697.
[http://dx.doi.org/10.1038/ncomms9697] [PMID: 26510376]

[72]   Gücüyener, C.; van den Bergh, J.; Gascon, J.; Kapteijn, F. Ethane/ethene separation turned on its head: selective ethane adsorption on the metal-organic framework ZIF-7 through a gate-opening mechanism. *J. Am. Chem. Soc.,* **2010**, *132*(50), 17704-17706.
[http://dx.doi.org/10.1021/ja1089765] [PMID: 21114318]

[73]   Liang, W.; Xu, F.; Zhou, X.; Xiao, J.; Xia, Q.; Li, Y.; Li, Z. Ethane selective adsorbent Ni(bdc)(ted)0.5 with high uptake and its significance in adsorption separation of ethane and ethylene. *Chem. Eng. Sci.,* **2016**, *148*, 275-281.
[http://dx.doi.org/10.1016/j.ces.2016.04.016]

[74]   Chen, Y.; Qiao, Z.; Wu, H.; Lv, D.; Shi, R.; Xia, Q.; Zhou, J.; Li, Z. An ethane-trapping MOF PCN-250 for highly selective adsorption of ethane over ethylene. *Chem. Eng. Sci.,* **2018**, *175*, 110-117.
[http://dx.doi.org/10.1016/j.ces.2017.09.032]

[75]   Li, L.; Lin, R.B.; Krishna, R.; Li, H.; Xiang, S.; Wu, H.; Li, J.; Zhou, W.; Chen, B. Ethane/ethylene separation in a metal-organic framework with iron-peroxo sites. *Science,* **2018**, *362*(6413), 443-446.
[http://dx.doi.org/10.1126/science.aat0586] [PMID: 30361370]

[76]   Kim, J.; Lin, L.C.; Martin, R.L.; Swisher, J.A.; Haranczyk, M.; Smit, B. Large-scale computational screening of zeolites for ethane/ethene separation. *Langmuir,* **2012**, *28*(32), 11914-11919.
[http://dx.doi.org/10.1021/la302230z] [PMID: 22784373]

[77]   Lin, R.B.; Wu, H.; Li, L.; Tang, X.L.; Li, Z.; Gao, J.; Cui, H.; Zhou, W.; Chen, B. Boosting ethane/ethylene separation within isoreticular ultramicroporous metal–organic frameworks. *J. Am. Chem. Soc.,* **2018**, *140*(40), 12940-12946.
[http://dx.doi.org/10.1021/jacs.8b07563] [PMID: 30216725]

[78]   Cui, X.; Chen, K.; Xing, H.; Yang, Q.; Krishna, R.; Bao, Z.; Wu, H.; Zhou, W.; Dong, X.; Han, Y.; Li, B.; Ren, Q.; Zaworotko, M.J.; Chen, B. Pore chemistry and size control in hybrid porous materials for acetylene capture from ethylene. *Science,* **2016**, *353*(6295), 141-144.
[http://dx.doi.org/10.1126/science.aaf2458] [PMID: 27198674]

[79]   Xue, Y.Y.; Bai, X.Y.; Zhang, J.; Wang, Y.; Li, S.N.; Jiang, Y.C.; Hu, M.C.; Zhai, Q.G. Precise Pore Space Partitions Combined with High-Density Hydrogen-Bonding Acceptors within Metal–Organic Frameworks for Highly Efficient Acetylene Storage and Separation. *Angew. Chem. Int. Ed.,* **2021**, *60*(18), 10122-10128.
[http://dx.doi.org/10.1002/anie.202015861] [PMID: 33533093]

[80]   Cadiau, A.; Adil, K.; Bhatt, P.M.; Belmabkhout, Y.; Eddaoudi, M. A metal-organic framework–based splitter for separating propylene from propane. *Science,* **2016**, *353*(6295), 137-140.
[http://dx.doi.org/10.1126/science.aaf6323] [PMID: 27387945]

[81]   Li, L.; Lin, R.B.; Krishna, R.; Wang, X.; Li, B.; Wu, H.; Li, J.; Zhou, W.; Chen, B. Flexible–Robust Metal–Organic Framework for Efficient Removal of Propyne from Propylene. *J. Am. Chem. Soc.,* **2017,** *139*(23), 7733-7736.
[http://dx.doi.org/10.1021/jacs.7b04268] [PMID: 28580788]

[82]   Kondo, A.; Noguchi, H.; Carlucci, L.; Proserpio, D.M.; Ciani, G.; Kajiro, H.; Ohba, T.; Kanoh, H.; Kaneko, K. Double-step gas sorption of a two-dimensional metal-organic framework. *J. Am. Chem. Soc.,* **2007,** *129*(41), 12362-12363.
[http://dx.doi.org/10.1021/ja073568h] [PMID: 17887754]

[83]   Kondo, A.; Kajiro, H.; Noguchi, H.; Carlucci, L.; Proserpio, D.M.; Ciani, G.; Kato, K.; Takata, M.; Seki, H.; Sakamoto, M.; Hattori, Y.; Okino, F.; Maeda, K.; Ohba, T.; Kaneko, K.; Kanoh, H. Super flexibility of a 2D Cu-based porous coordination framework on gas adsorption in comparison with a 3D framework of identical composition: framework dimensionality-dependent gas adsorptivities. *J. Am. Chem. Soc.,* **2011,** *133*(27), 10512-10522.
[http://dx.doi.org/10.1021/ja201170c] [PMID: 21671624]

[84]   Yang, S.; Sun, J.; Ramirez-Cuesta, A.J.; Callear, S.K.; David, W.I.F.; Anderson, D.P.; Newby, R.; Blake, A.J.; Parker, J.E.; Tang, C.C.; Schröder, M. Selectivity and direct visualization of carbon dioxide and sulfur dioxide in a decorated porous host. *Nat. Chem.,* **2012,** *4*(11), 887-894.
[http://dx.doi.org/10.1038/nchem.1457] [PMID: 23089862]

[85]   Wang, Q.; Ke, T.; Yang, L.; Zhang, Z.; Cui, X.; Bao, Z.; Ren, Q.; Yang, Q.; Xing, H. Separation of Xe from Kr with Record Selectivity and Productivity in Anion-Pillared Ultramicroporous Materials by Inverse Size-Sieving. *Angew. Chem. Int. Ed.,* **2020,** *59*(9), 3423-3428.
[http://dx.doi.org/10.1002/anie.201913245] [PMID: 31746086]

[86]   Sato, H.; Kosaka, W.; Matsuda, R.; Hori, A.; Hijikata, Y.; Belosludov, R.V.; Sakaki, S.; Takata, M.; Kitagawa, S.; Self-Accelerating, C.O. Self-accelerating CO sorption in a soft nanoporous crystal. *Science,* **2014,** *343*(6167), 167-170.
[http://dx.doi.org/10.1126/science.1246423] [PMID: 24336572]

[87]   DeCoste, J.B.; Weston, M.H.; Fuller, P.E.; Tovar, T.M.; Peterson, G.W.; LeVan, M.D.; Farha, O.K. Metal-organic frameworks for oxygen storage. *Angew. Chem. Int. Ed.,* **2014,** *53*(51), 14092-14095.
[http://dx.doi.org/10.1002/anie.201408464] [PMID: 25319881]

[88]   Zhou, Y.; Wei, L.; Yang, J.; Sun, Y.; Zhou, L. Adsorption of oxygen on superactivated carbon. *J. Chem. Eng. Data,* **2005,** *50*(3), 1068-1072.
[http://dx.doi.org/10.1021/je050036c]

[89]   Montoro, C.; Linares, F.; Quartapelle Procopio, E.; Senkovska, I.; Kaskel, S.; Galli, S.; Masciocchi, N.; Barea, E.; Navarro, J.A.R. Capture of nerve agents and mustard gas analogues by hydrophobic robust MOF-5 type metal-organic frameworks. *J. Am. Chem. Soc.,* **2011,** *133*(31), 11888-11891.
[http://dx.doi.org/10.1021/ja2042113] [PMID: 21761835]

[90]   Ma, F.J.; Liu, S.X.; Sun, C.Y.; Liang, D.D.; Ren, G.J.; Wei, F.; Chen, Y.G.; Su, Z.M. A sodalite-type porous metal-organic framework with polyoxometalate templates: adsorption and decomposition of dimethyl methylphosphonate. *J. Am. Chem. Soc.,* **2011,** *133*(12), 4178-4181.
[http://dx.doi.org/10.1021/ja109659k] [PMID: 21370859]

[91]   Lee, E.Y.; Suh, M.P. A robust porous material constructed of linear coordination polymer chains: reversible single-crystal to single-crystal transformations upon dehydration and rehydration. *Angew. Chem. Int. Ed.,* **2004,** *43*(21), 2798-2801.
[http://dx.doi.org/10.1002/anie.200353494] [PMID: 15150752]

[92]   Boesenberg, U.; Meirer, F.; Liu, Y.; Shukla, A.K.; Dell'Anna, R.; Tyliszczak, T.; Chen, G.; Andrews, J.C.; Richardson, T.J.; Kostecki, R.; Cabana, J. Mesoscale phase distribution in single particles of LiFePO$_4$ following lithium deintercalation. *Chem. Mater.,* **2013,** *25*(9), 1664-1672.
[http://dx.doi.org/10.1021/cm400106k] [PMID: 23745016]

[93]   Yang, C.; Kaipa, U.; Mather, Q.Z.; Wang, X.; Nesterov, V.; Venero, A.F.; Omary, M.A. Fluorous

metal-organic frameworks with superior adsorption and hydrophobic properties toward oil spill cleanup and hydrocarbon storage. *J. Am. Chem. Soc.*, **2011**, *133*(45), 18094-18097.
[http://dx.doi.org/10.1021/ja208408n] [PMID: 21981413]

[94]   Sepahpour, S.; Selamat, J.; Khatib, A.; Manap, M.Y.A.; Abdull Razis, A.F.; Hajeb, P. Inhibitory effect of mixture herbs/spices on formation of heterocyclic amines and mutagenic activity of grilled beef. *Food Addit. Contam. Part A Chem. Anal. Control Expo. Risk Assess.*, **2018**, *35*(10), 1911-1927.
[http://dx.doi.org/10.1080/19440049.2018.1488085] [PMID: 29913103]

[95]   Furukawa, H.; Gándara, F.; Zhang, Y.B.; Jiang, J.; Queen, W.L.; Hudson, M.R.; Yaghi, O.M. Water adsorption in porous metal-organic frameworks and related materials. *J. Am. Chem. Soc.*, **2014**, *136*(11), 4369-4381.
[http://dx.doi.org/10.1021/ja500330a] [PMID: 24588307]

[96]   Kim, H.; Yang, S.; Rao, S.R.; Narayanan, S.; Kapustin, E.A.; Furukawa, H.; Umans, A.S.; Yaghi, O.M.; Wang, E.N. Water harvesting from air with metal-organic frameworks powered by natural sunlight. *Science,* **2017**, *356*(6336), 430-434.
[http://dx.doi.org/10.1126/science.aam8743] [PMID: 28408720]

[97]   Kalmutzki, M.J.; Diercks, C.S.; Yaghi, O.M. Metal–organic frameworks for water harvesting from air. *Adv. Mater.,* **2018**, *30*(37), 1704304.
[http://dx.doi.org/10.1002/adma.201704304] [PMID: 29672950]

[98]   Kim, H.; Rao, S.R.; Kapustin, E.A.; Zhao, L.; Yang, S.; Yaghi, O.M.; Wang, E.N. Adsorption-based atmospheric water harvesting device for arid climates. *Nat. Commun.,* **2018**, *9*(1), 1191-1198.
[http://dx.doi.org/10.1038/s41467-018-03162-7] [PMID: 29568033]

[99]   Hanikel, N.; Prévot, M.S.; Fathieh, F.; Kapustin, E.A.; Lyu, H.; Wang, H.; Diercks, N.J.; Glover, T.G.; Yaghi, O.M. Rapid cycling and exceptional yield in a metal-organic framework water harvester. *ACS Cent. Sci.,* **2019**, *5*(10), 1699-1706.
[http://dx.doi.org/10.1021/acscentsci.9b00745] [PMID: 31660438]

[100]   Hanikel, N.; Prévot, M.S.; Yaghi, O.M. MOF water harvesters. *Nat. Nanotechnol.,* **2020**, *15*(5), 348-355.
[http://dx.doi.org/10.1038/s41565-020-0673-x] [PMID: 32367078]

[101]   Liu, C-H.; Nguyen, H-L.; Yaghi, O-M. *AsiaChem.,* **2020**, *1*, 18-25.

[102]   Xu, W.; Yaghi, O.M. Metal–Organic Frameworks for Water Harvesting from Air, Anywhere, Anytime. *ACS Cent. Sci.,* **2020**, *6*(8), 1348-1354.
[http://dx.doi.org/10.1021/acscentsci.0c00678] [PMID: 32875075]

[103]   Fathieh, F.; Kalmutzki, M.J.; Kapustin, E.A.; Waller, P.J.; Yang, J.; Yaghi, O.M. Practical water production from desert air. *Sci. Adv.,* **2018**, *4*(6), eaat3198.
[http://dx.doi.org/10.1126/sciadv.aat3198] [PMID: 29888332]

[104]   Dincă, M.; Long, J.R. High-enthalpy hydrogen adsorption in cation-exchanged variants of the microporous metal-organic framework $Mn_3[(Mn4Cl)_3(BTT)_8(CH_3OH)_{10}]_2$. *J. Am. Chem. Soc.,* **2007**, *129*(36), 11172-11176.
[http://dx.doi.org/10.1021/ja072871f] [PMID: 17705485]

[105]   Sheng, D.; Zhu, L.; Dai, X.; Xu, C.; Li, P.; Pearce, C.I.; Xiao, C.; Chen, J.; Zhou, R.; Duan, T.; Farha, O.K.; Chai, Z.; Wang, S. Successful Decontamination of $^{99}TcO_4^-$ in Groundwater at Legacy Nuclear Sites by a Cationic Metal-Organic Framework with Hydrophobic Pockets. *Angew. Chem. Int. Ed.,* **2019**, *58*(15), 4968-4972.
[http://dx.doi.org/10.1002/anie.201814640] [PMID: 30761705]

[106]   Han, Y.; Sheng, S.; Yang, F.; Xie, Y.; Zhao, M.; Li, J-R. Size-exclusive and coordination-induced selective dye adsorption in a nanotubular metal–organic framework. *J. Mater. Chem. A Mater. Energy Sustain.,* **2015**, *3*(24), 12804-12809.
[http://dx.doi.org/10.1039/C5TA00963D]

[107] Li, Y.; Yang, Z.; Wang, Y.; Bai, Z.; Zheng, T.; Dai, X.; Liu, S.; Gui, D.; Liu, W.; Chen, M.; Chen, L.; Diwu, J.; Zhu, L.; Zhou, R.; Chai, Z.; Albrecht-Schmitt, T.E.; Wang, S. A mesoporous cationic thorium-organic framework that rapidly traps anionic persistent organic pollutants. *Nat. Commun.,* **2017**, *8*(1), 1354.
[http://dx.doi.org/10.1038/s41467-017-01208-w] [PMID: 29116079]

[108] Alaerts, L.; Kirschhock, C.E.A.; Maes, M.; van der Veen, M.A.; Finsy, V.; Depla, A.; Martens, J.A.; Baron, G.V.; Jacobs, P.A.; Denayer, J.F.M.; De Vos, D.E. Selective adsorption and separation of xylene isomers and ethylbenzene with the microporous vanadium(IV) terephthalate MIL-47. *Angew. Chem. Int. Ed.,* **2007**, *46*(23), 4293-4297.
[http://dx.doi.org/10.1002/anie.200700056] [PMID: 17477460]

[109] Liu, Q.K.; Ma, J.P.; Dong, Y.B. Adsorption and separation of reactive aromatic isomers and generation and stabilization of their radicals within cadmium(II)-triazole metal-organic confined space in a single-crystal-to-single-crystal fashion. *J. Am. Chem. Soc.,* **2010**, *132*(20), 7005-7017.
[http://dx.doi.org/10.1021/ja101807c] [PMID: 20438078]

[110] Seo, J.S.; Whang, D.; Lee, H.; Jun, S.I.; Oh, J.; Jeon, Y.J.; Kim, K. A homochiral metal–organic porous material for enantioselective separation and catalysis. *Nature,* **2000**, *404*(6781), 982-986.
[http://dx.doi.org/10.1038/35010088] [PMID: 10801124]

[111] Peng, Y.; Gong, T.; Zhang, K.; Lin, X.; Liu, Y.; Jiang, J.; Cui, Y. *Nat. Commun.,* **2018**, *9*, 1191-1198.
[http://dx.doi.org/10.1038/s41467-018-03162-7] [PMID: 29568033]

[112] Evans, J.D.; Sumby, C.J.; Doonan, C.J. Post-synthetic metalation of metal–organic frameworks. *Chem. Soc. Rev.,* **2014**, *43*(16), 5933-5951.
[http://dx.doi.org/10.1039/C4CS00076E] [PMID: 24736674]

[113] Lin, X.M.; Li, T.T.; Wang, Y.W.; Zhang, L.; Su, C.Y. Two Zn(II) metal-organic frameworks with coordinatively unsaturated metal sites: structures, adsorption, and catalysis. *Chem. Asian J.,* **2012**, *7*(12), 2796-2804.
[http://dx.doi.org/10.1002/asia.201200601] [PMID: 23038041]

[114] Zhang, X.; Geng, L.; Zhang, Y.Z.; Zhang, D.S.; Zhang, R.; Fu, J.; Gao, J.; Carozza, J.C.; Zhou, Z.; Han, H. Construction of Cu-based MOFs with enhanced hydrogenation performance by integrating open electropositive metal sites. *CrystEngComm,* **2019**, *21*(36), 5382-5386.
[http://dx.doi.org/10.1039/C9CE01136F]

[115] Zhang, X.; Zhang, Y.Z.; Jin, Y.Q.; Geng, L.; Zhang, D.S.; Hu, H.; Li, T.; Wang, B.; Li, J.R. Pillar-Layered Metal–Organic Frameworks Based on a Hexaprismane [Co6(μ3-OH)6] Cluster: Structural Modulation and Catalytic Performance in Aerobic Oxidation Reaction. *Inorg. Chem.,* **2020**, *59*(16), 11728-11735.
[http://dx.doi.org/10.1021/acs.inorgchem.0c01611] [PMID: 32799465]

[116] Leus, K.; Vandichel, M.; Liu, Y.Y.; Muylaert, I.; Musschoot, J.; Pyl, S.; Vrielinck, H.; Callens, F.; Marin, G.B.; Detavernier, C.; Wiper, P.V.; Khimyak, Y.Z.; Waroquier, M.; Van Speybroeck, V.; Van Der Voort, P. The coordinatively saturated vanadium MIL-47 as a low leaching heterogeneous catalyst in the oxidation of cyclohexene. *J. Catal.,* **2012**, *285*(1), 196-207.
[http://dx.doi.org/10.1016/j.jcat.2011.09.014]

[117] Mao, J.; Yang, L.; Yu, P.; Wei, X.; Mao, L. Electrocatalytic four-electron reduction of oxygen with Copper (II)-based metal-organic frameworks. *Electrochem. Commun.,* **2012**, *19*, 29-31.
[http://dx.doi.org/10.1016/j.elecom.2012.02.025]

[118] Hasegawa, S.; Horike, S.; Matsuda, R.; Furukawa, S.; Mochizuki, K.; Kinoshita, Y.; Kitagawa, S. Three-dimensional porous coordination polymer functionalized with amide groups based on tridentate ligand: selective sorption and catalysis. *J. Am. Chem. Soc.,* **2007**, *129*(9), 2607-2614.
[http://dx.doi.org/10.1021/ja067374y] [PMID: 17288419]

[119] Roberts, J.M.; Fini, B.M.; Sarjeant, A.A.; Farha, O.K.; Hupp, J.T.; Scheidt, K.A. Urea metal-organic frameworks as effective and size-selective hydrogen-bond catalysts. *J. Am. Chem. Soc.,* **2012**, *134*(7),

3334-3337.
[http://dx.doi.org/10.1021/ja2108118] [PMID: 22296523]

[120] Feng, D.; Chung, W.C.; Wei, Z.; Gu, Z.Y.; Jiang, H.L.; Chen, Y.P.; Darensbourg, D.J.; Zhou, H.C. Construction of ultrastable porphyrin Zr metal-organic frameworks through linker elimination. *J. Am. Chem. Soc.,* **2013**, *135*(45), 17105-17110.
[http://dx.doi.org/10.1021/ja408084j] [PMID: 24125517]

[121] Liang, J.; Chen, R.P.; Wang, X.Y.; Liu, T.T.; Wang, X.S.; Huang, Y.B.; Cao, R. Postsynthetic ionization of an imidazole-containing metal–organic framework for the cycloaddition of carbon dioxide and epoxides. *Chem. Sci. (Camb.),* **2017**, *8*(2), 1570-1575.
[http://dx.doi.org/10.1039/C6SC04357G] [PMID: 28451286]

[122] Yoon, T.P.; Jacobsen, E.N. Privileged chiral catalysts. *Science,* **2003**, *299*(5613), 1691-1693.
[http://dx.doi.org/10.1126/science.1083622] [PMID: 12637734]

[123] Jiang, H.; Zhang, W.; Kang, X.; Cao, Z.; Chen, X.; Liu, Y.; Cui, Y. Topology-Based Functionalization of Robust Chiral Zr-Based Metal–Organic Frameworks for Catalytic Enantioselective Hydrogenation. *J. Am. Chem. Soc.,* **2020**, *142*(21), jacs.0c00637.
[http://dx.doi.org/10.1021/jacs.0c00637] [PMID: 32363868]

[124] Lions, M.; Tommasino, J.B.; Chattot, R.; Abeykoon, B.; Guillou, N.; Devic, T.; Demessence, A.; Cardenas, L.; Maillard, F.; Fateeva, A. Insights into the mechanism of electrocatalysis of the oxygen reduction reaction by a porphyrinic metal organic framework. *Chem. Commun. (Camb.),* **2017**, *53*(48), 6496-6499.
[http://dx.doi.org/10.1039/C7CC02113E] [PMID: 28569312]

[125] Sun, Y.; Huang, H.; Vardhan, H.; Aguila, B.; Zhong, C.; Perman, J.A.; Al-Enizi, A.M.; Nafady, A.; Ma, S. Facile Approach to Graft Ionic Liquid into MOF for Improving the Efficiency of CO$_2$ Chemical Fixation. *ACS Appl. Mater. Interfaces,* **2018**, *10*(32), 27124-27130.
[http://dx.doi.org/10.1021/acsami.8b08914] [PMID: 30016060]

[126] Dare, N.A.; Brammer, L.; Bourne, S.A.; Egan, T.J. Fe(III) Protoporphyrin IX Encapsulated in a Zinc Metal–Organic Framework Shows Dramatically Enhanced Peroxidatic Activity. *Inorg. Chem.,* **2018**, *57*(3), 1171-1183.
[http://dx.doi.org/10.1021/acs.inorgchem.7b02612] [PMID: 29308888]

[127] Yao, S.L.; Tian, X.M.; Li, L.Q.; Liu, S.J.; Zheng, T.F.; Chen, Y.Q.; Zhang, D.S.; Chen, J.L.; Wen, H.R.; Hu, T.L. A Cd^{II}-Based Metal-Organic Framework with *pcu* Topology as Turn-On Fluorescent Sensor for Al^{3+}. *Chem. Asian J.,* **2019**, *14*(20), 3648-3654.
[http://dx.doi.org/10.1002/asia.201900739] [PMID: 31276314]

[128] Sun, Y.; Christensen, M.; Johnsen, S.; Nong, N.V.; Ma, Y.; Sillassen, M.; Zhang, E.; Palmqvist, A.E.C.; Bøttiger, J.; Iversen, B.B. Low-cost high-performance zinc antimonide thin films for thermoelectric applications. *Adv. Mater.,* **2012**, *24*(13), 1693-1696.
[http://dx.doi.org/10.1002/adma.201104947] [PMID: 22388988]

[129] Chen, B.; Wang, L.; Zapata, F.; Qian, G.; Lobkovsky, E.B. A luminescent microporous metal-organic framework for the recognition and sensing of anions. *J. Am. Chem. Soc.,* **2008**, *130*(21), 6718-6719.
[http://dx.doi.org/10.1021/ja802035e] [PMID: 18452294]

[130] Chen, B.; Wang, L.; Xiao, Y.; Fronczek, F.R.; Xue, M.; Cui, Y.; Qian, G. A luminescent metal-organic framework with Lewis basic pyridyl sites for the sensing of metal ions. *Angew. Chem. Int. Ed.,* **2009**, *48*(3), 500-503.
[http://dx.doi.org/10.1002/anie.200805101] [PMID: 19072806]

[131] Lan, A.; Li, K.; Wu, H.; Olson, D.H.; Emge, T.J.; Ki, W.; Hong, M.; Li, J. A luminescent microporous metal-organic framework for the fast and reversible detection of high explosives. *Angew. Chem. Int. Ed.,* **2009**, *48*(13), 2334-2338.
[http://dx.doi.org/10.1002/anie.200804853] [PMID: 19180622]

[132] Pramanik, S.; Zheng, C.; Zhang, X.; Emge, T.J.; Li, J. New microporous metal-organic framework

demonstrating unique selectivity for detection of high explosives and aromatic compounds. *J. Am. Chem. Soc.,* **2011**, *133*(12), 4153-4155.
[http://dx.doi.org/10.1021/ja106851d] [PMID: 21384862]

[133] Banerjee, D.; Hu, Z.; Pramanik, S.; Zhang, X.; Wang, H.; Li, J. A luminescent cadmium metal–organic framework based on a triazolate–carboxylate ligand exhibiting selective gas adsorption and guest-dependent photoluminescence properties. *CrystEngComm,* **2013**, *15*, 9745-9750.
[http://dx.doi.org/10.1039/c3ce41680a]

[134] Hu, Z.; Pramanik, S.; Tan, K.; Zheng, C.; Liu, W.; Zhang, X.; Chabal, Y.J.; Li, J. Selective, Sensitive, and Reversible Detection of Vapor-Phase High Explosives *via* Two-Dimensional Mapping: A New Strategy for MOF-Based Sensors. *Cryst. Growth Des.,* **2013**, *13*(10), 4204-4207.
[http://dx.doi.org/10.1021/cg4012185]

[135] Pramanik, S.; Hu, Z.; Zhang, X.; Zheng, C.; Kelly, S.; Li, J. A systematic study of fluorescence-based detection of nitroexplosives and other aromatics in the vapor phase by microporous metal-organic frameworks. *Chemistry,* **2013**, *19*(47), 15964-15971.
[http://dx.doi.org/10.1002/chem.201301194] [PMID: 24123511]

[136] Takashima, Y.; Martínez, V.M.; Furukawa, S.; Kondo, M.; Shimomura, S.; Uehara, H.; Nakahama, M.; Sugimoto, K.; Kitagawa, S. Molecular decoding using luminescence from an entangled porous framework. *Nat. Commun.,* **2011**, *2*(1), 168-175.
[http://dx.doi.org/10.1038/ncomms1170] [PMID: 21266971]

[137] Xie, Z.; Ma, L.; deKrafft, K.E.; Jin, A.; Lin, W. Porous phosphorescent coordination polymers for oxygen sensing. *J. Am. Chem. Soc.,* **2010**, *132*(3), 922-923.
[http://dx.doi.org/10.1021/ja909629f] [PMID: 20041656]

[138] Ferrando-Soria, J.; Khajavi, H.; Serra-Crespo, P.; Gascon, J.; Kapteijn, F.; Julve, M.; Lloret, F.; Pasán, J.; Ruiz-Pérez, C.; Journaux, Y.; Pardo, E. Highly selective chemical sensing in a luminescent nanoporous magnet. *Adv. Mater.,* **2012**, *24*(41), 5625-5629.
[http://dx.doi.org/10.1002/adma.201201846] [PMID: 22887721]

[139] Wu, P.; Wang, J.; He, C.; Zhang, X.; Wang, Y.; Liu, T.; Duan, C. Luminescent metal-organic frameworks for selectively sensing nitric oxide in an aqueous solution and in living cells. *Adv. Funct. Mater.,* **2012**, *22*(8), 1698-1703.
[http://dx.doi.org/10.1002/adfm.201102157]

[140] Xu, H.; Rao, X.; Gao, J.; Yu, J.; Wang, Z.; Dou, Z.; Cui, Y.; Yang, Y.; Chen, B.; Qian, G. A luminescent nanoscale metal–organic framework with controllable morphologies for spore detection. *Chem. Commun. (Camb.),* **2012**, *48*(59), 7377-7379.
[http://dx.doi.org/10.1039/c2cc32346j] [PMID: 22627669]

[141] Liu, B.; Chen, Y. Responsive lanthanide coordination polymer for hydrogen sulfide. *Anal. Chem.,* **2013**, *85*(22), 11020-11025.
[http://dx.doi.org/10.1021/ac402651y] [PMID: 24191713]

[142] Zhu, X.; Zheng, H.; Wei, X.; Lin, Z.; Guo, L.; Qiu, B.; Chen, G. Metal–organic framework (MOF): a novel sensing platform for biomolecules. *Chem. Commun. (Camb.),* **2013**, *49*(13), 1276-1278.
[http://dx.doi.org/10.1039/c2cc36661d] [PMID: 23295434]

[143] Dou, Z.; Yu, J.; Cui, Y.; Yang, Y.; Wang, Z.; Yang, D.; Qian, G. Luminescent metal-organic framework films as highly sensitive and fast-response oxygen sensors. *J. Am. Chem. Soc.,* **2014**, *136*(15), 5527-5530.
[http://dx.doi.org/10.1021/ja411224j] [PMID: 24697214]

[144] Wang, G.Y.; Song, C.; Kong, D.M.; Ruan, W.J.; Chang, Z.; Li, Y. Two luminescent metal–organic frameworks for the sensing of nitroaromatic explosives and DNA strands. *J. Mater. Chem. A Mater. Energy Sustain.,* **2014**, *2*(7), 2213-2220.
[http://dx.doi.org/10.1039/C3TA14199C]

[145] Zhang, X.; Zhang, J.; Hu, Q.; Cui, Y.; Yang, Y.; Qian, G. Postsynthetic modification of metal–organic

framework for hydrogen sulfide detection. *Appl. Surf. Sci.,* **2015**, *355*, 814-819.
[http://dx.doi.org/10.1016/j.apsusc.2015.07.166]

[146] Cui, Y.; Song, R.; Yu, J.; Liu, M.; Wang, Z.; Wu, C.; Yang, Y.; Wang, Z.; Chen, B.; Qian, G. Dual-emitting MOF⊃dye composite for ratiometric temperature sensing. *Adv. Mater.,* **2015**, *27*(8), 1420-1425.
[http://dx.doi.org/10.1002/adma.201404700] [PMID: 25581401]

[147] Yao, Z.Q.; Xu, J.; Zou, B.; Hu, Z.; Wang, K.; Yuan, Y.J.; Chen, Y.P.; Feng, R.; Xiong, J.B.; Hao, J.; Bu, X.H. A Dual-Stimuli-Responsive Coordination Network Featuring Reversible Wide-Range Luminescence-Tuning Behavior. *Angew. Chem. Int. Ed.,* **2019**, *58*(17), 5614-5618.
[http://dx.doi.org/10.1002/anie.201900190] [PMID: 30779418]

[148] Wang, M.S.; Guo, S.P.; Li, Y.; Cai, L.Z.; Zou, J.P.; Xu, G.; Zhou, W.W.; Zheng, F.K.; Guo, G.C. A direct white-light-emitting metal-organic framework with tunable yellow-to-white photoluminescence by variation of excitation light. *J. Am. Chem. Soc.,* **2009**, *131*(38), 13572-13573.
[http://dx.doi.org/10.1021/ja903947b] [PMID: 19772357]

[149] He, J.; Zeller, M.; Hunter, A.D.; Xu, Z. White light emission and second harmonic generation from secondary group participation (SGP) in a coordination network. *J. Am. Chem. Soc.,* **2012**, *134*(3), 1553-1559.
[http://dx.doi.org/10.1021/ja2073559] [PMID: 22236070]

[150] Wibowo, A.C.; Vaughn, S.A.; Smith, M.D.; zur Loye, H.C. Novel bismuth and lead coordination polymers synthesized with pyridine-2,5-dicarboxylates: two single component "white" light emitting phosphors. *Inorg. Chem.,* **2010**, *49*(23), 11001-11008.
[http://dx.doi.org/10.1021/ic1014708] [PMID: 21043507]

[151] Sava, D.F.; Rohwer, L.E.S.; Rodriguez, M.A.; Nenoff, T.M. Intrinsic broad-band white-light emission by a tuned, corrugated metal-organic framework. *J. Am. Chem. Soc.,* **2012**, *134*(9), 3983-3986.
[http://dx.doi.org/10.1021/ja211230p] [PMID: 22339608]

[152] Xia, Y.P.; Wang, C.X.; An, L.C.; Zhang, D.S.; Hu, T.L.; Xu, J.; Chang, Z.; Bu, X.H. Utilizing an effective framework to dye energy transfer in a carbazole-based metal–organic framework for high performance white light emission tuning. *Inorg. Chem. Front.,* **2018**, *5*(11), 2868-2874.
[http://dx.doi.org/10.1039/C8QI00747K]

[153] Sun, C.Y.; Wang, X.L.; Zhang, X.; Qin, C.; Li, P.; Su, Z.M.; Zhu, D.X.; Shan, G.G.; Shao, K.Z.; Wu, H.; Li, J. Efficient and tunable white-light emission of metal–organic frameworks by iridium-complex encapsulation. *Nat. Commun.,* **2013**, *4*(1), 2717.
[http://dx.doi.org/10.1038/ncomms3717] [PMID: 24212250]

[154] Zhang, D.S.; Gao, Q.; Chang, Z.; Liu, X.T.; Zhao, B.; Xuan, Z.H.; Hu, T.L.; Zhang, Y.H.; Zhu, J.; Bu, X.H. Rational Construction of Highly Tunable Donor–Acceptor Materials Based on a Crystalline Host–Guest Platform. *Adv. Mater.,* **2018**, *30*(50), 1804715.
[http://dx.doi.org/10.1002/adma.201804715] [PMID: 30318756]

[155] Liu, X.T.; Wang, K.; Chang, Z.; Zhang, Y.H.; Xu, J.; Zhao, Y.S.; Bu, X.H. Engineering Donor–Acceptor Heterostructure Metal–Organic Framework Crystals for Photonic Logic Computation. *Angew. Chem. Int. Ed.,* **2019**, *58*(39), 13890-13896.
[http://dx.doi.org/10.1002/anie.201906278] [PMID: 31231920]

[156] Rieter, W.J.; Taylor, K.M.L.; An, H.; Lin, W.; Lin, W. Nanoscale metal-organic frameworks as potential multimodal contrast enhancing agents. *J. Am. Chem. Soc.,* **2006**, *128*(28), 9024-9025.
[http://dx.doi.org/10.1021/ja0627444] [PMID: 16834362]

[157] Rieter, W.J.; Taylor, K.M.L.; Lin, W. Surface modification and functionalization of nanoscale metal-organic frameworks for controlled release and luminescence sensing. *J. Am. Chem. Soc.,* **2007**, *129*(32), 9852-9853.
[http://dx.doi.org/10.1021/ja073506r] [PMID: 17645339]

[158] Taylor, K.M.L.; Jin, A.; Lin, W. Surfactant-assisted synthesis of nanoscale gadolinium metal-organic

frameworks for potential multimodal imaging. *Angew. Chem. Int. Ed.,* **2008**, *47*(40), 7722-7725.
[http://dx.doi.org/10.1002/anie.200802911] [PMID: 18767098]

[159] Taylor, K.M.L.; Rieter, W.J.; Lin, W. Manganese-based nanoscale metal-organic frameworks for magnetic resonance imaging. *J. Am. Chem. Soc.,* **2008**, *130*(44), 14358-14359.
[http://dx.doi.org/10.1021/ja803777x] [PMID: 18844356]

[160] Taylor-Pashow, K.M.L.; Della Rocca, J.; Xie, Z.; Tran, S.; Lin, W. Postsynthetic modifications of iron-carboxylate nanoscale metal-organic frameworks for imaging and drug delivery. *J. Am. Chem. Soc.,* **2009**, *131*(40), 14261-14263.
[http://dx.doi.org/10.1021/ja906198y] [PMID: 19807179]

[161] Foucault-Collet, A.; Gogick, K.A.; White, K.A.; Villette, S.; Pallier, A.; Collet, G.; Kieda, C.; Li, T.; Geib, S.J.; Rosi, N.L.; Petoud, S. Lanthanide near infrared imaging in living cells with Yb [3+] nano metal organic frameworks. *Proc. Natl. Acad. Sci. USA,* **2013**, *110*(43), 17199-17204.
[http://dx.doi.org/10.1073/pnas.1305910110] [PMID: 24108356]

[162] Horcajada, P.; Serre, C.; Vallet-Regí, M.; Sebban, M.; Taulelle, F.; Férey, G. Metal-organic frameworks as efficient materials for drug delivery. *Angew. Chem. Int. Ed.,* **2006**, *45*(36), 5974-5978.
[http://dx.doi.org/10.1002/anie.200601878] [PMID: 16897793]

[163] Wu, M.X.; Yang, Y.W. Metal-Organic Framework (MOF)-Based Drug/Cargo Delivery and Cancer Therapy. *Adv. Mater.,* **2017**, *29*(23), 1606134.
[http://dx.doi.org/10.1002/adma.201606134]

[164] Deria, P.; Mondloch, J.E.; Tylianakis, E.; Ghosh, P.; Bury, W.; Snurr, R.Q.; Hupp, J.T.; Farha, O.K. Perfluoroalkane functionalization of NU-1000 *via* solvent-assisted ligand incorporation: synthesis and CO_2 adsorption studies. *J. Am. Chem. Soc.,* **2013**, *135*(45), 16801-16804.
[http://dx.doi.org/10.1021/ja408959g] [PMID: 24175709]

[165] Pander, M.; Żelichowska, A.; Bury, W. Probing mesoporous Zr-MOF as drug delivery system for carboxylate functionalized molecules. *Polyhedron,* **2018**, *156*, 131-137.
[http://dx.doi.org/10.1016/j.poly.2018.09.006]

[166] Chen, Y.; Li, P.; Modica, J.A.; Drout, R.J.; Farha, O.K. Acid-resistant mesoporous metal–organic framework toward oral insulin delivery: protein encapsulation, protection, and release. *J. Am. Chem. Soc.,* **2018**, *140*(17), 5678-5681.
[http://dx.doi.org/10.1021/jacs.8b02089] [PMID: 29641892]

[167] Teplensky, M.H.; Fantham, M.; Li, P.; Wang, T.C.; Mehta, J.P.; Young, L.J.; Moghadam, P.Z.; Hupp, J.T.; Farha, O.K.; Kaminski, C.F.; Fairen-Jimenez, D. Temperature treatment of highly porous zirconium-containing metal–organic frameworks extends drug delivery release. *J. Am. Chem. Soc.,* **2017**, *139*(22), 7522-7532.
[http://dx.doi.org/10.1021/jacs.7b01451] [PMID: 28508624]

[168] Chen, S.; Chen, Q.; Dong, S.; Ma, J.; Yang, Y.W.; Chen, L.; Gao, H. Polymer brush decorated MOF nanoparticles loaded with AIEgen, anticancer drug, and supramolecular glue for regulating and in situ observing DOX release. *Macromol. Biosci.,* **2018**, *18*(12), 1800317.
[http://dx.doi.org/10.1002/mabi.201800317] [PMID: 30334359]

[169] Wu, M.X.; Yan, H.J.; Gao, J.; Cheng, Y.; Yang, J.; Wu, J.R.; Gong, B.J.; Zhang, H.Y.; Yang, Y.W. Multifunctional Supramolecular Materials Constructed from Polypyrrole@UiO-66 Nanohybrids and Pillararene Nanovalves for Targeted Chemophotothermal Therapy. *ACS Appl. Mater. Interfaces,* **2018**, *10*(40), 34655-34663.
[http://dx.doi.org/10.1021/acsami.8b13758] [PMID: 30226739]

[170] Jiang, K.; Zhang, L.; Hu, Q.; Zhao, D.; Xia, T.; Lin, W.; Yang, Y.; Cui, Y.; Yang, Y.; Qian, G. Pressure controlled drug release in a Zr-cluster-based MOF. *J. Mater. Chem. B Mater. Biol. Med.,* **2016**, *4*(39), 6398-6401.
[http://dx.doi.org/10.1039/C6TB01756H] [PMID: 32263448]

[171] Gandara-Loe, J.; Ortuño-Lizarán, I.; Fernández-Sanchez, L.; Alió, J.L.; Cuenca, N.; Vega-Estrada, A.;

Silvestre-Albero, J. Metal–Organic Frameworks as Drug Delivery Platforms for Ocular Therapeutics. *ACS Appl. Mater. Interfaces,* **2019,** *11*(2), 1924-1931.
[http://dx.doi.org/10.1021/acsami.8b20222] [PMID: 30561189]

[172] Jiang, K.; Zhang, L.; Hu, Q.; Yang, Y.; Lin, W.; Cui, Y.; Yang, Y.; Qian, G. A Biocompatible Ti-based metal-organic framework for pH responsive drug delivery. *Mater. Lett.,* **2018,** *225*, 142-144.
[http://dx.doi.org/10.1016/j.matlet.2018.05.006]

[173] Wu, M.X.; Gao, J.; Wang, F.; Yang, J.; Song, N.; Jin, X.; Mi, P.; Tian, J.; Luo, J.; Liang, F.; Yang, Y.W. Multistimuli responsive core–shell nanoplatform constructed from $Fe_3O_4@$ MOF equipped with pillar[6]arene nanovalves. *Small,* **2018,** *14*(17), 1704440.
[http://dx.doi.org/10.1002/smll.201704440]

[174] Lv, Y.; Ding, D.; Zhuang, Y.; Feng, Y.; Shi, J.; Zhang, H.; Zhou, T.L.; Chen, H.; Xie, R.J. Chromium-Doped Zinc Gallogermanate@Zeolitic Imidazolate Framework-8: A Multifunctional Nanoplatform for Rechargeable *In Vivo* Persistent Luminescence Imaging and pH-Responsive Drug Release. *ACS Appl. Mater. Interfaces,* **2019,** *11*(2), 1907-1916.
[http://dx.doi.org/10.1021/acsami.8b19172] [PMID: 30566326]

[175] Dong, H.; Yang, G.X.; Zhang, X.; Meng, X.B.; Sheng, J.L.; Sun, X.J.; Feng, Y.J.; Zhang, F.M. Folic Acid Functionalized Zirconium-Based Metal–Organic Frameworks as Drug Carriers for Active Tumor-Targeted Drug Delivery. *Chemistry,* **2018,** *24*(64), 17148-17154.
[http://dx.doi.org/10.1002/chem.201804153] [PMID: 30125400]

[176] Yang, J.C.; Chen, Y.; Li, Y.H.; Yin, X.B. Magnetic Resonance Imaging-Guided Multi-Drug Chemotherapy and Photothermal Synergistic Therapy with pH and NIR-Stimulation Release. *ACS Appl. Mater. Interfaces,* **2017,** *9*(27), 22278-22288.
[http://dx.doi.org/10.1021/acsami.7b06105] [PMID: 28616966]

[177] Zhang, H.; Jiang, W.; Liu, R.; Zhang, J.; Zhang, D.; Li, Z.; Luan, Y. Rational Design of Metal Organic Framework Nanocarrier-Based Codelivery System of Doxorubicin Hydrochloride/Verapamil Hydrochloride for Overcoming Multidrug Resistance with Efficient Targeted Cancer Therapy. *ACS Appl. Mater. Interfaces,* **2017,** *9*(23), 19687-19697.
[http://dx.doi.org/10.1021/acsami.7b05142] [PMID: 28530401]

[178] Hidalgo, T.; Giménez-Marqués, M.; Bellido, E.; Avila, J.; Asensio, M.C.; Salles, F.; Lozano, M.V.; Guillevic, M.; Simón-Vázquez, R.; González-Fernández, A.; Serre, C.; Alonso, M.J.; Horcajada, P. Chitosan-coated mesoporous MIL-100(Fe) nanoparticles as improved bio-compatible oral nanocarriers. *Sci. Rep.,* **2017,** *7*(1), 43099.
[http://dx.doi.org/10.1038/srep43099] [PMID: 28256600]

[179] Dong, Z.; Sun, Y.; Chu, J.; Zhang, X.; Deng, H. Multivariate Metal–Organic Frameworks for Dialing-in the Binding and Programming the Release of Drug Molecules. *J. Am. Chem. Soc.,* **2017,** *139*(40), 14209-14216.
[http://dx.doi.org/10.1021/jacs.7b07392] [PMID: 28898070]

[180] Wang, H.; Xu, J.; Zhang, D.S.; Chen, Q.; Wen, R.M.; Chang, Z.; Bu, X.H. Crystalline capsules: metal-organic frameworks locked by size-matching ligand bolts. *Angew. Chem. Int. Ed.,* **2015,** *54*(20), 5966-5970.
[http://dx.doi.org/10.1002/anie.201500468] [PMID: 25800154]

CHAPTER 2

Post-synthetic Modification and Engineering of Metal Nodes and Organic Ligands of MOFs for Catalytic Applications

Aleksander Ejsmont[1], Agata Chełmińska[1], Martyna Kotula[1], Anita Kubiak[1], Marcelina Kotschmarów[1], Aleksandra Galarda[1], Anna Olejnik[1] and Joanna Goscianska[1,*]

[1] *Adam Mickiewicz University in Poznań, Faculty of Chemistry, Department of Chemical Technology, Uniwersytetu Poznańskiego 8, 61-614 Poznań, Poland*

Abstract: Metal-organic frameworks (MOFs) emerged as adjustable and multipurpose materials, which are now intensively investigated worldwide. They are composed of a wide range of organic and inorganic building units which are a susceptible base for various post-synthetic modifications (PSMs). In the last years, altering MOFs composition has significantly contributed to their broad application in many fields, especially in heterogeneous catalysis. PSMs are employed to improve the physicochemical properties of MOFs such as stability or selectivity, but mostly to generate catalytically active sites. Here, we report diverse methods of metal- (exchange, doping, redox transformations) and ligand-based (functionalization, exchange, installation, removal) PSMs of MOFs, which can be effectively used for catalytic purposes. PSMs can either extend the MOF framework with catalytically active functionalities or contribute to defect engineering for open metal site formation. Moreover, combining different modifying procedures has been introduced as a tandem approach when various reactions prompt several changes in the framework. Epitaxial growth was also presented as PSM, which can govern catalytically beneficial features mostly for thin films, unattainable to achieve by conventional methods. Recent MOFs' PSM findings were reviewed to show new pathways and a continuously developing field of reticular chemistry which come across with the expectations for novel and more efficient catalysts.

Keywords: Coordination sites, Epitaxial growth, Ligand exchange, Ligand functionalization, Ligand installation, Ligand removal, Linkers, Metal exchange, Metal incorporation, Metal–organic frameworks, Metal nodes, MOF, MOF stability, Post-synthetic modification, PSM, Redox transformations, Transmetalation.

---

* **Corresponding author Joanna Goscianska:** Adam Mickiewicz University in Poznań, Faculty of Chemistry, Department of Chemical Technology, Uniwersytetu Poznańskiego 8, 61-614 Poznań, Poland; Tel: +48-618291607; E-mail: joanna.goscianska@amu.edu.pl

Junkuo Gao & Reza Abazari (Eds.)

1. INTRODUCTION

The search for new materials applied in the catalysis is grounded in the constant necessity for alternative and more efficient catalytic reaction pathways [1]. By any means, a proper catalyst does not influence the thermodynamic equilibrium, yet it provides opportune conditions for the substrate molecules to change the reaction kinetics [2]. In consequence, it lowers the energy barrier of the reaction, leading to its faster occurrence [3]. In heterogeneous catalysis, where the catalyst is usually in the form of a solid, catalytic reactions take place on the catalyst surface, thus it has to stand out with specific features [4, 5]. Firstly, the catalyst has to be stable enough during the proceeding reaction to only assist it, without creating interfering compounds, and in the end, it should remain unchanged. Moreover, it should be characterized by a well-developed surface area to provide a place for reagents, but above all, it must indicate abundance in catalytically active centers to allow substrates to interact [6, 7].

In the last few decades, intensive research devoted to MOFs showed that they meet the requirements of good catalysts [8 - 10]. Their extremely high porosity, the multiplicity of topologies, and improved stability created a wide spectrum of their applications in catalysis [11, 12]. On top of that, the possibility of post-synthetic modifications (PSMs) pushes MOFs utility even further [13 - 16]. It can translate into increasing their performance and selectivity or even giving them novel catalytic abilities [17, 18]. Usually, PSM is performed when the desired framework is particularly difficult to obtain in the direct synthesis. Despite it constitutes as an additional processing step, extended synthesis may be ultimately cost-effective [16].

Active sites in MOFs which provide catalytic abilities may occur in various forms [19]. Metal nodes exhibit catalytic activity if only they are accessible for substrates. When reagents are not able to approach nodes directly due to the steric hindrance, defect generation is required. For instance, linker displacement creates coordinatively unsaturated metal sites, known as open metal sites (OMSs), which mainly contribute to supporting catalytic reactions. The other types of active sites are outer surface terminating groups, which are exclusively located on the external side of the MOF particles. Therefore, decreasing particles size increases the number of active sites and catalytic activity. Moreover, terminating groups can constitute a base to attach different functional groups which are catalytically active. For instance, sulfonic or amino moieties may result in the formation of Brønsted acidic or basic centers, respectively [11]. They can also be generated by defect engineering, which does not always involve lattice constructions dislocation or modification. Besides organic linkers, inorganic nodes often are coordinated with solvent molecules or ligands which do not form the cage. Their removal at a specific temperature or by vacuum generates active sites which act as Lewis acids [20]. It all indicates that MOFs as materials constructed from inorganic and organic components, can be post-synthetically modified in many ways, due to the possibility of active sites generation on several levels (Fig. **1**). The first one is the metal-based PSM, where cations in metallic nodes can be substituted by different metals (transmetalation) [21], changed through redox transformations [22], incorporated with new species [23], or become more available through vacancies creation [24]. The second way is to interplay with organic struts *via* any reaction that will not damage the pristine framework unless intentionally for vacancies generation [25]. In order to increase catalytic activity and selectivity of MOFs, organic building units can be modified through ligand exchange, installation, and removal. Furthermore, MOFs owing to their large channels are capable of encapsulating catalytically active entities *i.e.* nanoparticles, molecules, or clusters *via* host-guest interactions [26]. Up to now, the most successful guest assemblage within framework cavities is conducted locally by impregnating MOFs with precursors, followed by thermal treatment, photochemical decomposition of precursor, or its redox reactions. This 'ship-in-a-bottle' approach enables confining and immobilizing guests without the risk of precursors aggregation if only the process is conducted in mild conditions and MOF maintains its structure [27].

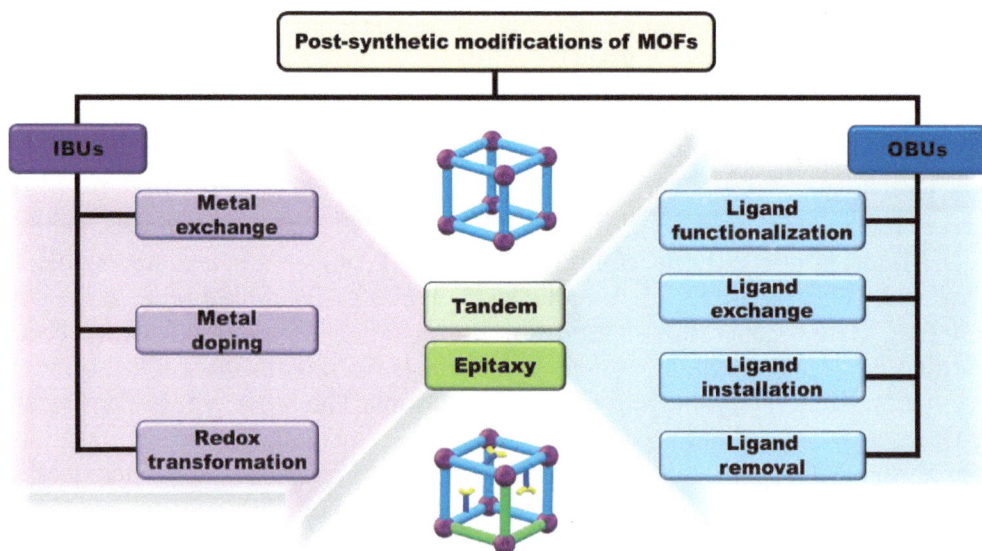

Fig. (1). Post-synthetic modifications of metal–organic frameworks (IBUs - Inorganic Building Units; OBUs - Organic Building Units).

Due to the plethora of reports concerning MOFs, in this work, we have compiled and analyzed the post-synthetic treatment methods of these materials, which allow the generation of new active sites showing a beneficial effect on their subsequent application in catalytic processes.

2. METAL-BASED POST-SYNTHETIC MODIFICATIONS OF MOFS

Considering inorganic-organic hybrid materials referred to as MOFs, in the context of catalytic applicability, the first component that is recognized as catalytically active is the inorganic building unit (IBU). Metal nodes take part in a variety of catalytic reactions *e.g.,* Co^{2+} OMSs in alcohol oxidation [28], Mn^{2+} in oxygen reduction [29], or Cu^{2+} in benzyl azide to phenylacetylene cycloaddition [30]. Also, metal complexes embedded within the framework exhibit catalytic activity *e.g.,* Mo^{6+} and V^{4+} complexes entrapped in Cu-based MOF in the epoxidation of olefins and allyl alcohols [31]. Interestingly, not all reactions when catalyzed by metal nodes require OMSs or metallic guests. In such cases, IBUs *i.e.,* Zr, Ti, Zn oxides can function as nanosized semiconductors coordinatively saturated with organic building units (OBUs) acting as antennae for electrons [19]. Then the catalysis proceeds on the outer sphere (OBUs), which serves as an electron acceptor/donor in redox type reactions *e.g.,* the photocatalytic benzyl alcohol [32] or propene [33] oxidation by Ti and Zn oxo clusters, respectively.

Metal nodes that comprise single metal, bimetallic systems, metal oxo clusters can be modulated towards enhanced catalytic activity in various ways. The replacement of one type of metal ions is most energetically demanding, because it involves overcoming large kinetic barriers in bond breaking and formation between OBUs and IBUs diffusion issues caused by steric hindrance of OBUs, and small pores [34]. However, the metal exchange process benefits if extra-framework metals lead to the charge balance or new metal has a more suitable valence electron configuration towards the required application *e.g.,* oxidation of *trans*-stilbene to stilbene oxide enabled by Cd-MOF transmetalation with Mn^{2+}/Cu^{2+} [35]. Metal metathesis PSM is also proceeded to increase MOF stability *via* forming stronger, more ionic bonds between metal nodes and organic linkers [36]. However, the change between IBUs and OBUs usually is only partial, since it is limited due to the initial low material stability under harsh PSM conditions.

In replacing one type of cations with another, crystal field stabilization, Jahn-Teller effect along with the radius of ions must be taken into account. To obtain MOF with increased stability, replaced cations should be smaller in size than the original ones. The type of the metal also has a stabilizing impact on the framework, for instance, transmetalation with the bivalent transition metals affects stability accordingly to the Irving Williams series: Mn(II) < Fe(II) < Co(II) < Ni(II) < Cu(II) > Zn(II) [37]. To avoid major structural disorder possible to occur during transmetalation, the change of metal oxidation state is an optional PSM, however, it is still in its infancy. Up to date, most oxidation reactions of IBUs were conducted for transition metals identified as soft Lewis acids *i.e.*, Cd, Co, Cu, Mn, Ni, and Zn, leading to more labile metal-ligand systems [34, 38]. An interesting approach suggested by Liu *et al.* [39] was to use as a template Mg-MOF with labile Mg–O. They exchanged Mg–O with Fe(II) and Cr(II) and subsequently oxidized them to highly water-stable Fe(III) and Cr(III)-MOFs. This post-synthetic oxidation process certainly suggests great potential for further research. It was especially proven for Mn(II)-based MOF, which after oxidation to Mn(III) revealed an outstanding catalytic performance in aerobic oxidation of alkylbenzenes [40]. In this chapter, we discuss what types of modifications can be made to the inorganic part of MOFs, and how they can affect the potential catalytic activity of materials.

2.1. Transmetalation (Metal Exchange)

Since 2007, when Mi and co-workers presented one of the earliest exchanges of cadmium to copper in PCP (porous coordination polymer), it was also suggested that other metals for instance Co or Ni can be replaced within the system [41]. Today it is a very promising PSM of MOFs especially applied for the design of novel catalysts whose framework and porosity are advantageous, whilst IBUs catalytic activity requires alteration [36]. In 2008, Zhao *et al.* [42] performed a tangible exchange of Cu(II) to Zn ions in PCP, which resulted in only a slight change of its activity in oxidative self-coupling catalysis of 2,6-di-*tert*-butylphenol. Although it only seems feasible that metal node swapping induces an anticipated change of the MOF topology, often the results are unexpected. The introduction of new metals may support favored and more typical geometrical formation, however, it does not have to be beneficial for catalysis [43]. Therefore, it is important to establish which PSM approaches are useful for the desired application. IBU replacement methods can be performed with different inorganic entities, *i.e.* metal or metal clusters, and also varying degrees (full or partial) of their quantitative exchange. Also, MOF single crystal as a core-shell structure can contain various metal ions with distinct radial distribution [34]. Presented reports on transmetalation confirm its huge advantages in the design of novel MOF-based catalysts constituting a promising direction for further research.

In 2012, Denysesko *et al.* [44] reported better MOF performance in gas conversion (CO to CO_2) after transmetalation. The isostructural replacement of Zn with Co in the MFU-4*l* (MFU = MOF Ulm University) resulted in the redox activity in CO oxidation, with maintained thermal stability (450 °C) of MOF. The MOF was subjected to the solvothermal treatment in the presence of $CoCl_2$ and LiCl in the DMF solution. Temperature programmed oxidation (TPO) and reduction (TPR) experiments showed that transmetalation increased accessibility to OMSs in the MOF. Moreover, the activation of modified MFU-4*l* before reaction was carried out at 200 °C. Whereas, the nanorods Co_3O_4 catalyst for CO oxidation requires high pre-treatment temperatures in the range 450-550 °C [45].

In 2013, Genna *et al.* [46] performed a heterogenization of homogeneous catalysis *via* modifying ZJU-28 (ZJU = Zhejiang University) with transition metals. The endogenous H_2NMe^{2+} was interchanged with metals such as Pd, Fe, Ir, Rh, and Ru. The Rh-MOF [Rh(dppe)(COD)]BF_4 (dppe = 1,2-bis(diphenylphosphino) ethane; COD = 1,5-cyclooctadiene) indicated better performance in catalyzing alkene hydrogenation than its homogeneous equivalent, and was easily regenerated under mild conditions. The reaction of converting 1-octene to *n*-octane after PSM was carried out 4-times without the loss of catalyst activity, whereas before modification, it was single-use.

Metal exchange in MOFs not always takes place in the nodes, forasmuch there are MOFs with linkers that contain metal ions *e.g.,* Werner complexes [47, 48]. A valid method of obtaining metal-based functionalities which are usually hardly available due to the tightly binding pincer ligands was presented by Wade and co-workers [49]. A highly stable Zr-MOF with a Co-based pincer complex functionalized with carboxylic acid as a linker was subjected to transmetalation with Rh and Pt complexes. It was observed that demetalation and exchange were possible to proceed upon the reduction of Co^{III} to Co^{II}. Moreover, a Pt complex [$PtCl_2(SMe_2)_2$] in the presence of a mild reducing agent NEt_3 almost entirely replaced Co^{II}. Such change could have a potential application for using many transition metals in high-temperature catalytic processes where phosphine-based pincer complexes ensure material stability and prolong its activity [50].

One type of MOF is able to catalyze many organic reactions which were noted in 2017 by Dutta and co-workers [51]. They observed new nitrate binding modes in porphyrin-based MOFs that caused the openness of the structure. It permitted high Zn replacement with Cu, from 70% up to 100% conducted at room temperature for 12-18 days. Most significantly, materials exhibited promising catalytic activity in chemo and regioselective reactions such as enamination of β-ketoesters, cycloadditions of terminal alkynes and organic azides, and synthesis of propargylamine derivatives.

The Zn replacement with Cu was also used by Salahshournia and co-workers [52] for TMU-17-NH_2 (TMU = Tarbiat Modares University). This Zn-MOF is based on linkers such as 4-bpdb standing for 1,4-bis(4-pyridyl)-2,3-diaza-1,3-butadiene and $H_2BDC-NH_2$, where H_2BDC is a terephthalic acid. The solvothermal reaction of these linkers with Cu salts does not result in obtaining $Cu-TMU-17-NH_2$, however, it is achievable *via* transmetalation (Fig. **2**). Nodes replacing procedure was a simple approach based on keeping Zn-MOF immersed for 7 days at room temperature in either $Cu(OAc)_2$ or $CuCl_2$ DMF solutions. The obtained Cu-MOF revealed the analogous framework with remained integrity and 95% of nodes interchange. Moreover, the material after PSM indicated high yields (~95%) in the [3 + 2] cycloaddition conducted in an environmental-friendly medium (PEG), resulting in 5-substituted-1*H*-tetrazoles. The same research group two years later presented another catalytic use of presented MOF additionally modified with Pd(II). A novel Cu/Pd bimetallic MOF with interchanged Zn nodes was prepared. The material was applied in the Suzuki-Miyaura cross-coupling reaction, and it revealed high catalytic efficiency, up to ~90% in green solvents (water/ethanol). Furthermore, in comparison to other Pd-based catalysts, modified MOF catalyzed the reaction in the shortest time (10 min) and exhibited the highest reusability (5 times) [53]. Bimetallic or multimetallic catalysts are characterized by higher selectivity, efficiency, and significant resistance to deactivation in comparison to monometallic catalysts [54]. Moreover, copper and its derivatives (*e.g.,* acetates, chlorides, cyanides, triflates) are cheaper and less damaging to the environment than noble metals [52, 55]. Due to their unique properties, multimetallic systems open new application possibilities, therefore as reports indicate, they will not be overlooked [56].

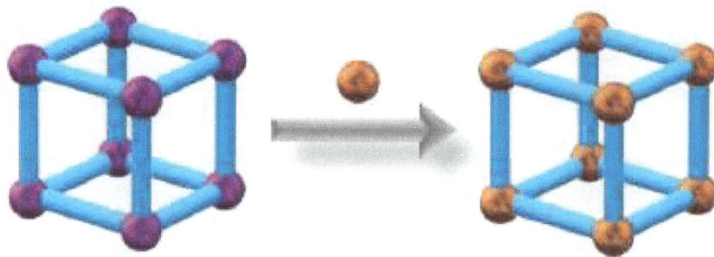

Fig. (2). Schematic example of transmetalation process of MOF, where one type of metal nodes are replaced with another.

2.2. Metal Incorporation (Metal Doping)

In recent years, increasingly more attention has been paid to the modification of MOFs that can have diverse functionalities and better properties than the parent MOFs. One type of PSM is the incorporation of metal ions or metal complexes into the framework (Fig. **3**). The metal doping can be accomplished either by incorporating metal ions into open channels of the framework or by direct coordination at a specific location in the framework. The vast number of reports on the catalytic activity of post-synthetically metalated MOFs signifies that this area has become the goal of many scientific studies on designing novel catalysts [37, 57].

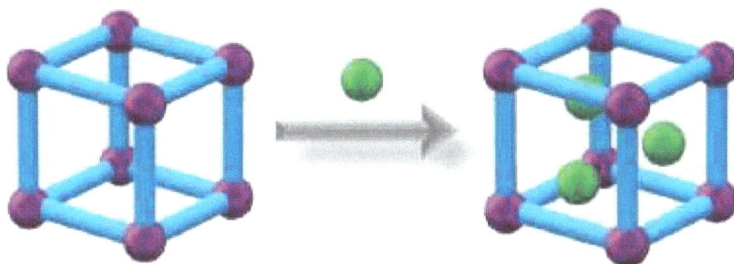

Fig. (3). Schematic example of the incorporation of metallic entities (*e.g.,* metal, metal clusters) within MOF.

Huang and co-workers [58] performed a post-synthetic modification of the aluminum-based metal–organic framework, Al-ATA-MOF (ATA = aluminum amino terephthalate) which is a photocatalyst for oxygen evolution. Incorporating Ni(II) cations into the amine groups of the ligand caused interactions between the benzene ring of ATA^{2-} and the Ni^{2+} ions. The modified MOF exhibits photocatalytic activity in overall water splitting. Li and co-workers [59] reported the UiO-type MOF (UiO = University of Oslo) with the 2,2'-bipyridyl moiety (bpy-UiO) and its post-synthetic metalation using Pd and Ir cations. The MOFs obtained not only demonstrated enhanced catalytic activity compared to their parent analog, but also had higher stability. They were used as catalysts in a wide range of organic transformations. The Pd-functionalized MOF (bpy-UiO-Ir) is a highly efficient catalyst of the borylation of aromatic C–H bonds and ortho-silylation of benzylic silyl ethers. Whereas, Ir-functionalized MOF (bpy-UiO-Pd) catalyzed the dehydrogenation of cyclohexenones. In addition, bpy-UiO-Ir maintained its catalytic activity and crystallinity after 20 catalytic cycles.

MOFs are a good platform to generate accessible OMSs, which often are Lewis acid sites with vacant orbitals in metal clusters or ion nodes. Furthermore, they can be introduced *via* the metalation approach into the framework, increasing the number of isolated catalytically active centers [60]. Zhang *et al.* [61] presented MOF-525-Co which is able to reduce CO_2 with high efficiency under visible-light irradiation. As a matrix for further modification, they selected MOF-525 $(Zr_6O_4OH)_4(TCPP-H_2)_3$ [TCPP = tetrakis(4-carboxyphenyl)porphyrin] which integrates Zr_6 clusters and porphyrin-based molecular units into a 3D network. Coordinatively unsaturated cobalt metal centers were incorporated into the porphyrin units by PSM to obtain a new composite. The presence of Co OMSs generates catalytically efficient active sites increasing CO_2 adsorption yield in comparison to porphyrins' active sites. Furthermore, the introduction of cobalt sites facilitated directional energy migration within the MOF by providing long-lived electrons to reduce CO_2 molecules adsorbed on the MOF. The results showed that porphyrin MOF comprises atomically dispersed catalytic centers allowing for higher photocatalytic reduction of CO_2 compared to the parent MOF, which corresponds to ~3-fold increase in CO evolution rate (200.6 μmol g^{-1} h^{-1}) and a ~6-fold enhancement in the CH_4 generation rate (36.76 μmol g^{-1} h^{-1}).

The suspension of metal clusters and metal nanoparticles into MOF scaffolds results in many interesting properties useful for catalytic applications. The presence of a large pore opening in the MOF provides the opportunity to introduce metal ions as guest species. Ji and co-workers [62] described the *in situ* encapsulation of ruthenium ions, which were subsequently reduced to a single Ru atom and triatomic Ru_3 cluster in functionalized ZIF-8 (ZIF = Zeolitic Imidazolate Framework). The modified MOF with the active Ru centers was applied as a chemoselective catalyst for semi-hydrogenation of alkynes to alkenes. The design and optimization of catalysts by combining efficient atomic sites and using MOFs as molecular sieves to achieve regioselectivity is a promising area of research, giving great possibilities for catalytic application in complex chemical reactions.

Zhou *et al.* [63] described an approach to construct photocatalytically active bifunctional MOF-253-Pt by PSM of 2,2'-bipyridine-based MOF with platinum ions. The modified MOF was used both as a photosensitizer and a photocatalyst for the reduction of water into H_2 under visible light irradiation. This work demonstrates that the concept of incorporating transition metal ions into MOF can afford new bifunctional materials for efficient photocatalytic H_2 evolution without additional catalyst and dye.

In 2020, Wei and co-workers [64] applied PSM to synthesize new CO_2 reduction photocatalyst - UiO-68-Fe-bpy based on a robust Zr(IV)-MOF platform with incorporated Fe(bpy)Cl$_3$ (bpy=4'-methyl-[2,2'-bipyridine]). Modified MOF indicated an enhanced light-absorbing ability and can reduce CO_2 to form CO with 100% selectivity under visible light irradiation. It was confirmed that PSM of MOFs is an effective strategy allowing to optimize the operation of the catalyst in photocatalytic processes.

2.3. Oxidation of the Coordination Site (Redox Transformations)

The wide range of applications requires robust MOFs constructed from high-valence metals. Compared with other porous materials, MOFs are crystalline structures that can be tuned at the atomic scale and functionalized by modifying linkers or selecting metal junctions. Their industrial application is hampered by the fact of their low chemical stability. From the hard-soft acid-base (HSAB) theory, hard Lewis acids such as Fe^{3+}, Cr^{3+}, Zr^{3+}, Ti^{4+} are good candidates for the MOF formation. These metals with carboxyl-based linkers (Lewis bases) can construct hydrolytically and chemically stable and robust MOFs. However, the synthesis of stable crystalline MOFs based on high-valence ions is difficult [39]. Therefore, an interesting option for obtaining high-valence MOFs has become post-synthetic metathesis and oxidation (PSMO), Fig. (**4**) [37].

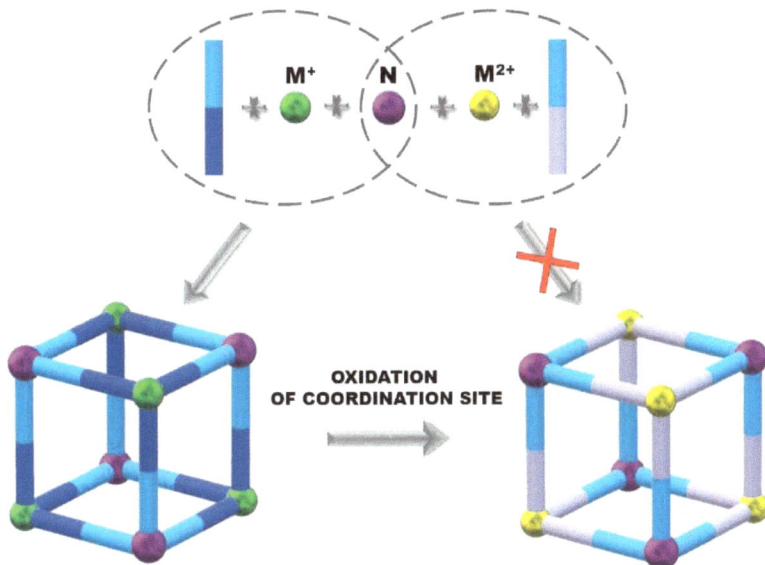

Fig. (4). Schematic example of post-synthetic oxidation of metal nodes in heterometallic metal–organic framework. M and N stand for different type of metal.

PSM of MOF allows for altering the structure for the desired application. However, numerous difficulties arise with PSM of coordination site due to structural breakdown or incompatibility between the newly formed coordination group and the original metallic site. In 2020, Han's group [57] developed an interesting concept by introducing "primary" and "secondary" nodes into Zr-based MOF $[Zr_6(\mu_3-OH)_8(OH)_8][Cu^I_4(L_1)_4]_2$ (1-SH-a, H_2L_1=6-mercaptopyridine-3-carboxylic acid). Post-synthetic oxidation at the coordination site occurs at "secondary" nodes $[Cu^I_4(L_1)_4]^{4-}$, while stabilization of the backbone is provided by "primary" nodes forming $[Zr_6(\mu_3-OH)_8(OH)_8][(Cu^I_{0,44}Cu^{II}_{0,56}(OH)_{0,56})_4(L_2)_4]_2$ (1-SO$_3$H, H_2L_2=6-sulfonatetinic acid). The obtained MOFs with unsaturated coordinated Cu(II) sites were effective catalysts in the ring-opening reaction of styrene oxide (regioselectivity of primary alcohol increases from 71% to 99%). This work provides information regarding the design of unique heterogeneous catalysts at the molecular level.

Liu and co-workers [39] acquired two metal–organic frameworks PCN-426-Cr(III) and PCN-426-Fe(III) (PCN = Porous Coordination Network) *via* PSMO of the metal nodes (Fig. **5**). Firstly, they synthesized Mg-based MOF (PCN-426-Mg) as a framework template. Next, the metal ions were exchanged with Cr(II) and Fe(II) to form the intermediate Cr(II)- and Fe(II)-MOFs. After air oxidation PCN-426-Cr(III) and PCN-426-Fe(III), were obtained in a single-crystal to single-crystal (SC-SC) transformation. Typically, MOFs based on high-valence metal ions occur in powder or amorphous form, hence this procedure aimed to use PSMO to synthesize the MOF in crystalline form. Additionally, the stability and porosity of the fabricated Cr(III)- and Fe(III)-MOFs were improved. In conclusion, the synthesis of the stable high-valence MOFs by direct metathesis of Mg^{2+}/Cr^{3+} or Mg^{2+}/Fe^{3+} is difficult to achieve, therefore PSMO strategy deserves more recognition and further study.

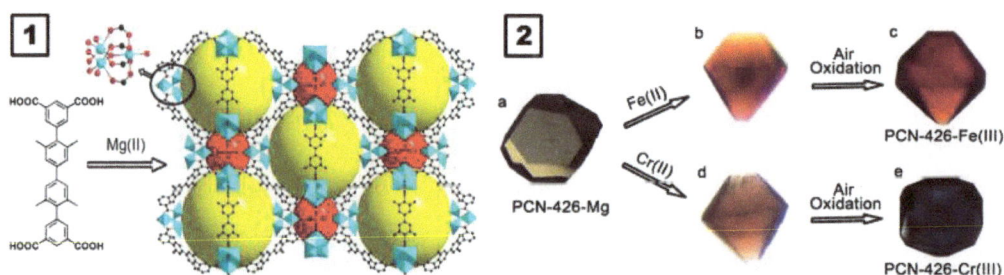

Fig. (5). 1) Synthesis of PCN-426-Mg, synthesized from TMQPTC ligands and Mg(II) ions. **2)** Optical microscope photographs of **(a)** PCN-426-Mg, **(b)** PCN-426-Mg after metathesis with FeCl$_2$, **(c)** PCN-426-Fe(III) after metal node oxidation, **(d)** PCN-426-Mg after metathesis with CrCl$_2$ and **(e)** PCN-426-Cr(III) after metal node oxidation (Reprinted with permission [39]. Copyright 2014, American Chemical Society).

Titanium-based MOFs (Ti-MOFs) are further examples of reticular materials difficult to synthesize directly. Due to their potential application in photocatalysis, much attention has been paid to their fabrication. Zou *et al.* [65] presented a synthetic strategy to obtain Ti(IV)-MOFs. Starting from appropriately selected matrices, MOF-74(Zn), MIL-100(Sc) (MIL = Materials Institute Lavoisier), PCN-333(Sc), and MOF-74(Mg), they prepared a series of porous photoactive titanium MOFs (MOF-74(Zn)-Ti, MIL-100(Sc)-Ti, PCN-333(Sc)-Ti, and MOF-74(Mg)-Ti). At first, the metathesis of metal ions Zn(II)/Mg(II) or Sc(III) to Ti(III) was carried out. The second step was based on the oxidation of metallic nodes to Ti(IV) in the air. The photocatalytic potential of the Ti-MOFs was investigated in the photodegradation of methylene blue (MB). All the obtained MOFs exhibited higher photocatalytic activity compared to TiO$_2$ and the parent MOFs. In the presence of TiO$_2$, MB degradation was slow, with less than 6% efficiency, whereas using PCN-333(Sc)-Ti and MIL-100(Sc)-Ti as catalysts, led to an increase in the process yield to 35% and 64%, respectively. Better photocatalytic performance was demonstrated by MOF-74(Zn)-Ti as well as MOF-74(Mg)-Ti, which achieved MB conversions up to 98% after 3 min, caused due to their ability to absorb a wider range of light irradiation. The high photocatalytic activity of titanium-based MOFs prepared by PSMO can solve the environmental issue due to the wide possibilities of their use in the photodegradation of organic pollutants (*e.g.,* rhodamine B and methylene blue) [66].

The Mn^{III} ions provide active catalytic sites for the oxidation of alkyl aromatic compounds. However, the instability of these ions causes them to be obtained by redox treatment of the more stable Mn^{II} precursors. Liao and co-workers [67] reported a porous metal azolate framework (MAF) based on Mn^{II} and rigid bis-triazolate ligand which has enabled oxidation to the Mn^{III} ion and its stabilization in solution. Catalytically active Mn^{III} sites present on the pore surface dramatically enhanced the activity of the oxidized MAF in the catalytic oxidation of alkylbenzenes. Shakya *et al.* [68] demonstrated a new approach for designing CuRhBTC (BTC^{3-} = 1,3,5-benzenetricarboxylate). By reducing the post-synthetically incorporated Rh^{3+}, catalytically active Rh^{2+} were obtained. It was observed that selective hydrogenation of propylene to propane at room temperature occurred only for the bimetallic MOF - CuRhBTC with Rh^{2+} as metal sites. No hydrogenation activity is exhibited by both CuBTC and CuMBTC (M = Co, Ir, Ni, Ru ions).

Presented reports demonstrate how the redox transformation of MOF inorganic components can be advantageous for different catalytic processes. It can increase the stability as well as catalytic efficiency of materials and direct their further applications.

3. LIGAND-BASED POST-SYNTHETIC MODIFICATIONS OF MOFS

Over the past few years, the approach to the role of organic ligands in MOF structure has changed. They are not only inert structural elements anymore, but could also give enormous benefits once their functionality and reactivity are utilized [69]. Organic ligands are significant parts of MOFs structure with regard to structural dimensionality and capability of binding with metal centers. Their length, size, and functional groups are necessary to establish such properties as porosity or active site availability in order to determine MOF's possible applications. Obtaining a new framework with improved properties requires some modifications. Attempts that have been made on ligand-based post-synthetic modification of MOFs include ligand functionalization, exchange, installation, and removal [37].

Throughout the years post-synthetically modified MOFs have found use in various fields. Their potential application as catalysts contributed to advances in the organic and organometallic chemistry of MOFs. Their similarity to heterogeneous catalysts in terms of high-performance and easy recovery as well as to homogeneous catalysts considering catalytic sites isolation and atomically defined structures is a significant benefit. Unsurprisingly, the PSM of MOFs is gaining more and more attention in the field of catalysis [69].

Numerous catalytic systems such as frustrated Lewis pairs or perrhenate anions with hydrosilanes as reductants are being used in the process of transforming carbon dioxide into methanol. However, only low yields (less than 10%) of desired products are obtained. MOF-based catalysts (*e.g.,* CuZn@UiO-bpy) can successfully reduce CO_2 to methanol as well, although high temperature (250 °C) and CO_2 pressure (4-15 MPa) are required. MOFs functionalized with nucleophilic N-heterocyclic carbenes (NHCs) have also gained interest, but producing free NHC moieties in aryl- or alkylimidazolium-based MOFs has been impossible because of material degradation. Zhang *et al.* [70] applied post-synthetic ligand exchange modification and incorporated a free NHC complex into a Zr^{IV}-based MOF. The obtained catalyst enables the quantitative transformation of silanes to methoxides using carbon dioxide under mild conditions (room temperature and 1 atm). Moreover, complete conversion of methoxide to methanol can be achieved through hydrolysis.

In 2018, Yuan and co-workers [71] performed post-synthetic modification in the form of ligand removal and ligand installation on a Zr-based MOF-PCN-160 (Fig. **6**). The incorporation of M-INA$_2$ (INA = isonicotinate) moieties led to the creation of *trans*-coordinated metal centers, which exhibit interesting properties in catalytic processes. Researchers used PCN-160-47%Ni MOF in the catalytic dimerization of ethylene. The material showed an intrinsic activity of 3360 h^{-1}, which is several times higher in comparison to other catalysts such as Ni-INA$_2$ or UiO-67-50%Ni. The equatorial positions of the open metal sites are exposed what facilitates ethylene binding and accelerates the reaction. Furthermore, there is a possibility to recycle PCN-160-47%Ni by centrifugation without major loss of its crystallinity and nickel content.

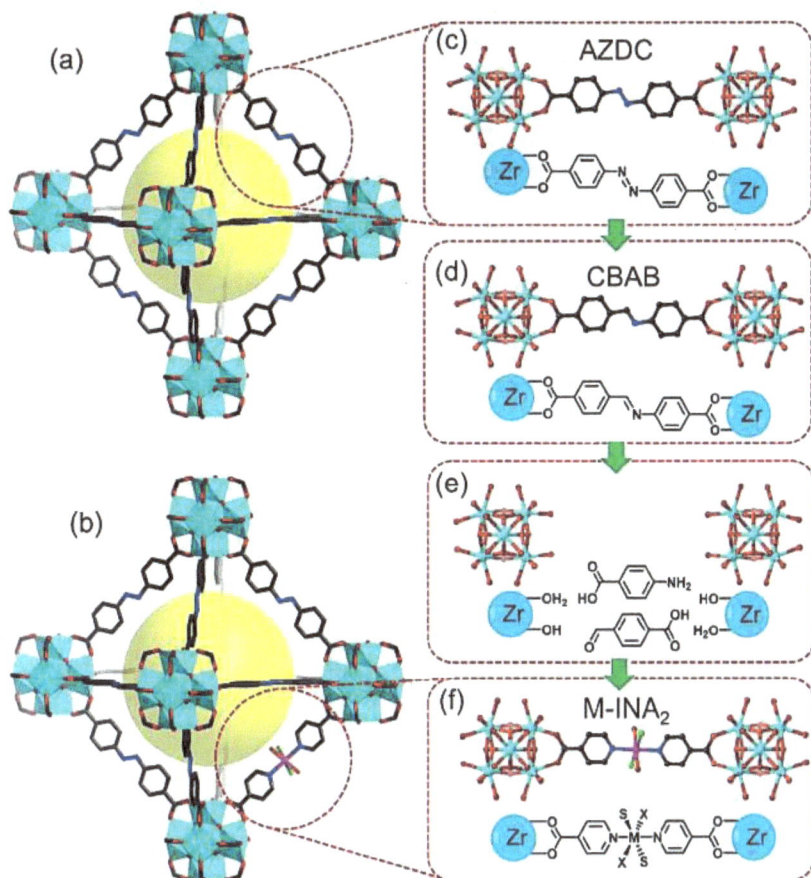

Fig. (6). The scheme represents the structures of PCN-160 **(a)** and PCN-160-R%M with trans-chelating ligands **(b)**. Modifications of PCN-160 ligand **(c)** by CBAB exchange **(d)**, ligand removal **(e)** and M-INA2 installation **(f)** are presented (Reprinted with permission [71]. Copyright 2018, Journal of the American Chemical Society).

The novel approach to prepare bifunctional Re-based MOF through post-synthetic ligand exchange (PSLE) was developed by Chen and co-workers [72]. PSLE is carried out at room temperature and it is a mild and efficient method. Due to high structural stability against water, the UiO-67 framework was selected for this modification. During PSM of UiO-67, ligand – 4,4'-biphenyl dicarboxylic acid (BPDC) was exchanged with dcpy ligand derived from tris-carbonyl-chloro(5,5'-dicarboxyl-2,2'-bipyridine)rhenium(I) – [Re(CO)$_3$(dcpy)Cl] complex. The obtained material (UiO-67-Re) can serve as a photosensitizer. Moreover, it catalyzes the photochemical reduction of carbon dioxide exhibiting high efficiency of carbon monoxide production. UiO-67-Re combines assets of a molecular catalyst with stability and orderliness of MOF substrate.

Sulfur organic compounds contained in fuels are a serious threat to the environment. The removal of thiophene derivatives *e.g.,* dibenzothiophene (DBT) can be achieved with oxidative desulfurization (ODS). In order to effectively conduct this reaction, a highly efficient catalyst is required. Ionic liquids (ILs) and MOFs have drawn interest as possible catalysts. However, ILs are difficult to reuse and MOF catalytic activity is usually low. So far, there have been a few attempts to combine MOFs with ionic liquids, although the composites turned out to be unstable and the MOF porous structure was not retained. Qi *et al.* [73] used post-synthetic ligand exchange to prepare a Fenton-like catalyst [mim(CH$_2$)$_3$COO]FeCl$_4$@UiO-66, which was exploited in dibenzothiophene oxidation. It showed good chemical, as well as thermal stability, and over 99% of DBT was removed. Furthermore, even after 6 recycle runs, [mim(CH$_2$)$_3$COO]FeCl$_4$@UiO-66 can maintain a high (>95%) removal efficiency.

The above examples have shown that MOFs are valuable materials that can be successfully used in the field of catalysis. Ligand-based post-synthetic modifications of MOFs can contribute to significant improvement of their catalytic activity. Hence, in the following parts of the chapter, we will focus on a detailed description of post-synthetic ligand functionalization, exchange, installation, and removal as ways for obtaining MOFs with desirable properties.

3.1. Ligand Functionalization

Preserving MOF's structural integrity is a large benefit of post-synthetic ligand modification (Fig. 7). Moreover, the newly created functional groups enhance MOF's properties so that they can be successfully used in various applications *e.g.*, in catalytic and adsorption processes [37].

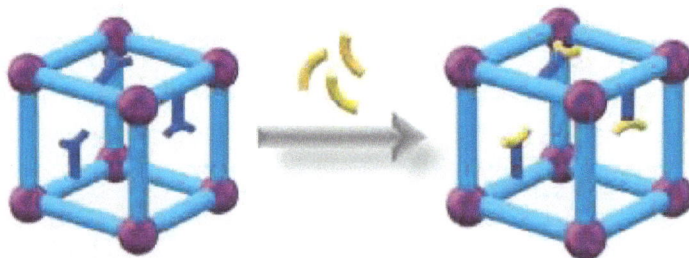

Fig. (7). Schematic view of post-synthetic ligand modification in metal–organic frameworks.

Post-synthetic modifications usually require high temperatures and are carried out as solid-liquid phase processes. They cause the limitation of the reaction progress because of reagents' diffusion in the framework. A solvent-less approach, in particular solid-gas phase reactions, may be applied to avoid the above-mentioned limitations in PSM. Albalad *et al.* [74] transformed olefin groups in a UiO-66-type MOF (ZrEBDC), in which 2-ethenylbenzene-1,4-dicarboxylic acid (H₂EBDC) was used as a ligand, into 1,2,4-trioxolane rings that were further converted to aldehydes or carboxylic acids. Due to the employment of controlled solid-gas phase ozonolysis, the modified MOF (ozo-ZrBDC) with stable 1,2,4-trioxolane moieties was obtained without single-crystallinity loss. Reducing or oxidizing the new groups of ozo-ZrBDC enables aldehydes or carboxylic acids' generation.

In 2020, Chen and co-workers [75] performed post-synthetic modification and grafted an anhydride generated by free carboxylic acid groups into a MOF with a hierarchical pore system. Al-based MOF (MIL-121) that contains 1,2,4,5-benzenetetracarboxylic acid (H₄BTEC) was selected by researchers. In order to form anhydride groups and create a hierarchical pore system in MIL-121 thermolysis was conducted. The synthesized material, which was termed HMIL-121, can immobilize organic molecules such as thiols, amines, or alcohols as well as inorganic Pt complex. Due to this ability, HMIL-121 may be used in drug delivery and for Pt-based catalysts fabrication.

Research concerning Zr^{IV}-based MOF applications as sorbents and catalysts in the degradation of chemical warfare agents (CWAs) has been already conducted. The catalytic activity of MOF can be enhanced by combining it with a suitable polymer. Nylon 6,6 (PA-66) is an easy to synthesize polyamide (PA) that is used to obtain covalent composites with MOFs through post-synthetic polymerization (PSP). Kalaj and co-workers [76] performed post-synthetic ligand modification on amine-functionalized MOF (UiO-66-NH₂) which was subsequently incorporated into PA-66. Numerous and widely used MOFs may be prepared from amine-functionalized ligands that can be transformed into amide groups. Due to the PSM of UiO-66-NH₂ with adipoyl chloride, reactive acyl chloride groups were formed and PSP reaction might have been carried out. The obtained hybrid material (PA-66-UiO-66-NH₂) exhibits high catalytic activity towards chemical warfare simulant – dimethyl-4-nitrophenyl phosphate (DMNP) degradation. Furthermore, its potential application in textile production is being investigated.

MOFs with pre-functionalized linkers can be further altered to increase their catalytic activity. An imidazolium functionalized UiO-66 (Im-UiO-66) was successfully modified with methyl groups (Me) into bifunctional (I⁻)Meim-UiO-66 *via* post-synthetic ionization. Readily fabricated Im-UiO-66 was treated with methyl iodide under reflux in acetonitrile. It resulted in 75% of the imidazole groups conversion to imidazolium and modification with methyl groups. Bifunctional materials having Brønsted acid sites and iodide ions revealed efficient catalytic activity in the solvent-free cycloaddition of CO₂ with epoxides.

Moreover, the process did not require the addition of co-catalysts that are typically used in this reaction. Catalytic cycloadditon was conducted in ambient conditions using MOF's nanoparticles characterized by the BET surface area of 560 m^2 g^{-1}, which yielded 93% of cyclocarbonate. Interestingly, smaller (I^-)Meim-UiO-66 nanoparticles (20-30 nm) indicated higher catalytic activity than nanospheres (100-300 nm) with a yield of 59%. Nanospheres had a higher degree of ionization and crystallinity, but less developed specific surface area (255 m^2 g^{-1}). As a result, the contact between the reactants and the catalytic active centers was reduced [77].

Encapsulating metal nanoparticles (NPs) in MOFs leads to the formation of NPs/MOF composites that can be used as catalysts in various chemical reactions. MOF's porous nature has an impact on catalyst selectivity and activity, therefore, different MOFs have to be selected for particular reactions to match the size of substrates and products. Modifying the organic ligands is a solution to adjust the pore size of the composites so that they can be utilized in various catalytic reactions. Liu *et al.* [78] described adjusting NPs/MOF pore sizes by post-synthetic modification. Researchers modified amino groups in ligands with alkyl chains of different lengths and employed the obtained composites as heterogeneous catalysts in olefins hydrogenation. Zr-based MOF (UiO-66-NH_2) and PVP-stabilized (PVP = polyvinylpyrrolidone) Pt NPs were chosen to synthesize Pt/UiO-66-NH_2 material. PSM was carried out with acetic, butyric, and hexanoic anhydrides. As a result, respectively Pt/UiO-66-AM1, Pt/UiO-66-AM3, and Pt/UiO-66-AM5 composites were obtained. Modified materials exhibited improved size selectivity in the hydrogenation of triphenylethylene, *trans*-stilbene, and cyclooctene. The results exposed the advantages of post-synthetic modified MOFs as supporting materials in catalysis.

In 2018, Kassymova and co-workers [79] applied PSM to synthesize NHC catalyst – Ag-NHC-MOF from azolium-based MOF. Integration of NHC into a framework led to the formation of numerous different catalytic sites which usually cannot be generated in catalytic systems simultaneously. A crystalline MOF used in this research consisted of Zn_8O clusters and ligands containing NHC with two aromatic carboxylates. During PSM, the ligand's carbon was deprotonated and a stable carbene was formed. The addition of metal in the next stage contributed to the development of a novel catalyst for the A^3-coupling reaction (aldehyde-alkyne-amine reaction). Post-synthetic ligand modification is one of the simplest methods of creating new materials with improved catalytic properties. Ag-NH--MOF showed better efficiency than previously reported catalysts (*e.g.,* thioether-based copper(I) Schiff base complex – Cu[(TS)PPh_3]X [80], nanosized Co_3O_4 [81], CuO nanoparticles [82]). Moreover, the A^3-coupling reaction may be conducted under mild conditions and the described catalyst is easy to separate and can be reused a few times without major activity loss.

Whilst Zr-based MOFs can be successfully applied in heterogeneous catalysis, their use might be limited if the reactant's particles exceed the pore size of the materials, thus some active centers can be unavailable for them. Ardila-Suárez *et al.* [83] utilized acetic acid to synthesize MOF-808 and then removed it to generate defects. With increasing the acetic acid concentration, more and wider mesopores were formed. Afterward, PSM with sulfuric acid was carried out and the obtained materials were activated under vacuum, which resulted in creating Lewis and Brønsted acid sites. This research may serve as a foundation for defects engineering towards acid catalysts synthesis.

Yaghi *et al.* [84] applied imidazolate-2-carboxyaldehyde (ICA) linker and synthesized ZIF-90, which was subsequently post-synthetically subjected to modifications. Due to its unique chemical and thermal stability, the reactions can be conducted under relatively hard conditions. $NaBH_4$ was used to perform the reduction of aldehyde to alcohol functionality. The surface area of the obtained ZIF-91 has slightly declined, whereas the porosity was maintained. Another PSM involved using ethanolamine to synthesize ZIF-92 through imine condensation. This material retained the crystalline structure of ZIF, which demonstrates that the beneficial organic transformations can be conducted without the alteration of the original structural integrity [15].

3.2. Ligand Exchange

Exchanging ligands in MOFs could lead to entirely new structures (Fig. 8). Frameworks formed by the new ligands can either resemble the parent structure or have different dimensions. The appropriate selection and replacement of a ligand is not a straightforward process. Several factors determine the success of the modification:

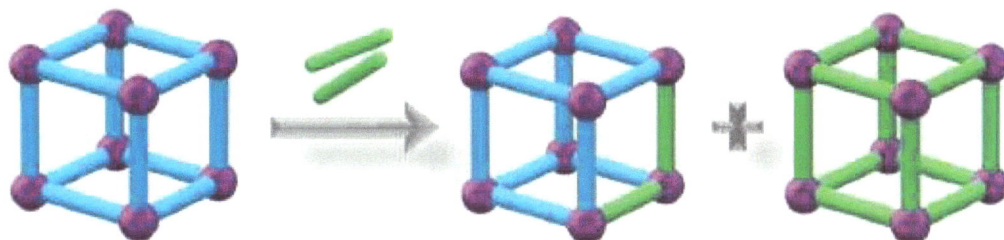

Fig. (8). Schematic view of general post-synthetic modification of metal–organic frameworks by ligand exchange.

• kinetic and thermodynamic stability of MOFs,

• incoming ligands' size and functionality,

• conditions (temperature, pressure, duration) under which the reaction takes place.

Post-synthetic ligand exchange can be conducted with new ligands or as a self-exchange reaction between the existing ones [37]. This modification provides excellent assets *e.g.,* in pore functionalization and introduction of catalytically active sites [85].

In 2019, Zhao *et al.* [86] performed post-synthetic ligand exchange on a Zr-based MOF – NU-1000 (NU = Northwestern University) which, has been broadly studied in terms of catalytic applications. This modification was applied in order to improve MOF stability. In the first step, a nanoscale NU-1000 was obtained. Afterward, a TCPP ligand in dimethylformamide (DMF) was introduced into the MOF by the PSLE method forming a modified MOF denoted as NT. TCPP partially replaced 1,3,6,8-tetrakis(p-benzoic acid)-pyrene (H_4TBAPy) in MOF structure. Employment of PSLE enabled the preservation of morphology and size of NU-1000.

The solvent-assisted ligand exchange (SALE) method was applied by Gharib and co-workers [87] to improve MOF efficiency in catalytic processes. N1,N3-di(pyridine-4-yl) malonamide ligand (S) has acidic hydrogen which catalyzes epoxide ring-opening reactions. Because the incorporation of N1,N3-di(pyridin--4-yl) malonamide ligand (S) cannot be conducted during MOF synthesis, SALE was used to successfully perform the post-synthetic modification of Co(II)-based MOFs – TMU-49, TMU-50, and TMU-51. The acylamide functional groups were exchanged by the new ligand leading to the formation of TMU-49S, TMU-50S, and TMU-51S MOFs. PSLE had a relevant impact on the catalytic activity of these metal–organic frameworks. Synthesized materials showed better efficiency in epoxide ring-opening reactions in comparison to parent MOFs. Furthermore, they did not exhibit a significant loss of catalytic activity after recycling.

In 2018, Pastore and co-workers [88] utilized PSLE to replace 1,4-benzene dicarboxylic acid in MOF-5 and ZIF-8 structures with poly(pyromellitic dianhydride-co-4,4'-oxydianiline) (PAA). The reaction was not affected by polymer concentration, but the rate depended on the number of exchangeable ligand sites. The product of this process was a cross-linked polymer-MOF composite which retained MOF porosity and was characterized by improved flexibility, stability, and strength. Obtaining hybrid PAA-MOF composites gives new possibilities for the future development of functional hybrid materials. So far polymer-MOF composites were successfully used as catalysts *e.g.,* in asymmetric aldol reaction [89].

Post-synthetic ligand exchange was also conducted by Kim *et al.* [90]. 2,2,6,6-tetramethylpiperidin (TEMPO) groups were anchored in UiO-67 to acquire a catalyst (UiO-67-TEMPO) for alcohol aerobic oxidation. For the purpose of improving its size selectivity, researchers obtained the surface-deactivated core-shell type MOF (UiO-67-TEMPO-CS) with the PSLE method. Deactivation of the surface was achieved with catalytically inactive ligand – (BPDC) without altering textural parameters and crystallinity. UiO-67-TEMPO-CS showed great size distinguishability of studied alcohols. Deactivating the surface using the PSLE strategy allows enhancing MOF catalytic properties.

As a result of the two-step PSM of ZIF-8 performed by [91] an acid-base bifunctional catalyst (ZIF8-A61-SO_3H) was synthesized. The first modification was carried out through the exchange of 2-methylimidazolate (2-mim) ligand with 3-amino-1,2,4-triazole (Atz). Afterward, the obtained ZIF8-A61 (with amine contents of 61%) was functionalized with sulfonic acid *via* a ring-opening reaction of 1,3-propanesultone with Atz NH_2 groups. ZIF8-A61-SO_3H was used as a heterogeneous catalyst in a one-pot deacetalization-Knoevenagel condensation with 100% conversion of benzaldehyde dimethylacetal and 98% selectivity toward benzylidenemalononitrile. Moreover, the material showed good stability and may be reused in the five consecutive cycles without a significant decrease in catalytic activity.

3.3. Ligand Installation

Even though PSLE enables to obtain new catalytically active materials, it has one significant disadvantage. The reaction occurs randomly with no site-specificity. The solution for this issue can be ligand installation (Fig. **9**). It is conducted as a stepwise installation of secondary ligands into prespecified sites without changing the original structure of MOFs. Although this modification is challenging due to the necessity to hold it in a tunable and predictable way, a few research works have already described it [37].

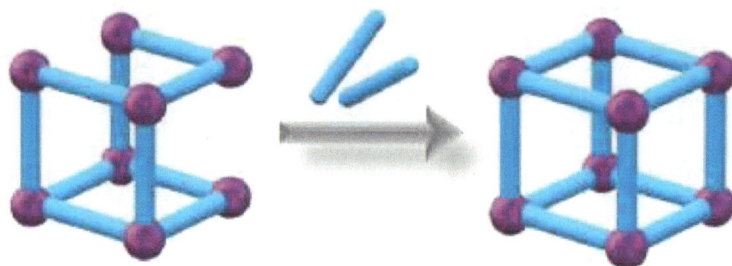

Fig. (9). Schematic representation of post-synthetic ligand installation in metal–organic frameworks.

Zr-based MOFs have recently gained much interest because of their excellent chemical and thermal stability. Zr^{4+} ion forms hexanuclear clusters – $[Zr_6O_4(OH)_4(COO)_{12}]$ when linear dicarboxylate ligands are entirely coordinated to $[Zr_6O_4(OH)_4]^{12+}$ which, results in the production of Zr-MOFs with 12-c fcu topology. One-pot synthesis or post-synthetic ligand exchange have been applied to prepare UiO-66-type MOFs containing two types of ligands of the same length. However, ligands are distributed at random, and placing them in designated positions within the structure remains a challenge. Post-synthetic ligand installation may unravel this problem. Inserting different carboxylate ligands into 8-c bcu Zr-MOFs can lead to the development of 10-c, 11-c, and 12-c Zr-MOFs. Kim *et al.* [92] described the symmetry-guided synthesis of Zr-based MOFs. Zigzag and linear dicarboxylate ligands that differ in length and symmetry were used and their precise positions in the framework were crystallographically defined. Post-synthetic installation of terephthalate derivatives of the same symmetry but different lengths, into 8-c bcu Zr-MOF (ZRN-bcu) resulted in the production of 12-c fcu MOFs with two types of ditopic heteroligands (Fig. **10**). Despite the fact that while post-synthetic ligand installation a structural change occurred, with single-crystal analysis the accurate locations of new ligands were defined. It is expected that the symmetry-guided post-synthetic ligand installation will become a common method of obtaining novel MOFs with desirable topologies as well as multiple functionalities in a well-defined and controlled way. Zr-based MOF with 12-c fcu topology (Zr-UiO-66) was also synthesized by Villoria-del-Álamo *et al.* [93]. This material has been investigated as a catalyst in a benzamide esterification reaction.

Multivariate heterogeneous catalysts *e.g.,* for tandem reactions possess multiple advantages but several limitations appear during their synthesis. One of the challenges is a simultaneous introduction of acidic and basic sites into the same catalyst. It results from the fact that preventing the neutralization of acid-base active sites during synthesis is hard. Furthermore, MOFs need to be resistant to acidic and basic conditions. Hu and co-workers [94] designed a dual-functional catalyst – PCN-700-AB by the installation of Brønsted acidic ligands – TPDC-

(COOH)$_2$ as well as basic ligands – BDC-NH$_2$ into a crystalline Zr-based MOF. The obtained catalyst showed great acid-base activity in a one-pot deacetalization-Knoevenagel condensation reaction. Brønsted acidic groups catalyzed the deacetalization reaction while Brønsted basic groups catalyzed the Knoevenagel condensation. Post-synthetic ligand installation is a new way of producing effective catalysts that contain bifunctional catalytically active sites and have increased active surface areas.

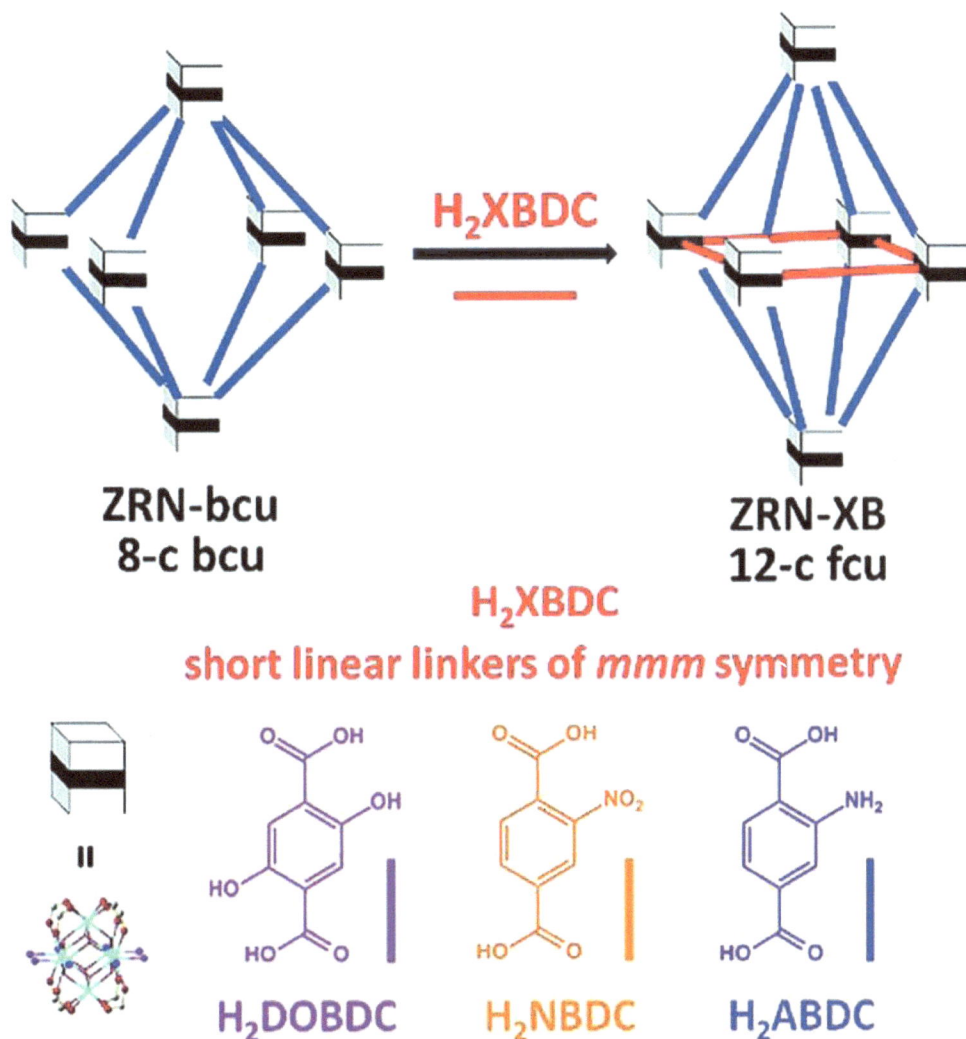

Fig. (10). Schematic view of 12-c fcu MOF preparation (Licensed under CC-BY [92]).

Investigating post-synthetic ligand installation resulted in discovering diverse tertiary as well as quaternary MOFs, which exhibit new properties in catalysis and gas storage. However, the design of a MOF matrix where more than three secondary ligands can be accommodated continues to be problematic. Asymmetric ligands that tend to form symmetry-reduced MOFs can be used to achieve the desired results. Pang and co-workers [95] obtained a carbazole-tetracarboxylate ligand to generate a matrix (Zr-MOF) with three coordination vacancies. Afterward, a sequential ligand installation was conducted to introduce three linear ligands (BDC, TPDC – 2',5'-dimethyl terphenyl-4,4"-dicarboxylate, BPDC) of various lengths and containing different functional groups into each vacancy. The resulting MOF (PCN-609), consisting of Zr_6 cluster and four different ligands (carbazole-tetracarboxylate ligand, BDC, TPDC, BPDC), is an excellent platform for incorporating multivariate functionalities inside a complex pore environment.

PCN-606-OH and PCN-606-OMe MOFs were chosen as platforms for ligand installation by Lollar *et al.* [96]. These materials contain tetratopic ligands as well as 8-connected Zr_6 clusters that exhibit high stability and possess open metal sites for post-synthetic ligand installation. 4,4'-(ethyne-1,2-diyl)dibenzoic acid (H_2EDDB) and Me_2TPDC ligands were inserted into PCN-606-OH and PCN-60--OMe to obtain PCN-606-OH-EDDB, PCN-606-OMe-EDDB, PCN-606- OH-TPDC, and PCN-606-OMe-TPDC MOFs. Since the TPDC ligand is longer and bulkier compared to the EDDB, a higher temperature was required to successfully conduct the installation. Post-synthetic ligand installation enabled placing ligands of proper topology and size inside pre-existing MOFs.

3.4. Ligand Removal

During ligand installation, the addition of new ligands to the MOF framework is conducted, while ligand removal is an exact opposite process (Fig. **11**). Saturated metal clusters, as well as non-labile and pro-labile ligands, are involved in this modification. The latter is then cleaved into a few elements to obtain a defective MOF. Such materials became the center of attention due to their numerous applications *e.g.,* in gas adsorption and catalysis [37].

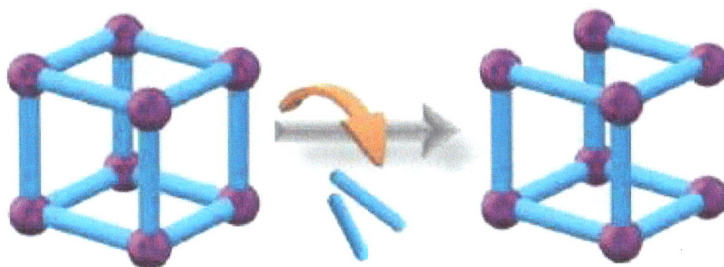

Fig. (11). Post-synthetic modification of metal–organic frameworks by ligand removal.

Large particle diffusion is impeded by microporosity in the majoriaty of MOFs. If such MOFs are applied in order to immobilize a large-sized catalyst or enzyme, their catalytic activity and access to the active sites will decline due to pore blocking. Hence, suitable pore size and stability, as well as hierarchical porosity, are necessary to utilize MOFs *e.g.,* in catalysis involving big molecules. Feng *et al.* [97] developed an effective approach to construct ultra-stable hierarchically porous Zr-MOFs (HP-UiO-66 MOFs) using a series of multivariate MOFs (MTV-UiO-66-NH$_2$-R%). Because of the instability of 2-amino-1-4-benzenedicarboxylate (BDC-NH$_2$) ligand, crystal defects were generated through thermolysis, and mesopores were formed. HP-MOFs chemical stability and crystallinity are maintained after removing thermolabile ligands from MTV-MOFs *via* the decarboxylation process. Moreover, due to thermolysis, ultra-small metal oxide (MO) nanoparticles (ZrO$_2$) immobilized in the MOF's structure were generated. As a result, a composite material (ZrO$_2$@HP-UiO-66), which displays high catalytic activity in reactions catalyzed by Lewis acid catalysts such as ZrO$_2$ and Zr-MOFs (*e.g.,* Meerwein-Ponndorf-Verley – MPV reaction), was synthesized.

In 2018, Guillerm and co-workers [98] performed solid-gas phase ozonolysis to cleave and remove particular ligands from MOFs. Organic ligands with non-terminal olefin groups were split into a few parts and then removed from the structure what provided a novel strategy for the post-synthetic fusion of micropores into mesopores. In the described work 4,4'-azobenzene dicarboxylic acid (H$_2$azo) and 4,4'-stilbene dicarboxylic acid (H$_2$sti) were exploited which resulted in obtaining MTV Zr-fcu-MOFs (Zr-fcu-azo and Zr-fcu-sti). Post-synthetic ligand removal enabled breaking the olefin bonds without affecting the remaining moieties. This method may be utilized in the selective removal and recovery of specific parts of frameworks *via* solid-gas manipulation.

Synthesizing MOFs with enlarged volumes and sizes of pores is the main goal in the research of these materials. Despite numerous attempts, synthesizing MOFs with adjusted pore apertures for particular applications remains a challenge. Yuan and co-workers [99] utilized ligand labilization to enhance porosity and pore sizes in MOFs. After the preparation of microporous MOFs, mesopores were formed as crystal defects by cleaving a pro-labile ligand and removing its fragments through acid treatment. Due to the maintenance of porosity as well as crystallinity after exposure to acid, researchers selected Zr-based MOFs for the ligand labilization strategy. 4-carboxybenzylidene-4-aminobenzate (CBAB) was chosen as a pro-labile ligand while azobenzene-4.4'-dicarboxylate (AZDC) acted as a robust ligand that supports the framework. Post-synthetic removal of CBAB ligand yielded in the fabrication of hierarchical porous structures within stable frameworks which can lead to improving MOFs efficiency, for instance in catalysis.

In 2020, Wang *et al.* [100] described a controlled ligand removal based on laser photolysis for generating mesopores inside microporous MOFs. UiO-66 framework with 1,4-benzenedicarbonate ligands and $[Zr_6O_4(OH)_4]$ clusters was used. TCPP was incorporated into the MOF structure to serve as a photolabile ligand. Hierarchically porous MOFs were obtained through selective elimination of TCPP ligands. Electrons' excitation or severe heat generated by the laser led to chemical bonds breaking which is the reason for organic ligands' removal and MOF cavities' enlargement. Post-synthetic ligand removal through photolysis enabled maintenance of the structural integrity of MOFs. Furthermore, this method can be used to create hierarchical MOF structures as well as in the production of complicated MOF patterns for catalytic applications *e.g.,* in cascade reactions.

In this chapter, we showed that ligands are important parts of MOFs' structure and their post-synthetic functionalization, exchange, installation, and removal can extend MOFs' catalytical applications. The presence of multiple types of organic ligands provides unlimited possibilities for obtaining frameworks with adjusted pore sizes and catalytically active sites while maintaining their stability as well as crystallinity. Moreover, if such materials are applied as catalysts, they can be reused without a significant activity loss. MOFs synthesized *via* ligand-based post-synthetic modifications are utilized in numerous chemical processes (*e.g.,* in carbon dioxide reduction, catalytic dimerization of ethylene, dibenzothiophene oxidation, A^3-coupling reaction, epoxide ring-opening reactions, and Meerwein-Ponndorf-Verley reaction) and are gaining more and more interest in the area of catalysis.

4. TANDEM PSM OF MOFS

The majority of MOF post-synthetic modifications are performed through one-step and one sort of reaction. However, conducting more than one reaction is possible because of the MOFs' robust structure that gives them enough stability. Combining various post-synthetic modification methods in a tandem way enables to obtain targeted structures with desirable functions and properties which cannot be achieved with single-step PSM (Fig. **12**) [13, 101]. Multiple structural factors such as shape, pores size, flexibility, and rigidity of the framework may be modified *via* tandem PSM. Two strategies for tandem post-synthetic modification have been described by Wang and Cohen [102]. In the first strategy, MOF active sites are partially modified by one reagent and afterward, the second reagent is used to modify the remaining ones. In another approach, a reagent with latent functional groups is employed to modify MOFs. Subsequently, these groups enter the structure and react with a second reagent. The introduction of functional groups into the framework has an impact on MOF applications in catalytical processes such as α-chlorination of butyraldehyde [103] and tandem deacetalization-Knoevenagel reaction [104].

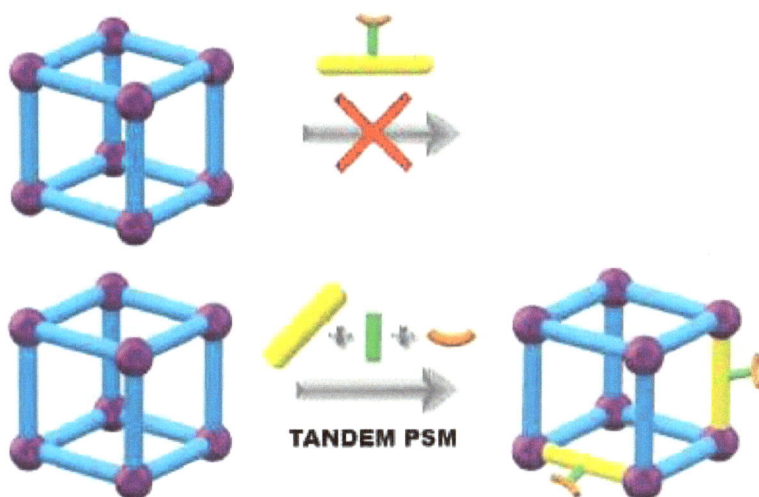

Fig. (12). Schematic example of tandem post-synthetic modification of metal–organic framework. Several functionalities can be introduced separately into the framework.

Karagiaridi *et al.* [105] performed a tandem PSM on ZIF-8 and obtained a Zn(im)$_2$ polymorph (SALEM-2) with SOD (sodalite) topology. Solvent-assisted linker exchange was used to replace a 2-methylimidazolate (2-mim) in a parent MOF with imidazolate (im) linkers and subsequently, the material was treated with n-butyllithium. The deprotonated SALEM-2 exhibited similar behavior to NHC-like-catalyst, although both alkyl groups were substituted with transition-metal cations. The received product may be applied as Brønsted base catalyst in the conjugate addition of methanol or benzyl alcohol to the α,β-unsaturated ketone(4-

hex-en-3-one), while ZIF-8 is catalytically inactive regarding the mentioned reaction.

Fei and Cohen [60] introduced 2,3-dimercaptoterephthalate (TCAT) into UiO-66 MOF followed by metalation of thiocatechol group with Pd, which resulted in obtaining Pd-mono(thiocatecholato) sites through a tandem approach (Fig. **13**). The prepared UiO-66-PdTCAT MOF can act as a reusable and effective heterogeneous catalyst in alcohol oxidation reactions. Pd-metalated material oxidizes aromatic reagents what contribute to the conversion of sp^2 C–H bonds into ethers and aryl halides. UiO-66-PdTCAT has been utilized in the oxidation of benzo[h]quinoline in order to install alkoxy groups on the C–H single bond. Methoxy-functionalized benzo[h]quinoline was synthesized in almost quantitative yield within 6 hours. Despite the fact that the use of homogeneous Pd(II) catalyst allowed to complete the reaction in 3 hours, UiO-66PdTCAT showed great recyclability while maintaining high yields (over 90%).

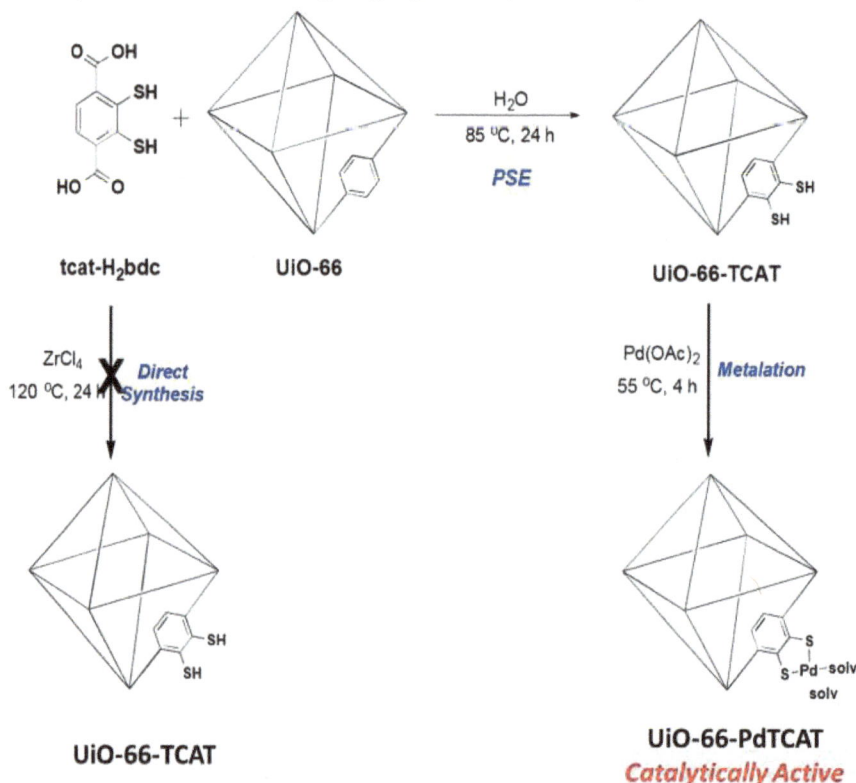

Fig. (13). Synthesis of UiO-66-TCAT and UiO-66-PdTCAT (Reprinted with permission [60]. Copyright 2015, American Chemical Society).

Heterogeneous acidic catalysts based on solid supports possess many advantages *e.g.,* eco-friendliness, non-corrosive nature, easy separation, and recovery. Furthermore, they are an excellent alternative for mineral acids such as H_2SO_4 or HCl. A great candidate for catalytic usage is MIL-101(Cr) which exhibits thermal and chemical stability and has a high surface area (over 3000 $m^2 \ g^{-1}$). In 2020, Mortazavi *et al.* [106] used a tandem post-synthetic modification strategy to synthesize two types of MIL-101(Cr) functionalized with sulfonic acid. The obtained MOFs have subsequently been applied as highly active and efficient Brønsted acid catalysts for styrene oxide methanolysis. Researchers coordinated piperazine or 1,4-diazabicyclo[2.2.2]octane (DABCO) to MIL-101(Cr) open metal sites. Subsequently, uncoordinated nitrogen atoms of N-heterocyclic compounds were covalently attached to 1,3-propane sultone. As a result, -SO_3H groups were generated in the framework, and MIL-pip-SO_3H as well as MIL-DABCO-SO_3H were synthesized. Compared to pure MIL-101(Cr), the reaction yield in styrene oxide methanolysis increased from less than 1% to approximately 99%. Moreover, the catalysts were easily separated from other reactants without a significant loss of catalytic activity. Tandem post-synthetic modification allows incorporating –SO_3H groups inside MIL-101(Cr) under mild conditions and is considered a desirable strategy for improving MOFs catalytic activities.

IRMOFs are isoreticular metal–organic frameworks that have a similar network topology, however, they are made of different components. IRMOF-3 $[(Zn_4O)(O_2C–C_6H_3(NH_2)–CO_2)_3]_n$ is the material which after proper modification is widely used in various catalytic processes *i.e.,* synthesis of cyclic carbonates [107], the cycloaddition of allyl glycidyl ether and CO_2 [107], Knoevenagel condensation [108], or Paal-Knorr [109], Heck [110], unsymmetrical Hantzsch coupling reactions [111]. Such broad applicability is the result of its susceptibility to modifications that are rarely capable of destroying the original structure topology. In terms of tandem PSM, an inverse-electron-demand Diels-Alder reaction was used on mixed linkers (BDC/BDC-NH_2) of IRMOF-3 by the Cohen group [112]. It was established that the proper ratio of tagged linker BDC-NH_2 to BDC was essential to efficiently conduct modification on IRMOF-3. A "tag" stands for functionality that is stable and non-structure-defining during framework formation [113]. Firstly, a tagged IRMOF-3 was modified with pentenoic anhydride (AMPent) to obtain an olefin group rich in electrons. Alkenes and alkynes readily react with electron-deficient dienes *e.g.,* *s*-tetrazines and their derivatives. Therefore, in the next step IRMOF-3(AMPent)$_n$ was subjected to reaction with dimethyl-1,2,4,5-tetrazine-3,6-dicarboxylate (TDC) in $CHCl_3$. It led to a "click" reaction and synthesis of Diels-Alder cycloaddition product IRMOF-3-(AMPent-TDC)$_n$. The same reaction was performed for UMCM-NH_2 (UMCM = University of Michigan Crystalline Material), where all linkers are tagged with –NH_2. After PSM, the obtained UMCM-1-AMPent-TDC maintained its high crystallinity, stability, and microporosity. This method allows the introduction of bulky organic molecules which are usually low-soluble *e.g.,* tetrazines. After click reaction, tetrazine-based MOFs are more stable at room temperature in the humid air [114]. Further extending tandem PSM for IRMOF-3 was again presented by Cohen and co-workers [14]. They found that IRMOF-3 can be modified with many sequential reagents, for instance with decanoic anhydride (–AM9), allyl isocyanate (–URAl), propyl isocyanate (–UR3), and crotonic anhydride (–AMCrot) (Fig. **14**). To avoid loss of IRMOF-3 crystallinity, the modification has to be restricted to ~50% of the amine sites. Such stepwise and precisely composed modification results in a profound alteration of IRMOF properties, which could be highly valuable for enantioselective or oxidation catalysis. Rosseinsky and co-workers [115] reported catalytic activity of IRMOF-3-Vsal$_{0.4}$ (sal = salicylaldehyde) in the oxidation of cyclohexene with ^{t}BuOOH. They conducted a first dual tandem covalent and coordinate covalent PSM of IRMOF-3. At first, the material required impregnation with toluene (tol) to decrease its moisture sensitivity. It permitted further treatment of IRMOF-3-tol under ambient conditions and its functionalization with salicylaldehyde. In consequence, free amines decorating IRMOF-3 were converted into Shiff base ligands, which as chelators were able to bind metal in vanadium derivative VO(acac)$_2$.

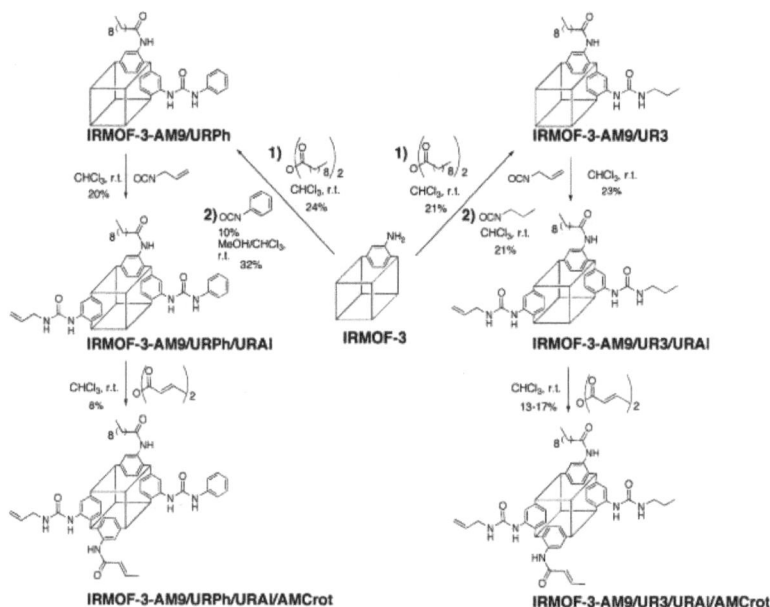

Fig. (14). Tandem PSM of IRMOF-3 with various modifiers: phenyl urea (URPh), decanoic anhydride (AM9), allyl isocyanate (URAl), propyl isocyanate (UR3), and crotonic anhydride (AMCrot) (Reprinted with permission [116]. Copyright 2009, American Chemical Society).

Through tandem PSM, MOFs with numerous functionalities, which are impossible to acquire by direct synthesis methods, can be obtained. Moreover, this type of functionalization gives the opportunity to observe complex reactions between MOF components and modifiers so that designing a given material for a specific application can be easily achieved. Coupling various reactions during the modification process can contribute not only to the generation of catalytically active centers but also to increase the stability and selectivity of MOFs.

5. EPITAXIAL GROWTH ON MOFS

The previously mentioned methods (*e.g.,* transmetalation or ligand functionalization) often do not provide a complete transition to the desired product. Epitaxial growth is the next major PSM relying on uniform distribution of one MOF onto another using methods such as dipping, spraying, pumping, flowing, or spin-coating layer by layer [117 - 119]. The surface of the core MOFs supports the evolution of single-shell crystals, where MOFs' pillar ligands are able to interchange [120]. Distorted interfacial structures formed between different hybridized solid phases, in consequence, may contribute to novel material properties *e.g.,* increased catalytic activity in tandem acetalization/oxidation [121] or CO_2 hydrogenation reactions [122]. Moreover, materials obtained by epitaxy can be subjected to further treatment *e.g.,* carbonization. It should be mentioned that MOF-derived carbons demonstrate good catalytic activity in the degradation of dyes or reduction of aryl nitro compounds [123].

Non-catalytically active MOF can be treated as a matrix for additional metal ions or complexes which exhibit catalytic activity. It is an alternative PSM where incorporation is conducted through metal inclusion in the open channels or direct coordination to metal sites. Incorporation can be carried out by the straightforward infiltration of preformed particles dispersed in a surfactant solution. The second way requires several steps, where MOF is impregnated with a proper metal precursor, followed by thermal treatment and conversion within the framework. For each modification, spatial confinement impedes particle growth if only the framework remains inert. This procedure allows the introduction of different types of metals, which if constitute metal nodes can create unstable MOF topology. Therefore, hetero-metallic doping of MOFs can be highly beneficial for catalysis *e.g.,* manganese salen (sal = salicylaldehyde, en = ethylenediamine), which is a system where Mn^{IV} is coordinated by tetradentate C_2-symmetric ligand. It caused improved catalyst stability during olefin epoxidation and enabled its reuse [124]. In turn, Ce^{3+}/Ce^{4+} incorporation promoted photocatalytic H_2 production from ammonia borane [125]. It is worth mentioning that the most oxo-bridged metals, especially Ti clusters incorporated into MOFs, can contribute to increased their performance in redox reactions commonly applied in photocatalysis [126].

Fabrication of heterointerfaces of which at least one phase is MOF enables creating superstructures with micro- or nanometer length scales along with tunable mechanical and electronic properties [120, 127]. MOFs can be obtained on numerous surfaces such as polymers, silica, metals, or even other MOFs. Such materials are expected to exhibit catalytic activity due to their unique alignment [128]. Therefore, in recent years, much attention has been paid to the stability of MOF hybrid materials and their efficiency in catalysis [23]. For example, MOF-on-MOF core-shell UiO-68 with various linkers has potential application in photoswitchable catalysis, which is an evolving field offering dynamic control of catalyst selectivity and activity. The outer layer with 2',5'-dimethyl-2,2"-bi-[4-(2,2',6,6'-tetrafluoroazobenzene)]-[1,1':4',1"-terphenyl]-4,4"-dicarboxylic acid linker, acts as a light-responsive gate for molecules of pyrene-derivatives. It could increase the selectivity, whilst implementing MOF cores with 3,3"-dihydrox--2',5'-dimethyl-[1,1':4',1"-terphenyl]-4,4"-dicarboxylic acid linker having catalytic properties is beneficial for photoswitchable catalysis [129].

The goal in a layer-by-layer approach is to obtain a specific material in the form of a film, therefore the reactants have to be kept apart. During synthesis, especially of MOF-on-MOF films, several factors of the parent MOFs have to be taken into account. For instance, to obtain MOF unit cell with sensible dimensions and symmetry, substrate termination, growth orientation, diffusion behavior, and chirality have to be considered [117]. It is necessary because mixing linkers differing in chemical nature or size will not always allow successful surface-mounted MOF (SURMOF) synthesis (Fig. **15**). A valuable side of this approach is that dependently on a technique, the thickness of the aimed film product can be altered *e.g.,* by the frequency of immersion cycles. It provides high control over MOF crystallites nucleation. Such procedures can eventually resolve kinetic and mass-transfer limitations during MOF syntheses caused by steric hindrance and small pores. In consequence, it enables precise MOF fabrication using a broad variety of substrates with the facilitation of urgently coveted size-selective catalysis *via* diffusion control.

A good example of improving catalyst properties was presented by Silvestre *et al.* [130] which, combined catalytically active magnetic particles with MOF. Magnetic particles mixed with silica are known for their catalytic application [131 - 133]. The group reported magnetic/MOF core/shell particles with the increased bulk surface (from 17 m^2 g^{-1} to 1150 m^2 g^{-1}) to improve catalytic performance of material. Using the liquid phase epitaxy (LPE) approach facilitated the deposition of Cu-MOF with a well-defined thickness on a wafer of COOH-terminated silica magnetic particles (MagPrep® Silica).

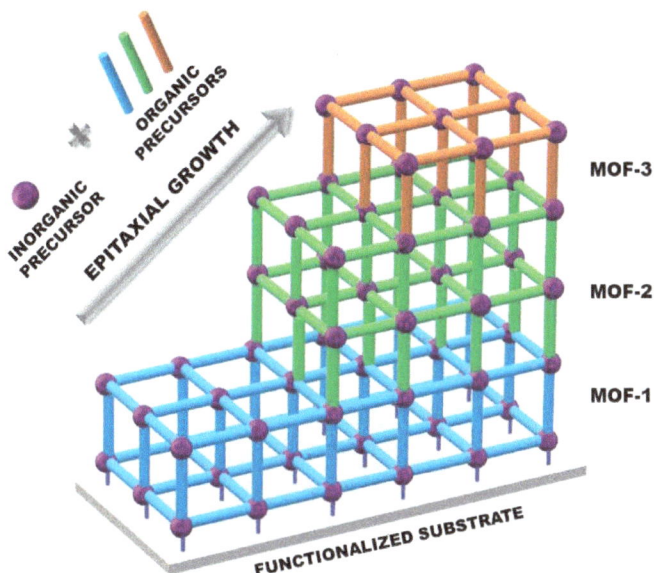

Fig. (15). Schematic example of the epitaxy approach as metal–organic framework post-synthetic modification. Inorganic and organic precursors may vary and construct MOF@MOF heterointerfaces.

The LPE procedure also allows encapsulating pre-synthesized particles within the MOF network *e.g.,* Pd-Cs guest incorporation in metalloporphyrin MOF PIZA-1 (porphyrinic Illinois zeolite analog no. 1) [121]. The spray method with the etanolic solutions of substrates provided high dispersion of material. The following substrates were employed: $Co(OAc)_2$ was used as IBUs, TCPP as OBUs to construct host MOF PIZA-1, and for guest – $Pd[P(C_6H_5)_3]_4$. Owing to the oxidation abilities of Pd-Cs, and Lewis acidic sites in PIZA-1, it resulted in the well-defined bifunctional catalyst. The material is able to catalyze benzyl alcohol oxidation, followed by acetalization of obtained benzaldehyde with ethylene glycol. Interestingly, the powder mixture of catalytically active components (PIZA-1 and Pd-Cs) under the same conditions exhibited a low substrate conversion of 55.3%, whereas using encapsulated material increased it up to 85%.

It is worth mentioning that the epitaxial approach for catalytically active hybrid material not necessarily always requires many steps. Usually, complex synthesis is not associated with green chemistry due to the many reagents and solvents which are involved in various steps. The Zhou group [134] reported the one-pot synthesis of core-shell PCN-222@UiO-67 with mismatching lattices for olefin epoxidation. During the solvothermal reaction, $ZrCl_4$ acted as a metal cations source, whereas TCPP and BPDC were applied as linkers to form different MOFs. These linkers indicated different nucleation kinetics under the same reaction conditions. TCPP binds to the metal center quicker and more strongly than BPDC, which ultimately led to the faster PCN-222 needle-like crystals formation. The PCN-222 was a template for the second MOF UiO-67, and it also had an accelerating effect on its nucleation. Moreover, it was possible to incorporate other metals during the synthesis such as Fe, Mn, Ni, and Zn *via* various metalloporphyrins.

Designing multimetallic MOF materials has been particularly investigated for ZIFs and their various core-shell structures. Singh and co-workers [135] presented the seed-mediated methodology to create bimetallic core-shell materials ZIF-8@ZIF-67 and ZIF-67@ZIF-8. Once the MOF seeds acting as the core are synthesized in the solution, a linker and salt of another metal are introduced in the next step to build them up on the core surface. Zhang *et al.* [136] extended such epitaxial growth and created multilayered ZIF-67@ZIF-8@ZIF-67 and ZIF-8@ZIF-67@ZIF-8. A key aspect of successful layer development is a homological feature between particles of these MOFs despite different metals. Carrying out the process at a low nucleation rate and long shell growth time prevents the generation of free-standing, monometallic MOFs in solution. Epitaxial post-synthetic modifications enable control of the quantitative ratio of one metal to another and may contribute to various catalytic applications of these materials. For instance, Zn/Co-ZIFs have been applied in the catalytic conversion of CO_2 to carbonates, the semihydrogenation of acetylene, or photocatalytic dyes degradation [137, 138]. Moreover, Zn/Co-ZIF-derived carbons and composites were tested in catalytically driven oxygen reduction reactions [139]. It indicates that MOF PSMs are unlimited and give ample opportunity to create new materials with catalytically active sites. The alteration of modification parameters allows the adjustment of the catalytic properties of the materials to the desired reaction.

CONCLUDING REMARKS

In recent years, inspiring progress in methods for the covalent and coordinative post-synthetic modification of MOF materials has been achieved. They are one of the effective tools to tailor the structure and physicochemical properties of MOFs. The internal or external surface of the crystals, pore apertures, sizes, and shapes, as well as framework rigidity and flexibility, can be precisely altered during PSM. A tandem PSM combining different reactions provides unexpected and at the same time also very desirable MOF complexity and multifunctionality. Moreover, post-synthetic modification approaches based on inorganic and organic components of MOFs help to address specific challenges related to creating new catalytically active centers and improving framework stability.

The proper selection of controllable PSM is essential from the point of view of designing MOF materials with high catalytic activity and selectivity expanding their full potential as catalysts in various types of catalytic reactions. The transfer of the applied post-synthetic modification methods from the laboratory to the industrial scale will demand the implementation of green chemistry principles. Then, the newly developed MOF catalysts can be an environmental-friendly alternative to conventional catalysts.

LIST OF ABBREVIATIONS

AM9	Decanoic anhydride
AMCrot	Crotonic anhydride
AMPent	Pentenoic anhydride
ATA	Aluminum amino terephthalate
Atz	3-amino-1,2,4-triazole
AZDC	Azobenzene-4.4'-dicarboxylate
BDC	Terephtalate
BET	Brunauer–Emmett–Teller
bpdb	1,4-bis(4-pyridyl)-2,3-diaza-1,3-butadiene
BPDC	4,4'-biphenyl dicarboxylate
bpy	2,2'-bipyridyl
BTC	1,3,5-benzenetricarboxylate
BTEC 1	1,2,4,5-benzenetetracarboxylate
CBAB	4-carboxybenzylidene-4-aminobenzate
COD	1,5-cyclooctadiene
CWA	Chemical warfare agent
DABCO	4-diazabicyclo[2.2.2]octane
DBT	Dibenzothiophene
dcpy	(2,2'-bipyridine)-5,5'-dicarboxylate
DMF	Dimethylformamide
DMNP	Dimethyl-4-nitrophenyl phosphate
dppe	1,2-bis(diphenylphosphino)ethane
EBDC	2-ethenylbenzene-1,4-dicarboxylate
EDDB	4,4'-(ethyne-1,2-diyl)dibenzoate
en	Ethylenediamine

HMIL	Hierarchically porous MIL
HP	Hierarchically porous
HSAB	Hard-soft acid-base theory
IBU	Inorganic building unit
ICA	Imidazolate-2-carboxyaldehyde
IL	Ionic liquid
im	Imidazolate
IRMOF	Isoreticular MOF
LPE	Liquid phase epitaxy
MAF	Metal azolate framework
MB	Methylene blue
MFU	MOF Ulm University
MIL	Materials Institute Lavoisier
	Methylimidazolate
MOF	Metal-organic framework
MPV	Meerwein-Ponndorf-Verley
MTV	Multivariate MOF
NHC	N-heterocyclic carbine
NP	Nanoparticles
NU	Northwestern University
OBU	Organic building unit
ODS	Oxidative desulfurization
OMS	Open-metal site
PA	Polyamide
PAA	Poly(pyromellitic dianhydride-co-4,4'-oxydianiline)
PCN	Porous coordination network
PCP	Porous coordination polymer
PEG	Poly(ethylene glycol)
PIZA	Porphyrinic Illinois zeolite analog
PSLE	Post-synthetic ligand exchange
PSM	Post-synthetic modification
PSMO	Post-synthetic metathesis and oxidation
PSP	Post-synthetic polymerization
PVP	Polyvinylpyrrolidone
sal	Salicylaldehyde

SALE	Solvent-assisted ligand exchange
SALEM-2	Zn(im)$_2$ polymorph
SOD	Sodalite
sti	4,4'-stilbene dicarboxylate
SURMOF	Surface-mounted MOF
TBAPy	1,3,6,8-tetrakis(p-benzoic acid)-pyren
TCAT 2	3-dimercaptoterephthalate
TCPP	tetrakis(4-carboxyphenyl)porphyrin
TDC	Dimethyl-1,2,4,5-tetrazine-3,6-dicarboxylate
TEMPO 2	2,6,6-tetramethylpiperidin
TMU	Tarbiat Modares University
TPDC	2',5'-dimethyl terphenyl-4,4''-dicarboxylate
TPO	Temperature programmed oxidation
TPR	Temperature programmed reduction
UiO	University of Oslo
UMCM	University of Michigan Crystalline Material
UR3	Propyl isocyanate
URAl	Allyl isocyanate
ZIF	Zeolitic Imidazolate Framework
ZJU	Zhejiang University

CONSENT FOR PUBLICATION

Not applicable.

CONFLICT OF INTEREST

The author declares no conflict of interest, financial or otherwise.

ACKNOWLEDGEMENTS

Declared none.

REFERENCES

[1]   Isaeva, V.I.; Kustov, L.M. The application of metal-organic frameworks in catalysis (Review). *Petrol. Chem.,* **2010**, *50*(3), 167-180.
[http://dx.doi.org/10.1134/S0965544110030011]

[2]   Nørskov, J.K.; Studt, F.; Abild-Pedersen, F.; Bligaard, T. *Fundamental Concepts in Heterogeneous Catalysis*; John Wiley & Sons, **2014**.
[http://dx.doi.org/10.1002/9781118892114]

[3]   Roduner, E. Understanding catalysis. *Chem. Soc. Rev.,* **2014**, *43*(24), 8226-8239.
      [http://dx.doi.org/10.1039/C4CS00210E] [PMID: 25311156]

[4]   Waclawek, S.; Vellora, T.P.V.; Černík, M. Major advances and challenges in heterogeneous catalysis
      for environmental applications: a review. *Ecol Chem Eng S,* **2018**, *25*(1).
      [http://dx.doi.org/10.1515/eces-2018-0001]

[5]   García-García, P.; Müller, M.; Corma, A. MOF catalysis in relation to their homogeneous counterparts
      and conventional solid catalysts. *Chem. Sci. (Camb.),* **2014**, *5*(8), 2979-3007.
      [http://dx.doi.org/10.1039/c4sc00265b]

[6]   Ejsmont, A.; Andreo, J.; Lanza, A.; Galarda, A.; Macreadie, L.; Wuttke, S.; Canossa, S.; Ploetz, E.;
      Goscianska, J. Applications of reticular diversity in metal–organic frameworks: An ever-evolving state
      of the art. *Coord. Chem. Rev.,* **2021**, *430*, 213655.
      [http://dx.doi.org/10.1016/j.ccr.2020.213655]

[7]   Sun, S.; Li, H.; Xu, Z.J. Impact of surface area in evaluation of catalyst activity. *Joule,* **2018**, *2*(6),
      1024-1027.
      [http://dx.doi.org/10.1016/j.joule.2018.05.003]

[8]   Bavykina, A.; Kolobov, N.; Khan, I.S.; Bau, J.A.; Ramirez, A.; Gascon, J. Metal–Organic Frameworks
      in Heterogeneous Catalysis: Recent Progress, New Trends, and Future Perspectives. *Chem. Rev.,* **2020**,
      120(16), 8468-8535.
      [http://dx.doi.org/10.1021/acs.chemrev.9b00685] [PMID: 32223183]

[9]   Pascanu, V.; González Miera, G.; Inge, A.K.; Martín-Matute, B. Metal–organic frameworks as
      catalysts for organic synthesis: A critical perspective. *J. Am. Chem. Soc.,* **2019**, *141*(18), 7223-7234.
      [http://dx.doi.org/10.1021/jacs.9b00733] [PMID: 30974060]

[10] Freund, R.; Zaremba, O.; Arnauts, G.; Ameloot, R.; Skorupskii, G.; Dincă, M.; Bavykina, A.; Gascon,
      J.; Ejsmont, A.; Goscianska, J.; Kalmutzki, M.; Lächelt, U.; Ploetz, E.; Diercks, C.S.; Wuttke, S. The
      Current Status of MOF and COF Applications. *Angew. Chem. Int. Ed.,* **2021**, *60*(45), 23975-24001.
      [http://dx.doi.org/10.1002/anie.202106259] [PMID: 33989445]

[11] Alhumaimess, M.S. Metal–organic frameworks and their catalytic applications. *J. Saudi Chem. Soc.,*
      2020, *24*(6), 461-473.
      [http://dx.doi.org/10.1016/j.jscs.2020.04.002]

[12] Ding, M.; Cai, X.; Jiang, H.L. Improving MOF stability: approaches and applications. *Chem. Sci.
      (Camb.),* **2019**, *10*(44), 10209-10230.
      [http://dx.doi.org/10.1039/C9SC03916C] [PMID: 32206247]

[13] Yin, Z.; Wan, S.; Yang, J.; Kurmoo, M.; Zeng, M.H. Recent advances in post-synthetic modification
      of metal–organic frameworks: New types and tandem reactions. *Coord. Chem. Rev.,* **2019**, *378*, 500-
      512.
      [http://dx.doi.org/10.1016/j.ccr.2017.11.015]

[14] Tanabe, K.K.; Cohen, S.M. Postsynthetic modification of metal–organic frameworks—a progress
      report. *Chem. Soc. Rev.,* **2011**, *40*(2), 498-519.
      [http://dx.doi.org/10.1039/C0CS00031K] [PMID: 21103601]

[15] Wang, Z.; Cohen, S.M. Postsynthetic modification of metal–organic frameworks. *Chem. Soc. Rev.,*
      2009, *38*(5), 1315-1329.
      [http://dx.doi.org/10.1039/b802258p] [PMID: 19384440]

[16] Kalaj, M.; Cohen, S.M. Postsynthetic Modification: An Enabling Technology for the Advancement of
      Metal–Organic Frameworks. *ACS Cent. Sci.,* **2020**, *6*(7), 1046-1057.
      [http://dx.doi.org/10.1021/acscentsci.0c00690] [PMID: 32724840]

[17] Goetjen, T.A.; Liu, J.; Wu, Y.; Sui, J.; Zhang, X.; Hupp, J.T.; Farha, O.K. Metal–organic framework
      (MOF) materials as polymerization catalysts: a review and recent advances. *Chem. Commun. (Camb.),*
      2020, *56*(72), 10409-10418.

[http://dx.doi.org/10.1039/D0CC03790G] [PMID: 32745156]

[18]   Wang, Z.; Tanabe, K.K.; Cohen, S.M. Accessing postsynthetic modification in a series of metal-organic frameworks and the influence of framework topology on reactivity. *Inorg. Chem.*, **2009**, *48*(1), 296-306.
[http://dx.doi.org/10.1021/ic801837t] [PMID: 19053339]

[19]   Valvekens, P.; Vermoortele, F.; De Vos, D. Metal–organic frameworks as catalysts: the role of metal active sites. *Catal. Sci. Technol.*, **2013**, *3*(6), 1435-1445.
[http://dx.doi.org/10.1039/c3cy20813c]

[20]   Hu, Z.; Zhao, D. Metal–organic frameworks with Lewis acidity: synthesis, characterization, and catalytic applications. *CrystEngComm*, **2017**, *19*(29), 4066-4081.
[http://dx.doi.org/10.1039/C6CE02660E]

[21]   Chen, W.; Zhang, H.; Qiao, A.; Tao, H. Impact of solvent substitution on kinetically controlled transmetalation behaviours in a MOF. *New J. Chem.*, **2020**, *44*(34), 14679-14685.
[http://dx.doi.org/10.1039/D0NJ02935A]

[22]   D'Alessandro, D.M. Exploiting redox activity in metal–organic frameworks: concepts, trends and perspectives. *Chem. Commun. (Camb.)*, **2016**, *52*(58), 8957-8971.
[http://dx.doi.org/10.1039/C6CC00805D] [PMID: 26988560]

[23]   Chen, L.; Xu, Q. Metal-Organic Framework Composites for Catalysis. *Matter*, **2019**, *1*(1), 57-89.
[http://dx.doi.org/10.1016/j.matt.2019.05.018]

[24]   Wu, Y.; Li, Y.; Gao, J.; Zhang, Q. Recent advances in vacancy engineering of metal-organic frameworks and their derivatives for electrocatalysis. *SusMat*, **2021**, *1*(1), 66-87.
[http://dx.doi.org/10.1002/sus2.3]

[25]   Zhang, Y.; Yang, X.; Zhou, H.C. Synthesis of MOFs for heterogeneous catalysis *via* linker design. *Polyhedron*, **2018**, *154*, 189-201.
[http://dx.doi.org/10.1016/j.poly.2018.07.021]

[26]   Ranocchiari, M.; Bokhoven, J.A. Catalysis by metal–organic frameworks: fundamentals and opportunities. *Phys. Chem. Chem. Phys.*, **2011**, *13*(14), 6388-6396.
[http://dx.doi.org/10.1039/c0cp02394a] [PMID: 21234497]

[27]   Wang, T.; Gao, L.; Hou, J.; Herou, S.J.A.; Griffiths, J.T.; Li, W.; Dong, J.; Gao, S.; Titirici, M.M.; Kumar, R.V.; Cheetham, A.K.; Bao, X.; Fu, Q.; Smoukov, S.K. Rational approach to guest confinement inside MOF cavities for low-temperature catalysis. *Nat. Commun.*, **2019**, *10*(1), 1340.
[http://dx.doi.org/10.1038/s41467-019-08972-x] [PMID: 30902984]

[28]   Wang, J.C.; Ding, F.W.; Ma, J.P.; Liu, Q.K.; Cheng, J.Y.; Dong, Y.B. Co (II)-MOF: a highly efficient organic oxidation catalyst with open metal sites. *Inorg. Chem.*, **2015**, *54*(22), 10865-10872.
[http://dx.doi.org/10.1021/acs.inorgchem.5b01938] [PMID: 26497909]

[29]   Gonen, S.; Lori, O.; Cohen-Taguri, G.; Elbaz, L. Metal organic frameworks as a catalyst for oxygen reduction: an unexpected outcome of a highly active Mn-MOF-based catalyst incorporated in activated carbon. *Nanoscale*, **2018**, *10*(20), 9634-9641.
[http://dx.doi.org/10.1039/C7NR09081A] [PMID: 29756623]

[30]   Luz, I.; Llabrés i Xamena, F.X.; Corma, A.; Corma, A. Bridging homogeneous and heterogeneous catalysis with MOFs: Cu-MOFs as solid catalysts for three-component coupling and cyclization reactions for the synthesis of propargylamines, indoles and imidazopyridines. *J. Catal.*, **2012**, *285*(1), 285-291.
[http://dx.doi.org/10.1016/j.jcat.2011.10.001]

[31]   Zamani, S.; Abbasi, A.; Masteri-Farahani, M. Post-synthetic modification of porous [Cu3(BTC)2] (BTC = benzene-1,3,5-tricarboxylate) metal organic framework with molybdenum and vanadium complexes for the epoxidation of olefins and allyl alcohols. *React. Kinet. Mech. Catal.*, **2021**, *132*(1), 235-250.

[http://dx.doi.org/10.1007/s11144-020-01912-7]

[32] Dan-Hardi, M.; Serre, C.; Frot, T.; Rozes, L.; Maurin, G.; Sanchez, C.; Férey, G. A new photoactive crystalline highly porous titanium(IV) dicarboxylate. *J. Am. Chem. Soc.,* **2009**, *131*(31), 10857-10859.
[http://dx.doi.org/10.1021/ja903726m] [PMID: 19621926]

[33] Khajavi, H.; Gascon, J.; Schins, J.M.; Siebbeles, L.D.A.; Kapteijn, F. Unraveling the Optoelectronic and Photochemical Behavior of Zn $_4$ O-Based Metal Organic Frameworks. *J. Phys. Chem. C,* **2011**, *115*(25), 12487-12493.
[http://dx.doi.org/10.1021/jp201760s]

[34] Lalonde, M.; Bury, W.; Karagiaridi, O.; Brown, Z.; Hupp, J.T.; Farha, O.K. Transmetalation: routes to metal exchange within metal–organic frameworks. *J. Mater. Chem. A Mater. Energy Sustain.,* **2013**, *1*(18), 5453-5468.
[http://dx.doi.org/10.1039/c3ta10784a]

[35] Zhang, Z.; Zhang, L.; Wojtas, L.; Nugent, P.; Eddaoudi, M.; Zaworotko, M.J. Templated synthesis, postsynthetic metal exchange, and properties of a porphyrin-encapsulating metal-organic material. *J. Am. Chem. Soc.,* **2012**, *134*(2), 924-927.
[http://dx.doi.org/10.1021/ja209643b] [PMID: 22191602]

[36] Brozek, C.K.; Dincă, M. Cation exchange at the secondary building units of metal–organic frameworks. *Chem. Soc. Rev.,* **2014**, *43*(16), 5456-5467.
[http://dx.doi.org/10.1039/C4CS00002A] [PMID: 24831234]

[37] Mandal, S.; Natarajan, S.; Mani, P.; Pankajakshan, A. Post-Synthetic Modification of Metal–Organic Frameworks Toward Applications. *Adv. Funct. Mater.,* **2021**, *31*(4), 2006291.
[http://dx.doi.org/10.1002/adfm.202006291]

[38] Brozek, C.K.; Cozzolino, A.F.; Teat, S.J.; Chen, Y.S.; Dincă, M. Quantification of site-specific cation exchange in metal-organic frameworks using multi-wavelength anomalous X-ray dispersion. *Chem. Mater.,* **2013**, *25*(15), 2998-3002.
[http://dx.doi.org/10.1021/cm400858d]

[39] Liu, T.F.; Zou, L.; Feng, D.; Chen, Y.P.; Fordham, S.; Wang, X.; Liu, Y.; Zhou, H.C. Stepwise synthesis of robust metal-organic frameworks *via* postsynthetic metathesis and oxidation of metal nodes in a single-crystal to single-crystal transformation. *J. Am. Chem. Soc.,* **2014**, *136*(22), 7813-7816.
[http://dx.doi.org/10.1021/ja5023283] [PMID: 24840498]

[40] Liao, P.Q.; Li, X.Y.; Bai, J.; He, C.T.; Zhou, D.D.; Zhang, W.X.; Zhang, J.P.; Chen, X.M. Drastic enhancement of catalytic activity *via* post-oxidation of a porous MnII triazolate framework. *Chemistry,* **2014**, *20*(36), 11303-11307.
[http://dx.doi.org/10.1002/chem.201403123] [PMID: 25043981]

[41] Mi, L.; Hou, H.; Song, Z.; Han, H.; Xu, H.; Fan, Y.; Ng, S.W. Rational construction of porous polymeric cadmium ferrocene-1,1′- disulfonates for transition metal ion exchange and sorption. *Cryst. Growth Des.,* **2007**, *7*(12), 2553-2561.
[http://dx.doi.org/10.1021/cg070468e]

[42] Zhao, J.; Mi, L.; Hu, J.; Hou, H.; Fan, Y. Cation exchange induced tunable properties of a nanoporous octanuclear Cu(II) wheel with double-helical structure. *J. Am. Chem. Soc.,* **2008**, *130*(46), 15222-15223.
[http://dx.doi.org/10.1021/ja8007227] [PMID: 18939798]

[43] Noori, Y.; Akhbari, K. Post-synthetic ion-exchange process in nanoporous metal–organic frameworks; an effective way for modulating their structures and properties. *RSC Advances,* **2017**, *7*(4), 1782-1808.
[http://dx.doi.org/10.1039/C6RA24958B]

[44] Denysenko, D.; Werner, T.; Grzywa, M.; Puls, A.; Hagen, V.; Eickerling, G.; Jelic, J.; Reuter, K.; Volkmer, D. Reversible gas-phase redox processes catalyzed by Co-exchanged MFU-4l(arge). *Chem. Commun. (Camb.),* **2012**, *48*(9), 1236-1238.

[http://dx.doi.org/10.1039/C2CC16235K] [PMID: 22179398]

[45]   Xie, X.; Li, Y.; Liu, Z.Q.; Haruta, M.; Shen, W. Low-temperature oxidation of CO catalysed by Co3O4 nanorods. *Nature,* **2009,** *458*(7239), 746-749.
[http://dx.doi.org/10.1038/nature07877] [PMID: 19360084]

[46]   Genna, D.T.; Wong-Foy, A.G.; Matzger, A.J.; Sanford, M.S. Heterogenization of homogeneous catalysts in metal-organic frameworks *via* cation exchange. *J. Am. Chem. Soc.,* **2013,** *135*(29), 10586-10589.
[http://dx.doi.org/10.1021/ja402577s] [PMID: 23837970]

[47]   Dhakshinamoorthy, A.; Asiri, A.M.; Garcia, H. Mixed-metal or mixed-linker metal organic frameworks as heterogeneous catalysts. *Catal. Sci. Technol.,* **2016,** *6*(14), 5238-5261.
[http://dx.doi.org/10.1039/C6CY00695G]

[48]   Diercks, C.S.; Kalmutzki, M.J.; Diercks, N.J.; Yaghi, O.M. Conceptual Advances from Werner Complexes to Metal–Organic Frameworks. *ACS Cent. Sci.,* **2018,** *4*(11), 1457-1464.
[http://dx.doi.org/10.1021/acscentsci.8b00677] [PMID: 30555897]

[49]   Kassie, A.A.; Duan, P.; McClure, E.T.; Schmidt-Rohr, K.; Woodward, P.M.; Wade, C.R. Postsynthetic Metal Exchange in a Metal–Organic Framework Assembled from Co(III) Diphosphine Pincer Complexes. *Inorg. Chem.,* **2019,** *58*(5), 3227-3236.
[http://dx.doi.org/10.1021/acs.inorgchem.8b03318] [PMID: 30762343]

[50]   van der Boom, M.E.; Milstein, D. Cyclometalated phosphine-based pincer complexes: mechanistic insight in catalysis, coordination, and bond activation. *Chem. Rev.,* **2003,** *103*(5), 1759-1792.
[http://dx.doi.org/10.1021/cr960118r] [PMID: 12744693]

[51]   Dutta, G.; Jana, A.K.; Natarajan, S. Assembling Porphyrins into Extended Network Structures by Employing Aromatic Dicarboxylates: Synthesis, Metal Exchange, and Heterogeneous Catalytic Studies. *Chemistry,* **2017,** *23*(37), 8932-8940.
[http://dx.doi.org/10.1002/chem.201700985] [PMID: 28422333]

[52]   Salahshournia, B.; Hamadi, H.; Nobakht, V. Engineering a Cu-MOF Nano-Catalyst by using Post-Synthetic Modification for the Preparation of 5-Substituted 1 *H* -Tetrazoles. *Appl. Organomet. Chem.,* **2018,** *32*(8), e4416.
[http://dx.doi.org/10.1002/aoc.4416]

[53]   Salahshournia, B.; Hamadi, H.; Nobakht, V. Designing a bifunctional metal-organic framework by tandem post-synthetic modifications; an efficient and recyclable catalyst for Suzuki-Miyaura cross-coupling reaction. *Polyhedron,* **2020,** *189*, 114749.
[http://dx.doi.org/10.1016/j.poly.2020.114749]

[54]   Liang, Y.; Liu, Y.; Deng, J.; Zhang, K.; Hou, Z.; Zhao, X.; Zhang, X.; Zhang, K.; Wei, R.; Dai, H. Coupled Palladium–Tungsten Bimetallic Nanosheets/TiO 2 Hybrids with Enhanced Catalytic Activity and Stability for the Oxidative Removal of Benzene. *Environ. Sci. Technol.,* **2019,** *53*(10), 5926-5935.
[http://dx.doi.org/10.1021/acs.est.9b00370] [PMID: 31035751]

[55]   Yan, X.; Yang, X.; Xi, C. Recent progress in copper-catalyzed electrophilic amination. *Catal. Sci. Technol.,* **2014,** *4*(12), 4169-4177.
[http://dx.doi.org/10.1039/C4CY00773E]

[56]   Rostamnia, S.; Alamgholiloo, H.; Liu, X. Pd-grafted open metal site copper-benzene-1,4-dicarboxylate metal organic frameworks (Cu-BDC MOF's) as promising interfacial catalysts for sustainable Suzuki coupling. *J. Colloid Interface Sci.,* **2016,** *469*, 310-317.
[http://dx.doi.org/10.1016/j.jcis.2016.02.021] [PMID: 26897567]

[57]   Han, Y.; Sinnwell, M.A.; Surbella, R.G., III; Xue, W.; Huang, H.; Zheng, J.; Peng, B.; Verma, G.; Yang, Y.; Liu, L.; Ma, S.; Thallapally, P.K. Postsynthetic Oxidation of the Coordination Site in a Heterometallic Metal–Organic Framework: Tuning Catalytic Behaviors. *Chem. Mater.,* **2020,** *32*(12), 5192-5199.
[http://dx.doi.org/10.1021/acs.chemmater.0c01267]

[58]   An, Y.; Liu, Y.; An, P.; Dong, J.; Xu, B.; Dai, Y.; Qin, X.; Zhang, X.; Whangbo, M.H.; Huang, B. Ni ^{II} Coordination to an Al-Based Metal-Organic Framework Made from 2-Aminoterephthalate for Photocatalytic Overall Water Splitting. *Angew. Chem. Int. Ed.,* **2017**, *56*(11), 3036-3040.
[http://dx.doi.org/10.1002/anie.201612423] [PMID: 28170148]

[59]   Manna, K.; Zhang, T.; Lin, W. Postsynthetic metalation of bipyridyl-containing metal-organic frameworks for highly efficient catalytic organic transformations. *J. Am. Chem. Soc.,* **2014**, *136*(18), 6566-6569.
[http://dx.doi.org/10.1021/ja5018267] [PMID: 24758529]

[60]   Fei, H.; Cohen, S.M. Metalation of a thiocatechol-functionalized Zr(IV)-based metal-organic framework for selective C-H functionalization. *J. Am. Chem. Soc.,* **2015**, *137*(6), 2191-2194.
[http://dx.doi.org/10.1021/ja5126885] [PMID: 25650584]

[61]   Zhang, H.; Wei, J.; Dong, J.; Liu, G.; Shi, L.; An, P.; Zhao, G.; Kong, J.; Wang, X.; Meng, X.; Zhang, J.; Ye, J. Efficient Visible-Light-Driven Carbon Dioxide Reduction by a Single-Atom Implanted Metal-Organic Framework. *Angew. Chem. Int. Ed.,* **2016**, *55*(46), 14310-14314.
[http://dx.doi.org/10.1002/anie.201608597] [PMID: 27736031]

[62]   Ji, S.; Chen, Y.; Zhao, S.; Chen, W.; Shi, L.; Wang, Y.; Dong, J.; Li, Z.; Li, F.; Chen, C.; Peng, Q.; Li, J.; Wang, D.; Li, Y. Atomically Dispersed Ruthenium Species Inside Metal-Organic Frameworks: Combining the High Activity of Atomic Sites and the Molecular Sieving Effect of MOFs. *Angew. Chem. Int. Ed.,* **2019**, *58*(13), 4271-4275.
[http://dx.doi.org/10.1002/anie.201814182] [PMID: 30730605]

[63]   Du, Y.; Hong, J.; Wang, Y.; Zhang, W.; Zhou, T.; Borgna, A.; Han, J.; Xu, R. Metal Insertion in Metal-Organic Frameworks Forming a Bifunctional hotocatalyst for Hydrogen Production Oxide on Energy Application View project Post-synthesis modification of a metal-organic framework to construct a bifunctional photocatalyst for hydroge.

[64]   Wei, Y.P.; Yang, S.; Wang, P.; Guo, J.H.; Huang, J.; Sun, W.Y. Iron(III)-bipyridine incorporated metal-organic frameworks for photocatalytic reduction of CO $_2$ with improved performance. *Dalton Trans.,* **2021**, *50*(1), 384-390.
[http://dx.doi.org/10.1039/D0DT03500A] [PMID: 33320135]

[65]   Zou, L.; Feng, D.; Liu, T.F.; Chen, Y.P.; Yuan, S.; Wang, K.; Wang, X.; Fordham, S.; Zhou, H.C. A versatile synthetic route for the preparation of titanium metal-organic frameworks. *Chem. Sci. (Camb.),* **2016**, *7*(2), 1063-1069.
[http://dx.doi.org/10.1039/C5SC03620H] [PMID: 29896371]

[66]   Gao, J.; Miao, J.; Li, P.Z.; Teng, W.Y.; Yang, L.; Zhao, Y.; Liu, B.; Zhang, Q. A p-type Ti(IV)-based metal-organic framework with visible-light photo-response. *Chem. Commun. (Camb.),* **2014**, *50*(29), 3786-3788.
[http://dx.doi.org/10.1039/C3CC49440C] [PMID: 24522830]

[67]   Liao, P.Q.; Li, X.Y.; Bai, J.; He, C.T.; Zhou, D.D.; Zhang, W.X.; Zhang, J.P.; Chen, X.M. Drastic enhancement of catalytic activity *via* post-oxidation of a porous MnII triazolate framework. *Chemistry,* **2014**, *20*(36), 11303-11307.
[http://dx.doi.org/10.1002/chem.201403123] [PMID: 25043981]

[68]   Shakya, D.M.; Ejegbavwo, O.A.; Rajeshkumar, T.; Senanayake, S.D.; Brandt, A.J.; Farzandh, S.; Acharya, N.; Ebrahim, A.M.; Frenkel, A.I.; Rui, N.; Tate, G.L.; Monnier, J.R.; Vogiatzis, K.D.; Shustova, N.B.; Chen, D.A. Selective Catalytic Chemistry at Rhodium(II) Nodes in Bimetallic Metal-Organic Frameworks. *Angew. Chem. Int. Ed.,* **2019**, *58*(46), 16533-16537.
[http://dx.doi.org/10.1002/anie.201908761] [PMID: 31529667]

[69]   Chen, T.H.; Popov, I.; Kaveevivitchai, W.; Miljanić, O.Š. Metal-organic frameworks: Rise of the ligands. *Chem. Mater.,* **2014**, *26*(15), 4322-4325.
[http://dx.doi.org/10.1021/cm501657d]

[70]   Zhang, X.; Sun, J.; Wei, G.; Liu, Z.; Yang, H.; Wang, K.; Fei, H. *In Situ* Generation of an N-

Heterocyclic Carbene Functionalized Metal–Organic Framework by Postsynthetic Ligand Exchange: Efficient and Selective Hydrosilylation of CO $_2$. *Angew. Chem. Int. Ed.,* **2019**, *58*(9), 2844-2849.
[http://dx.doi.org/10.1002/anie.201813064] [PMID: 30609209]

[71]   Yuan, S.; Zhang, P.; Zhang, L.; Garcia-Esparza, A.T.; Sokaras, D.; Qin, J.S.; Feng, L.; Day, G.S.; Chen, W.; Drake, H.F.; Elumalai, P.; Madrahimov, S.T.; Sun, D.; Zhou, H.C. Exposed Equatorial Positions of Metal Centers *via* Sequential Ligand Elimination and Installation in MOFs. *J. Am. Chem. Soc.,* **2018**, *140*(34), 10814-10819.
[http://dx.doi.org/10.1021/jacs.8b04886] [PMID: 30089362]

[72]   Chen, X.H.; Wei, Q.; Hong, J.D.; Xu, R.; Zhou, T.H. Bifunctional metal–organic frameworks toward photocatalytic CO2 reduction by post-synthetic ligand exchange. *Rare Met.,* **2019**, *38*(5), 413-419.
[http://dx.doi.org/10.1007/s12598-019-01259-6]

[73]   Qi, Z.; Qiu, T.; Wang, H.; Ye, C. Synthesis of ionic-liquid-functionalized UiO-66 framework by post-synthetic ligand exchange for the ultra-deep desulfurization. *Fuel,* **2020**, *268*, 117336.
[http://dx.doi.org/10.1016/j.fuel.2020.117336]

[74]   Albalad, J.; Xu, H.; Gándara, F.; Haouas, M.; Martineau-Corcos, C.; Mas-Ballesté, R.; Barnett, S.A.; Juanhuix, J.; Imaz, I.; Maspoch, D. Single-Crystal-to-Single-Crystal Postsynthetic Modification of a Metal–Organic Framework *via* Ozonolysis. *J. Am. Chem. Soc.,* **2018**, *140*(6), 2028-2031.
[http://dx.doi.org/10.1021/jacs.7b12913] [PMID: 29364654]

[75]   Chen, S.; Song, Z.; Lyu, J.; Guo, Y.; Lucier, B.E.G.; Luo, W.; Workentin, M.S.; Sun, X.; Huang, Y. Anhydride Post-Synthetic Modification in a Hierarchical Metal–Organic Framework. *J. Am. Chem. Soc.,* **2020**, *142*(9), 4419-4428.
[http://dx.doi.org/10.1021/jacs.9b13414] [PMID: 32037827]

[76]   Kalaj, M.; Denny, M.S., Jr; Bentz, K.C.; Palomba, J.M.; Cohen, S.M. Nylon–MOF Composites through Postsynthetic Polymerization. *Angew. Chem. Int. Ed.,* **2019**, *58*(8), 2336-2340.
[http://dx.doi.org/10.1002/anie.201812655] [PMID: 30511412]

[77]   Liang, J.; Chen, R.P.; Wang, X.Y.; Liu, T.T.; Wang, X.S.; Huang, Y.B.; Cao, R. Postsynthetic ionization of an imidazole-containing metal–organic framework for the cycloaddition of carbon dioxide and epoxides. *Chem. Sci. (Camb.),* **2017**, *8*(2), 1570-1575.
[http://dx.doi.org/10.1039/C6SC04357G] [PMID: 28451286]

[78]   Liu, Y.; Shen, Y.; Zhang, W.; Weng, J.; Zhao, M.; Zhu, T.; Chi, Y.R.; Yang, Y.; Zhang, H.; Huo, F. Engineering channels of metal–organic frameworks to enhance catalytic selectivity. *Chem. Commun. (Camb.),* **2019**, *55*(78), 11770-11773.
[http://dx.doi.org/10.1039/C9CC06061H] [PMID: 31513185]

[79]   Kassymova, M.; de Mahieu, A.; Chaemchuen, S.; Demeyere, P.; Mousavi, B.; Zhuiykov, S.; Yusubov, M.S.; Verpoort, F. Post-synthetically modified MOF for the A ^{3} -coupling reaction of aldehyde, amine, · and alkyne. *Catal. Sci. Technol.,* **2018**, *8*(16), 4129-4140.
[http://dx.doi.org/10.1039/C8CY00662H]

[80]   Naeimi, H.; Moradian, M. Thioether-based copper(I) Schiff base complex as a catalyst for a direct and asymmetric A3-coupling reaction. *Tetrahedron Asymmetry,* **2014**, *25*(5), 429-434.
[http://dx.doi.org/10.1016/j.tetasy.2014.02.002]

[81]   Bhatte, K.D.; Sawant, D.N.; Deshmukh, K.M.; Bhanage, B.M. Nanosize Co3O4 as a novel, robust, efficient and recyclable catalyst for A3-coupling reaction of propargylamines. *Catal. Commun.,* **2011**, *16*(1), 114-119.
[http://dx.doi.org/10.1016/j.catcom.2011.09.012]

[82]   Nasrollahzadeh, M.; Mohammad Sajadi, S.; Rostami-Vartooni, A. Green synthesis of CuO nanoparticles by aqueous extract of Anthemis nobilis flowers and their catalytic activity for the A3 coupling reaction. *J. Colloid Interface Sci.,* **2015**, *459*, 183-188.
[http://dx.doi.org/10.1016/j.jcis.2015.08.020] [PMID: 26291574]

[83]   Ardila-Suárez, C.; Díaz-Lasprilla, A.M.; Díaz-Vaca, L.A.; Balbuena, P.B.; Baldovino-Medrano, V.G.;

Ramírez-Caballero, G.E. Synthesis, characterization, and post-synthetic modification of a micro/mesoporous zirconium–tricarboxylate metal–organic framework: towards the addition of acid active sites. *CrystEngComm,* **2019**, *21*(19), 3014-3030.
[http://dx.doi.org/10.1039/C9CE00218A]

[84]   Morris, W.; Doonan, C.J.; Furukawa, H.; Banerjee, R.; Yaghi, O.M. Crystals as molecules: postsynthesis covalent functionalization of zeolitic imidazolate frameworks. *J. Am. Chem. Soc.,* **2008**, *130*(38), 12626-12627.
[http://dx.doi.org/10.1021/ja805222x] [PMID: 18754585]

[85]   Tu, M.; Wannapaiboon, S.; Fischer, R.A. Inter-conversion between zeolitic imidazolate frameworks: a dissolution–recrystallization process. *J. Mater. Chem. A Mater. Energy Sustain.,* **2020**, *8*(27), 13710-13717.
[http://dx.doi.org/10.1039/D0TA02975K]

[86]   Zhao, X.; Zhang, Z.; Cai, X.; Ding, B.; Sun, C.; Liu, G.; Hu, C.; Shao, S.; Pang, M. Postsynthetic Ligand Exchange of Metal–Organic Framework for Photodynamic Therapy. *ACS Appl. Mater. Interfaces,* **2019**, *11*(8), 7884-7892.
[http://dx.doi.org/10.1021/acsami.9b00740] [PMID: 30698413]

[87]   Gharib, M.; Esrafili, L.; Morsali, A.; Retailleau, P. Solvent-assisted ligand exchange (SALE) for the enhancement of epoxide ring-opening reaction catalysis based on three amide-functionalized metal–organic frameworks. *Dalton Trans.,* **2019**, *48*(24), 8803-8814.
[http://dx.doi.org/10.1039/C9DT00941H] [PMID: 31134242]

[88]   Pastore, V.J.; Cook, T.R.; Rzayev, J. Polymer–MOF Hybrid Composites with High Porosity and Stability through Surface-Selective Ligand Exchange. *Chem. Mater.,* **2018**, *30*(23), 8639-8649.
[http://dx.doi.org/10.1021/acs.chemmater.8b03881]

[89]   Dong, X.W.; Yang, Y.; Che, J.X.; Zuo, J.; Li, X.H.; Gao, L.; Hu, Y.Z.; Liu, X.Y. Heterogenization of homogeneous chiral polymers in metal–organic frameworks with enhanced catalytic performance for asymmetric catalysis. *Green Chem.,* **2018**, *20*(17), 4085-4093.
[http://dx.doi.org/10.1039/C8GC01323C]

[90]   Kim, S.; Lee, J.; Jeoung, S.; Moon, H.R.; Kim, M. Surface-Deactivated Core–Shell Metal–Organic Framework by Simple Ligand Exchange for Enhanced Size Discrimination in Aerobic Oxidation of Alcohols. *Chemistry,* **2020**, *26*(34), 7568-7572.
[http://dx.doi.org/10.1002/chem.202000933] [PMID: 32096306]

[91]   Lee, Y.R.; Do, X.H.; Hwang, S.S.; Baek, K.Y. Dual-functionalized ZIF-8 as an efficient acid-base bifunctional catalyst for the one-pot tandem reaction. *Catal. Today,* **2021**, *359*, 124-132.
[http://dx.doi.org/10.1016/j.cattod.2019.06.076]

[92]   Kim, H.; Kim, D.; Moon, D.; Choi, Y.N.; Baek, S.B.; Lah, M.S. Symmetry-guided syntheses of mixed-linker Zr metal–organic frameworks with precise linker locations. *Chem. Sci. (Camb.),* **2019**, *10*(22), 5801-5806.
[http://dx.doi.org/10.1039/C9SC01301F] [PMID: 31293768]

[93]   Villoria-del-Álamo, B.; Rojas-Buzo, S.; García-García, P.; Corma, A. Zr-MOF-808 as Catalyst for Amide Esterification. *Chemistry,* **2021**, *27*(14), 4588-4598.
[http://dx.doi.org/10.1002/chem.202003752] [PMID: 33026656]

[94]   Hu, X.J.; Li, Z.X.; Xue, H.; Huang, X.; Cao, R.; Liu, T.F. Designing a Bifunctional Brønsted Acid–Base Heterogeneous Catalyst Through Precise Installation of Ligands on Metal–Organic Frameworks. *CCS Chemistry,* **2020**, *2*(1), 616-622.
[http://dx.doi.org/10.31635/ccschem.019.201900040]

[95]   Pang, J.; Yuan, S.; Qin, J.; Wu, M.; Lollar, C.T.; Li, J.; Huang, N.; Li, B.; Zhang, P.; Zhou, H.C. Enhancing Pore-Environment Complexity Using a Trapezoidal Linker: Toward Stepwise Assembly of Multivariate Quinary Metal–Organic Frameworks. *J. Am. Chem. Soc.,* **2018**, *140*(39), 12328-12332.
[http://dx.doi.org/10.1021/jacs.8b07411] [PMID: 30227706]

[96] Lollar, C.T.; Pang, J.; Qin, J.; Yuan, S.; Powell, J.A.; Zhou, H.C. Thermodynamically Controlled Linker Installation in Flexible Zirconium Metal–Organic Frameworks. *Cryst. Growth Des.,* **2019,** *19*(4), 2069-2073.
[http://dx.doi.org/10.1021/acs.cgd.8b01637]

[97] Feng, L.; Yuan, S.; Zhang, L.L.; Tan, K.; Li, J.L.; Kirchon, A.; Liu, L.M.; Zhang, P.; Han, Y.; Chabal, Y.J.; Zhou, H.C. Creating Hierarchical Pores by Controlled Linker Thermolysis in Multivariate Metal–Organic Frameworks. *J. Am. Chem. Soc.,* **2018,** *140*(6), 2363-2372.
[http://dx.doi.org/10.1021/jacs.7b12916] [PMID: 29345141]

[98] Guillerm, V.; Xu, H.; Albalad, J.; Imaz, I.; Maspoch, D. Postsynthetic Selective Ligand Cleavage by Solid–Gas Phase Ozonolysis Fuses Micropores into Mesopores in Metal–Organic Frameworks. *J. Am. Chem. Soc.,* **2018,** *140*(44), 15022-15030.
[http://dx.doi.org/10.1021/jacs.8b09682] [PMID: 30351020]

[99] Yuan, S.; Zou, L.; Qin, J.S.; Li, J.; Huang, L.; Feng, L.; Wang, X.; Bosch, M.; Alsalme, A.; Cagin, T.; Zhou, H.C. Construction of hierarchically porous metal–organic frameworks through linker labilization. *Nat. Commun.,* **2017,** *8*(1), 15356.
[http://dx.doi.org/10.1038/ncomms15356] [PMID: 28541301]

[100] Wang, K.Y.; Feng, L.; Yan, T.H.; Wu, S.; Joseph, E.A.; Zhou, H.C. Rapid Generation of Hierarchically Porous Metal–Organic Frameworks through Laser Photolysis. *Angew. Chem. Int. Ed.,* **2020,** *59*(28), 11349-11354.
[http://dx.doi.org/10.1002/anie.202003636] [PMID: 32243687]

[101] Yu, Q.; Li, Z.; Cao, Q.; Qu, S.; Jia, Q. Advances in luminescent metal-organic framework sensors based on post-synthetic modification. *Trends Analyt. Chem.,* **2020,** *129*, 115939.
[http://dx.doi.org/10.1016/j.trac.2020.115939]

[102] Wang, Z.; Cohen, S.M. Tandem modification of metal-organic frameworks by a postsynthetic approach. *Angew. Chem. Int. Ed.,* **2008,** *47*(25), 4699-4702.
[http://dx.doi.org/10.1002/anie.200800686] [PMID: 18442157]

[103] Fracaroli, A.M.; Siman, P.; Nagib, D.A.; Suzuki, M.; Furukawa, H.; Toste, F.D.; Yaghi, O.M. Seven Post-synthetic Covalent Reactions in Tandem Leading to Enzyme-like Complexity within Metal–Organic Framework Crystals. *J. Am. Chem. Soc.,* **2016,** *138*(27), 8352-8355.
[http://dx.doi.org/10.1021/jacs.6b04204] [PMID: 27346625]

[104] Liu, H.; Xi, F.G.; Sun, W.; Yang, N.N.; Gao, E.Q. Amino- and Sulfo-Bifunctionalized Metal–Organic Frameworks: One-Pot Tandem Catalysis and the Catalytic Sites. *Inorg. Chem.,* **2016,** *55*(12), 5753-5755.
[http://dx.doi.org/10.1021/acs.inorgchem.6b01057] [PMID: 27254287]

[105] Karagiaridi, O.; Lalonde, M.B.; Bury, W.; Sarjeant, A.A.; Farha, O.K.; Hupp, J.T. Opening ZIF-8: a catalytically active zeolitic imidazolate framework of sodalite topology with unsubstituted linkers. *J. Am. Chem. Soc.,* **2012,** *134*(45), 18790-18796.
[http://dx.doi.org/10.1021/ja308786r] [PMID: 23088345]

[106] Mortazavi, S.S.; Abbasi, A.; Masteri-Farahani, M. Influence of SO_3H groups incorporated as Brønsted acidic parts by tandem post-synthetic functionalization on the catalytic behavior of MIL-101(Cr) MOF for methanolysis of styrene oxide. *Colloids Surf. A Physicochem. Eng. Asp.,* **2020,** *599*, 124703.
[http://dx.doi.org/10.1016/j.colsurfa.2020.124703]

[107] Zhou, X.; Zhang, Y.; Yang, X.; Zhao, L.; Wang, G. Functionalized IRMOF-3 as efficient heterogeneous catalyst for the synthesis of cyclic carbonates. *J. Mol. Catal. Chem.,* **2012,** *361-362*, 12-16.
[http://dx.doi.org/10.1016/j.molcata.2012.04.008]

[108] Llabrés i Xamena, F.X.; Cirujano, F.G.; Corma, A. An unexpected bifunctional acid base catalysis in IRMOF-3 for Knoevenagel condensation reactions. *Microporous Mesoporous Mater.,* **2012,** *157*, 112-117.

[http://dx.doi.org/10.1016/j.micromeso.2011.12.058]

[109] Phan, N.T.S.; Nguyen, T.T.; Luu, Q.H.; Nguyen, L.T.L. Paal–Knorr reaction catalyzed by metal–organic framework IRMOF-3 as an efficient and reusable heterogeneous catalyst. *J. Mol. Catal. Chem.*, **2012**, *363-364*, 178-185.
[http://dx.doi.org/10.1016/j.molcata.2012.06.007]

[110] Nuri, A.; Vucetic, N.; Smått, J.H.; Mansoori, Y.; Mikkola, J.P.; Murzin, D.Y. Pd Supported IRMOF-3: Heterogeneous, Efficient and Reusable Catalyst for Heck Reaction. *Catal. Lett.*, **2019**, *149*(7), 1941-1951.
[http://dx.doi.org/10.1007/s10562-019-02756-0]

[111] Rostamnia, S.; Xin, H. Basic isoreticular metal-organic framework (IRMOF-3) porous nanomaterial as a suitable and green catalyst for selective unsymmetrical Hantzsch coupling reaction. *Appl. Organomet. Chem.*, **2014**, *28*(5), 359-363.
[http://dx.doi.org/10.1002/aoc.3136]

[112] Chen, C.; Allen, C.A.; Cohen, S.M. Tandem postsynthetic modification of metal-organic frameworks using an inverse-electron-demand Diels-Alder reaction. *Inorg. Chem.*, **2011**, *50*(21), 10534-10536.
[http://dx.doi.org/10.1021/ic2017598] [PMID: 21985297]

[113] Burrows, A.D.; Frost, C.G.; Mahon, M.F.; Richardson, C. Post-synthetic modification of tagged metal-organic frameworks. *Angew. Chem. Int. Ed.*, **2008**, *47*(44), 8482-8486.
[http://dx.doi.org/10.1002/anie.200802908] [PMID: 18825761]

[114] Vinu, M.; Sivasankar, K.; Prabu, S.; Han, J.L.; Lin, C.H.; Yang, C.C.; Demel, J. Tetrazine-Based Metal-Organic Frameworks as Scaffolds for Post-Synthetic Modification by the Click Reaction. *Eur. J. Inorg. Chem.*, **2020**, *2020*(5), 461-466.
[http://dx.doi.org/10.1002/ejic.201901230]

[115] Ingleson, M.J.; Perez Barrio, J.; Guilbaud, J.B.; Khimyak, Y.Z.; Rosseinsky, M.J. Framework functionalisation triggers metal complex binding. *Chem. Commun. (Camb.)*, **2008**, (23), 2680-2682.
[http://dx.doi.org/10.1039/b718367d] [PMID: 18535706]

[116] Garibay, S.J.; Wang, Z.; Tanabe, K.K.; Cohen, S.M. Postsynthetic modification: a versatile approach toward multifunctional metal-organic frameworks. *Inorg. Chem.*, **2009**, *48*(15), 7341-7349.
[http://dx.doi.org/10.1021/ic900796n] [PMID: 19580256]

[117] Gu, Z.G.; Zhang, J. Epitaxial growth and applications of oriented metal–organic framework thin films. *Coord. Chem. Rev.*, **2019**, *378*, 513-532.
[http://dx.doi.org/10.1016/j.ccr.2017.09.028]

[118] Ikigaki, K.; Okada, K.; Takahashi, M. Epitaxial Growth of Multilayered Metal–Organic Framework Thin Films for Electronic and Photonic Applications. *ACS Appl. Nano Mater.*, **2021**, *4*(4), 3467-3475.
[http://dx.doi.org/10.1021/acsanm.0c03462]

[119] Hwang, J.; Ejsmont, A.; Freund, R.; Goscianska, J.; Schmidt, B.V.K.J.; Wuttke, S. Controlling the morphology of metal–organic frameworks and porous carbon materials: metal oxides as primary architecture-directing agents. *Chem. Soc. Rev.*, **2020**, *49*(11), 3348-3422.
[http://dx.doi.org/10.1039/C9CS00871C] [PMID: 32249855]

[120] Furukawa, S.; Hirai, K.; Nakagawa, K.; Takashima, Y.; Matsuda, R.; Tsuruoka, T.; Kondo, M.; Haruki, R.; Tanaka, D.; Sakamoto, H.; Shimomura, S.; Sakata, O.; Kitagawa, S. Heterogeneously hybridized porous coordination polymer crystals: fabrication of heterometallic core-shell single crystals with an in-plane rotational epitaxial relationship. *Angew. Chem. Int. Ed.*, **2009**, *48*(10), 1766-1770.
[http://dx.doi.org/10.1002/anie.200804836] [PMID: 19072803]

[121] Vohra, M.I.; Li, D.J.; Gu, Z.G.; Zhang, J. Insight into the epitaxial encapsulation of Pd catalysts in an oriented metalloporphyrin network thin film for tandem catalysis. *Nanoscale*, **2017**, *9*(23), 7734-7738.
[http://dx.doi.org/10.1039/C7NR02284K] [PMID: 28574075]

[122] Pan, X.; Xu, H.; Zhao, X.; Zhang, H. Metal–Organic Framework-Membranized Bicomponent Core–Shell Catalyst HZSM-5@UIO-66-NH$_2$/Pd for CO$_2$ Selective Conversion. *ACS Sustain. Chem.& Eng.*, **2020**, *8*(2), 1087-1094.
[http://dx.doi.org/10.1021/acssuschemeng.9b05912]

[123] Gu, Z.G.; Zhang, D.X.; Fu, W.Q.; Fu, Z.H.; Vohra, M.I.; Zhang, L.; Wöll, C.; Zhang, J. Facile Synthesis of Metal-Loaded Porous Carbon Thin Films *via* Carbonization of Surface-Mounted Metal–Organic Frameworks. *Inorg. Chem.*, **2017**, *56*(6), 3526-3531.
[http://dx.doi.org/10.1021/acs.inorgchem.6b03140] [PMID: 28267315]

[124] Cho, S.H.; Ma, B.; Nguyen, S.T.; Hupp, J.T.; Albrecht-Schmitt, T.E. A metal–organic framework material that functions as an enantioselective catalyst for olefin epoxidation. *Chem. Commun. (Camb.)*, **2006**, (24), 2563-2565.
[http://dx.doi.org/10.1039/B600408C] [PMID: 16779478]

[125] Wen, M.; Kuwahara, Y.; Mori, K.; Zhang, D.; Li, H.; Yamashita, H. Synthesis of Ce ions doped metal–organic framework for promoting catalytic H$_2$ production from ammonia borane under visible light irradiation. *J. Mater. Chem. A Mater. Energy Sustain.*, **2015**, *3*(27), 14134-14141.
[http://dx.doi.org/10.1039/C5TA02320C]

[126] Wang, A.; Zhou, Y.; Wang, Z.; Chen, M.; Sun, L.; Liu, X. Titanium incorporated with UiO-66(Zr-)-type Metal–Organic Framework (MOF) for photocatalytic application. *RSC Advances*, **2016**, *6*(5), 3671-3679.
[http://dx.doi.org/10.1039/C5RA24135A]

[127] Wang, Z.; Liu, J.; Lukose, B.; Gu, Z.; Weidler, P.G.; Gliemann, H.; Heine, T.; Wöll, C. Nanoporous designer solids with huge lattice constant gradients: multiheteroepitaxy of metal-organic frameworks. *Nano Lett.*, **2014**, *14*(3), 1526-1529.
[http://dx.doi.org/10.1021/nl404767k] [PMID: 24512342]

[128] Xie, Z.; Liu, Z.; Wang, Y.; Yang, Q.; Xu, L.; Ding, W. An overview of recent development in composite catalysts from porous materials for various reactions and processes. *Int. J. Mol. Sci.*, **2010**, *11*(5), 2152-2187.
[http://dx.doi.org/10.3390/ijms11052152] [PMID: 20559508]

[129] Mutruc, D.; Goulet-Hanssens, A.; Fairman, S.; Wahl, S.; Zimathies, A.; Knie, C.; Hecht, S. Modulating guest uptake in core–shell MOFs with visible light. *Angew. Chem. Int. Ed.*, **2019**, *58*(37), 12862-12867.
[http://dx.doi.org/10.1002/anie.201906606] [PMID: 31183909]

[130] Silvestre, M.E.; Franzreb, M.; Weidler, P.G.; Shekhah, O.; Wöll, C. Magnetic Cores with Porous Coatings: Growth of Metal-Organic Frameworks on Particles Using Liquid Phase Epitaxy. *Adv. Funct. Mater.*, **2013**, *23*(9), 1210-1213.
[http://dx.doi.org/10.1002/adfm.201202078]

[131] Salgueiriño-Maceira, V.; Correa-Duarte, M.A.; Spasova, M.; Liz-Marzán, L.M.; Farle, M. Cover Picture: Composite Silica Spheres with Magnetic and Luminescent Functionalities (Adv. Funct. Mater. 4/2006). *Adv. Funct. Mater.*, **2006**, *16*(4), NA.
[http://dx.doi.org/10.1002/adfm.200690014]

[132] Govan, J.; Gun'ko, Y. Recent Advances in the Application of Magnetic Nanoparticles as a Support for Homogeneous Catalysts. *Nanomaterials (Basel)*, **2014**, *4*(2), 222-241.
[http://dx.doi.org/10.3390/nano4020222] [PMID: 28344220]

[133] Phan, N.T.S.; Jones, C.W. Highly accessible catalytic sites on recyclable organosilane-functionalized magnetic nanoparticles: An alternative to functionalized porous silica catalysts. *J. Mol. Catal. Chem.*, **2006**, *253*(1-2), 123-131.
[http://dx.doi.org/10.1016/j.molcata.2006.03.019]

[134] Yang, X.; Yuan, S.; Zou, L.; Drake, H.; Zhang, Y.; Qin, J.; Alsalme, A.; Zhou, H.C. One-Step Synthesis of Hybrid Core–Shell Metal–Organic Frameworks. *Angew. Chem. Int. Ed.*, **2018**, *57*(15),

3927-3932.
[http://dx.doi.org/10.1002/anie.201710019] [PMID: 29451952]

[135] Panchariya, D.K.; Rai, R.K.; Anil Kumar, E.; Singh, S.K. Core–shell zeolitic imidazolate frameworks for enhanced hydrogen storage. *ACS Omega,* **2018**, *3*(1), 167-175.
[http://dx.doi.org/10.1021/acsomega.7b01693] [PMID: 31457885]

[136] Zhang, J.; Zhang, T.; Xiao, K.; Cheng, S.; Qian, G.; Wang, Y.; Feng, Y. Novel and Facile Strategy for Controllable Synthesis of Multilayered Core–Shell Zeolitic Imidazolate Frameworks. *Cryst. Growth Des.,* **2016**, *16*(11), 6494-6498.
[http://dx.doi.org/10.1021/acs.cgd.6b01161]

[137] Huang, Z.; Zhou, J.; Zhao, Y.; Cheng, H.; Lu, G.; Morawski, A.W.; Yu, Y. Stable core–shell ZIF-8@ZIF-67 MOFs photocatalyst for highly efficient degradation of organic pollutant and hydrogen evolution. *J. Mater. Res.,* **2021**, *36*(3), 602-614.
[http://dx.doi.org/10.1557/s43578-021-00117-5]

[138] Ejsmont, A.; Jankowska, A.; Goscianska, J. Insight into the Photocatalytic Activity of Cobalt-Based Metal–Organic Frameworks and Their Composites. *Catalysts,* **2022**, *12*(2), 110.
[http://dx.doi.org/10.3390/catal12020110]

[139] Li, K.; Zhang, Y.; Wang, P.; Long, X.; Zheng, L.; Liu, G.; He, X.; Qiu, J. Core-Shell ZIF-67@ZIF-8-derived multi-dimensional cobalt-nitrogen doped hierarchical carbon nanomaterial for efficient oxygen reduction reaction. *J. Alloys Compd.,* **2022**, *903*, 163701.
[http://dx.doi.org/10.1016/j.jallcom.2022.163701]

CHAPTER 3

MOFs and Their Composites as Catalysts for Organic Reactions

Anna Olejnik[1], Aleksandra Galarda[1], Anita Kubiak[1], Marcelina Kotschmarów[1], Aleksander Ejsmont[1], Agata Chełmińska[1], Martyna Kotula[1], Simona M. Coman[2] and Joanna Goscianska[1,*]

[1] *Adam Mickiewicz University in Poznań, Faculty of Chemistry, Department of Chemical Technology, Uniwersytetu Poznańskiego 8, 61-614 Poznań, Poland*

[2] *University of Bucharest, Faculty of Chemistry, Department of Organic Chemistry, Biochemistry and Catalysis, Bd. Regina Elisabeta, 4-12, 030018 Bucharest, Romania*

Abstract: In recent years, metal-organic frameworks (MOFs) have significantly contributed to broadening the frontiers of science. Due to their distinctive properties including well-developed surface area, high porosity, multifarious composition, tunable and uniform pore structures, and comprehensive functionality, they were applied in different fields such as separation, drug delivery, fuel storage, chemical sensing, and catalysis. The application of pristine MOFs as materials that speed up the reaction rate could be restricted mainly because of the limited number of active sites and their low mechanical and thermal stability. In order to enhance their catalytic properties, metal-organic frameworks can be functionalized or integrated with a variety of materials to obtain composites or hybrids. The review outlines the state of art concerning the application of MOFs and their composites as catalysts in various organic transformation processes. A particular focus was given to the oxidation of alkanes, cycloalkanes, alkylbenzenes, alcohols, thiols, sulfides. Furthermore, the role of metal-organic frameworks in hydrogenation and C–C coupling reactions were also presented.

Keywords: Alkanes, Alkenes, Brønsted acid sites, Catalysis, C–C coupling reactions, Composites, Hybrids, Hydrogenation, Lewis acid sites, Linkers, Metal nodes, Metal–organic frameworks, MOF, MOF active sites, MOF stability, Oxidation, Reduction, Sulfides, Thiols.

1. INTRODUCTION

In recent times, metal-organic frameworks (MOFs) have attracted great attention of scientists from different fields. These materials are classified as hybrids being

[*] **Corresponding author Joanna Goscianska**: Adam Mickiewicz University in Poznań, Faculty of Chemistry, Department of Chemical Technology, Uniwersytetu Poznańskiego 8, 61-614 Poznań, Poland; Tel.:+48-618291607, E-mail: joanna.goscianska@amu.edu.pl

Junkuo Gao & Reza Abazari (Eds.)

composed of inorganic parts (metal nodes) connected by coordination bonds to organic linkers. The most common representatives of MOFs are MOF-5, UiO-66, MIL-101, HKUST-1, PCN-14 and ZIF-8 [1 - 4]. They exhibit distinctive properties including well-developed surface area, high porosity, multifarious composition, tunable and uniform pore structures, and comprehensive functionality [5, 6]. As porous materials, MOFs fill the gap between zeolites (possessing small pore size) and silicate (having larger pore size). So far metal-organic frameworks with micropores to mesopores have been described in the literature [7, 8]. Due to the presented above unique features, MOFs have the potential to be applied in different fields such as separation [9], fuel storage [10], drug delivery [11, 12], chemical sensing [13, 14], and environmental remediation [15]. These porous polymers have also been recommended as catalysts in different processes [16 - 18]. The catalytic activity of MOF originates either from metal ions or functional groups attached to the organic linkers of the framework. Additionally, high density and spatially separated active sites of metal-organic frameworks are crucial features that support their function as catalysts [19]. Furthermore, the post-synthetic modification or *in situ* processes enable to adjust the metal-organic framework structure and introduce additional functional acid-base groups ($-SO_3H$, $-NH_2$, *etc.*) or embed active metal nanoparticles (Pt, Ru, Cu, *etc.*) and metal complexes inside the MOF cages (or anchored on their surface) that are beneficial to design an appropriate catalyst for the target application. The high porosity and permeable channels facilitate the delivery of reactants to catalytic sites. Therefore, MOFs and their composites represent a new class of recyclable heterogeneous catalysts owing remarkable properties. Although their application as catalysts is still at the developing phase, a series of studies have been performed in this area so far, showing that MOFs exhibit considerable catalytic properties including high activity, appropriate stability, and reusability [20 - 22].

Despite the fact that the application of MOFs and MOF-derived materials for various catalytic reactions has been documented in recent reviews [23 - 25], in this fast-growing research field, we would like to provide an overview of the catalytic behavior and advantages of the MOFs and their composites in different valuable organic processes such as oxidation, hydrogenation, and C–C coupling reactions.

2. MOFS AND THEIR STABILITY

Stability is the fundamental factor in the development of materials. In this context, it should be highlighted that although MOFs have unique properties, their full potential application is limited, mainly due to their low chemical, mechanical, thermal, and hydrothermal stabilities [16]. Chemical stability refers to the resistance to different solvents, acids, bases, and solutions with strongly coordinating anions (for example, phosphate anion), while the mechanical and thermal stability is related to the capability of materials to preserve their structure under exposure to pressure, heat, and vacuum. The chemical stability of metal-organic frameworks is related to the strength of the metal-ligand bond [26]. It should be noted that mostly the thermodynamic factors have an influence on the metal-ligand coordination bond strength [27]. Therefore, the stronger the coordination bonds, the more stable MOFs can be created [28]. It was proved that the chemical stability of UiO-66 and SUMOF-7 series declined with the lengthening of the linker and an increase of pore sizes [1, 29]. This is due to the kinetic factors, that are associated with the coordination number, the rigidity of the linker, and surface hydrophobicity.

The structural framework of MOFs can be decomposed both in acidic solutions, which can accelerate the formation of a protonated linker, and in basic solutions facilitating the formation of a hydroxide ligated node [5, 30]. However, the chemical stability can be enhanced by using high valence metal ions including Zr^{4+}, Fe^{3+} or Cr^{3+} and by the interactions with different ligands (imidazolates and triazolates). Outstanding stability in water is therefore achieved for azolate MOFs thanks to the strong metal–nitrogen bonds [31]. Nevertheless, despite recent progress in understanding and improving the chemical stability of MOFs, it was proved that MOFs fabricated by the traditional methods are less competitive compared to commercial catalysts when applied in a harsh reactive environment [32].

It should be added that the nature of reagents can also have an influence on the stability of MOFs. Timofeeva *et al.* observed the destruction of MAF-6(S) structure due to the leaching of Zn^{2+} caused by the polar reagents [33]. In turn, Linder-Patton *et al.* detected that the surface of ZIF-8 was unstable in the catalytic processes in the presence of hydrophobic reagents with polar functional groups [34].

The mechanical stability of MOFs decreased with the increase in porosity. However, it was shown that when metal–organic frameworks are filled with solvent, they were more mechanically stable than the same materials with empty pores [35 - 38].

It should be underlined that the thermal, chemical, and hydrothermal stabilities are crucial in terms of MOF's characterization and application as catalysts in various reactions. When metal-organic frameworks undergo decomposition, it is hard to determine their structure by using X-ray diffraction. The thermal stability of

MOF-5 is ca. 300 °C, moreover, it is not stable in aqueous solutions and it decomposed after long contact with humid air [39]. The thermal degradation of metal–organic frameworks is caused by node-linker bond breakage after which the linker combustions occurred [26]. In the thermal treatment, degradation of MOF's structure can cause its melting [40], amorphization [41], linker dehydrogenation, and graphitization [42]. These processes occur gradually throughout the heating step or when temperature above the decomposition limit is achieved. It was proved that the thermal stability of MOFs could be improved by using oxy-anion-terminated linkers with higher valency metal centers [43, 44]. On the other hand, hydrothermal stability indicated that the material remains unchanged in the presence of moisture at higher temperatures. The studies demonstrated that intermolecular or intramolecular forces, hydrophobic functional groups, and perfluorinated linkers within metal–organic frameworks could enhance hydrothermal stability. Therefore, MOFs were combined with different functional materials to obtain novel composites with enhanced stability properties.

Since MOFs are generally not very stable, scientists carry out cyclic tests to check if they can be reused without losing their catalytic activity. The examples presented in this chapter show that different types of MOFs can be used as catalysts several times.

3. ACTIVE SITES IN MOF AND MOF COMPOSITE

The effectiveness of catalysts is associated with the interaction between active sites and microenvironments. The open metal sites (OMS) present in metal-organic frameworks suggest that these materials might be the promising Lewis acid catalysts [45, 46]. It was proved that the application of pristine MOFs as catalysts could be restricted not only due to their low mechanical and thermal stability but mainly due to the limited number of active sites [47]. In order to enhance their catalytic properties, metal–organic frameworks can be functionalized or integrated with a variety of materials (metal or metal oxides nanoparticles [48, 49], carbon materials, polyoxometalates [50], silica, quantum dots, ionic liquids [51], enzymes and molecular species) to obtain composites or hybrids [5, 52]. Consequently, the unique catalytical properties of novel materials can be achieved by inducing the synergic effect. The catalytic activity can also originate from the functional groups which are combined with linkers of the framework [53]. Furthermore, catalysts including metal complexes, metal nanoparticles (NPs), and even biomolecules can be built in the MOF cages or attached to their surface. In simple terms, three types of active sites in metal–organic frameworks can be distinguished. The first one is coordinatively unsaturated metal centers that interact with the substrate. The second one engages the incorporation of active catalytic species inside the pores of metal–organic frameworks. The third one involves the functionalized ligands. According to Huang *et al.* [54], active sites include metal nodes, functional organic linkers and guest species located inside the pores (Fig. **1**). A great variety of metal nods in metal–organic frameworks such as aluminum(III), chromium(III), cobalt(II), copper(II), iron(III), magnesium(II), manganese(II), nickel(II), palladium(II), silver(I), titanium(III), zinc(II), and zirconium(II) have been applied in the broad range of organic reactions including acetalization [55], Friedel-Crafts acylation [56], cyclopropanation [57], Aza-Michael reaction [58], 1,3-dipolar cycloaddition [59], and esterification [60].

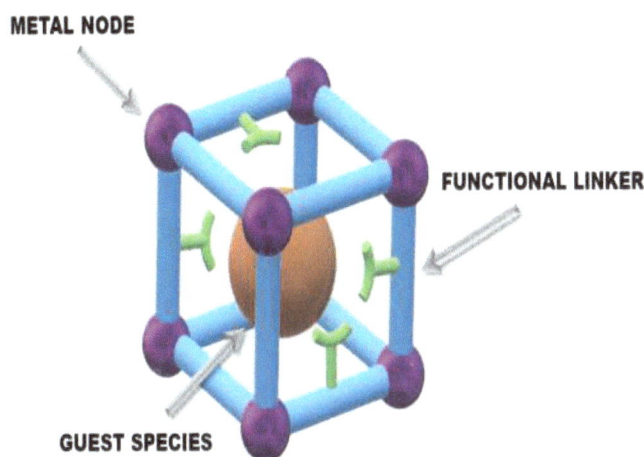

METAL NODE

FUNCTIONAL LINKER

GUEST SPECIES

Fig. (1). Different types of metal-organic framework active sites.

As far as functional groups are concerned, sulfonic acid groups and dicarboxylic acid groups can be introduced to the linker as acidic sites, while amino, pyrrolidine, amide groups can serve as basic sites. These types of groups exhibit great catalytic performance in different types of organic reactions such as esterification [61], alcoholysis of epoxides [62], aldol condensation [63], Michael addition [64], and acetalization [65].

When MOF-metal nanoparticle composites are applied in catalytic reactions, metal functions act as active sites, while metal–organic frameworks are used as a stabilizer and activity enhancer for metal nanoparticles [66, 67]. It should be highlighted that metal–organic frameworks play an important role in the reaction process, because they can organize molecules, stabilize the transition states, and introduce another active site. The functional organic groups in MOF linkers can act as active sites to get along with the metal nanoparticles to support the synergistic catalysis in photocatalytic and tandem reactions [68]. For instance, Tang *et al.* [69] obtained core-shell Pd@IRMMOF-3 material that was applied as multifunctional catalysts in tandem Knoevenagel condensation-hydrogenation reaction. By using this type of MOF composite, 86% selectivity to the target product for a total conversion of the substrate was reached. The amino groups of the IRMOF-3 play the role of active sites in the condensation process to fabricate 2-(4-nitrobenzylidene)malononitrile, while the palladium NP cores catalyzed the hydrogenation reaction to get the target product as 2-(-aminobenzylidene)malononitrile. Toyao *et al.* [70] applied MIL-101(Al)-NH$_2$ as bifunctional acid-base catalyst in a tandem deacetalization-Knoevenagel condensation reaction. The Brønsted acid sites (carboxylic acid groups) and the Lewis acid sites (aluminum) catalyzed benzaldehyde dimethylacetal deacetalization, while the following condensation of benzaldehyde with malononitrile was supported by coordinatively unsaturated aluminum Lewis acid sites and the amine basic sites. The excellent selectivity was achieved when MIL-101(Al)-NH$_2$ was applied in tandem Meinwald rearrangement-Knoevenagel condensation reaction [71]. In the first reaction, the aldehyde was obtained in high yield *via* a tertiary benzylic carbocation intermediate, whereas aluminum Lewis acid sites promoted the epoxide ring-opening reaction. Afterward, the amino groups catalyzed the Knoevenagel reaction of an aldehyde with methylene nucleophile.

4. MOFS AS CATALYSTS IN OXIDATION PROCESSES

4.1. Oxidation of Alkanes, Cycloalkanes and Alkylbenzenes

Oxidation reactions are one of the most significant parts of chemical conversions in organic chemistry and industry. In nature, they play a key role in delivering essential compounds or fine chemicals such as vitamins, hormones, and fragrances [72]. The development of methods for catalytic oxidation of alkanes and alkenes is a valid scientific subject with significant technological capability. However, it should be stressed that these processes must be environmentally friendly, non-toxic, and economic. Additionally, they should be performed under mild conditions [73]. Oxidation reactions with the majority of commonly employed oxidants frequently generate large amounts of waste and economic loss. Another issue in this topic is selectivity to desired products. Due to the inert nature of substrates, they require extremely reactive oxidants and hard reaction conditions which generate a large number of side reactions [74]. The structure of metal–organic frameworks makes them promising materials as catalysts in oxidation reactions of alkanes [75]. Some of the raw organic compounds like alkanes and alkenes are relevant to industrial interest. However, they suffer from low conversion and the formation of undesirable by-products [76]. Examples of MOF applied as catalysts in oxidation reactions of alkanes, cycloalkanes, and alkylbenzenes are presented in Table **1**.

Table 1. Overview of MOFs applied as catalysts in oxidation reactions of alkane, cycloalkanes and alkylbenzenes.

MOF Catalyst	Substrate	Oxidant	Conditions of Reaction	Product	Conversion (%)	Ref.
Mn(TPP)Cl@Im-MIL-101	cyclooctane	NaIO$_4$	CH$_3$CN/ H$_2$O, RT	cyclooctanol/ cyclooctanone	75	[82]
Mn(TPP)Cl@Im-MIL-101	cyclohexane	NaIO$_4$	CH$_3$CN/ H$_2$O, RT	cyclohexanol/cyclohexanone (K/A oil)	58	[82]
NHPI/Fe(BTC)	cyclooctane	O$_2$	120°C, 6 h	cyclooctanone	28	[83]
NHPI/Fe(BTC)	tetralin	O$_2$	120°C, 3 h	tetranol/tetralone	33	[83]
MIL-101(Cr) immobilized on monolithic structure	tetralin	tert-butyl hydroperoxide	PhCl, 80°C, 21 h	tetralone	90	[84]
CZJ-1	cyclohexane	iodosylbenzene (PhIO)	CH$_3$CN, RT, 6 h	cyclohexanone/ cyclohexanol (K/A oil)	94	[85]
Cu-NU-1000	methane	O$_2$	1 bar methane, 150 °C, 3 h	methanol	90	[86]
Au/MIL-53(Cr)	cyclohexane	O$_2$	130 °C, 6h	cyclohexanone/ cyclohexanol (K/A oil)	31	[87]
Au−Pd/MIL-101	cyclohexane	O$_2$	150 °C, 4 h	cyclohexanone/ cyclohexanol (K/A oil)	40	[88]
Cu-Schiff base complex@MIL-101	styrene	H$_2$O$_2$	Tris-HCl buffer, 30 °C	benzaldehyde	93.5	[89]
ZC-700	styrene	air	DMF, 6 h	styrene oxide	91	[90]
Co(II) MOF	styrene	TBHP	CH$_3$CN, 60 °C, 22-24 h	styrene oxide	99	[91]

Selective oxidation of C−H bonds is a fundamental and essential process to obtain high-value chemicals. One of the most interesting reactions is the oxidation of alkylbenzenes to aromatic ketones which are widely applied in medicine and pesticide production. Recently, Li *et al.* [77] have used two copper-containing Keggin-type polyoxometalate-based metal−organic frameworks (POMOFs), *i.e.*, [Cu$^I_6$(trz)$_6${PW$_{12}$O$_{40}$}$_2$] (HENU-2, HENU = Henan University; trz = 1,2,4- triazole) and [CuI$_3$(trz)$_3${PMo$_{12}$O$_{40}$}] (HENU-3) as catalysts for oxidation of diphenylmethane. Polyoxometalate-based metal−organic frameworks (POMOFs) consisted of POM units and holding open network structures were considered as a promising heterogeneous catalytic platform for various organic reactions. These two POMOFs were highly effective heterogeneous catalysts that facilitate the oxidation of diphenylmethane to ketone in the presence of tert-butyl hydroperoxide (TBHP) (Fig. **2**). When HENU-2 and HENU-3 have been applied as catalysts, 100% selectivity toward benzophenone was obtained for conversion of diphenylmethane of 95.2% (HENU-2) and 92.8% (HENU-3), respectively. Among the reported POM-based catalysts containing non-noble metals, HENU-2 exhibited the best catalytic activity in diphenylmethane oxidation. HENU-2 and HENU-3 were stable and could be easily recovered from the reaction system. The results indicated that HENU-2 could be effectively recycled three times with almost preserved high catalytic efficiency. Oxidation of different alkylbenzenes to the corresponding ketone products was also examined. Oxidation of ethylbenzene occurred with the excellent catalytic activity of MOF materials (conversion - 90.6−100%, selectivity - 100%), while the conversions of substituted ethylbenzene derivatives were lower (70.6−91.4%).

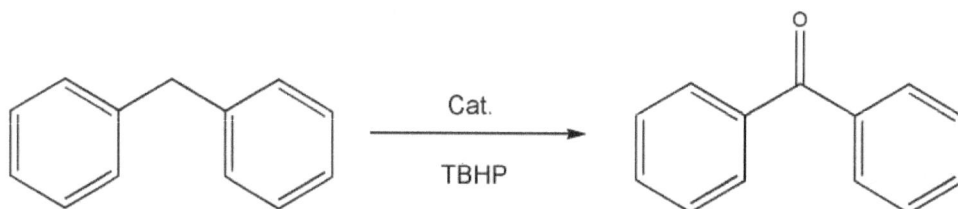

Fig. (2). Scheme of the synthesis of benzophenone by oxidation of diphenylmethane.

Another type of reaction in which MOFs were applied as catalysts is selective aerobic oxidation of cumene. This process is important in industry because it is one of the steps in phenol production [78]. Moreover, cumene hydroperoxide (CHP) is also utilized in the Sumitomo process for the large-scale production of propylene oxide [79]. The reaction scheme is presented in Fig. (**3**). Nowacka *et al.* [80] prepared Co−Ni and Mn−Ni bimetallic trimesate MOF by a fast aqueous synthesis and investigated its catalytic activity in the selective aerobic oxidation of cumene to CHP. The results have shown that the isolation of Co^{2+} (or Mn^{2+}) in an inert Ni-BTC framework is a facile strategy to achieve more than 90% CHP selectivity. The reaction conditions were optimized to reach the maximum CHP selectivity *via* using mixed-metal Co_5Ni-BTC (Co: Ni ratio of 5:95). A high selectivity (91%) to CHP for a 30% conversion of cumene was obtained at 90 °C and after 7 h of reaction. Co_5Ni-BTC maintained its catalytic performance for at least five catalytic cycles with the same CHP selectivity.

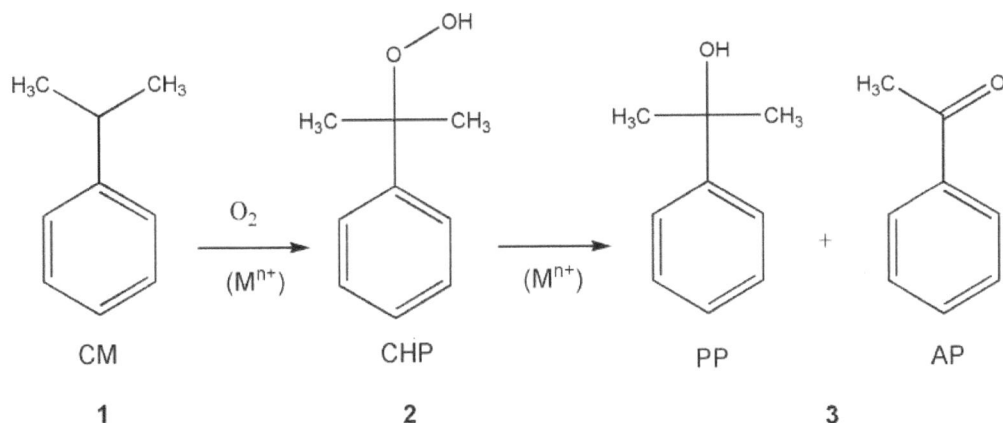

Fig. (3). Aerobic oxidation of cumene (CM) to cumene hydroperoxide (CHP). 1. Autocatalytic or catalyzed by metal ions (M^{n+}). 2. CHP decomposition catalyzed by M^{n+} ions. 3. Formation of 2-phenyl-2-propanol (PP) and acetophenone (AP) as the main by-products.

Selective oxidation of alkylbenzenes to aldehydes is one of the most significant and valuable chemical reactions because aldehydes (especially benzaldehydes) have been widely used to produce dyes, pharmaceuticals, perfumes, and fine chemicals. In this aspect, Ma *et al.* [81] reported two new polyoxometalates-based metal–organic frameworks (POMOFs) $[Co(BBTZ)_2][H_3BW_{12}O_{40}]10H_2O$ and $[Co_3(H_2O)_6(BBTZ)_4][BW_{12}O_{40}] \cdot NO_3 \cdot 4H_2O$ (BBTZ=1,4-bis(1,2,4-triazol-1- ylmethyl)benzene). Materials were utilized in the selective oxidation of styrene to aldehyde (Fig. **4**). The first material enabled a selectivity of 96% to benzaldehyde for a total conversion of styrene within only 4 h. Additionally, it was reused seven times without a significant loss of catalytic activity. The second structure required 6 h to afford a selectivity of 93% to benzaldehyde for a total conversion of styrene. It proved that both polyoxometalates-based metal–organic frameworks possessed similar abilities to catalyze the oxidation of styrene to benzaldehyde.

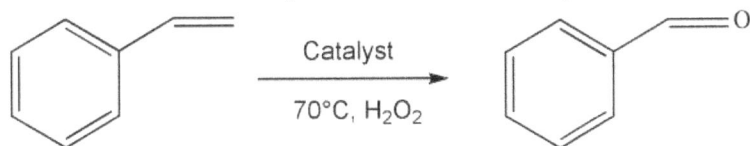

Fig. (4). The oxidation of styrene to benzaldehyde.

Gascon and co-workers [92] synthesized Fe-containing MOFs - MIL-53(Al, Fe) which were applied in the oxidation of methane to methanol under mild aqueous conditions. Two different synthetic pathways were used for the incorporation of Fe units into the MIL-53(Al) framework. The first one involved post-synthetic cation exchange using hydrothermally synthesized MIL-53(Al) and $FeCl_3$ solution

(HTS). The second one was *in situ* Fe incorporation *via* the addition of $FeCl_3$ aliquots during electrochemical synthesis of MIL-53(Al) from an Al electrode and a terephthalic acid solution (ECS). The material included spatially isolated oxo-bridged Fe_2 units similar to the carboxylate-bridging di-iron active site of methane monooxygenase (MMO). In nature, methane monooxygenase enzymes can directly and selectively oxidize methane to methanol under mild aqueous conditions (Fig. **5**). The ECS samples exhibited a higher activity in comparison to those obtained through post-synthetic ion exchange. It was demonstrated that the hybrid MIL-53(Al, Fe) material is able to catalyze methane oxidation with H_2O_2, producing methanol as the main oxidation product with selectivity toward oxygenates of ca. 80% and turnover number (TON) up to 350. The results indicated that MOFs could be remarkable platforms for the development of model catalysts mimicking full enzymatic cycles.

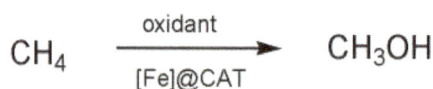

$$CH_4 \xrightarrow[\text{[Fe]@CAT}]{\text{oxidant}} CH_3OH$$

Fig. (5). Reaction scheme of the direct catalytic methane oxidation.

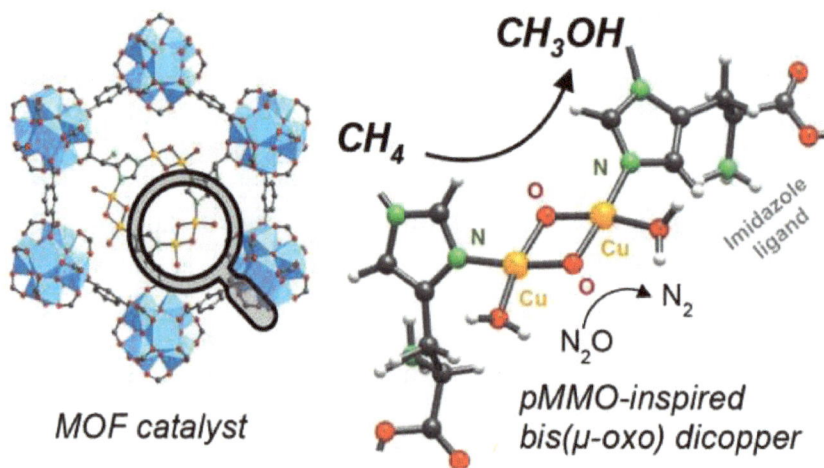

Fig. (6). Bioinspired metal–organic framework catalysts for selective methane oxidation to methanol by installing imidazole moieties and incorporation of reactive copper–oxygen complexes within the framework Reprinted with permission [93]. Copyright 2018, American Chemical Society.

Baek and co-workers [93] performed similar studies. They synthesized metal-organic framework catalysts inspired by particulate methane monooxygenase (pMMO) for selective methane oxidation to methanol. MOF was used as a backbone for the formation of an enzyme-like active sites by installing biologically significant imidazole moieties (L-histidine (His), 4-imidazoleacrylic acid (Iza), and 5-benzimidazolecarboxylic acid (Bzz)). Then, the ligands were metalating to incorporate reactive copper–oxygen complexes within the framework (Fig. **6**).

Among different MOFs, MOF-808 can be used to host and stabilize highly active copper-oxygen complexes. The highest methanol performance exhibits MOF - 808-Bzz-Cu - 71.8 ± 23.4 µmol/g under isothermal conditions at 150 °C. This strategy could serve as an inspiration for different biological catalysts in MOFs. Taking into account hydrocarbon oxidation transformations, the oxidation of cyclohexane in the presence of molecular oxygen without organic solvents and additives is considered to be one of the most valuable industrial reactions for cyclohexanone and cyclohexanol (ketone-alcohol oil, KA) production. However, in the traditional industrial cyclohexane autocatalytic oxidation process, the conversion of cyclohexane has only reached 3–5% and the selectivity to KA oil is 80%. Due to this issue, exploring efficient catalysts for cyclohexane oxidation under mild conditions is desirable. For this reason, Wang *et al.* [94] synthesized mixed-valence $\{V_{16}\}$ cluster-based metal–organic framework ([Ni(4,4`-bpy)$_2$]$_2$ [V$^{IV}_7$ V$^V_9$O$_{38}$Cl]·(4,4`-bpy)·6H$_2$O, named as NENU-MV-1 and used it as a catalyst in the oxidation of cyclohexane. They tuned the average particle size of NENU-MV-1 from 500 nm to 25 µm through changes in the reaction temperature, concentration, and molar ratio of metal source to ligand during synthesis. It was considered that the catalyst with mixed oxidation states (V^{4+} or V^{5+}) had catalytically active sites associated with the fast electrons transfer. One of the catalysts, NENU-MV-1 with the smallest particle size (500 nm) achieved higher conversion of cyclohexane (24.6%) and selectivity of the mixture of cyclohexanone and cyclohexanol (99%) than that of bulk crystalline NENU-M--1. Turnover frequency (TOF) calculations and kinetic analysis revealed that more active sites were available on the surface of smaller-sized catalysts, and as a result, full contact with the cyclohexane was possible. This remarkable catalytic activity could be related to the presence of both V^{IV} and V^V centers in the catalytic system. V^{IV} centers might constitute suitable sites for increased O$_2$ activation and adsorption, while V^V centers are preferable to the adsorption and oxidation of cyclohexane. Furthermore, NENU-MV-1-500 nm showed good recyclability after five cycles of being regenerated by ethanol. Oxidation of cyclohexane to cyclohexanol and cyclohexanone was used as a model reaction to examine the effect of a hydrophobic pore environment on product selectivity and catalyst stability in a series of iron-based frameworks (Fig. **7**). Xiao and co-workers [95] synthesized the terphenyl derivative of Fe-MOF-74 - Fe$_2$(dotpdc) (H$_4$dotpdc = 4,4″- dihydroxy-[1,1′:4′,1″-terphenyl]-3,3″-dicarboxylic acid), and three modified derivatives in which the central ring was replaced with tetrafluoro-, tetramethyl-, or di-tert-butylaryl groups (R= F, CH$_3$, and tBu). It was found that the insertion of simple nonpolar groups significantly increased the stability and selectivity of framework-embedded iron sites for the oxidation of cyclohexane. At the same time, it did not directly affect the reactivity of the iron centers and structure. In contrast to the similarity of the pore size, surface area, and unit cell parameters of the four Fe$_2$(dotpdc^R) frameworks, there was a big variation in alcohol: ketone (A:K) selectivity and TONs increasing in the order H < F < CH$_3$ < tBu. The differences in the A:K selectivities (an increase from 2.8 for Fe$_2$(dotpdc) to 8.4 for Fe$_2$(dotpdc^{tBu})) were associated with the various enthalpies of cyclohexane adsorption on these frameworks.

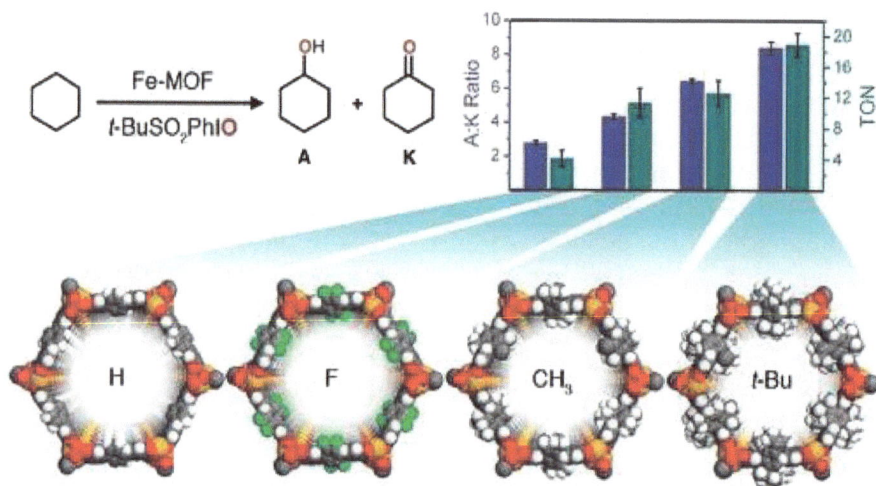

Fig. (7). Iron based frameworks as catalysts in the oxidation of cyclohexane. Reprinted with permission [95]. Copyright 2016, American Chemical Society.

4.2. Oxidation of Alkenes and Cycloalkenes

An important type of reaction in the industry is the oxidation of C=C double bonds. Oxidation of alkenes can be performed by various methods which led to dihydroxylation, epoxidation, aminohydroxylation, and carbonylation of C=C double bond [72]. The reactions that can take place during the oxidation of alkenes are shown in Fig. (8). Nowadays, fossil reserves are depleting, emission of greenhouse gases is rising, hence, the progress of catalytic and sustainable processes of renewable raw materials has become an important topic, particularly the synthesis of oxygen-containing compounds [96]. Examples of MOF applied as catalysts in oxidation reactions of alkenes and cycloalkenes in recent years are presented in Table **2**.

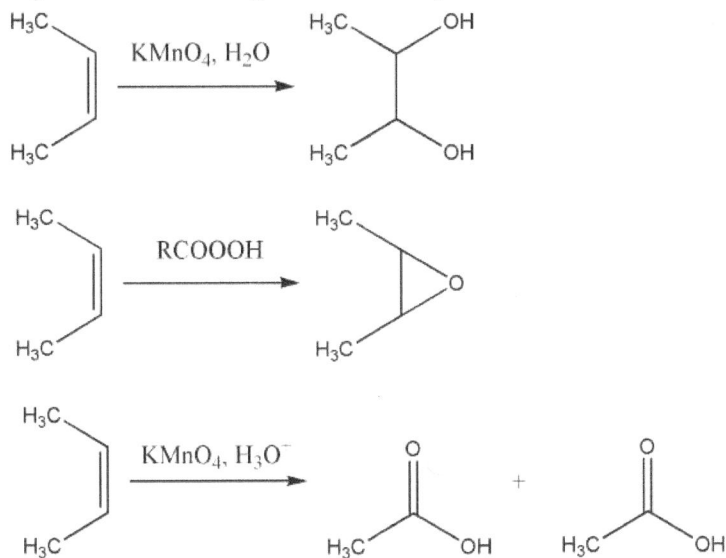

Fig. (8). Scheme of the possible reactions during alkenes oxidation.

Table 2. Overview of MOFs applied as catalysts in oxidation reactions of alkenes and cycloalkenes.

MOF Catalyst	Substrate	Oxidant	Conditions of Reaction	Product	Conversion (%)	Ref.
Mn(TPP)Cl@Im-MIL-101	cyclooctene	NaIO$_4$	CH$_3$CN/ H$_2$O, RT	cyclooctene oxide	96	[82]
Mn(TPP)Cl@Im-MIL-101	cyclohexene	NaIO$_4$	CH$_3$CN/ H$_2$O, RT	cyclohexene oxide	92	[82]
Co^{II} Schiff-base@Cr-M-L-101-P2I, isobutyraldehyd (co-catalyst)	cyclooctene	air	CH$_3$CN, 35 °C, 5 h	cyclooctene oxide	99	[99]
IRMOF-3[Mn]	cyclohexene	O$_2$	Toluene, 40 °C, 6 h	cyclohexene oxide	67.5	[100]
MIL-101	cyclopentene	H$_2$O$_2$	CH$_3$CN, 8 h	glutaric acid	75	[101]
MIL-101	1-octene	H$_2$O$_2$	CH$_3$CN, 8 h	heptanoic acid	90	[101]
MIL-101	indene	H$_2$O$_2$	CH$_3$CN, 3 h	homophthalic acid	100	[101]
Cu-MOF-2	cyclododecene	O$_2$	(CH$_3$)$_3$ CCHO, CH$_3$CN, 40 °C, 6 h	1,2-epoxycyclododecane	>99	[102]

MOF Catalyst	Substrate	Oxidant	Conditions of Reaction	Product	Conversion (%)	Ref.
Cu-MOF-2	hexene	O_2	$(CH_3)_3$ CCHO, CH_3CN, 40 °C, 6 h	1,2-epoxyhexane	75	[102]

Sun *et al.* [97] partially substituted Ni^{2+} in Ni-MOF-74 by active Co^{2+} (61%) *via* a post-synthetic metal-exchange method to enhance the catalytic performance in the reaction of cyclohexene oxidation (Fig. **9**). Newly created Co/Ni-MOF-74 exhibited improved catalytic activity compared to Ni-MOF-74 and even to pure Co-MOF-74 containing a similar amount of Co^{2+}. The conversion of cyclohexene reached 54.7%, and the compound has been converted mainly to 2-cyclohexen-1-ol (20.7%) and 2-cyclohexen-1-one (30.4%), and only 0.6% to cyclohexene hydroperoxide as an undesired by-product. The material showed a high total selectivity of 93.4% to cyclohexen-1-ol and 2-cyclohexen-1-one.

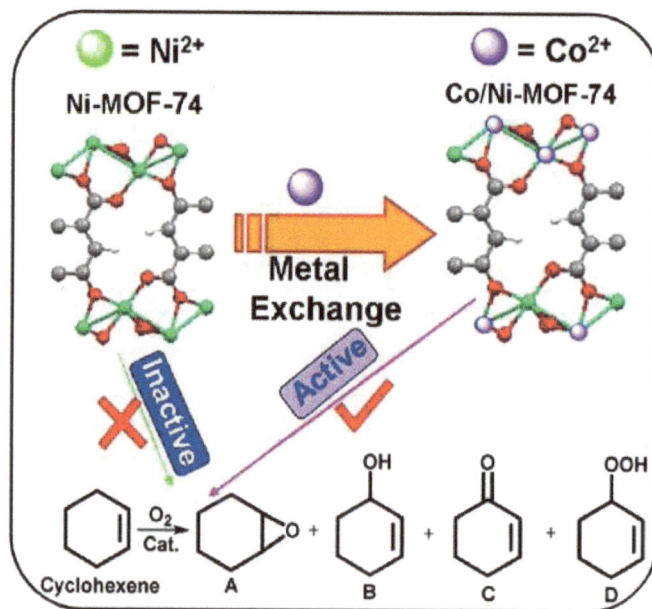

Fig. (9). Co^{2+} substitution during aerobic oxidation of cyclohexene over inactive Ni-MOF-74. Reprinted with permission [97]. Copyright 2015 American Chemical Society.

Allylic C–H oxidation of alkenes to the adequate α,β-unsaturated enones or 1,4-enediones is crucial in the drug precursor synthesis. Many soluble metal complexes demonstrate the catalytic activity in oxidation reactions, however, several limitations *e.g.,* corrosion, difficulty in recovery, deposition on reactor walls, or separation of the catalyst from the reaction media are still a problem. Metal–organic frameworks based on lanthanides (Ln-MOFs) are attractive Lewis acid catalysts in alkenes oxidation. This property has been exploited by Tran and co-workers [98] who obtained two isostructural lanthanide metal–organic frameworks (Ln-MOF-589, Ln = La^{3+}, Ce^{3+}) constructed from benzoimide-phenanthroline tetracarboxylic acid (H_4BIPA-TC) as a linker.

The materials were used as catalysts in the oxidation of olefins using anhydrous tert-butyl hydroperoxide (TBHP) as an oxidant. The two Ln-MOF-589 exhibited excellent catalytic activity in the oxidation of styrene and cyclohexene under mild conditions (0.7 mol% of MOF catalyst, 75 °C, 10 h, solvent-free, and N_2 environment). The presence of Ce-MOF-589 material results in remarkable conversion of styrene (94%) and cyclohexene (90%). Selectivity in the formation of styrene oxide from styrene and 2-cyclohexen-1-one from cyclohexene reached 85 and 95%, respectively. Transformation of cyclohexene by La-MOF-589 was performed with high selectivity (78%) to 2-cyclohexen-1-one but with lower conversion (60%). The higher catalytic efficiency of Ce-MOF-589 could be explained by the easy formation of the stable (+IV) oxidation state of the Ce ion according to its vacant *f*-shell. Moreover, Ce-MOF-589 was recovered and reused up to six times without any significant decrease in catalytic activity. Complexes of transition metals, especially molybdenum complexes (oxomolybdenum species), exhibited high catalytic activity in the epoxidation of alkenes. Unfortunately, they suffer from limited applicability because of poor catalyst separation and reusability. The solution to this problem is the immobilization of the catalyst onto various supports. This process combines advantages of recovery and easy separation of heterogeneous catalysts with the high catalytic activity of soluble transition metal complexes.

Hence, Afzali *et al.* [103] synthesized and functionalized UiO-66, known for its chemical and thermal stability, *via* immobilizing $Mo(CO)_3$. UiO-66-$Mo(CO)_3$-L material (L - synthesized by a large-scale procedure under ultrasonic conditions) was examined for the epoxidation of *cis*-cyclooctene as a model reaction and other alkenes with TBHP. The highest yield in *cis*-cyclooctene oxidation (98%) and selectivity (100%) were achieved with 100 mg of catalyst at 50 °C in the presence of TBHP as oxygen donor, and with acetonitrile as a solvent. The catalyst was stable under the reaction conditions and could be reused for four consecutive cycles. The results exhibited that the UiO-66-$Mo(CO)_3$-L catalyst was very active in the epoxidation of a wide range of alkenes, both linear and cyclic. Conversion of linear alkenes *e.g.,* octene and 1-decene reached 75% and 56%, respectively, with selectivities to 1,2-epoxyoctane and 1-decene oxide at the levels of 96 and 93%, respectively. Cyclic olefins such as cyclohexene were converted efficiently and selectively to cyclohexene oxide with a yield of 95%. Due to the epoxy efficiency, selectivity, reaction time, and straightforward preparation, UiO-66-Mo $(CO)_3$-L is a promising catalyst in alkenes oxidation.

It is widely considered that titanium-containing molecular sieves belong to the best catalysts for selective oxidation with hydroperoxides [104]. Titanium dioxide, TiO_2, is currently recognized as one of the most successful photocatalysts owing to its chemical stability, high efficiency, abundance, and low toxicity. Hence, titanium-based MOFs are perceived also as attractive photocatalysts, due to their promising optoelectronic and photocatalytic properties [105, 106]. Maksimchuk *et al.* [107] evaluated the catalytic activity of titanium-based metal–organic framework MIL-125 in the selective oxidation of cyclohexene (CyH) in acetonitrile with environmentally friendly oxidants, as hydrogen peroxide (H_2O_2) and tert-butyl hydroperoxide ('BuOOH). The aim of presented work was to explain the factors that influence the catalytic performance of this MOF. To increase the catalytic efficiency, alkene conversion, and change the product distribution, a source of protons ($HClO_4$ or CF_3SO_3H) was added in the amount of 1 molar equivalent (relative to Ti^{n+}). The effect of protons addition on the epoxidation selectivity was more significant for the reaction conducted with H_2O_2 than with 'BuOOH. One proton's equivalent ($HClO_4$) increased the cyclohexene conversion to 42% in 45 min *vs.* 26% in 60 min without acid. CF_3SO_3H addition caused a similar effect, but also enhanced the ratio of diol relative to epoxide. Total selectivity towards diol and epoxide with CF_3SO_3H achieved 75–80% in comparison to only 35% without acid additives. Drop in temperature from 50 to 30 °C reduced the maximum conversion of cyclohexene and changed the ratio between the heterolytic oxidation products, refraining the diol formation. MIL-125 catalyst could be readily recovered from the reaction mixture by filtration and regenerated by washing it with methanol to remove the reaction products from micropores. The material can be reused in at least four following runs without the loss of its catalytic activity.

Monoterpenes like α- and β-pinene are the key components of gum turpentine, a by-product obtained from the pulp and paper industry. They constitute inexpensive starting materials for the synthesis of multiple fragrances, flavors, agrochemicals, and therapeutically active substances. During the autoxidation of α-pinene, fine chemicals such as pinene oxide, verbenone, and verbenol are produced. They could be employed as precursors for taxol, menthol, citral, and sandalwood fragrance or vitamins A and E. The challenge in this field was using a greener reaction procedure, *i.e.* using oxygen from the air as the sole oxidant and avoiding the application of reagents like toxic organic hydroperoxides, which produce huge amounts of waste. Therefore, Raupp *et al.* [96] attempted to perform the immobilization of Mn complexes on MIL-53NH$_2$(Al)- MIXMIL-5-

-NH$_2$(50) *via* post-synthetic modification with maleic anhydride and Mn(III) acetate. This novel Mn-containing MOF catalyst was compared to Mn(III) acetate – a homogeneous catalyst in the oxidation of α-pinene. The reaction of α-pinene with molecular oxygen in diethyl carbonate/dimethylformamide (DEC/DMF) resulted in 31% conversion with high selectivity to pinene oxide (54%) as the main product, and smaller amounts of verbenol (3%) and verbenone (8%) in a mixture. The results were similar to those obtained in α-pinene oxidation with homogeneous Mn(III) acetate (conversion of 31% and the pinene oxide formation with 55% selectivity) after 6 h. However, the turnover number (TON) of the Mn-containing MOF was higher (166) compared to the homogeneous catalytic system (23). Additionally, the heterogeneous MOF catalyst was stable under the applied reaction conditions over at least five catalytic cycles and can be recycled.

Products of alkanes and alkenes oxidation are the most valuable industrial compounds. They are important in various fields, including medicine, pharmacy, or petrochemistry. Metal–organic frameworks are promising platforms in the oxidation of alkanes and alkenes. With the presence of MOF catalyst, conversion of substrates in oxidation reactions is higher and selectivity is enhanced towards desirable products. Oxidants and reaction conditions are often milder, and the production of toxic wastes is not observed. Metal–organic frameworks are stable under reaction conditions and could be easily recovered from the reaction system.

4.3. Oxidation of Alcohols

One of the fundamental processes in organic synthesis of significant industrial importance is the oxidation of alcohols to carbonyl compounds. In the last two decades, various heterogeneous catalytic systems using air or oxygen as the oxidant [108] and metal-based catalysts containing palladium [109], gold [110], or ruthenium [111] have been developed. Unfortunately, many of them require harsh reaction conditions and the selectivity to desired products is difficult to control. Moreover, all these metals fall in the so-called "critical elements" with reduced availability and, consequently, a high price. In this context, over the past few years, the oxidation of alcohols using MOFs as catalysts has proved to be extremely valuable [108]. It can be carried out with the use of molecular oxygen, hydroxide, or other oxidizing reagents either thermally or under the presence of light [112].

The catalytic oxidation of alcohols is highly influenced by the used conditions, including reaction temperature, substrate/catalyst molar ratio, nature and amount of base, type of solvent, and proportion of co-catalysts. The goal is to use as small amounts of metal as possible to carry out the reaction without solvents and at room temperature. Particular attention should be paid to the type and concentration of the base, which is often required to promote the oxidation of alcohols. The mechanism of the reaction is based on shifting the alcohol/alkoxide dissociation equilibrium towards the alkoxide by proton cleavage. The presence of a base increases the concentration of the alcoholate, which binds to metal ions more strongly than alcohol, and this promotes oxidation. The choice of the base affects not only catalytic activity but also structural stability of MOF [112]. Unfortunately, many MOFs are not stable in the presence of bases, thus, for this reason, but also taking into account aspects of green chemistry it would be much better to avoid or minimize the base concentration. The MOF network relies heavily on metal–ligand coordination bonds and bases can disrupt this interaction by coordinating more with metal ions than with ligands. In the absence of a base, alcohols bind to the metal cation instead of alkoxides. This interaction is weaker due to the lack of a negative charge and the lower electron density of the alcohols [113].

It is also worth considering the scope of application of MOFs as catalysts in the oxidation of alcohols. In general, the reactivity of secondary acyclic alcohols is lower than benzylic alcohols and therefore requires harsh reaction conditions. Secondary benzylic alcohols tend to react more slowly than primary alcohols due to a hindrance around the hydroxyl group. However, secondary alcohols can be oxidized with radicals and other more stable intermediate reaction products. They are formed with less activation energy than the corresponding products of primary alcohol oxidation, therefore secondary alcohols react faster. Thus, when selecting a suitable MOF as an oxidation catalyst, its stability, alcohol reactivity, range, and reaction conditions *i.e.,* temperature and the presence of a base must be considered. However, not all MOFs structures have suitable pore size and sufficient stability. Moreover, many of them are difficult to synthesize or they have undesirable structural or physicochemical characteristics [112]. Hence, it appears that there are some preferred MOFs commonly used as catalysts in the oxidation of alcohols, as presented in Table **3**.

Table 3. Overview of MOFs applied as catalysts in oxidation reactions of alcohols.

MOF Catalyst	Substrate	Oxidant	Conditions of Reaction	Product	Conversion (%)	Ref.
Ru@MOF-253	1°/2° alcohols	PhI(Oac)$_2$	22~40 °C, 1.5~5 h in CH$_2$Cl$_2$	ketone or aldehyde	55-99	[108]
TEMPO/Cu$_3$(BTC)$_2$	benzyl alcohol	O$_2$	40~75 °C, 22~44 h in CH$_3$CN+ Na$_2$CO$_3$	benzaldehyde	90	[114]

(Table 3) cont.....

MOF Catalyst	Substrate	Oxidant	Conditions of Reaction	Product	Conversion (%)	Ref.
TEMPO/ PEG/Cu$_3$(BTC)$_2$	benzyl alcohol	O$_2$	75 °C, 3~8 h in CH$_3$CN	benzaldehyde	100	[115]
RuCl$_3$@ MIL101(Cr)-TPY	benzyl alcohol	H$_2$O$_2$	100 °C, 6 h in H$_2$O	benzaldehyde	>99	[116]
UiO-66-CrCAT	2-heptanol	TBHP	70.1 °C, 24 h in C$_6$H$_5$Cl	2-heptanone	100	[117]
Pt@MOF-177	allylic & aliphatic alcohols	air	solvent- and base free, RT, 24 h	ketone or aldehyde	>99	[118]
0.35%Pd /MIL-101(Cr)	cinnamyl alcohol	O$_2$	80 °C, 0.5 h in toluene	cinnamyl aldehyde	97	[119]
TEMPO/ CuCl$_2$ @MOF-NH$_2$	benzyl alcohol	air	70 °C, 6 h in CH$_3$CN	benzaldehyde	100	[120]
Au@ZIF-8 and Au@ZIF-90	benzyl alcohol	O$_2$	80 °C, 24 h in MeOH at 5 bar	methyl benzoate	13-81	[121]
STA-12(Fe)	benzyl alcohol	TBHP	RT, in EtOAc+Na$_2$S$_2$O$_4$	benzaldehyde	100	[122]
[Pd(2-pymo)$_2$]$_n$	3-phenyl-2-propen-1-ol	air	90 °C, 20 h in toluene	3-phenylprop-2-enal	>99	[123]
Au@MIL-101	benzylic, allylic and aliphatic alcohols	O$_2$	80 °C, 1 atm, 1~35 h in toluene	aldehyde or ketone	23-99	[124]
Co/C–N700	1-phenyl-ethanol	air	110 °C, 48 h in H$_2$O	acetophenone	98	[125]
UoB-3	1°/2° alcohols	TBHP	65 °C, 45-90 min solvent free	ketone or aldehyde	75-95	[126]
[Co(L^1)(bdc)(MeOH)]$_n$	benzyl alcohol	TBHP	60 °C, 22 h in DMF	benzaldehyde	88	[91]
[Co$_3$(BTC)$_2$]$_n$	benzyl alcohol	O$_2$	95 °C, 10 h in DMF	benzaldehyde	92.9	[127]

HKUST-1 (Hong Kong University of Science and Technology) is an example of MOFs that has been employed in catalytic oxidation of benzylic alcohols [128]. It consists of Cu ions connected with benzene-1,3,5-tricarboxylate (BTC). $Cu_3(BTC)_2$ is stable up to 240 °C, so it can be applied as a solid catalyst in a wide range of liquid phase reactions [129]. Using this type of MOF, oxidation of benzyl alcohol was carried out with the addition of co-catalyst 2,2,6,6-tetramethylpiperidine-1-oxyl (TEMPO) in the presence of Na_2CO_3 and acetonitrile medium (Fig. **10**). $Cu_3(BTC)_2$ showed high catalytic activity with a yield to benzaldehyde of 89%.

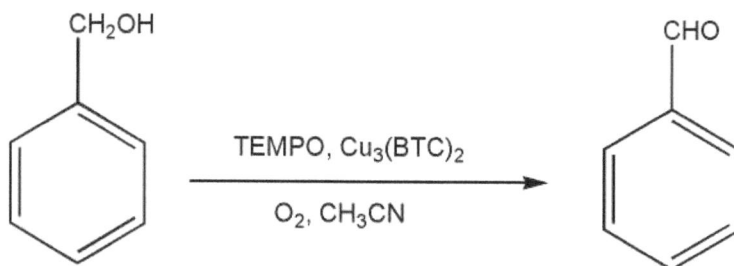

Fig. (10). Aerobic oxidation of benzyl alcohol catalyzed by $Cu_3(BTC)_2$ hosting piperidinyl oxyl (TEMPO) [114].

The X-ray diffraction (XRD) analysis of the spent catalyst showed changes in the crystal structure compared with the fresh catalyst. This suggests that the conditions required for the reaction are inconsistent with the stability of the used MOF. It has also been revealed that benzoic acid formed during the reaction can lead to catalyst deactivation by coordination with free metal centers [114].

Later, a mesoporous $Cu_3(BTC)_2$ structure was obtained and its activity in the aerobic oxidation of benzyl alcohol to benzaldehyde was investigated. Polyethylene glycol (PEG) was immobilized in the pores of MOF by hydrogen bonding with BTC, which prevented the collapse of the mesopores. PEG was also used as the reaction medium to induce the crystallization of MOF at room temperature, while the micelles of P104 the triblock copolymer $(EO_{27}PO_{61}EO_{27})$ acted as a matrix for the formation of mesopores. After removal of the solvent, PEG-stabilized mesoporous MOF nanocrystals were obtained. The catalytic activity of this MOF was high and complete conversion of benzyl alcohol in the presence of oxygen and TEMPO in an acetonitrile medium at 75 °C was achieved. This revealed that the PEG/$Cu_3(BTC)_2$ activity significantly exceeded the activity of commercial $Cu_3(BTC)_2$ in aerobic oxidation. Smaller particle size and mesoporous structure of PEG/$Cu_3(BTC)_2$ promote the diffusion of substrates and products and increase the availability of catalytic active sites. The post-reaction XRD analysis of the catalyst showed no differences in the structure between the fresh and the four times used sample. The heterogeneity and stability of the catalyst were evaluated by the hot filtration test. No further substrate conversion was detected, confirming that oxidation took place on the catalyst surface [115].

Another modified MOF that was used in the oxidation of alcohols is chromium(III) terephthalate, MIL101(Cr) (Matériaux de l'Institut Lavoisier) with the molecular formula $Cr_3(F,OH)-(H_2O)_2O[(O_2C)-C_6H_4-(CO_2)]_3 \times nH_2O$ (n=25) [2]. The terpyridyl moiety (TPY), a tridentate oligopyridine o-donor chelator, was incorporated into the porous MIL-101(Cr), using click post-synthetic functionalization. MOF with an open terpyridyl unit can be employed as a platform for the synthesis of various heterogeneous single active site metal catalysts. To generate the appropriate complex on the tridentate chelator, $RuCl_3$ was added to form the $RuCl_3$@MIL101(Cr)-TPY. The catalyst obtained was used to oxidize benzyl alcohol to benzaldehyde at 100 °C with hydrogen peroxide as an oxidant and exhibited a high yield (99%) after 6 h. The oxidation with $RuCl_3 \times nH_2O$ showed only 49% yield under identical conditions. $RuCl_3$@MIL101(Cr)-TPY maintained its activity for five cycles without significant loss of activity [116].

The next type of MOF commonly applied in catalysis is the UiO-66 (University of Oslo), formed by the reaction of 1,4-benzenedicarboxylate (BDC) with zirconium salt [130]. UiO-66 is widely used due to its high thermal resistance and chemical stability [1]. For the oxidation of 2-heptanol to 2-heptanone using tert-butylhydroperoxide (TBHP) as an oxidant in chlorobenzene at 70 °C, UiO-66 catalysts modified with monocatecholate metal groups (UiO-66-CrCAT, 1 mol% Cr) afforded 100% yield to 2-heptanone. For the attachment of monocatecholate metal groups to UiO-66, two post-synthetic strategies were applied: post-synthetic exchange (PSE) and post-synthetic deprotection (PSD). The first method is a more efficient and convenient approach to obtain the UiO-66 catalyst and the subsequent metalation of the catechol functional group attached to the MOF structure allowed to obtain UiO-66-CrCAT. A significant increase in the reaction rate with a high yield to 2-heptanone was achieved by using only 0.5 mol% Cr loading under solvent-free conditions, which are very beneficial for environmental protection due to the avoidance of chlorinated solvents. Furthermore, unmodified UiO-66 and UiO-66-CAT achieved 8% and 12% yields of 2-heptanone, respectively. In addition, several cyclic, aromatic and secondary aliphatic alcohols have been converted to ketones in chlorobenzene and under solvent-free conditions with high yields. UiO-66-CrCAT was reused five times without yield loss, and a hot filtration test showed no catalytically active Cr compounds in the reaction medium. The XRD analysis of the spent catalyst showed high crystallinity without structural changes. Extended X-ray absorption fine structure (EXAFS) and X-ray photoelectron spectroscopy (XPS) exhibited that after one oxidation cycle there is no significant change with the Cr coordination environment, suggesting high resistance of this catalyst [117].

The employment of MOFs as catalysts in alcohol oxidation reactions that have industrial interest should be further explored. In particular, the production of fine chemicals such as flavoring agents, fragrances, and pharmaceuticals is very important. So far, soluble transition metals have been mainly used as homogeneous catalysts. However, properly designed MOFs may show sufficiently higher catalytic activity and stability to be used in industrial processes.

4.4. Oxidation of Thiols and Sulfides

Environmental pollution has become the main problem around the world, mainly because it has a negative influence on human health. Pollutants may disturb metabolic activity, they are toxic and carcinogenic. The human activity led, for instance, to a significant accumulation of sulfur-based compound contaminants in the environment. In order to overcome this problem, organic transformations such as oxidation have become a widely accepted alternative to traditional physical treatment technologies to remove these pollutants [72]. In recent years, there has been a growing interest in the oxidation of thiols to disulfides, which are involved in many biochemical and industrial processes. They are widely used in the synthesis of agrochemicals, pharmaceuticals, and vulcanization of rubber [131, 132]. The production of disulfides by thiols oxidation is an efficient industrial method due to the simplicity of the procedure and the high availability of many thiols [133]. A common limitation in the oxidation of thiols is the wide variety of the obtained by-products in addition to target disulfides, such as sulfoxides, sulfones, and sulfonic acids. Therefore, the development of a selective oxidation process of thiols to disulfides, avoiding the excessive formation of by-products, has become the target of many researchers in the field [134].

One of the MOFs used for the aerobic oxidation of thiols to disulfides is the Fe-BTC [134]. In the presence of this solid redox catalyst, the thiophenol was oxidized with a conversion of 98%, by using molecular oxygen, in an acetonitrile medium and under mild reaction conditions (70 °C) (Fig. **11**). Much poorer efficiency was obtained by carrying out the same reaction with a homogeneous catalyst based on iron(III) nitrate, despite the use of a larger amount of it. Further advantages of using the solid Fe-BTC redox catalyst were demonstrated by comparing it with other MOFs, namely $Cu_3(BTC)_2$ and $Al_2(BDC)_3$, which showed lower catalytic activity with thiophenol conversions of 28 and 4%, respectively.

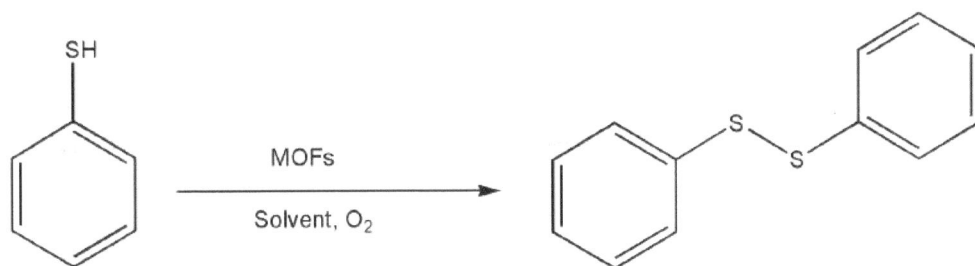

Fig. (11). Aerobic oxidation of thiophenol to diphenyldisulfide [134].

Dalapati *et al.* [135] synthesized cerium-based MOF material containing 3,4-dimethylthieno[2,3-b]thiophene-2,5-dicarboxylic acid as ligand under solvo thermal conditions. In the presence of this catalyst the thiophenol conversion observed in four oxidation cycles, at 70 °C and in methanol solvent, was 100%, 98%, 97%, and 96%, respectively. Moreover, the crystallinity of the catalyst recovered after the fourth cycle was very similar to the fresh one, which was confirmed by the XRD analysis. In addition, the versatility of this catalyst in the oxidation reaction with substituted thiophenols was also investigated. The reaction was performed at 70 °C for 12 h, using molecular oxygen as an oxidant. Cyclohexanethiol was oxidized with a 26% yield to the corresponding disulfide while thiophenols with the attached electron-withdrawing groups, such as the halogen groups at positions 2, 3, and 4, led to disulfides in 80-90% yields. On the other hand, thiophenols with electron-donating substituents, such as amino, methyl, and methoxy groups at positions 2 and 4, led to 60-70% yields of the corresponding disulfide.

Song *et al.* [136] investigated the aerobic oxidation of thiols to disulfides catalyzed by POM-MOF. They synthesized a composite by incorporating Keggin-type polyoxometalate (POM) $[CuPW11O39]^{5-}$ into the pores of MOF-199 (HKUST-1). The new material catalyzed the fast chemical and shape-selective oxidation of thiols to disulfides. The highest conversion (95%) was obtained for 2-hydroxyethanethiol. Conversions decrease with increasing the carbon atoms of thiols. This is coherent with the oxidation of thiol by POM units incorporated in MOF: sterically larger thiols are less accessible to the POM units and hence are oxidized at a slower rate. It should be mentioned that the pristine MOF or POM were inactive or had very low catalytic activity. It was also investigated whether the size of the MOF crystals influenced the thiol oxidation reaction. Dhakshinamoorthy *et al.* [137] used MIL-100 (Fe) as a catalyst for the oxidation of thiophenol. The reaction was carried out at 70 °C, the conversion was 100% after 3 hours and the main product was diphenyldisulfide. After using it three times, MIL-100 (Fe) was still stable and active. For small molecules like thiophenol, it has been found that the reaction took place in the pores of the MOF and did not occur on its external surface.

The oxidation of sulfides to sulfur oxides has also gained increasing interest in recent years [138]. These compounds are especially undesirable in sewerage, causing the corrosion of the pipes and stimulating the metabolic activity of sulfate-reducing bacteria [139]. Sulfide ion is also one of the main pollutants in wastewater from oil and gas extraction [140]. Organic sulfur oxides, on the other hand, are useful intermediates and active ingredients in biological applications, organic synthesis, and the petroleum industry [141]. A subject of great interest in green chemistry is the aerobic photooxidation of sulfides in water to sulfur oxides [142]. In this context, Wei *et al.* [143] synthesized three highly stable Ir-Zr MOFs using Ir(III) complexes as linkers and the Zr6 cluster as nodes: Zr6 – Irphen (phen is 1,10-phenanthroline), Zr6 – Irbpy (bpy is 2,2'-bipyridine), Zr6 – IrbpyOMe (bpyOMe is 4,4'-dimethoxy-2,2'-bipyridine). Their photocatalytic activity was tested in the oxidation of sulfides to sulfoxides in water, at room temperature and under irradiation with visible light (Fig. **12**). In the presence of Zr6-Irphen catalyst, sulfoxide was obtained with a yield of 100%, after 6 h, while Zr6-Irbpy and Zr6-IrbpyOMe catalysts led to yields of 57% and 82%, respectively. Powder X-ray diffraction (PXRD) patterns of the reused catalysts showed that MOFs could be recycled and reused at least 10 times without losing catalytic activity. These studies provide a new strategy for green sulfoxide synthesis under ambient conditions.

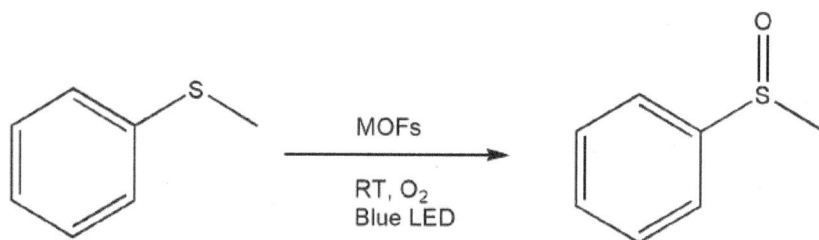

Fig. (12). Aerobic photooxidation of sulfide to sulfoxide [143].

Johnson *et al.* [144] reported the first example of controllable *in situ* metalation of an anionic indium porphyrinic metal−organic framework UNLPF-10 (University of Nebraska−Lincoln porous framework). The extent of the metalation of the porphyrin macrocycles in UNLPF-10 was tuned by changing the In/L **ratio** (indium to ligand) during MOF synthesis. The oxidation of sulfide under irradiation with a blue LED lamp revealed the high photocatalytic activity of UNLPF-10. Therefore, within 8 hours of photo-induced oxidation, the thioanisole was completely and selectively converted to sulfoxide. Moreover, the catalyst displayed high stability, and a yield of 97% to sulfoxide being obtained after five catalytic cycles. It has also been shown that the number of indium-metalated porphyrin sites (In-porph) affects the rate of thioanisole photooxidation. Thus, in the presence of UNLPF-10 with 25% In-porph sites, full conversion of thioanisole was reached after 40 h, while in the presence of the UNLPF-10 with 98% In-porph sites the same conversion level was achieved within only 8 h. In addition, MOF displayed an enhanced photocatalytic activity for photo-induced sulfide oxidation compared to the homogeneous tetraphenylporphyrin analog or its indium-metalated derivative.

The oxidation of methylsulfobenzene to sulfoxides takes place selectively in the presence of $[Yb(C_4H_4O_4)1.5]$ catalyst under mild reaction conditions (*i.e.*, 6 h, 60 °C, H_2O_2 as oxidant) [145]. However, the conversion of methylsulfobenzene was rather modest (50%). Contrarily, very high conversions (99%) of methylphenyl sulfide were obtained in the presence of $[Yb_4(OH)10(H_2O)_4][2,6-AQDS]$ and $[Yb(OH)(2,6-AQDS)(H_2O)]$ catalysts, under similar reaction conditions [146].

High selectivities to sulfoxides (95-100%) were also reported not long ago by Haddadi *et al.* [141] for thioanisole oxidation in the presence of another two MOF-based catalysts, namely $[Cu(Phen) (4.4'-bpy)(H_2O)]_2[PW_{12}O_{40}](4.4'-bpy)$ (PW-MOF) and $[Cu_3(4.4'-bpy)_3] [PMo_{12}O_{40}] (C_5H_6N_2)\square O\square 5H_2O$ (PMo-MOF). Reactions were carried out in the presence of 1 mmol of hydrogen peroxide and 0.01 mmol of catalyst, at room temperature for 60 minutes. As a result, products were obtained with a yield of 94–99.9% and very high selectivity to sulfoxides (95-100%). Moreover, the PW-MOF catalyst could be easily recovered and reused five times without losing its catalytic activity. The same catalytic systems have been shown to be effective for a wide range of aromatic and aliphatic sulfides which were selectively oxidized to the corresponding sulfoxides (90–99% yields) under optimized reaction conditions. The developed catalytic system is safe, easy to use, cheap, environmentally friendly, and reproducible. The mild reaction conditions make the catalytic procedure an ecological alternative that includes a recyclable solid catalyst, a green oxidizer, and a halogen-free solvent.

5. MOFS AS CATALYSTS FOR C–C COUPLING REACTIONS

The development of metal-catalyzed cross-coupling reactions over the past three decades has revolutionized the way carbon-carbon bonds between sp and sp^2 carbon atoms are formed. Many natural products, building blocks for supramolecular chemistry and self-assembly, organic materials, polymers, and compounds in medicinal chemistry were synthesized by this methodology for the first time or in a much more efficient way than before. The 2010 Nobel prize in chemistry for E. Negishi, R. Heck, and A. Suzuki underlines the importance of direct C–C bond formation.

Metal-catalyzed cross-coupling reactions involve the formation of organometallic intermediates by Lewis acid sites allowing access to stereospecific and regioisomeric products [147]. The homogeneous organometallic palladium complexes have gained huge relevance in most common coupling reactions such as Suzuki-Miyaura, Mizoroki-Heck, Sonogashira, Stille, Ulmann and Buchwald-Hartwig. However, homogeneous catalysis has several drawbacks, in particular, the problem of recycling the catalyst or even the lack of reuse of the catalyst. This leads to a loss of expensive metal and ligands and thus to impurities in the reaction products.

Therefore, in recent years much attention has been paid to transition from the homogenous to heterogeneous catalysts [148]. Heterogeneous Pd fixed to a solid support with large porosity is a promising option. Therefore, metal–organic frameworks were proposed as a host for palladium nanoparticles and MOF composites have been applied as catalysts in the formation of C–C bonds in a simple and reliable way [149]. Table **4** presents an overview of MOF catalysts applied in C–C coupling reactions.

Fe_3O_4@PDA-PD@[Cu(btc)$_2$] exhibited very high efficiency in the Suzuki-Miyaura coupling reaction between aryl halides and arylboronic acids [150]. Moreover, the catalyst could be reused and recycled without its decomposition. Corma *et al.* [123] have also studied the activity of another type of catalysts such as Pd-MOF (Pd(2-pymo)$_2$]$_n$·3H$_2$O (2-pymo - 2-hydroxypyrimidinolate) in the Suzuki-Miyuara coupling reaction between phenylboronic acid and 4-bromoanisole (Fig. **13**). The reaction took place with a selectivity of 99% to *p*-methoxybiphenyl for a conversion of 85%, at 150 °C and after 5 h.

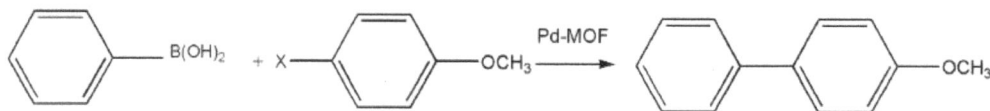

Fig. (13). Suzuki-Miyaura coupling catalyzed by Pd-MOF [123].

In another study MIL-101, with a large total surface area (4000 m^2/g) [2] and pore system formed by both hydrophilic and hydrophobic networks [151], have been used as a support for palladium nanoparticles (NPs). In the presence of Pd/MIL-101 catalyst the Suzuki-Miyuara coupling reaction between 3-chloroanisole and phenylboronic acid takes place with a yield of 82% [152]. Amino-functionalized MIL-101 with embedded palladium nanoparticles has also been used in the reaction between 4-bromoanisole and phenylboronic acid [153].

Table 4. Overview of MOFs applied as catalysts in C–C coupling reactions.

Type of Coupling Reaction	MOF Catalyst	Active Site	Conversion (%)	Number of Cycles in Recyclability Test	Ref.
Suzuki-Miyaura	Pd@MIL-101 (Cr)	Palladium nanoparticles encapsulated in MOF pores	99	5	[154]
	IRMOF-3-PL-Pd	Palladium anchored to walls of modified IRMOF-3	99	5	[155]
	Pd@UiO-66	Immobilized Pd nanoparticles on zirconium MOF surface	99	5	[156]
	UiO-67-3-Pl-Pd	Palladium nanoclusters encapsulated inside the cavity of MOF	95	10	[157]
	UiO-67-bpydc	Pd(bipyridine)Cl$_2$ complexed on metal surface	89	3	[158]
Mizoroki-Heck	Pd@MIL-101(Cr)	Palladium nanoparticles encapsulated in MOF pores	99	5	[154]
	UiO-67-Pd-NHDC	Pd-NHDC complex doped onto MOF surface	99	5	[159]
	UiO-67-3-Pl-Pd	Pd nanoclusters Encapsulated into MOF pores	95	10	[157]
Sonogashira	Pd-TPOP-1	Palladium nanoparticles dispersed on organic polymer matrix surface	70	4	[160]
	Pd@UiO-67-bpdy	Palladium immobilized on MOF surface	99	5	[161]
	Pd(II)-NHC@Zn-Azolium MOF	Palladium attached to azolium containing MOF surface	98	4	[162]

(Table 4) cont.....

Type of Coupling Reaction	MOF Catalyst	Active Site	Conversion (%)	Number of Cycles in Recyclability Test	Ref.
Stille	Pd@IRMOF-3	Palladium anchored to walls of modified IRMOF-3	90	5	[155]

Interestingly, in the Suzuki cross-coupling reaction, Pd nanoparticles supported on scandium-MOFs exhibited higher catalytic activity in comparison to activated carbon-supported palladium catalysts, as recently reported by Zhang *et al.* [163]. Furthermore, the metal leaching was irrelevant, and the catalytic performance was meaningful in a few runs. In turn, in the Friedel-Crafts alkylation of toluene with benzyl chloride, the use of MOF-5@SiO$_2$@Fe$_3$O$_4$ led to a conversion of 97% in solvent-free conditions only by using high amounts of catalyst (26.8 wt.%), as Li *et al.* [164] reported. However, its excellent stability and recyclability represent important advantages. The performance of MIL-53(Al)@SiO$_2$@Fe$_3$O$_4$ as a catalyst was assessed in the Friedel-Crafts acylation of 2-methylindole with benzoyl chloride [165]. The MOF composite with 38.8 wt.% MIL-53(Al) led to a high selectivity (81%) to 3-acetylindole for a 98% conversion of 2-methylindole. The catalyst could be reused for five runs. Metal–organic frameworks were also successfully applied in Ullmann-type reactions [166]. For instance, in the reaction of phenols with aryl iodides to produce diaryl ethers, MOF-199 catalyst led to a yield of 82% at 120 °C and after 6 h [167]. It is worth mentioning that MOF-199 was reusable without the significant loss of its catalytic activity. Chen *et al.* [168] reported the use of MOF-253 catalyst in Ullmann homocoupling and Suzuki-Miyaura cross-coupling. The material was prepared through a hydrothermal methodology followed by the immobilization of palladium(II) chloride precursor on its surface. It was showed that the obtained MOF-253·PdCl$_2$ exhibited higher activity in comparison to homogeneous Pd(bpy)Cl$_2$ and PdCl$_2$(CH$_3$CN)$_2$ species most probably due to the charge transfer between Pd and neighboring ligands in the catalyst. The supplement of the Suzuki-Miyaura reaction is the Stille coupling of aryl halide with alkyl or aryl stannane [169]. In this type of reaction, IRMOF-3-PI-Pd was used as a catalyst (Fig. **14**) [155]. In the coupling between bromobenzene and tributylphenyltin the best yield (88%) was obtained when lithium chloride was added to ethanol.

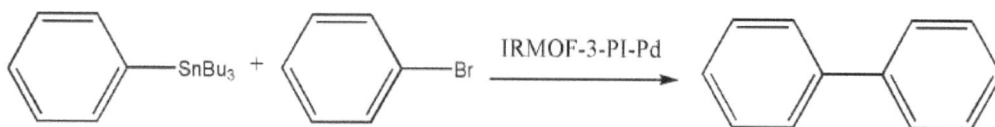

Fig. (14). Stille coupling catalyzed by Pd-MOF [155].

Among different aryl halides tested in the presence of IRMOF-3-PI-Pd, high product yields were achieved only when bromobenzene was used [155]. The catalytic recycling revealed no changes in the yield after five runs. Moreover, the powder XRD and FT-IR data remained unchanged for both recovered and parent IRMOF-3-PI-Pd demonstrating the high stability of the material.

A crucial role in modern organic synthesis plays the Mizoroki-Heck coupling reaction between aryl/vinyl halides and alkenes. MIL-101 modified with 3-aminopropyltrialkoxysilane (APS-MIL-101) and ethylenediamine (ED-MIL-101), and subsequently loaded with palladium were utilized as catalysts in the reaction between iodobenzene and acrylic acid [170]. The palladium leaching was not observed in the recycling test. The efficiency of Pd/MIL-53-(Al)-NH$_2$ as a catalyst in the reaction between bromobenzene and styrene was also tested (Fig. **15**) [171]. The highest yield (81%) was reached with 0.05 mol% of MOF-based catalyst in the presence of Et$_3$N in DMF solvent. The results proved that the catalytic efficiency of Pd/MIL-53-(Al)-NH$_2$ was much better compared to the Pd/C in the presence of which the yield was only 49%. Moreover, the Pd/MIL-53-(Al)-NH$_2$ catalyst could be recycled five times with a slight decrease in the yield. After the 5th run, the yield was at the level of 76%.

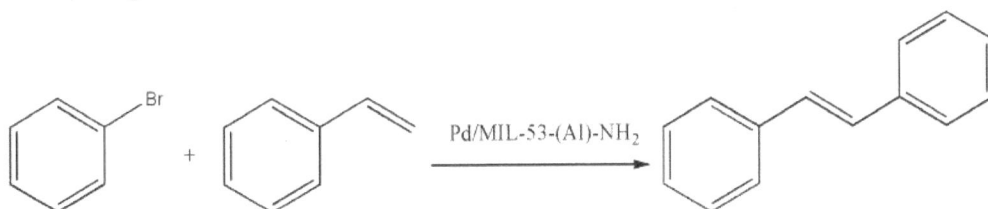

Fig. (15). Mizoroki-Heck coupling between bromobenzene and styrene [171].

Pd^{2+}@UiO-67 has been developed as an efficient catalyst for Mizoroki-Heck reaction between 4-choloracetophenone and styrene (Fig. **16**) [172]. When the process was performed in DMF with the presence of tetrabutylammonium bromide and potassium carbonate, 99% conversion of 3-chloroacetophenone with more than 99% selectivity to the desired product (4-aceto-1,1'-biphenyl) was achieved. Other results proved that the catalytic efficiency of Pd^{2+}@UiO-67 was much better compared to less porous catalysts (Ln$_2$[Pd(bpydc) Cl$_2$]$_2$[Pd(Hbpydc)Cl$_2$]$_2$). Furthermore, the structural integrity of the MOF material was retained after performing the reactions [173].

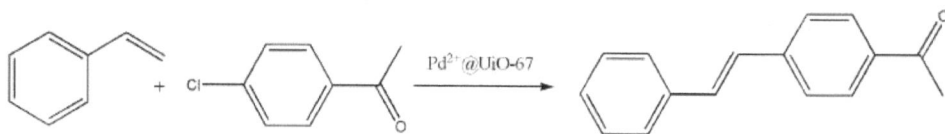

Fig. (16). Reaction between styrene and 4-chloroacetophenone catalyzed by Pd^{2+}@UiO-67 [172].

MOFs were also employed in the Sonogashira coupling of aryl halides and terminal alkynes to aryl-substituted alkynes [174]. The obtained products are valuable intermediates in the synthesis of essential pharmaceuticals [175]. The catalytic performance of Pd/MOF-5 was studied in the coupling that occurred between iodobenzene and phenylacetylene to obtained diphenylacetylene in high yield (98%) (Fig. **17**) [176]. Furthermore, the same catalyst exhibited great activity for coupling aliphatic alkynes and aryl iodide. Nevertheless, Pd/MOF-5 showed some limitations. Notably, when ortho-substituted iodobenzenes were used as the substrate, it revealed lower yields compared to the para-isomer. Furthermore, when the aryl halide was altered to chloro or bromo derivative, no conversion was registered. Not least, the catalytic recycling tests demonstrated that Pd/MOF-5 lost its activity after the third cycle, most probably due to the oxidation of palladium.

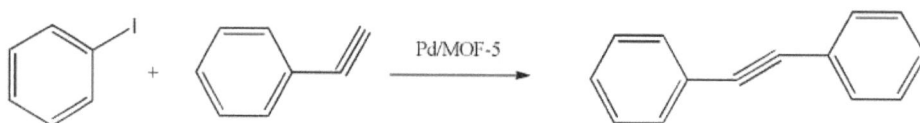

Fig. (17). Sonogashira coupling of iodobenzene and phenylacetylene [176].

Truong *et al.* [177] used a crystalline porous metal-organic framework, namely Ni$_2$(bdc)$_2$(DABCO), as a catalyst in the coupling reaction of phenylboronic acid and phenylacetylene. This material showed a better catalytic activity (86% conversion) than Ni$_3$(btc)$_2$ and Ni(btc)(bpy) (conversions of 51 and 59%, respectively). In the leaching test, no active species were detected in the solution. Furthermore, any decrease in conversion was observed when the catalyst was reused five times. The XRD analysis of the spent catalyst showed that Ni$_2$(bdc)$_2$(DABCO) remained still highly crystalline.

In this subsection, it was shown that metal-organic frameworks can be used to promote cross-coupling reactions. One approach was to apply MOFs supported palladium catalysts. Another one is based on including small metal nanoparticles within the pores of MOFs. For this purpose, the ligand can be functionalized with amino groups to interact with NPs. The main aim is to develop catalysts for cross-coupling reactions that can be reusable and stable under different reaction conditions. All the presented results proved that MOFs have the potential to replace the current catalysts that are used in an industrial scale.

6. MOFS AS CATALYSTS IN HYDROGENATION REACTIONS

Catalytic hydrogenation has a long tradition in organic synthesis dating back to the days of Sabatier [178] who received the 1912 Nobel Prize in chemistry for his pioneering work in this area. Today the catalytic hydrogenation is a mature technology that is widely applied in the industrial organic synthesis of different products, such as fine chemicals, medicals, or polymers. Yet, new applications carry on to appear, and the scope of catalytic hydrogenations continues to be extended to more difficult reductions, such as the direct conversion of carboxylic acids to aldehydes.

A broad variety of compounds such as cinnamaldehyde, aldehydes, alkenes, alkynes, or furfural can selectively and efficiently undergo this process, under mild conditions, in which pairs of hydrogen atoms are being added to a molecule, resulting in its reduction or saturation. Hydrogen atoms, which are further being transferred to unsaturated compounds, are generally acquired from molecular hydrogen or another hydrogenation source, such as *iso*-propanol, formic acid or hydrazine. Many hydrogenation catalysts have been developed over time, from homogeneous to heterogeneous [179]. Despite the fact that commonly used hydrogenation catalysts, such as metal oxides, metal particles, or metal sulfides have a notable impact on improving selectivity and efficiency of the reaction, they have diverse critical drawbacks, which foster the urge for optimization. Then, in order to improve the stabilization of these traditional catalysts, some supports like zeolites or silicas have been used, but their poor tunability became another limiting factor in obtaining high selectivities and conversions. MOF-based catalysts as opposed to the traditional catalysts, have more adjustable properties and superior tunability [180]. MOFs have a broad spectrum of potential applications in hydrogenation reactions due to their unique features, such as: *i)* the excellent hydrogen storage characteristics which can improve the catalytic activity; *ii)* their surface areas largely exceeding those of traditional porous materials (*e.g.,* zeolites and activated carbons) able to accommodate high amounts of highly dispersed active sites, and *iii)* easy functionalization. The overview of some commonly used MOF-based catalysts in hydrogenation reactions is shown in Table **5**. When served as templates/precursors, MOFs can be converted, by pyrolysis, to MOF-derived porous carbon materials, which are much more stable and conductive than pristine MOFs but, at the same time, with inherited characters of pristine MOFs such as the composition diversity and dispersion, large surface area, and tailored porosity [181]. Therefore, these materials may afford better catalytic performances than their parent MOFs. Not least, besides single active sites in pristine MOF materials, these catalysts can have other active sites as a result of coating with functional materials or encapsulation of inorganic and organic guests. Properties of the active phase such as distribution and size can be controlled, which has a significant impact on the selectivity and activity of hydrogenation reaction. High specific surface areas of MOFs cannot only enhance adsorption of reductants and reactants, increasing their local concentrations at the same time, but also positively affect the quantity of the active sites. Both factors influence catalytic activity. By the use of easily tunable MOF-based catalysts, the hydrogenation process can be controlled. Adjustable sizes of pore channels and windows allow an effective way of transferring the reagents to the active centers in the diffusion process [6].

Table 5. Overview of MOFs applied as catalysts in hydrogenation reactions.

MOF-based Catalyst	Type of Hydrogenation Reaction	Product	Conversion /Yield(%)	Ref.
Pd-MIL-101 (Cr)	reduction	hydrocinnamaldehyde	99.8/n.a	[182]
Pd-MIL-53 (Al)	reduction	hydrocinnamaldehyde	99.9/n.a	[182]
Pd-ZIF-8 (1%)	reduction	hydrocinnamaldehyde	>99.9/n.a.	[182]
Pd@MOF-1	reduction of nitrobenzene	aniline	-/98	[183]
Ni/MIL-53(Al)	reduction of nitrobenzene	aniline	92/n.a.	[184]
Fe- and Co-functionalized MOFs	olefin hydrogenation	cyclohexane	100/n.a.	[185]
Pd nanocubes@ZIF-8	olefin hydrogenation	hexane	n.a./66	[186]
Pd/H-UiO-66	furfural hydrogenation	furfuryl alcohol	100/>99	[187]
Pt/H-UiO-66	furfural hydrogenation	furfuryl alcohol	78/49	[187]

Although this type of catalysts has many worthwhile properties, they also have many drawbacks that impede their practical usage. The main obstacle is their poor stability which strongly limits their catalytic efficiency, particularly under harsh conditions such as organic solvents, humid air, high pressure, and temperature. The other two major disadvantages are their higher costs than the traditional catalysts and their poor reusability. MOF-based catalysts can be used in hydrogenation reactions of various compounds such as alkynes, alkenes, aldehydes, olefins, esters, nitro groups, alcohols and phenols, and others [188]

One of the most stable and selective catalysts was constructed by encapsulation of platinum nanoclusters in an amino-functionalized Zr-terephthalated metal–organic framework (UiO-66–NH$_2$). It was demonstrated that by using this MOF in chemoselective hydrogenation of cinnamaldehyde (Fig. **18**) not only its high conversion (over 85%) but also high selectivity (over 90%) of cinnamyl alcohol can be reached. To achieve this effect, Pt nanocluster cannot be presented on the outside surface of MOF but must be embedded in MOF functionalized with amines. The major benefit of this catalyst is the fact that it can be recycled numerous times with no impact on its selectivity and activity [189].

Due to the porous crystalline system, ZIF (zeolitic imidazolate framework) has both MOF and zeolites required properties. Comparing to other MOFs, this catalyst exhibits higher chemical, thermal and hydrothermal stability. ZIF-8 can be used as support for platinum, silver, gold, ruthenium, palladium, and gold-nickel bimetallic nanoparticles and be applied in amino-carbonylation, acetophenone asymmetric hydrogenation, hydrogenation of 1,4-butynediol, and alkene hydrogenation in the gas phase [190].

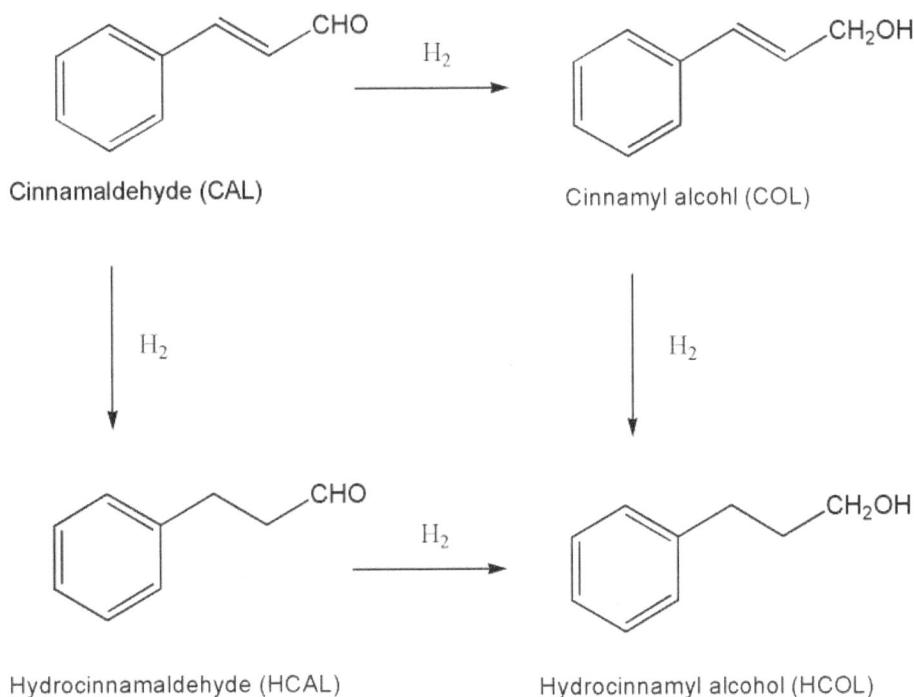

Fig. (18). The hydrogenation reaction of cinnamaldehyde [182].

The hydrogenation reaction of cinnamaldehyde to acquire saturated aldehydes and unsaturated alcohols is very important in chemical industries such as perfume, medicine, and food additives. Comparing to traditional catalysts made with oxide supports (Al$_2$O$_3$, SiO$_2$), these catalysts demonstrate better activity and selectivity (over 90%) in the hydrogenation reaction [182]. Nowadays, hydrogenation of nitroarenes is becoming more and more popular because of resulting in anilines, which are vital raw chemical compounds with varied applications in chemical industries, such as medicines, polymers, dyes or fine chemicals. This process consists of breaking of hydrogen bond, reduction of nitro groups, condensation of hydroxylamine and creating the nitro compounds. However, besides the reduction of nitro functionality, other functional groups from the substrate like carbonyl, or halide can also be reduced, being the major limitation in selective nitroarenes hydrogenation [191]. In order to produce aniline selectively, lanthanides (Pr, La, Nd, Sm) based MOF (Ln-MOF, where Ln is La or Pr) were synthesized by hydrothermal approach and applied in nitroarenes hydrogenation.

Because of their unique catalytic efficiency, transition–metal phosphides are becoming more popular, especially cobalt phosphide. In this context, ZIF-67, a metal–organic framework with a high surface area, was prepared by solvothermal method, at a temperature higher than 100 °C in methanol solvent. By encapsulating it with red phosphorus followed by pyrolysis, phosphide-based Co_2P/CNx nanocubes were produced with high catalytic efficiency in selective nitroarenes hydrogenation to aniline (Fig. **19**) [190].

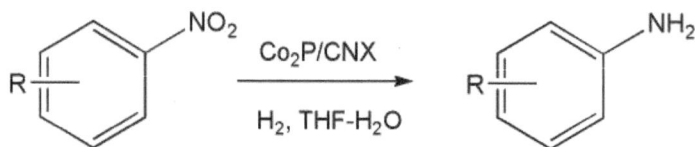

Fig. (19). The hydrogenation reaction of substituted nitroarenes [192].

The selective hydrogenation of phenol led to cyclohexanol (Fig. **20**), significant raw material in the chemical industry. Nitrogen-doped carbon prepared from MOF with encapsulated nickel-cobalt alloy nanoparticles (Co-Ni@NC) displays a high catalytic efficiency [193]. Conversion of phenol reached over 99% in the presence of the optimized catalyst (*e.g.,* the alloy composition and the temperature of the pyrolysis), a level much higher than for individual Ni and Co N-doped carbon catalysts. Comparing to Cu, Ni-Cu alloys, and Co-Cu alloys (N-doped carbon), Co-Ni@NC catalyst also displays the best catalytic performance, the single-step pathway of cyclohexanol formation preventing the formation of other intermediate by-products of phenol hydrogenation (shown in Fig 20). Its magnetic separation and recyclability represent other major advantages of this catalytic material. Not least, it can be used in the hydrogenation of a broad spectrum of phenol derivatives generating a large spectrum of compounds with important industrial applications [193].

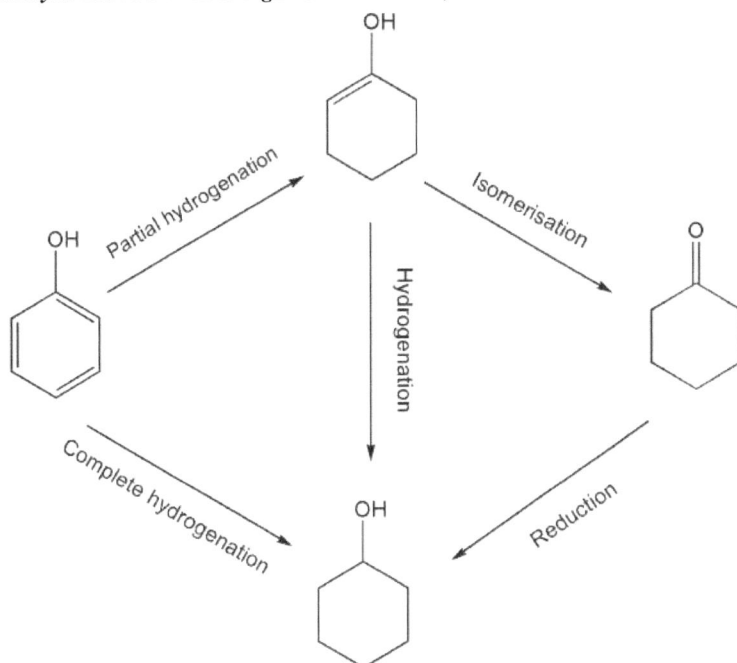

Fig. (20). Main hydrogenation pathways of phenol [193].

A vast range of heterogeneous and homogenous catalysts are synthesized to support the hydrogenation of alkenes, which is a significant process in organic chemistry. MOF-based catalysts, such as Ca-MOF (AEPF-1), UiO-66 with encapsulated Pt nanoparticles, and Zn/Co-ZIF, were found to be efficient for the hydrogenation of alkenes and alkynes, and their catalytic potential can be increased by adjusting the porosity. For instance, it was investigated that using methanol in the synthesis affects linkage of ZIF-67 in the presence of Co^{2+} resulting in initialization of phase conversion under solvothermal conditions. That conversion leads to creation of great-defined hollow Zn/Co ZIF composite, which coated with ZIF-8 increases better porous encapsulation and gas storage what enables acetylene semi-hydrogenation with excellent selectivity and activity [194]. ZIF-8 with encapsulated iridium metal NPs was also identified as a highly selective catalyst for the hydrogenation of phenylacetylene, but also by-products formed *via* excessive hydrogenation were detected. Pd–Ag catalyst encapsulated in MOF framework exhibited efficiency for the partial hydrogenation of phenylacetylene (Fig. **21**). The selectivity to hydrogenation products could be increased by inserting selective agents along with metal nanoparticles into MOFs, for instance, insertion of nitrogen compounds was reported as a decisive factor that improves selectivity and activity of semi-hydrogenation catalyst [195].

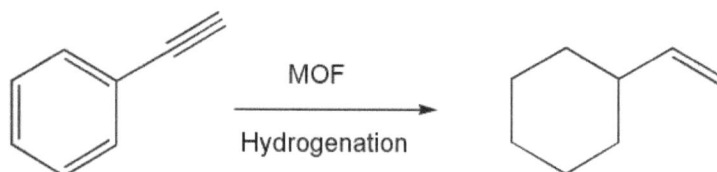

Fig. (21). Selective hydrogenation of phenylacetylene [196].

Metal-organic framework-based catalysts possess unique properties that make them a potential candidate for various hydrogenation reactions. Their tunable structure, high surface area and the presence of active centers assist in increasing activities and selectivities during the hydrogenation of chemical compounds such as cinnamaldehyde, alkenes, phenol, *etc.*

CONCLUDING REMARKS

Metal–organic frameworks (MOFs) and their composites represent promising heterogeneous catalysts for a large number of organic reactions that are relevant for the pharmaceutical, textile, and agrochemical industries. They can replace homogeneous Lewis and Brønsted acid catalysts or complex organometallic catalysts in oxidation, reduction, and C–C coupling reactions. Their capacity to accommodate high amounts of acid-base and the redox functional sites enable them to catalyze these reactions efficiently and selectively. In addition, the distinctive features, such as tunable porosity, high specific surface area, and rich surface chemistry, may allow synergistic interactions, and hence improved catalytic activities.

However, despite the significant progress that has been made to increase MOFs catalytic efficiency some further improvements to ensure real sustainable processes need to be developed. In this context, it is important to consider the chemical, thermal and mechanical stabilities of MOFs as a major issue to ensure their reusability. Nowadays, many MOFs have been moved from "delicate and moisture-sensitive" materials to nearly robust ones but more focuses on testifying the catalyzed reactions under scale-up conditions are needed. Moreover, many of the presented reactions are still carried out using either organic solvents or additives with moderate or even high toxicity. Intensive research should be focused on procedures using environmental-friendly solvents, avoiding the use of additives, and designing reactions in an atom-economically way to reduce waste.

LIST OF ABBREVIATIONS

2-pymo	2-pyrimidinolate
AEPF	Alkaline-earth polymer framework

AP	Acetophenone
APS	3-aminopropyltrialkoxysilane
AQDS - Anthraquinone-2	6-disulfonate
BBTZ - 1	4-bis(1,2,4-triazol-1-ylmethyl)benzene
bdc - 1	4-benzenedicarboxylate
Bpy - 2	2'-Bipyridine
Bpydc - 2	2-bipyridine-5,5'-dicarboxylic acid
bpyOMe - 4	4'-dimethoxy-2,2'-bipyridine
BTC - 1	3,5-benzenetricarboxylate
'BuOOH	Tert-butyl hydroperoxide
Bzz	5-benzimidazolecarboxylic acid
CAL	Cinnamalaldehyde
CAT	Catechol functional group
CHP	Cumene hydroperoxide
CM	Cumene
Co/C	N - N-doped carbon composite
COL	Cinnamyl alcohol
CyH	Cyclohexene
CZJ	Chemistry Department of Zhejiang University
DABCO - 1	4-diazabicyclo[2.2.2]octane
DEC/DMF	Diethyl carbonate/dimethylformamide
DMF	Dimethylformamide
ECS	Electrochemical synthesis
ED	Ethylenediamine
EO	Ethylene oxide
EtOAc	Ethyl acetate
EXAFS	Extended X-ray absorption fine structure
H$_4$BIPA-TC	Benzoimidephenanthroline tetracarboxylic acid
H$_4$dotpdc - 4	4''- dihydroxy-[1,1':4',1''-terphenyl]-3,3''-dicarboxylic acid
HCAL	Hydrocinnamalaldehyde
HCOL	Hydrocinnamal alcohol
HENU	Henan University
His - L	histidine
HKUST	Hong Kong University of Science and Technology
HTS	Hydrothermally synthesized solution

Im	Imidazole
In-porph	Indium-metalated porphyrin sites
Iza	4-imidazoleacrylic acid
IRMOF	Isoreticular MOF
K/A oil	Ketone-alcohol oil
L	Ligand
LED	Light-emitting diode
Ln-MOFs - Metal	organic frameworks based on lanthanides
MeOH	Methanol
MIL	Matériaux de l'Institut Lavoisier
MMO	Methane monooxygenase
MOFs - Metal	organic frameworks
NENU-MV-1 - mixed-valence {V16} cluster-based metal	organic framework
NHC	N-heterocyclic carbenes
NHPI	N-hydroxyphthalimide
NPs	Nanoparticles
NU	Northwestern University
OMS	Open Metal Sites
P104	$(EO)_{27}(PO)_{61}(EO)_{27}$
PCN	Porous coordination network
PDA	Polydopamine
Pd-NHDC	Palladium N-heterocyclic bis-carbene
PEG	Poly(ethylene glycol)
phen - 1	10-phenanthroline
PhIO	Iodosylbenzene
PhI(OAc)$_2$-	Phenyliododiacetate
PI	2-pyridyl-imine
pMMO	Particulate methane monooxygenase
PO	Propylene oxide
POMOFs	Polyoxometalate-based metal–organic frameworks
PP	2-phenyl-2-propanol
PSD	Post-synthetic deprotection
PSE	Post-synthetic exchange
PXRD	Powder X-ray diffraction

STA-12	St. Andrews microporous material number 12
TBHP	Tert-butyl hydroperoxide
tBu	Di-tert-butylaryl
TEMPO - 2	6,6-tetra-methyl-piperidine-1-oxy
THF	Tetrahydrofuran
TOF	Turnover frequency
TON	Turnover number
TPOP-1	Triazine functionalized porphyrin based porous organic polymer
TPP	Tetraphenylporphyrin
Tpy - 2	2':6',2''-terpyridine
trz - 1	2,4- triazole
UiO-66	Universitetet i Oslo
UoB-3 - Co-MOF nanostructures from coordination assembly of cobalt acetate with 4	4'-[benzene-1,4-diylbis (methylylidenenitrilo)]dibenzoic acid (H$_2$bdda)
UNLPF-10	University of Nebraska–Lincoln porous framework
XRD	X-ray diffraction
XPS	X-ray photoelectron spectroscopy
ZC - ZIF-67 and their corresponding C	N–Co nanocomposite materials
ZIF	Zeolitic Imidazolate Framework

CONSENT FOR PUBLICATION

Not applicable.

CONFLICT OF INTEREST

The author declares no conflict of interest, financial or otherwise.

ACKNOWLEDGEMENTS

Declared none.

REFERENCES

[1]   Cavka, J.H.; Jakobsen, S.; Olsbye, U.; Guillou, N.; Lamberti, C.; Bordiga, S.; Lillerud, K.P. A new zirconium inorganic building brick forming metal organic frameworks with exceptional stability. *J. Am. Chem. Soc.*, **2008**, *130*(42), 13850-13851.
[http://dx.doi.org/10.1021/ja8057953] [PMID: 18817383]

[2]   Férey, G.; Mellot-Draznieks, C.; Serre, C.; Millange, F.; Dutour, J.; Surblé, S.; Margiolaki, I. A chromium terephthalate-based solid with unusually large pore volumes and surface area. *Science (80-.)*, **2005**, *309*(5743), 2040-2042.

[3]   Chui, S.S.-Y.; Lo, S.M.-F.; Charmant, J.P.H.; Orpen, A.G.; Williams, I.D. A chemically functionalizable nanoporous material [Cu3 (TMA) 2 (H2O) 3] n. *Science (80-.)*, **1999**, *283*(5405), 1148-1150.

[4]   Jin, C.X.; Shang, H.B. Synthetic methods, properties and controlling roles of synthetic parameters of zeolite imidazole framework-8: A review. *J. Solid State Chem.*, **2021**, *297*, 122040.
[http://dx.doi.org/10.1016/j.jssc.2021.122040]

[5]   Chen, L.; Xu, Q. Metal-organic framework composites for catalysis. *Matter*, **2019**, *1*(1), 57-89.
[http://dx.doi.org/10.1016/j.matt.2019.05.018]

[6]   Furukawa, H.; Cordova, K.E.; O'Keeffe, M.; Yaghi, O.M. The chemistry and applications of metal-organic frameworks. *Science (80-.)*, **2013**, *341*(6149), 1230444-1230444.

[7]   Xuan, W.; Zhu, C.; Liu, Y.; Cui, Y. Mesoporous metal–organic framework materials. *Chem. Soc. Rev.*, **2012**, *41*(5), 1677-1695.
[http://dx.doi.org/10.1039/C1CS15196G] [PMID: 22008884]

[8]   Lu, J.; Wu, J.K.; Jiang, Y.; Tan, P.; Zhang, L.; Lei, Y.; Liu, X.Q.; Sun, L.B. Fabrication of Microporous Metal–Organic Frameworks in Uninterrupted Mesoporous Tunnels: Hierarchical Structure for Efficient Trypsin Immobilization and Stabilization. *Angew. Chem. Int. Ed.*, **2020**, *59*(16), 6428-6434.
[http://dx.doi.org/10.1002/anie.201915332] [PMID: 32017320]

[9]   Zhao, X.; Wang, Y.; Li, D.S.; Bu, X.; Feng, P. Metal–organic frameworks for separation. *Adv. Mater.*, **2018**, *30*(37), 1705189.
[http://dx.doi.org/10.1002/adma.201705189] [PMID: 29582482]

[10]   He, Y.; Chen, F.; Li, B.; Qian, G.; Zhou, W.; Chen, B. Porous metal–organic frameworks for fuel storage. *Coord. Chem. Rev.*, **2018**, *373*, 167-198.
[http://dx.doi.org/10.1016/j.ccr.2017.10.002]

[11]   Sun, Y.; Zheng, L.; Yang, Y.; Qian, X.; Fu, T.; Li, X.; Yang, Z.; Yan, H.; Cui, C.; Tan, W. Metal–organic framework nanocarriers for drug delivery in biomedical applications. *Nano-Micro Lett.*, **2020**, *12*(1), 103.
[http://dx.doi.org/10.1007/s40820-020-00423-3] [PMID: 34138099]

[12]   Wang, L.; Zheng, M.; Xie, Z. Nanoscale metal–organic frameworks for drug delivery: a conventional platform with new promise. *J. Mater. Chem. B Mater. Biol. Med.*, **2018**, *6*(5), 707-717.
[http://dx.doi.org/10.1039/C7TB02970E] [PMID: 32254257]

[13]   Kreno, L.E.; Leong, K.; Farha, O.K.; Allendorf, M.; Van Duyne, R.P.; Hupp, J.T. Metal-organic framework materials as chemical sensors. *Chem. Rev.*, **2012**, *112*(2), 1105-1125.
[http://dx.doi.org/10.1021/cr200324t] [PMID: 22070233]

[14]   Xia, T.; Song, T.; Zhang, G.; Cui, Y.; Yang, Y.; Wang, Z.; Qian, G. A terbium metal–organic framework for highly selective and sensitive luminescence sensing of Hg2+ ions in aqueous solution. *Chemistry*, **2016**, *22*(51), 18429-18434.
[http://dx.doi.org/10.1002/chem.201603531] [PMID: 27747951]

[15]   Wen, J.; Fang, Y.; Zeng, G. Progress and prospect of adsorptive removal of heavy metal ions from aqueous solution using metal–organic frameworks: A review of studies from the last decade. *Chemosphere*, **2018**, *201*, 627-643.
[http://dx.doi.org/10.1016/j.chemosphere.2018.03.047] [PMID: 29544217]

[16]   Farrusseng, D.; Aguado, S.; Pinel, C. Metal-organic frameworks: opportunities for catalysis. *Angew. Chem. Int. Ed.*, **2009**, *48*(41), 7502-7513.
[http://dx.doi.org/10.1002/anie.200806063] [PMID: 19691074]

[17]   Lee, J.; Farha, O.K.; Roberts, J.; Scheidt, K.A.; Nguyen, S.T.; Hupp, J.T. Metal–organic framework materials as catalysts. *Chem. Soc. Rev.,* **2009**, *38*(5), 1450-1459.
[http://dx.doi.org/10.1039/b807080f] [PMID: 19384447]

[18]   Corma, A.; García, H.; Llabrés i Xamena, F.X. Engineering metal organic frameworks for heterogeneous catalysis. *Chem. Rev.,* **2010**, *110*(8), 4606-4655.
[http://dx.doi.org/10.1021/cr9003924] [PMID: 20359232]

[19]   Yadav, S.; Dixit, R.; Sharma, S.; Dutta, S.; Solanki, K.; Sharma, R.K. Magnetic metal–organic framework composites: structurally advanced catalytic materials for organic transformations. *Mater. Adv,* **2021**.

[20]   Wu, C.D.; Zhao, M. Incorporation of molecular catalysts in metal–organic frameworks for highly efficient heterogeneous catalysis. *Adv. Mater.,* **2017**, *29*(14), 1605446.
[http://dx.doi.org/10.1002/adma.201605446] [PMID: 28256748]

[21]   Dhakshinamoorthy, A.; Asiri, A.M.; Garcia, H. Metal organic frameworks as versatile hosts of Au nanoparticles in heterogeneous catalysis. *ACS Catal.,* **2017**, *7*(4), 2896-2919.
[http://dx.doi.org/10.1021/acscatal.6b03386]

[22]   Wang, C.C.; Du, X.D.; Li, J.; Guo, X.X.; Wang, P.; Zhang, J. Photocatalytic Cr(VI) reduction in metal-organic frameworks: A mini-review. *Appl. Catal. B,* **2016**, *193*, 198-216.
[http://dx.doi.org/10.1016/j.apcatb.2016.04.030]

[23]   Yang, S.; Peng, L.; Bulut, S.; Queen, W.L. Recent Advances of MOFs and MOF-Derived Materials in Thermally Driven Organic Transformations. *Chemistry,* **2019**, *25*(9), 2161-2178.
[http://dx.doi.org/10.1002/chem.201803157] [PMID: 30114320]

[24]   Chen, Y.Z.; Zhang, R.; Jiao, L.; Jiang, H.L. Metal–organic framework-derived porous materials for catalysis. *Coord. Chem. Rev.,* **2018**, *362*, 1-23.
[http://dx.doi.org/10.1016/j.ccr.2018.02.008]

[25]   Konnerth, H.; Matsagar, B.M.; Chen, S.S.; Prechtl, M.H.G.; Shieh, F.K.; Wu, K.C.W. Metal-organic framework (MOF)-derived catalysts for fine chemical production. *Coord. Chem. Rev.,* **2020**, *416*(1), 213319.
[http://dx.doi.org/10.1016/j.ccr.2020.213319]

[26]   Remya, V.R.; Kurian, M. Synthesis and catalytic applications of metal–organic frameworks: a review on recent literature. *Int. Nano Lett.,* **2019**, *9*(1), 17-29.
[http://dx.doi.org/10.1007/s40089-018-0255-1]

[27]   Burtch, N.C.; Jasuja, H.; Walton, K.S. Water stability and adsorption in metal-organic frameworks. *Chem. Rev.,* **2014**, *114*(20), 10575-10612.
[http://dx.doi.org/10.1021/cr5002589] [PMID: 25264821]

[28]   Bunzen, H. *Chemical Stability of Metal-organic Frameworks for Applications in Drug Delivery.,* **2021**, *7*, 998-1007.

[29]   Yao, Q.; Bermejo Gómez, A.; Su, J.; Pascanu, V.; Yun, Y.; Zheng, H.; Chen, H.; Liu, L.; Abdelhamid, H.N.; Martín-Matute, B.; Zou, X. Series of Highly Stable Isoreticular Lanthanide Metal–Organic Frameworks with Expanding Pore Size and Tunable Luminescent Properties. *Chem. Mater.,* **2015**, *27*(15), 5332-5339.
[http://dx.doi.org/10.1021/acs.chemmater.5b01711]

[30]   Deng, X.; Li, Z.; García, H. Visible Light Induced Organic Transformations Using Metal-Organi--Frameworks (MOFs). *Chemistry,* **2017**, *23*(47), 11189-11209.
[http://dx.doi.org/10.1002/chem.201701460] [PMID: 28503763]

[31]   Bosch, M.; Zhang, M.; Zhou, H.-C. Increasing the stability of metal-organic frameworks. *Adv. Chem,* **2014**, (182327.10), 1155.

[32]   Ma, Y.J.; Jiang, X.X.; Lv, Y.K. Recent advances in preparation and applications of magnetic

framework composites. *Chem. Asian J.,* **2019**, *14*(20), 3515-3530.
[http://dx.doi.org/10.1002/asia.201901139] [PMID: 31553124]

[33]   Timofeeva, M.N.; Lukoyanov, I.A.; Panchenko, V.N.; Bhadra, B.N.; Gerasimov, E.Y.; Jhung, S.H. Effect of MAF-6 crystal size on its physicochemical and catalytic properties in the cycloaddition of CO_2 to propylene oxide. *Catalysts,* **2021**, *11*(9), 1061.
[http://dx.doi.org/10.3390/catal11091061]

[34]   Linder-Patton, O.M.; de Prinse, T.J.; Furukawa, S.; Bell, S.G.; Sumida, K.; Doonan, C.J.; Sumby, C.J. Influence of nanoscale structuralisation on the catalytic performance of ZIF-8: a cautionary surface catalysis study. *CrystEngComm,* **2018**, *20*(34), 4926-4934.
[http://dx.doi.org/10.1039/C8CE00746B]

[35]   Li, W.; Thirumurugan, A.; Barton, P.T.; Lin, Z.; Henke, S.; Yeung, H.H.M.; Wharmby, M.T.; Bithell, E.G.; Howard, C.J.; Cheetham, A.K. Mechanical tunability *via* hydrogen bonding in metal-organic frameworks with the perovskite architecture. *J. Am. Chem. Soc.,* **2014**, *136*(22), 7801-7804.
[http://dx.doi.org/10.1021/ja500618z] [PMID: 24815319]

[36]   ul Qadir, N; Said, S.A.M; Bahaidarah, HM Structural stability of metal organic frameworks in aqueous media–controlling factors and methods to improve hydrostability and hydrothermal cyclic stability. *Microporous Mesoporous Mater,* **2015**, *201*, 60-90.

[37]   Howarth, A.J.; Liu, Y.; Li, P.; Li, Z.; Wang, T.C.; Hupp, J.T.; Farha, O.K. Chemical, thermal and mechanical stabilities of metal–organic frameworks. *Nat. Rev. Mater.,* **2016**, *1*(3), 15018.
[http://dx.doi.org/10.1038/natrevmats.2015.18]

[38]   Yang, J.; Grzech, A.; Mulder, F.M.; Dingemans, T.J. Methyl modified MOF-5: a water stable hydrogen storage material. *Chem. Commun. (Camb.),* **2011**, *47*(18), 5244-5246.
[http://dx.doi.org/10.1039/c1cc11054c] [PMID: 21451855]

[39]   Li, H.; Shi, W.; Zhao, K.; Li, H.; Bing, Y.; Cheng, P. Enhanced hydrostability in Ni-doped MOF-5. *Inorg. Chem.,* **2012**, *51*(17), 9200-9207.
[http://dx.doi.org/10.1021/ic3002898] [PMID: 22900917]

[40]   Bennett, T.D.; Yue, Y.; Li, P.; Qiao, A.; Tao, H.; Greaves, N.G.; Richards, T.; Lampronti, G.I.; Redfern, S.A.T.; Blanc, F.; Farha, O.K.; Hupp, J.T.; Cheetham, A.K.; Keen, D.A. Melt-quenched glasses of metal–organic frameworks. *J. Am. Chem. Soc.,* **2016**, *138*(10), 3484-3492.
[http://dx.doi.org/10.1021/jacs.5b13220] [PMID: 26885940]

[41]   Smedskjaer, M.M.; Mauro, J.C.; Yue, Y. Prediction of glass hardness using temperature-dependent constraint theory. *Phys. Rev. Lett.,* **2010**, *105*(11), 115503.
[http://dx.doi.org/10.1103/PhysRevLett.105.115503] [PMID: 20867584]

[42]   Salunkhe, R.R.; Kaneti, Y.V.; Kim, J.; Kim, J.H.; Yamauchi, Y. Nanoarchitectures for metal–organic framework-derived nanoporous carbons toward supercapacitor applications. *Acc. Chem. Res.,* **2016**, *49*(12), 2796-2806.
[http://dx.doi.org/10.1021/acs.accounts.6b00460] [PMID: 27993000]

[43]   Mondloch, J.E.; Bury, W.; Fairen-Jimenez, D.; Kwon, S.; DeMarco, E.J.; Weston, M.H.; Sarjeant, A.A.; Nguyen, S.T.; Stair, P.C.; Snurr, R.Q.; Farha, O.K.; Hupp, J.T. Vapor-phase metalation by atomic layer deposition in a metal-organic framework. *J. Am. Chem. Soc.,* **2013**, *135*(28), 10294-10297.
[http://dx.doi.org/10.1021/ja4050828] [PMID: 23829224]

[44]   Devic, T.; Serre, C. High valence 3p and transition metal based MOFs. *Chem. Soc. Rev.,* **2014**, *43*(16), 6097-6115.
[http://dx.doi.org/10.1039/C4CS00081A] [PMID: 24947910]

[45]   Yadav, A.; Kanoo, P. Metal-Organic Frameworks as Platform for Lewis-Acid-Catalyzed Organic Transformations. *Chem. Asian J.,* **2019**, *14*(20), 3531-3551.
[http://dx.doi.org/10.1002/asia.201900876] [PMID: 31509343]

[46]   Panchenko, V.N.; Timofeeva, M.N.; Jhung, S.H. Acid-base properties and catalytic activity of metal-organic frameworks: A view from spectroscopic and semiempirical methods. *Catal. Rev., Sci. Eng.,* **2016**, *58*(2), 209-307.
[http://dx.doi.org/10.1080/01614940.2016.1128193]

[47]   Burtch, N.C.; Heinen, J.; Bennett, T.D.; Dubbeldam, D.; Allendorf, M.D. Mechanical properties in metal–organic frameworks: emerging opportunities and challenges for device functionality and technological applications. *Adv. Mater.,* **2018**, *30*(37), 1704124.
[http://dx.doi.org/10.1002/adma.201704124] [PMID: 29149545]

[48]   Yang, Q.; Xu, Q.; Jiang, H.L. Metal–organic frameworks meet metal nanoparticles: synergistic effect for enhanced catalysis. *Chem. Soc. Rev.,* **2017**, *46*(15), 4774-4808.
[http://dx.doi.org/10.1039/C6CS00724D] [PMID: 28621344]

[49]   Falcaro, P.; Ricco, R.; Yazdi, A.; Imaz, I.; Furukawa, S.; Maspoch, D.; Ameloot, R.; Evans, J.D.; Doonan, C.J. Application of metal and metal oxide nanoparticles@MOFs. *Coord. Chem. Rev.,* **2016**, *307*, 237-254.
[http://dx.doi.org/10.1016/j.ccr.2015.08.002]

[50]   Du, D.Y.; Qin, J.S.; Li, S.L.; Su, Z.M.; Lan, Y.Q. Recent advances in porous polyoxometalate-based metal–organic framework materials. *Chem. Soc. Rev.,* **2014**, *43*(13), 4615-4632.
[http://dx.doi.org/10.1039/C3CS60404G] [PMID: 24676127]

[51]   Fujie, K.; Otsubo, K.; Ikeda, R.; Yamada, T.; Kitagawa, H. Low temperature ionic conductor: ionic liquid incorporated within a metal–organic framework. *Chem. Sci. (Camb.),* **2015**, *6*(7), 4306-4310.
[http://dx.doi.org/10.1039/C5SC01398D] [PMID: 29218200]

[52]   Zhu, Q.L.; Xu, Q. Metal–organic framework composites. *Chem. Soc. Rev.,* **2014**, *43*(16), 5468-5512.
[http://dx.doi.org/10.1039/C3CS60472A] [PMID: 24638055]

[53]   Hu, M.L.; Safarifard, V.; Doustkhah, E.; Rostamnia, S.; Morsali, A.; Nouruzi, N.; Beheshti, S.; Akhbari, K. Taking organic reactions over metal-organic frameworks as heterogeneous catalysis. *Microporous Mesoporous Mater.,* **2018**, *256*, 111-127.
[http://dx.doi.org/10.1016/j.micromeso.2017.07.057]

[54]   Huang, Y.B.; Liang, J.; Wang, X.S.; Cao, R. Multifunctional metal–organic framework catalysts: synergistic catalysis and tandem reactions. *Chem. Soc. Rev.,* **2017**, *46*(1), 126-157.
[http://dx.doi.org/10.1039/C6CS00250A] [PMID: 27841411]

[55]   Dhakshinamoorthy, A.; Alvaro, M.; Garcia, H. Metal organic frameworks as solid acid catalysts for acetalization of aldehydes with methanol. *Adv. Synth. Catal.,* **2010**, *352*(17), 3022-3030.
[http://dx.doi.org/10.1002/adsc.201000537]

[56]   Phan, N.T.S.; Le, K.K.A.; Phan, T.D. MOF-5 as an efficient heterogeneous catalyst for Friedel–Crafts alkylation reactions. *Appl. Catal. A Gen.,* **2010**, *382*(2), 246-253.
[http://dx.doi.org/10.1016/j.apcata.2010.04.053]

[57]   Falkowski, J.M.; Wang, C.; Liu, S.; Lin, W. Actuation of asymmetric cyclopropanation catalysts: reversible single-crystal to single-crystal reduction of metal-organic frameworks. *Angew. Chem. Int. Ed.,* **2011**, *50*(37), 8674-8678.
[http://dx.doi.org/10.1002/anie.201104086] [PMID: 21826773]

[58]   Nguyen, L.T.L.; Nguyen, T.T.; Nguyen, K.D.; Phan, N.T.S. Metal–organic framework MOF-199 as an efficient heterogeneous catalyst for the aza-Michael reaction. *Appl. Catal. A Gen.,* **2012**, *425-426*, 44-52.
[http://dx.doi.org/10.1016/j.apcata.2012.02.045]

[59]   Jing, X.; He, C.; Dong, D.; Yang, L.; Duan, C. Homochiral crystallization of metal-organic silver frameworks: asymmetric [3+2] cycloaddition of an azomethine ylide. *Angew. Chem. Int. Ed.,* **2012**, *51*(40), 10127-10131.
[http://dx.doi.org/10.1002/anie.201204530] [PMID: 22962040]

[60]   de la Iglesia, Ó.; Sorribas, S.; Almendro, E.; Zornoza, B.; Téllez, C.; Coronas, J. Metal-organic framework MIL-101(Cr) based mixed matrix membranes for esterification of ethanol and acetic acid in a membrane reactor. *Renew. Energy,* **2016**, *88*, 12-19.
[http://dx.doi.org/10.1016/j.renene.2015.11.025]

[61]   Jiang, J.; Gándara, F.; Zhang, Y.B.; Na, K.; Yaghi, O.M.; Klemperer, W.G. Superacidity in sulfated metal-organic framework-808. *J. Am. Chem. Soc.,* **2014**, *136*(37), 12844-12847.
[http://dx.doi.org/10.1021/ja507119n] [PMID: 25157587]

[62]   Zhou, Y.X.; Chen, Y.Z.; Hu, Y.; Huang, G.; Yu, S.H.; Jiang, H.L. MIL-101-SO3H: a highly efficient Brønsted acid catalyst for heterogeneous alcoholysis of epoxides under ambient conditions. *Chemistry,* **2014**, *20*(46), 14976-14980.
[http://dx.doi.org/10.1002/chem.201404104] [PMID: 25291973]

[63]   Lun, D.J.; Waterhouse, G.I.N.; Telfer, S.G. A general thermolabile protecting group strategy for organocatalytic metal-organic frameworks. *J. Am. Chem. Soc.,* **2011**, *133*(15), 5806-5809.
[http://dx.doi.org/10.1021/ja202223d] [PMID: 21443248]

[64]   Beheshti, S.; Morsali, A. Post-modified anionic nano-porous metal–organic framework as a novel catalyst for solvent-free Michael addition reactions. *RSC Advances,* **2014**, *4*(70), 37036-37040.
[http://dx.doi.org/10.1039/C4RA05226A]

[65]   Jin, Y.; Shi, J.; Zhang, F.; Zhong, Y.; Zhu, W. Synthesis of sulfonic acid-functionalized MIL-101 for acetalization of aldehydes with diols. *J. Mol. Catal. Chem.,* **2014**, *383-384*, 167-171.
[http://dx.doi.org/10.1016/j.molcata.2013.12.005]

[66]   Li, X.; Goh, T.W.; Li, L.; Xiao, C.; Guo, Z.; Zeng, X.C.; Huang, W. Controlling Catalytic Properties of Pd Nanoclusters through Their Chemical Environment at the Atomic Level Using Isoreticular Metal–Organic Frameworks. *ACS Catal.,* **2016**, *6*(6), 3461-3468.
[http://dx.doi.org/10.1021/acscatal.6b00397]

[67]   Chen, L.; Chen, H.; Luque, R.; Li, Y. Metal–organic framework encapsulated Pd nanoparticles: towards advanced heterogeneous catalysts. *Chem. Sci. (Camb.),* **2014**, *5*(10), 3708-3714.
[http://dx.doi.org/10.1039/C4SC01847H]

[68]   Chen, Y.Z.; Wang, Z.U.; Wang, H.; Lu, J.; Yu, S.H.; Jiang, H.L. Singlet oxygen-engaged selective photo-oxidation over Pt nanocrystals/porphyrinic MOF: the roles of photothermal effect and Pt electronic state. *J. Am. Chem. Soc.,* **2017**, *139*(5), 2035-2044.
[http://dx.doi.org/10.1021/jacs.6b12074] [PMID: 28103670]

[69]   Zhao, M.; Deng, K.; He, L.; Liu, Y.; Li, G.; Zhao, H.; Tang, Z. Core-shell palladium nanoparticle@metal-organic frameworks as multifunctional catalysts for cascade reactions. *J. Am. Chem. Soc.,* **2014**, *136*(5), 1738-1741.
[http://dx.doi.org/10.1021/ja411468e] [PMID: 24437922]

[70]   Toyao, T.; Fujiwaki, M.; Horiuchi, Y.; Matsuoka, M. Application of an amino-functionalised metal–organic framework: an approach to a one-pot acid–base reaction. *RSC Advances,* **2013**, *3*(44), 21582-21587.
[http://dx.doi.org/10.1039/c3ra44701d]

[71]   Srirambalaji, R.; Hong, S.; Natarajan, R.; Yoon, M.; Hota, R.; Kim, Y.; Ho Ko, Y.; Kim, K. Tandem catalysis with a bifunctional site-isolated Lewis acid–Brønsted base metal–organic framework, NH2-MIL-101(Al). *Chem. Commun. (Camb.),* **2012**, *48*(95), 11650-11652.
[http://dx.doi.org/10.1039/c2cc36678a] [PMID: 23104231]

[72]   Mohammadian, R.; Karimi Alavijeh, M.; Kamyar, N.; Amini, M.M.; Shaabani, A. Metal–organic frameworks as a new platform for molecular oxygen and aerobic oxidation of organic substrates: Recent advances. *Polyhedron,* **2018**, *156*, 174-187.
[http://dx.doi.org/10.1016/j.poly.2018.09.029]

[73]   Nodzewska, A.; Wadolowska, A.; Watkinson, M. Recent advances in the catalytic oxidation of alkene

and alkane substrates using immobilized manganese complexes with nitrogen containing ligands. *Coord. Chem. Rev.,* **2019**, *382*, 181-216.
[http://dx.doi.org/10.1016/j.ccr.2018.12.004]

[74] Olivo, G.; Cussó, O.; Borrell, M.; Costas, M. Oxidation of alkane and alkene moieties with biologically inspired nonheme iron catalysts and hydrogen peroxide: from free radicals to stereoselective transformations. *J. Biol. Inorg. Chem.,* **2017**, *22*(2-3), 425-452.
[http://dx.doi.org/10.1007/s00775-016-1434-z] [PMID: 28124122]

[75] Barona, M.; Ahn, S.; Morris, W.; Hoover, W.; Notestein, J.M.; Farha, O.K.; Snurr, R.Q. Computational Predictions and Experimental Validation of Alkane Oxidative Dehydrogenation by Fe$_2$ M MOF Nodes. *ACS Catal.,* **2020**, *10*(2), 1460-1469.
[http://dx.doi.org/10.1021/acscatal.9b03932]

[76] Carabineiro, S.A.C. Supported Gold Nanoparticles as Catalysts for the Oxidation of Alcohols and Alkanes. *Front Chem.,* **2019**, *7*, 702.
[http://dx.doi.org/10.3389/fchem.2019.00702] [PMID: 31750289]

[77] Li, D.; Ma, X.; Wang, Q.; Ma, P.; Niu, J.; Wang, J. Copper-Containing Polyoxometalate-Based Metal–Organic Frameworks as Highly Efficient Heterogeneous Catalysts toward Selective Oxidation of Alkylbenzenes. *Inorg. Chem.,* **2019**, *58*(23), 15832-15840.
[http://dx.doi.org/10.1021/acs.inorgchem.9b02189] [PMID: 31721567]

[78] Zakoshansky, V.M. The Cumene Process for Phenol—Acetone Production. *Petrol. Chem.,* **2007**, *47*(4), 307-307.
[http://dx.doi.org/10.1134/S0965544107040135]

[79] Tsuji, J. Development of New Propylene Oxide Process. **2006**.

[80] Nowacka, A.; Briantais, P.; Prestipino, C.; Llabrés i Xamena, F.X. Selective Aerobic Oxidation of Cumene to Cumene Hydroperoxide over Mono- and Bimetallic Trimesate Metal–Organic Frameworks Prepared by a Facile "Green" Aqueous Synthesis. *ACS Sustain. Chem.& Eng.,* **2019**, *7*(8), 7708-7715.
[http://dx.doi.org/10.1021/acssuschemeng.8b06472]

[81] Ma, Y.; Peng, H.; Liu, J.; Wang, Y.; Hao, X.; Feng, X.; Khan, S.U.; Tan, H.; Li, Y. Polyoxometalate-based metal-organic frameworks for selective oxidation of aryl alkenes to aldehydes. *Inorg. Chem.,* **2018**, *57*(7), 4109-4116.
[http://dx.doi.org/10.1021/acs.inorgchem.8b00282] [PMID: 29533068]

[82] Zadehahmadi, F.; Tangestaninejad, S.; Moghadam, M.; Mirkhani, V.; Mohammadpoor-Baltork, I.; Khosropour, A.R.; Kardanpour, R. Synthesis and characterization of mangenese(III) porphyrin supported on imidazole modified chloromethylated MIL-101(Cr): A heterogeneous and reusable catalyst for oxidation of hydrocarbons with sodium periodate. *J. Solid State Chem.,* **2014**, *218*, 56-63.
[http://dx.doi.org/10.1016/j.jssc.2014.05.016]

[83] Dhakshinamoorthy, A.; Alvaro, M.; Garcia, H. Atmospheric-pressure, liquid-phase, selective aerobic oxidation of alkanes catalysed by metal-organic frameworks. *Chemistry,* **2011**, *17*(22), 6256-6262.
[http://dx.doi.org/10.1002/chem.201002664] [PMID: 21495095]

[84] Ramos-Fernandez, E.V.; Garcia-Domingos, M.; Juan-Alcañiz, J.; Gascon, J.; Kapteijn, F. MOFs meet monoliths: Hierarchical structuring metal organic framework catalysts. *Appl. Catal. A Gen.,* **2011**, *391*(1-2), 261-267.
[http://dx.doi.org/10.1016/j.apcata.2010.05.019]

[85] Xie, M.H.; Yang, X.L.; He, Y.; Zhang, J.; Chen, B.; Wu, C.D. Highly efficient C-H oxidative activation by a porous Mn(III)-porphyrin metal-organic framework under mild conditions. *Chemistry,* **2013**, *19*(42), 14316-14321.
[http://dx.doi.org/10.1002/chem.201302025] [PMID: 24038207]

[86] Zheng, J.; Ye, J.; Ortuño, M.A.; Fulton, J.L.; Gutiérrez, O.Y.; Camaioni, D.M.; Motkuri, R.K.; Li, Z.; Webber, T.E.; Mehdi, B.L.; Browning, N.D.; Penn, R.L.; Farha, O.K.; Hupp, J.T.; Truhlar, D.G.; Cramer, C.J.; Lercher, J.A. Selective Methane Oxidation to Methanol on Cu-Oxo Dimers Stabilized by

Zirconia Nodes of an NU-1000 Metal–Organic Framework. *J. Am. Chem. Soc.*, **2019**, *141*(23), 9292-9304.
[http://dx.doi.org/10.1021/jacs.9b02902] [PMID: 31117650]

[87]   Sun, Z.; Li, G.; Liu, L.; Liu, H. Au nanoparticles supported on Cr-based metal-organic framework as bimetallic catalyst for selective oxidation of cyclohexane to cyclohexanone and cyclohexanol. *Catal. Commun.*, **2012**, *27*, 200-205.
[http://dx.doi.org/10.1016/j.catcom.2012.07.017]

[88]   Long, J.; Liu, H.; Wu, S.; Liao, S.; Li, Y. Selective oxidation of saturated hydrocarbons using Au-Pd alloy nanoparticles supported on metal-organic frameworks. *ACS Catal.*, **2013**, *3*(4), 647-654.
[http://dx.doi.org/10.1021/cs300754k]

[89]   Wu, Y.; Wang, W.; Liu, L.; Zhu, S.; Wang, X.; Hu, E.; Hu, K. Novel Synthesis of Cu-Schiff Base Complex@Metal-Organic Framework MIL-101 *via* a Mild Method: A Comparative Study for Rapid Catalytic Effects. *ChemistryOpen*, **2019**, *8*(3), 333-338.
[http://dx.doi.org/10.1002/open.201900032] [PMID: 30976473]

[90]   Yu, G.; Sun, J.; Muhammad, F.; Wang, P.; Zhu, G. Cobalt-based metal organic framework as precursor to achieve superior catalytic activity for aerobic epoxidation of styrene. *RSC Advances*, **2014**, *4*(73), 38804-38811.
[http://dx.doi.org/10.1039/C4RA03746D]

[91]   Wang, J.C.; Ding, F.W.; Ma, J.P.; Liu, Q.K.; Cheng, J.Y.; Dong, Y.B. Co(II)-MOF: A Highly Efficient Organic Oxidation Catalyst with Open Metal Sites. *Inorg. Chem.*, **2015**, *54*(22), 10865-10872.
[http://dx.doi.org/10.1021/acs.inorgchem.5b01938] [PMID: 26497909]

[92]   Osadchii, D.Y.; Olivos-Suarez, A.I.; Szécsényi, Á.; Li, G.; Nasalevich, M.A.; Dugulan, I.A.; Crespo, P.S.; Hensen, E.J.M.; Veber, S.L.; Fedin, M.V.; Sankar, G.; Pidko, E.A.; Gascon, J. Isolated fe sites in metal organic frameworks catalyze the direct conversion of methane to methanol. *ACS Catal.*, **2018**, *8*(6), 5542-5548.
[http://dx.doi.org/10.1021/acscatal.8b00505]

[93]   Baek, J.; Rungtaweevoranit, B.; Pei, X.; Park, M.; Fakra, S.C.; Liu, Y.S.; Matheu, R.; Alshmimri, S.A.; Alshehri, S.; Trickett, C.A.; Somorjai, G.A.; Yaghi, O.M. Bioinspired Metal–Organic Framework Catalysts for Selective Methane Oxidation to Methanol. *J. Am. Chem. Soc.*, **2018**, *140*(51), 18208-18216.
[http://dx.doi.org/10.1021/jacs.8b11525] [PMID: 30525562]

[94]   Wang, S.; Sun, Z.; Zou, X.; Zhang, Z.; Fu, G.; Li, L.; Zhang, X.; Luo, F. Enhancing catalytic aerobic oxidation performance of cyclohexane *via* size regulation of mixed-valence V_{16} cluster-based metal–organic frameworks. *New J. Chem.*, **2019**, *43*(36), 14527-14535.
[http://dx.doi.org/10.1039/C9NJ03614H]

[95]   Xiao, D.J.; Oktawiec, J.; Milner, P.J.; Long, J.R. Pore Environment Effects on Catalytic Cyclohexane Oxidation in Expanded Fe_2(dobdc) Analogues. *J. Am. Chem. Soc.*, **2016**, *138*(43), 14371-14379.
[http://dx.doi.org/10.1021/jacs.6b08417] [PMID: 27704846]

[96]   Raupp, Y.S.; Yildiz, C.; Kleist, W.; Meier, M.A.R. Aerobic oxidation of α-pinene catalyzed by homogeneous and MOF-based Mn catalysts. *Appl. Catal. A Gen.*, **2017**, *546*, 1-6.
[http://dx.doi.org/10.1016/j.apcata.2017.07.047]

[97]   Sun, D.; Sun, F.; Deng, X.; Li, Z. Mixed-Metal Strategy on Metal–Organic Frameworks (MOFs) for Functionalities Expansion: Co Substitution Induces Aerobic Oxidation of Cyclohexene over Inactive Ni-MOF-74. *Inorg. Chem.*, **2015**, *54*(17), 8639-8643.
[http://dx.doi.org/10.1021/acs.inorgchem.5b01278] [PMID: 26288128]

[98]   Tran, Y.B.N.; Nguyen, P.T.K. Lanthanide metal–organic frameworks for catalytic oxidation of olefins. *New J. Chem.*, **2021**, *45*(4), 2090-2102.
[http://dx.doi.org/10.1039/D0NJ05685E]

[99] Wang, J.; Yang, M.; Dong, W.; Jin, Z.; Tang, J.; Fan, S.; Lu, Y.; Wang, G. Co(II) complexes loaded into metal–organic frameworks as efficient heterogeneous catalysts for aerobic epoxidation of olefins. *Catal. Sci. Technol.,* **2016**, *6*(1), 161-168.
[http://dx.doi.org/10.1039/C5CY01099C]

[100] Bhattacharjee, S.; Yang, D.A.; Ahn, W.S. A new heterogeneous catalyst for epoxidation of alkenes *via* one-step post-functionalization of IRMOF-3 with a manganese(ii) acetylacetonate complex. *Chem. Commun. (Camb.),* **2011**, *47*(12), 3637-3639.
[http://dx.doi.org/10.1039/c1cc00069a] [PMID: 21290069]

[101] Saedi, Z.; Tangestaninejad, S.; Moghadam, M.; Mirkhani, V.; Mohammadpoor-Baltork, I. MIL-101 metal–organic framework: A highly efficient heterogeneous catalyst for oxidative cleavage of alkenes with H2O2. *Catal. Commun.,* **2012**, *17*, 18-22.
[http://dx.doi.org/10.1016/j.catcom.2011.10.005]

[102] Qi, Y.; Luan, Y.; Yu, J.; Peng, X.; Wang, G. Nanoscaled copper metal-organic framework (MOF) based on carboxylate ligands as an efficient heterogeneous catalyst for aerobic epoxidation of olefins and oxidation of benzylic and allylic alcohols. *Chemistry,* **2015**, *21*(4), 1589-1597.
[http://dx.doi.org/10.1002/chem.201405685] [PMID: 25430789]

[103] Afzali, N.; Tangestaninejad, S.; Moghadam, M.; Mirkhani, V.; Mechler, A.; Mohammadpoor-Baltork, I.; Kardanpour, R.; Zadehahmadi, F. Oxidation reactions catalysed by molybdenum(VI) complexes grafted on UiO-66 metal-organic framework as an elegant nanoreactor. *Appl. Organomet. Chem.,* **2018**, *32*(1), e3958.
[http://dx.doi.org/10.1002/aoc.3958]

[104] Cavani, F. Liquid Phase Oxidation *via* Heterogeneous Catalysis. Organic Synthesis and Industrial Applications. In: *Angew. Chemie Int*; Mario, G. Clerici; Oxana, A. Kholdeeva, Eds.; , **2014**; 53, pp. (30)7707-7707.

[105] Sun, D.; Li, Z. Robust Ti- and Zr-Based Metal-Organic Frameworks for Photocatalysis. *Chin. J. Chem.,* **2017**, *35*(2), 135-147.
[http://dx.doi.org/10.1002/cjoc.201600647]

[106] Zhu, J.; Li, P.Z.; Guo, W.; Zhao, Y.; Zou, R. Titanium-based metal–organic frameworks for photocatalytic applications. *Coord. Chem. Rev.,* **2018**, *359*, 80-101.
[http://dx.doi.org/10.1016/j.ccr.2017.12.013]

[107] Maksimchuk, N.; Lee, J.; Ayupov, A.; Chang, J.S.; Kholdeeva, O. Cyclohexene Oxidation with H2O2 over Metal-Organic Framework MIL-125(Ti): The Effect of Protons on Reactivity. *Catalysts,* **2019**, *9*(4), 324.
[http://dx.doi.org/10.3390/catal9040324]

[108] Carson, F.; Agrawal, S.; Gustafsson, M.; Bartoszewicz, A.; Moraga, F.; Zou, X.; Martín-Matute, B. Ruthenium complexation in an aluminium metal-organic framework and its application in alcohol oxidation catalysis. *Chemistry,* **2012**, *18*(48), 15337-15344.
[http://dx.doi.org/10.1002/chem.201200885] [PMID: 23042715]

[109] Layek, K.; Maheswaran, H.; Arundhathi, R.; Kantam, M.L.; Bhargava, S.K. Nanocrystalline Magnesium Oxide Stabilized Palladium(0): An Efficient Reusable Catalyst for Room Temperature Selective Aerobic Oxidation of Alcohols. *Adv. Synth. Catal.,* **2011**, *353*(4), 606-616.
[http://dx.doi.org/10.1002/adsc.201000591]

[110] Liu, P.; Guan, Y.; Santen, R.A.; Li, C.; Hensen, E.J.M. Aerobic oxidation of alcohols over hydrotalcite-supported gold nanoparticles: the promotional effect of transition metal cations. *Chem. Commun. (Camb.),* **2011**, *47*(41), 11540-11542.
[http://dx.doi.org/10.1039/c1cc15148g] [PMID: 21952125]

[111] Zhang, Y.; Wang, J.; Zhang, T. Novel Ca-doped CePO4 supported ruthenium catalyst with superior catalytic performance for aerobic oxidation of alcohols. *Chem. Commun. (Camb.),* **2011**, *47*(18), 5307-5309.

[http://dx.doi.org/10.1039/c1cc10626k] [PMID: 21461414]

[112] Dhakshinamoorthy, A.; Asiri, A.M.; Garcia, H. Tuneable nature of metal organic frameworks as heterogeneous solid catalysts for alcohol oxidation. *Chem. Commun. (Camb.)*, **2017**, *53*(79), 10851-10869.
[http://dx.doi.org/10.1039/C7CC05927B] [PMID: 28936534]

[113] Ding, M.; Cai, X.; Jiang, H.L. Improving MOF stability: approaches and applications. *Chem. Sci. (Camb.)*, **2019**, *10*(44), 10209-10230.
[http://dx.doi.org/10.1039/C9SC03916C] [PMID: 32206247]

[114] Dhakshinamoorthy, A.; Alvaro, M.; Garcia, H. Aerobic oxidation of benzylic alcohols catalyzed by metal-organic frameworks assisted by TEMPO. *ACS Catal.*, **2011**, *1*(1), 48-53.
[http://dx.doi.org/10.1021/cs1000703]

[115] Xue, Z.; Zhang, J.; Peng, L.; Han, B.; Mu, T.; Li, J.; Yang, G. Poly(ethylene glycol) stabilized mesoporous metal-organic framework nanocrystals: efficient and durable catalysts for the oxidation of benzyl alcohol. *ChemPhysChem*, **2014**, *15*(1), 85-89.
[http://dx.doi.org/10.1002/cphc.201300809] [PMID: 24285509]

[116] Wu, S.; Chen, L.; Yin, B.; Li, Y. "Click" post-functionalization of a metal–organic framework for engineering active single-site heterogeneous Ru(III) catalysts. *Chem. Commun. (Camb.)*, **2015**, *51*(48), 9884-9887.
[http://dx.doi.org/10.1039/C5CC02741A] [PMID: 25995141]

[117] Fei, H.; Shin, J.; Meng, Y.S.; Adelhardt, M.; Sutter, J.; Meyer, K.; Cohen, S.M. Reusable oxidation catalysis using metal-monocatecholato species in a robust metal-organic framework. *J. Am. Chem. Soc.*, **2014**, *136*(13), 4965-4973.
[http://dx.doi.org/10.1021/ja411627z] [PMID: 24597832]

[118] Proch, S.; Herrmannsdörfer, J.; Kempe, R.; Kern, C.; Jess, A.; Seyfarth, L.; Senker, J. Pt@MOF-177: synthesis, room-temperature hydrogen storage and oxidation catalysis. *Chemistry*, **2008**, *14*(27), 8204-8212.
[http://dx.doi.org/10.1002/chem.200801043] [PMID: 18666269]

[119] Chen, G.; Wu, S.; Liu, H.; Jiang, H.; Li, Y. Palladium supported on an acidic metal–organic framework as an efficient catalyst in selective aerobic oxidation of alcohols. *Green Chem.*, **2013**, *15*(1), 230-235.
[http://dx.doi.org/10.1039/C2GC36618E]

[120] Taher, A.; Kim, D.W.; Lee, I.M. Highly efficient metal organic framework (MOF)-based copper catalysts for the base-free aerobic oxidation of various alcohols. *RSC Advances*, **2017**, *7*(29), 17806-17812.
[http://dx.doi.org/10.1039/C6RA28743C]

[121] Esken, D.; Turner, S.; Lebedev, O.I.; Van Tendeloo, G.; Fischer, R.A. Au@ZIFs: Stabilization and Encapsulation of Cavity-Size Matching Gold Clusters inside Functionalized Zeolite Imidazolate Frameworks, ZIFs. *Chem. Mater.*, **2010**, *22*(23), 6393-6401.
[http://dx.doi.org/10.1021/cm102529c]

[122] Farrokhi, A.; Jafarpour, M.; Alipour, M. Highly selective and efficient oxidation of benzylic alcohols with sulfate radical over metal-organic frameworks. *J. Organomet. Chem.*, **2019**, *903*, 120995.
[http://dx.doi.org/10.1016/j.jorganchem.2019.120995]

[123] Llabres, I.; Xamena, F.; Abad, A.; Corma, A.; Garcia, H. MOFs as catalysts: Activity, reusability and shape-selectivity of a Pd-containing MOF. *J. Catal.*, **2007**, *250*(2).

[124] Liu, H.; Liu, Y.; Li, Y.; Tang, Z.; Jiang, H. Metal−Organic Framework Supported Gold Nanoparticles as a Highly Active Heterogeneous Catalyst for Aerobic Oxidation of Alcohols. *J. Phys. Chem. C*, **2010**, *114*(31), 13362-13369.
[http://dx.doi.org/10.1021/jp105666f]

[125] Bai, C.; Li, A.; Yao, X.; Liu, H.; Li, Y. Efficient and selective aerobic oxidation of alcohols catalysed by MOF-derived Co catalysts. *Green Chem.,* **2016**, *18*(4), 1061-1069.
[http://dx.doi.org/10.1039/C5GC02082D]

[126] Aryanejad, S.; Bagherzade, G.; Moudi, M. Design and development of novel Co-MOF nanostructures as an excellent catalyst for alcohol oxidation and Henry reaction, with a potential antibacterial activity. *Appl. Organomet. Chem.,* **2019**, *33*(6), e4820.
[http://dx.doi.org/10.1002/aoc.4820]

[127] Peng, L.; Wu, S.; Yang, X.; Hu, J.; Fu, X.; Li, M.; Bai, L.; Huo, Q.; Guan, J. Oxidation of benzyl alcohol over metal organic frameworks M-BTC (M = Co, Cu, Fe). *New J. Chem.,* **2017**, *41*(8), 2891-2894.
[http://dx.doi.org/10.1039/C7NJ00588A]

[128] Lin, K.S.; Adhikari, A.K.; Ku, C.N.; Chiang, C.L.; Kuo, H. Synthesis and characterization of porous HKUST-1 metal organic frameworks for hydrogen storage. *Int. J. Hydrogen Energy,* **2012**, *37*(18), 13865-13871.
[http://dx.doi.org/10.1016/j.ijhydene.2012.04.105]

[129] Chui, S.S.Y.; Lo, S.M.F.; Charmant, J.P.H.; Orpen, A.G.; Williams, I.D. A chemically functionalizable nanoporous material [Cu3(TMA)2 (H2O)3](n). *Science (80-.),* **1999**, *283*(5405), 1148-1150.

[130] Winarta, J.; Shan, B.; Mcintyre, S.M.; Ye, L.; Wang, C.; Liu, J.; Mu, B. A Decade of UiO-66 Research: A Historic Review of Dynamic Structure, Synthesis Mechanisms, and Characterization Techniques of an Archetypal Metal–Organic Framework. *Cryst. Growth Des.,* **2020**, *20*(2), 1347-1362.
[http://dx.doi.org/10.1021/acs.cgd.9b00955]

[131] Saxena, A.; Kumar, A.; Mozumdar, S. Ni-nanoparticles: An efficient green catalyst for chemo-selective oxidative coupling of thiols. *J. Mol. Catal. Chem.,* **2007**, *269*(1-2), 35-40.
[http://dx.doi.org/10.1016/j.molcata.2006.12.042]

[132] Xiang, H.P.; Rong, M.Z.; Zhang, M.Q. Self-healing, Reshaping, and Recycling of Vulcanized Chloroprene Rubber: A Case Study of Multitask Cyclic Utilization of Cross-linked Polymer. *ACS Sustain. Chem.& Eng.,* **2016**, *4*(5), 2715-2724.
[http://dx.doi.org/10.1021/acssuschemeng.6b00224]

[133] Ogawa, A.; Nishiyama, Y.; Kambe, N.; Murai, S.; Sonoda, N. Selenium, carbon monoxide and water as a new reduction system: Reductive cleavage of disulfides and diselenides to thiols and selenols. *Tetrahedron Lett.,* **1987**, *28*(28), 3271-3274.
[http://dx.doi.org/10.1016/S0040-4039(00)95490-X]

[134] Dhakshinamoorthy, A.; Alvaro, M.; Garcia, H. Aerobic oxidation of thiols to disulfides using iron metal–organic frameworks as solid redox catalysts. *Chem. Commun. (Camb.),* **2010**, *46*(35), 6476-6478.
[http://dx.doi.org/10.1039/c0cc02210a] [PMID: 20714572]

[135] Dalapati, R.; Sakthivel, B.; Ghosalya, M.K.; Dhakshinamoorthy, A.; Biswas, S. A cerium-based metal–organic framework having inherent oxidase-like activity applicable for colorimetric sensing of biothiols and aerobic oxidation of thiols. *CrystEngComm,* **2017**, *19*(39), 5915-5925.
[http://dx.doi.org/10.1039/C7CE01053B]

[136] Song, J.; Luo, Z.; Britt, D.K.; Furukawa, H.; Yaghi, O.M.; Hardcastle, K.I.; Hill, C.L. A multiunit catalyst with synergistic stability and reactivity: a polyoxometalate-metal organic framework for aerobic decontamination. *J. Am. Chem. Soc.,* **2011**, *133*(42), 16839-16846.
[http://dx.doi.org/10.1021/ja203695h] [PMID: 21913693]

[137] Dhakshinamoorthy, A.; Alvaro, M.; Hwang, Y.K.; Seo, Y.K.; Corma, A.; Garcia, H. Intracrystalline diffusion in Metal Organic Framework during heterogeneous catalysis: Influence of particle size on the activity of MIL-100 (Fe) for oxidation reactions. *Dalton Trans.,* **2011**, *40*(40), 10719-10724.
[http://dx.doi.org/10.1039/c1dt10826c] [PMID: 21879084]

[138]　Caliari, P.C.; Pacheco, M.J.; Ciríaco, L.F.; Lopes, A.M.C. Anodic oxidation of sulfide to sulfate: Effect of the current density on the process kinetics. *J. Braz. Chem. Soc.,* **2017**, *28*(4), 557-566.

[139]　Jiang, G.; Sun, J.; Sharma, K.R.; Yuan, Z. Corrosion and odor management in sewer systems. *Curr. Opin. Biotechnol.,* **2015**, *33*, 192-197.
[http://dx.doi.org/10.1016/j.copbio.2015.03.007] [PMID: 25827114]

[140]　Altaş, L.; Büyükgüngör, H. Sulfide removal in petroleum refinery wastewater by chemical precipitation. *J. Hazard. Mater.,* **2008**, *153*(1-2), 462-469.
[http://dx.doi.org/10.1016/j.jhazmat.2007.08.076] [PMID: 17913353]

[141]　Haddadi, H.; Hafshejani, S.M.; Farsani, M.R. Selective and Reusable Oxidation of Sulfides to Sulfoxides with Hydrogen Peroxide Catalyzed by Organic–Inorganic Polyoxometalate-Based Frameworks. *Catal. Lett.,* **2015**, *145*(11), 1984-1990.
[http://dx.doi.org/10.1007/s10562-015-1595-3]

[142]　Zhao, W.; Yang, C.; Huang, J.; Jin, X.; Deng, Y.; Wang, L.; Su, F.; Xie, H.; Wong, P.K.; Ye, L. Selective aerobic oxidation of sulfides to sulfoxides in water under blue light irradiation over $Bi_4O_5Br_2$. *Green Chem.,* **2020**, *22*(15), 4884-4889.
[http://dx.doi.org/10.1039/D0GC01930E]

[143]　Wei, L.Q.; Ye, B.H. Cyclometalated Ir–Zr Metal–Organic Frameworks as Recyclable Visible-Light Photocatalysts for Sulfide Oxidation into Sulfoxide in Water. *ACS Appl. Mater. Interfaces,* **2019**, *11*(44), 41448-41457.
[http://dx.doi.org/10.1021/acsami.9b15646] [PMID: 31604013]

[144]　Johnson, J.A.; Zhang, X.; Reeson, T.C.; Chen, Y.S.; Zhang, J. Facile control of the charge density and photocatalytic activity of an anionic indium porphyrin framework *via in situ* metalation. *J. Am. Chem. Soc.,* **2014**, *136*(45), 15881-15884.
[http://dx.doi.org/10.1021/ja5092672] [PMID: 25341191]

[145]　Bernini, M.C.; Gándara, F.; Iglesias, M.; Snejko, N.; Gutiérrez-Puebla, E.; Brusau, E.V.; Narda, G.E.; Monge, M.Á. Reversible breaking and forming of metal-ligand coordination bonds: temperature-triggered single-crystal to single-crystal transformation in a metal-organic framework. *Chemistry,* **2009**, *15*(19), 4896-4905.
[http://dx.doi.org/10.1002/chem.200802385] [PMID: 19322845]

[146]　Gándara, F.; Puebla, E.G.; Iglesias, M.; Proserpio, D.M.; Snejko, N.; Monge, M.Á. Controlling the structure of arenedisulfonates toward catalytically active materials. *Chem. Mater.,* **2009**, *21*(4), 655-661.
[http://dx.doi.org/10.1021/cm8029517]

[147]　Dhakshinamoorthy, A.; Asiri, A.M.; Garcia, H. Metal–organic frameworks catalyzed C–C and C–heteroatom coupling reactions. *Chem. Soc. Rev.,* **2015**, *44*(7), 1922-1947.
[http://dx.doi.org/10.1039/C4CS00254G] [PMID: 25608717]

[148]　De Vos, D.E.; Dams, M.; Sels, B.F.; Jacobs, P.A. Ordered mesoporous and microporous molecular sieves functionalized with transition metal complexes as catalysts for selective organic transformations. *Chem. Rev.,* **2002**, *102*(10), 3615-3640.
[http://dx.doi.org/10.1021/cr010368u] [PMID: 12371896]

[149]　Dhakshinamoorthy, A.; Garcia, H. Catalysis by metal nanoparticles embedded on metal–organic frameworks. *Chem. Soc. Rev.,* **2012**, *41*(15), 5262-5284.
[http://dx.doi.org/10.1039/c2cs35047e] [PMID: 22695806]

[150]　Ma, R.; Yang, P.; Ma, Y.; Bian, F. Facile Synthesis of Magnetic Hierarchical Core-Shell Structured Fe_3O_4@PDA-Pd@MOF Nanocomposites: Highly Integrated Multifunctional Catalysts. *ChemCatChem,* **2018**, *10*(6), 1446-1454.
[http://dx.doi.org/10.1002/cctc.201701693]

[151]　Kawano, M.; Kawamichi, T.; Haneda, T.; Kojima, T.; Fujita, M. The modular synthesis of functional

porous coordination networks. *J. Am. Chem. Soc.,* **2007**, *129*(50), 15418-15419.
[http://dx.doi.org/10.1021/ja0752540] [PMID: 18031041]

[152]   Yuan, B.; Pan, Y.; Li, Y.; Yin, B.; Jiang, H. A highly active heterogeneous palladium catalyst for the Suzuki-Miyaura and Ullmann coupling reactions of aryl chlorides in aqueous media. *Angew. Chem. Int. Ed.,* **2010**, *49*(24), 4054-4058.
[http://dx.doi.org/10.1002/anie.201000576] [PMID: 20432496]

[153]   Pascanu, V.; Yao, Q.; Bermejo Gómez, A.; Gustafsson, M.; Yun, Y.; Wan, W.; Samain, L.; Zou, X.; Martín-Matute, B. Sustainable catalysis: rational Pd loading on MIL-101Cr-NH2 for more efficient and recyclable Suzuki-Miyaura reactions. *Chemistry,* **2013**, *19*(51), 17483-17493.
[http://dx.doi.org/10.1002/chem.201302621] [PMID: 24265270]

[154]   Shang, N.; Gao, S.; Zhou, X.; Feng, C.; Wang, Z.; Wang, C. Palladium nanoparticles encapsulated inside the pores of a metal–organic framework as a highly active catalyst for carbon–carbon cross-coupling. *RSC Advances,* **2014**, *4*(97), 54487-54493.
[http://dx.doi.org/10.1039/C4RA10065D]

[155]   Saha, D.; Sen, R.; Maity, T.; Koner, S. Anchoring of palladium onto surface of porous metal-organic framework through post-synthesis modification and studies on Suzuki and Stille coupling reactions under heterogeneous condition. *Langmuir,* **2013**, *29*(9), 3140-3151.
[http://dx.doi.org/10.1021/la304147j] [PMID: 23373729]

[156]   Pourkhosravani, M.; Dehghanpour, S.; Farzaneh, F. Palladium nanoparticles supported on zirconium metal organic framework as an efficient heterogeneous catalyst for the Suzuki–Miyaura coupling reaction. *Catal. Lett.,* **2016**, *146*(2), 499-508.
[http://dx.doi.org/10.1007/s10562-015-1674-5]

[157]   Sun, D.; Li, Z. Double-solvent method to Pd nanoclusters encapsulated inside the cavity of NH2–Uio-66 (Zr) for efficient visible-light-promoted suzuki coupling reaction. *J. Phys. Chem. C,* **2016**, *120*(35), 19744-19750.
[http://dx.doi.org/10.1021/acs.jpcc.6b06710]

[158]   Fei, H.; Cohen, S.M. A robust, catalytic metal–organic framework with open 2,2′-bipyridine sites. *Chem. Commun. (Camb.),* **2014**, *50*(37), 4810-4812.
[http://dx.doi.org/10.1039/C4CC01607F] [PMID: 24687158]

[159]   Wei, Y.L.; Li, Y.; Chen, Y.Q.; Dong, Y.; Yao, J.J.; Han, X.Y.; Dong, Y.B. Pd(II)-NHD-Functionalized UiO-67 Type MOF for Catalyzing Heck Cross-Coupling and Intermolecular Benzyne–Benzyne–Alkene Insertion Reactions. *Inorg. Chem.,* **2018**, *57*(8), 4379-4386.
[http://dx.doi.org/10.1021/acs.inorgchem.7b03271] [PMID: 29617122]

[160]   Modak, A.; Pramanik, M.; Inagaki, S.; Bhaumik, A. A triazine functionalized porous organic polymer: excellent CO$_2$ storage material and support for designing Pd nanocatalyst for C–C cross-coupling reactions. *J. Mater. Chem. A Mater. Energy Sustain.,* **2014**, *2*(30), 11642-11650.
[http://dx.doi.org/10.1039/C4TA02150A]

[161]   Bai, C.; Jian, S.; Yao, X.; Li, Y. Carbonylative Sonogashira coupling of terminal alkynes with aryl iodides under atmospheric pressure of CO using Pd(II)@MOF as the catalyst. *Catal. Sci. Technol.,* **2014**, *4*(9), 3261-3267.
[http://dx.doi.org/10.1039/C4CY00488D]

[162]   Ezugwu, C.I.; Mousavi, B.; Asrafa, M.A.; Mehta, A.; Vardhan, H.; Verpoort, F. An N-heterocyclic carbene based MOF catalyst for Sonogashira cross-coupling reaction. *Catal. Sci. Technol.,* **2016**, *6*(7), 2050-2054.
[http://dx.doi.org/10.1039/C5CY01944C]

[163]   Zhang, L.; Su, Z.; Jiang, F.; Zhou, Y.; Xu, W.; Hong, M. Catalytic palladium nanoparticles supported on nanoscale MOFs: a highly active catalyst for Suzuki–Miyaura cross-coupling reaction. *Tetrahedron,* **2013**, *69*(44), 9237-9244.
[http://dx.doi.org/10.1016/j.tet.2013.08.059]

[164] Li, Q.; Jiang, S.; Ji, S.; Shi, D.; Li, H. Synthesis of magnetically recyclable MOF-5@SiO2@Fe3O4 catalysts and their catalytic performance of Friedel–Crafts alkylation. *J. Porous Mater.,* **2015**, *22*(5), 1205-1214.
[http://dx.doi.org/10.1007/s10934-015-9997-6]

[165] Jiang, S.; Yan, J.; Habimana, F.; Ji, S. Preparation of magnetically recyclable MIL-53(Al)@SiO2@Fe3O4 catalysts and their catalytic performance for Friedel–Crafts acylation reaction. *Catal. Today,* **2016**, *264*, 83-90.
[http://dx.doi.org/10.1016/j.cattod.2015.10.003]

[166] Mondal, S. Recent advancement of Ullmann-type coupling reactions in the formation of C–C bond. *ChemTexts.,* **2016**, *2*(4), 17.
[http://dx.doi.org/10.1007/s40828-016-0036-2]

[167] Phan, N.T.S.; Nguyen, T.T.; Nguyen, C.V.; Nguyen, T.T. Ullmann-type coupling reaction using metal-organic framework MOF-199 as an efficient recyclable solid catalyst. *Appl. Catal. A Gen.,* **2013**, *457*, 69-77.
[http://dx.doi.org/10.1016/j.apcata.2013.02.005]

[168] Chen, L.; Gao, Z.; Li, Y. Immobilization of Pd(II) on MOFs as a highly active heterogeneous catalyst for Suzuki–Miyaura and Ullmann-type coupling reactions. *Catal. Today,* **2015**, *245*, 122-128.
[http://dx.doi.org/10.1016/j.cattod.2014.03.074]

[169] Stille, J.K. The palladium-catalyzed cross-coupling reactions of organotin reagents with organic electrophiles. *Angew. Chem. Int. Ed. Engl.,* **1986**, *25*(6), 508-524. [new synthetic methods (58)].
[http://dx.doi.org/10.1002/anie.198605081]

[170] Hwang, Y.K.; Hong, D.Y.; Chang, J.S.; Jhung, S.H.; Seo, Y.K.; Kim, J.; Vimont, A.; Daturi, M.; Serre, C.; Férey, G. Amine grafting on coordinatively unsaturated metal centers of MOFs: consequences for catalysis and metal encapsulation. *Angew. Chem. Int. Ed.,* **2008**, *47*(22), 4144-4148.
[http://dx.doi.org/10.1002/anie.200705998] [PMID: 18435442]

[171] Huang, Y.; Gao, S.; Liu, T.; Lü, J.; Lin, X.; Li, H.; Cao, R. Palladium Nanoparticles Supported on Mixed-Linker Metal-Organic Frameworks as Highly Active Catalysts for Heck Reactions. *ChemPlusChem,* **2012**, *77*(2), 106-112.
[http://dx.doi.org/10.1002/cplu.201100021]

[172] Chen, L.; Rangan, S.; Li, J.; Jiang, H.; Li, Y. A molecular Pd(II) complex incorporated into a MOF as a highly active single-site heterogeneous catalyst for C–Cl bond activation. *Green Chem.,* **2014**, *16*(8), 3978-3985.
[http://dx.doi.org/10.1039/C4GC00314D]

[173] Huang, S.L.; Jia, A.Q.; Jin, G.X. Pd(diimine)Cl2 embedded heterometallic compounds with porous structures as efficient heterogeneous catalysts. *Chem. Commun. (Camb.),* **2013**, *49*(24), 2403-2405.
[http://dx.doi.org/10.1039/c3cc38714c] [PMID: 23417107]

[174] Sonogashira, K. Development of Pd–Cu catalyzed cross-coupling of terminal acetylenes with sp2-carbon halides. *J. Organomet. Chem.,* **2002**, *653*(1-2), 46-49.
[http://dx.doi.org/10.1016/S0022-328X(02)01158-0]

[175] King, A.O.; Yasuda, N. Palladium-catalyzed cross-coupling reactions in the synthesis of pharmaceuticals. In: *Organometallics in Process Chemistry*; Springer, **2004**; pp. 205-245.
[http://dx.doi.org/10.1007/b94551]

[176] Gao, S.; Zhao, N.; Shu, M.; Che, S. Palladium nanoparticles supported on MOF-5: A highly active catalyst for a ligand- and copper-free Sonogashira coupling reaction. *Appl. Catal. A Gen.,* **2010**, *388*(1-2), 196-201.
[http://dx.doi.org/10.1016/j.apcata.2010.08.045]

[177] Truong, T.; Nguyen, C.K.; Tran, T.V.; Nguyen, T.T.; Phan, N.T.S. Nickel-catalyzed oxidative coupling of alkynes and arylboronic acids using the metal–organic framework Ni $_2$ (BDC) $_2$ (DABCO)

as an efficient heterogeneous catalyst. *Catal. Sci. Technol.,* **2014**, *4*(5), 1276-1285.
[http://dx.doi.org/10.1039/C3CY01053H]

[178]   Sabatier, P.; Senderens, J.B. Direct hydrogenation of oxides of carbon in presence of various finely divided metals. *CR Acad Sci,* **1902**, *134*(1), 689-691.

[179]   Chen, Z.; Chen, J.; Li, Y. Metal–organic-framework-based catalysts for hydrogenation reactions. *Chin. J. Catal.,* **2017**, *38*(7), 1108-1126.
[http://dx.doi.org/10.1016/S1872-2067(17)62852-3]

[180]   Sahoo, T.; Panda, J.; Sahu, J.R.; Sahu, R. Application of Metal Organic Framework and Derived Material in Hydrogenation Catalysis. *Appl. Met. Fram. Their Deriv. Mater,* **2020**, 193-213.

[181]   Chughtai, A.H.; Ahmad, N.; Younus, H.A.; Laypkov, A.; Verpoort, F. Metal–organic frameworks: versatile heterogeneous catalysts for efficient catalytic organic transformations. *Chem. Soc. Rev.,* **2015**, *44*(19), 6804-6849.
[http://dx.doi.org/10.1039/C4CS00395K] [PMID: 25958955]

[182]   Zhao, Y.; Liu, M.; Fan, B.; Chen, Y.; Lv, W.; Lu, N.; Li, R. Pd nanoparticles supported on ZIF-8 as an efficient heterogeneous catalyst for the selective hydrogenation of cinnamaldehyde. *Catal. Commun.,* **2014**, *57*, 119-123.
[http://dx.doi.org/10.1016/j.catcom.2014.08.015]

[183]   Bao, L.; Yu, Z.; Fei, T.; Yan, Z.; Li, J.; Sun, C.; Pang, S. Palladium supported on metal–organic framework as a catalyst for the hydrogenation of nitroarenes under mild conditions. *Appl. Organomet. Chem.,* **2020**, *34*(6), e5607.
[http://dx.doi.org/10.1002/aoc.5607]

[184]   Jun, J.I.A.N.G.; Gang, L.I.; Ling-Hao, K.O.N.G. Hydrogenation of nitrobenzene catalyzed by metal-organic framework-supported Ni nanoparticles. *Wuli Huaxue Xuebao,* **2015**, *31*(1), 137-144.
[http://dx.doi.org/10.3866/PKU.WHXB201411171]

[185]   Manna, K.; Zhang, T.; Carboni, M.; Abney, C.W.; Lin, W. Salicylaldimine-based metal-organic framework enabling highly active olefin hydrogenation with iron and cobalt catalysts. *J. Am. Chem. Soc.,* **2014**, *136*(38), 13182-13185.
[http://dx.doi.org/10.1021/ja507947d] [PMID: 25187995]

[186]   Yang, Q.; Xu, Q.; Yu, S.H.; Jiang, H.L. Pd Nanocubes@ZIF-8: Integration of Plasmon-Driven Photothermal Conversion with a Metal-Organic Framework for Efficient and Selective Catalysis. *Angew. Chem. Int. Ed.,* **2016**, *55*(11), 3685-3689.
[http://dx.doi.org/10.1002/anie.201510655] [PMID: 26799948]

[187]   Fang, R.; Chen, L.; Shen, Z.; Li, Y. Efficient hydrogenation of furfural to fufuryl alcohol over hierarchical MOF immobilized metal catalysts. *Catal. Today,* **2021**, *368*, 217-223.
[http://dx.doi.org/10.1016/j.cattod.2020.03.019]

[188]   Chen, J.; Li, Y. The Road to MOF-Related Functional Materials and Beyond: Desire, Design, Decoration and Development. **2016**, 1456-1476.

[189]   Guo, Z.; Xiao, C.; Maligal-ganesh, R. V; Zhou, L.; Goh, T.W.; Li, X.; Tesfagaber, D.; Thiel, A.; Huang, W. Pt Nanoclusters Con fi ned within Metal – Organic Framework Cavities for Chemoselective Cinnamaldehyde Hydrogenation. **2014**.

[190]   Zhang, M.; Yang, Y.; Li, C.; Liu, Q.; Williams, C.T.; Liang, C. Catalysis Science & Technology catalysts for selective hydrogenation of. 8 (c). **2014**, 329-332.

[191]   Reis, P.M.; Royo, B. Chemoselective hydrogenation of nitroarenes and deoxygenation of pyridine N-oxides with H2 catalyzed by MoO2Cl2. *Tetrahedron Lett.,* **2009**, *50*(8), 949-952.
[http://dx.doi.org/10.1016/j.tetlet.2008.12.038]

[192]   Yang, A.S.; Peng, L.; Oveisi, E.; Bulut, S.; Daniel, T.; Asgari, M.; Trukhina, O.; Queen, W.L. Accepted Article.

[193] Huang, G.; Yang, Q.; Xu, Q.; Yu, S.; Jiang, H. Zuschriften Metal – Organic Framework Composites Polydimethylsiloxane Coating for a Palladium / MOF Composite : Highly Improved Catalytic Performance by Surface Hydrophobization Zuschriften. **2016**, 7505-7509.

[194] Yang, J.; Zhang, F.; Lu, H.; Hong, X.; Jiang, H.; Wu, Y.; Li, Y. Hollow Zn/Co ZIF Particles Derived from Core-Shell ZIF-67@ZIF-8 as Selective Catalyst for the Semi-Hydrogenation of Acetylene. *Angew. Chem. Int. Ed.,* **2015**, *54*(37), 10889-10893.
[http://dx.doi.org/10.1002/anie.201504242] [PMID: 26333054]

[195] Peng, L.; Zhang, J.; Yang, S.; Han, B.; Sang, X.; Liu, C.; Yang, G. The ionic liquid microphase enhances the catalytic activity of Pd nanoparticles supported by a metal–organic framework. *Green Chem.,* **2015**, *17*(8), 4178-4182.
[http://dx.doi.org/10.1039/C5GC01333J]

[196] Chen, L.; Chen, X.; Liu, H.; Li, Y. Encapsulation of mono- or bimetal nanoparticles inside metal-organic frameworks *via in situ* incorporation of metal precursors. *Small,* **2015**, *11*(22), 2642-2648.
[http://dx.doi.org/10.1002/smll.201403599] [PMID: 25644718]

CHAPTER 4

Metal-Organic Frameworks and Their Derived Structures for Biomass Upgrading

Yushan Wu[1], Yanfei Xu[1], Chuan Qin[1] and Mingyue Ding[1,*]

[1] *School of Power and Mechanical Engineering, Hubei International Scientific and Technological Cooperation Base of Sustainable Resource and Energy, Hubei Province Key Laboratory of Accoutrement Technique in Fluid Machinery & Power Engineering, Wuhan University, Wuhan 430072, China*

Abstract: Biomass valorization is receiving increasing attention over the past years with the consumption of traditional fossil fuels as well as the deterioration of the global environment. The transformation of biomass into highly value-added chemicals and important feedstocks will be of keen interest and great impact. The conversion process of biomass requires efficient and durable catalysts with high selectivity and stable structures. This chapter focuses on the employment of metal-organic frameworks (MOFs), MOF composites (metal, metal oxide, or polyoxometalates combined with MOFs), and MOF-derived materials (carbon, carbon-supported metal or metal oxide by using MOF as precursors) as solid catalysts for the upgrading of biomass into important fine chemicals. First, we will give a short introduction of biomass and MOFs, and then the brief biomass valorization reactions by MOFs and MOF-based catalysts based on the types of substrates. The last segment is summary of the state of the art, challenges, as well as prospects of MOFs and MOFs-derived structures for biomass transformation.

Keywords: Biomass, Heterogeneous catalysts, Metal-organic frameworks, Platform chemicals, Porous materials, Transformation.

1. INTRODUCTION

1.1. Biomass

Conventional fossil resources, crude oil, coal, and natural gas are used to satisfy our energy consumption, improve the national economy and people's livelihood. However, severe environmental issues occurred due to the high consumption of

---

* **Corresponding author Mingyue Ding:** School of Power and Mechanical Engineering, Hubei International Scientific and Technological Cooperation Base of Sustainable Resource and Energy, Hubei Province Key Laboratory of Accoutrement Technique in Fluid Machinery & Power Engineering, Wuhan University, Wuhan 430072, China; E-mail: dingmy@whu.edu.cn

Junkuo Gao & Reza Abazari (Eds.)

traditional fossil energy, such as the greenhouse effect, acid rain, atmosphere and marine pollution [1, 2]. At the same time, these conventional fossil sources are not renewable and will be depleted one day. Hence, it is necessary and urgent to develop alternative renewable energy resources such as biomass, solar energy, wind energy, as well as hydrogen energy owing to their features of being rich in sources, wide distribution, renewable and environmental-friendly. Amongst, biomass is regarded as the only potential reliable candidate for the replacement of conventional fossil energy sources to produce fuels and fine chemicals [3 - 7].

Biomass can be obtained from an inexpensive and wide range of substances, such as forest products (green plants, forestry, woods, logging residues), crops (agricultural and husbandry wastes as well as breeding), domestic and industrial wastes (kitchen waste, biodiesel or bioethanol production, and paper industry and so on), so as to avoid the competition with human food.

Lignocellulose, in general, is the most abundant form of biomass. The main construction of lignocellulosic biomass is lignin (20%-30%), cellulose (40%-50%), and hemicellulose (18%-28%) (Fig. **1**) [8, 9]. Lignin is composed of an aromatic polymer with methoxylated phenylpropane units, cellulose is a crystalline homopolysaccharide polymer and made up of glucose units, while hemicellulose is a complex polymer in amorphous state (xylose monomer unit is the main component) [10 - 13].

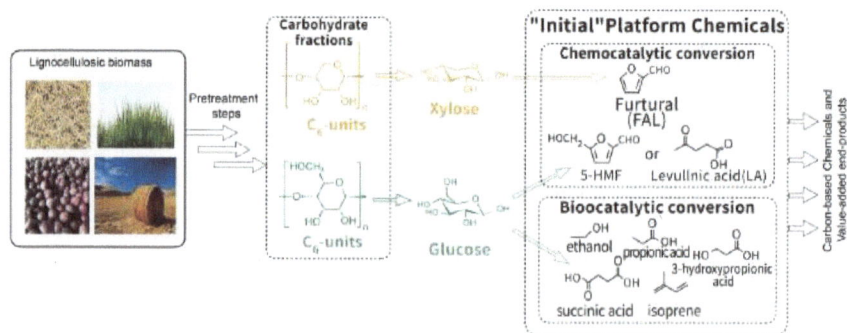

Fig. (1). Lignocellulosic-derived initial platform chemicals. Reprinted from ref. 14 with permission from American Chemical Society, Copyright 2018 [14].

Cellulose is a homopolymer consisting of glucose units with a high degree of polymerization (10000 to 15000 in wood and cotton) [15]. It is hard to be hydrolyzed ascribed to the existence of intramolecular and intermolecular hydrogen bonding between the units of anhydro glucan [16]. The bottom and top part of the cellulose chains are entirely hydrophobic, and both ends of the chains are usually hydrophilic. This long range ordered hydrogen bonding feature makes cellulose a highly crystalline and robust material toward chemical reactivity. Through strong acid hydrolysis process, cellulose can be degraded into glucose tetramer, glucose trimer, glucose dimer, and even glucose (Fig. **2**) [17]. Further dehydration of glucose obtains a very important chemical intermediate of 5-hydroxymethylfurfural (HMF) [18], which can be widely used for the production of various indispensable industrial feedstocks such as 5-formyl-2-furancarboxylic acid and 5-hydroxymethyl-2-furancarboxylic acid intermediates, 2,5-furandicarboxylic acid *via* furan-2,5-diformylfuran (DFF) [19], 2,5-dihydroxymethyl-tetrahydrofuran (DHMTHF) through di(hydroxymethyl) furfural (DHMF) intermediate [20]. Besides, other important products like levulinic acid (LA), and γ-valerolactone (GVL) can be also obtained from HMF [21, 22].

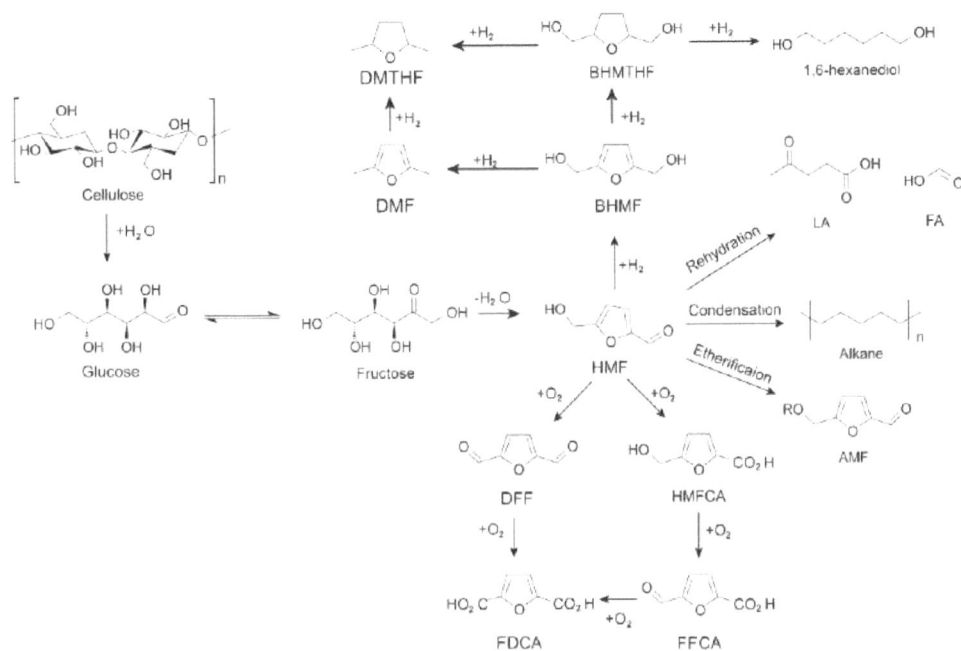

Fig. (2). Conversion of cellulose into diverse chemicals. Reprinted from ref. 23 with permission from American Chemical Society, Copyright 2018 [22].

Hemicellulose is sugar polymer which is made up of five and six-carbon sugars: this raw material can be transformed to C5 sugar monomers (Fig. **3**) [24]. The most abundant constitute of hemicellulose is xylan, which is usually composed of xylose polymer. Hemicellulose is amorphous due to its branched feature, at the same time, it is facile to be hydrolyzed into its basic monomer in comparison to cellulose. In addition, the xylan can yield arabinose and xylose, and can be transformed into kinds of high valuable intermediates, such as furfural and xylitol [25, 26]. In general, furfural is a very important platform chemical, which can be further converted into liquid fuel additives, fuels, and significant derivatives like furfuryl alcohol (FA), 2-methyl tetrahydrofuran (MTHF), 2-methyl furan (MF), tetrahydrofurfuryl alcohol (THFOL), cyclopentanone (CPO) and so on [27 - 30].

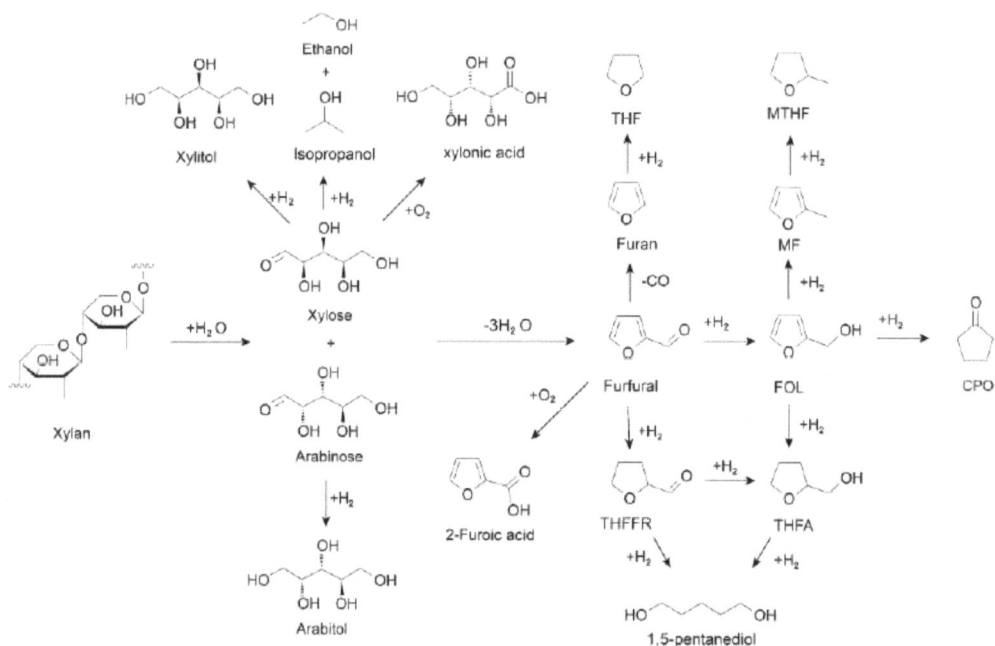

Fig. (3). The route of hemicellulose conversion [23]. Reprinted from ref. 23 with permission from American Chemical Society, Copyright 2018.

Lignin, consisted of phenylpropane units with disordered structures and linked by carbon-carbon bonds to form three dimensional frameworks, making up a quarter in biomass, is a potential candidate to supply aromatic components *via* simple process [31 - 33]. It is originated from the precursors of three aromatic alcohol, namely β-coumarol, coniferol and myrosinol, corresponding to three types of lignin, that is p-hydroxyphenyl lignin, syringyl lignin and guaiacyl lignin, respectively. Compared with the pyrolysis process of cellulose, it seems much difficult for the conversion of lignin because of its complicated structure, which is composed of oxygen-rich component such as coniferyl, phenol, and sinapyl alcohols [34, 35]. Lignin can be oxidized, and reduced, and also undergo hydrolysis, alcoholysis, carboxylation, condensation or graft copolymerization due to the presence of phenolic hydroxyl, aromatic group, alcohol hydroxyl, and carbon-based conjugated double bond in the molecular structures.

1.2. MOFs and MOFs-based Structures

Metal-organic frameworks (MOFs), as a new class of porous crystalline materials, are consisted of organic linkers and metal oxides by strong bonds with high range orders, where the metal ions and oxygen/carbon atoms are connected at atomic or molecular level [36 - 40]. Until now, there are thousands of MOFs designed and synthesized, almost the majority of the metals in elemental table were introduced to the structures of MOFs. And the organic linkers are usually dicarboxylate acid, tricarboxylate caid, tetra-carboxylate acid, or nitrogen-containing linkers [36, 41 - 44]. MOFs have drawn increasing concern over the past decades due to their intrinsic properties of ultrahigh specific surface areas, permanent porosity, adjustable pore size and channel environment, as well as controllable structures and topology. Indeed, these unique features endow their wide applications in the fields of gas sorption/separation, drug delivery, catalysis, energy storage, electron device and so on [45 - 50].

For catalytic reactions, the active sites are inevitably indispensable to the performance. The insoluble nature of MOFs guarantees their easy recycling for heterogeneous catalysis, and the open channels with a rich pore environment facilitate the fast diffusion and transport of guest species. As for pristine MOFs, the catalytic ability was originated from the open metal sites as Lewis acid sites or defects as active sites [51 - 54], and Brønsted acid originated from the organic linkers with dangling acid sites [55 - 58]. Besides, some ligands with inherently catalytic ability [59] (porphyrin or metalloporphyrin, and Schiff-base complexes) are engineered into the skeleton of MOFs [60, 61], which could modulate the electronic properties of the MOFs that allowed them to catalyze related reactions. So far, MOF-based catalysts are still less or similarly active compared with the inorganic solid-acid or homogeneous catalysts. Of course, the pristine MOFs have limited applications in the field of catalysis just by the restricted active sites in frameworks. Therefore, it is necessary to expand and enrich the scopes of these porous materials to overcome the disadvantages of pristine MOFs. MOFs can be served as a host or support for the encapsulation of guest species (metals, metal oxides, polymers, quantum dots, enzymes, polyoxometalates (POMs), and organic molecules) by one-pot synthesis or post-modification method [62 - 68]. These supported catalytic species have the ability to complete different reactions efficiently with the aid of MOFs, which can prevent the active species from aggregation, leaching, deactivation and decomposition. Moreover, MOFs can be applied as precursors for the fabrication of supported metal (dual metal or alloy), metal oxide, or porous carbon materials (graphene, carbon nanotube, amorphous carbon) through high-temperature calcination or laser treatment [69 - 76]. The obtained metal nanoparticles are usually evenly distributed on or encapsulated in the porous carbon species with uniform particle size due to the atomic contact of metal and carbon atoms in the structures of MOFs (Fig. **4**). Additionally, MOF-derived structures possess high conductivity, high specific surface areas, high thermal and chemical stability, thus endowing them excellent catalytic performance in many fields.

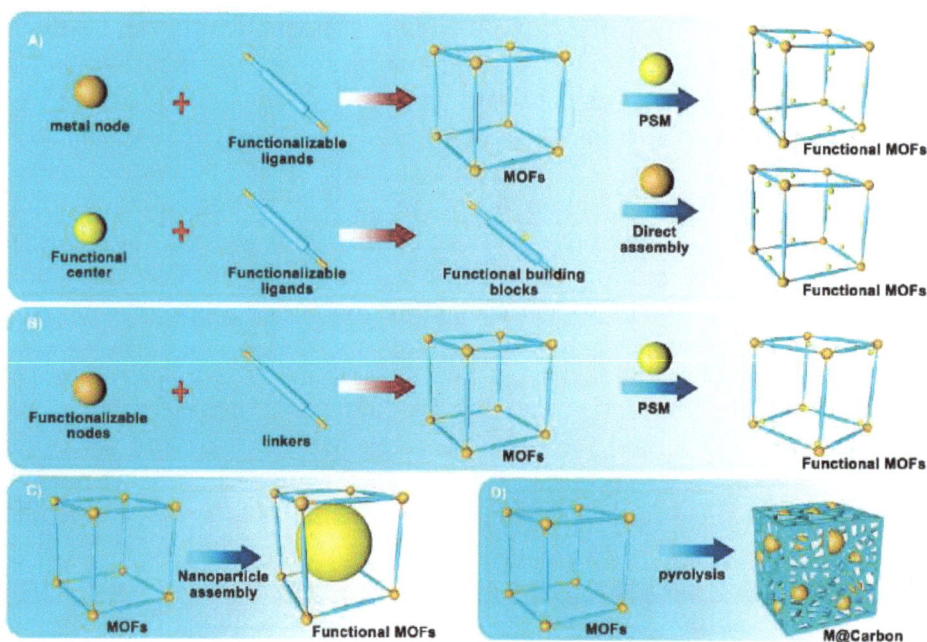

Fig. (4). The assembling processes of functional MOFs. Reprinted from ref. 76 with permission from American Chemical Society, Copyright 2020.

This chapter is focused on the MOFs and MOFs-derived materials in the applications of biomass conversion. At this point, the brief methods for the synthesis of MOFs and their derived structures will not be discussed here, and the relevant content about it could be found in the former chapters.

1.3. Scope of this Chapter

This Chapter focuses on the applications of MOFs and MOF-derived structures (MOF composites, functionalized MOFs, and metal/metal oxide derived from MOFs) for biomass transformation. Great efforts have been devoted to the improvement of biomass valorization by using MOFs and MOFs-derived materials. At the same time, some critical reviews have been published and summarized the biomass conversion over MOF-based catalysts. For example, Janiak *et al.* reviewed MOFs as catalysts for biomass upgrading in comparison with other catalysts like zeolites [77]. Wu *et al.* summarized the conversion of lignocellulose by MOF-derived composites as efficient heterogeneous catalysts [23]. Besides, Heynderickx *et al.* reviewed the conversion of sugars into platform chemicals employing MOFs as catalysts [78]. Sels *et al.* summarized the sustainable biomass transformation over functionalized heterogeneous catalysts including MOFs and MOFs-derived structures [79]. Garcia and co-authors reviewed the synthesis of MOF-derived composites for the upgrading of biomass [80]. Here, we concentrate on the MOFs and MOF-derived structures for biomass transformation, as well as the representation of structure–activity relationship. We will summarize biomass upgrading reactions catalyzed by MOFs and their derived materials based on previous literature. The contents have been classified on the basis of catalytic reaction types, starting with those literatures involving cellulose, hemicellulose or lignin as the raw materials, then up to the use of glucose or fructose as substrates, and followed by the use of platform chemicals (hydroxymethylfurfural, furfural, levulinic acid, *etc*) catalyzed by MOFs and their derived structures.

2. BIOMASS CONVERSION OVER MOFS AND THEIR DERIVED STRUCTURES

2.1. Transformation of Cellulose Biomass

Cellulose presents one of the most abundant renewable sources, which can be transformed into various platform chemicals and feedstocks. The pioneering work of cellulose hydrolysis by MOFs was conducted by Kitagawa and co-authors [81], they synthesized an ultrahigh stable MIL-101-SO_3H MOF with strong Brønsted acid sites in a large pore surface (Fig. **5**). The resultant catalyst displayed only 1.4, 2.6, and 1.2% yield of glucose, xylose, and cellobiose, respectively, and total 5.3% yield for mono- or disaccharides in water due to the inefficient mass transfer. What's more, MIL-101-SO_3H exhibited excellent stability during the 13 recycling tests, and no deactivation was observed, indicating MOF was a promising candidate for the conversion of biomass.

Besides, MIL-53(Al), constructed from aluminum nodes and terephthalic acid linkers with ultrahigh chemical and thermal stability, was also investigated as a heterogeneous catalyst for carboxymethyl cellulose (CMC) upgrading [82]. As a result, a high yield of 40.4% for 5-hydroxymethyl-furaldehyde (5-HMF) from

CMC was obtained over MIL-53 in the water phase compared with other MOFs with different metal anodes (Table **1**). MIL-53(Al) with weak Brønsted acid sites was beneficial to the conversion of cellulose. On the other hand, the -COOH groups in MIL-53 can adsorb the CMC and water molecules through hydrogen bonding. This catalyst displayed about 38.4% yield to 5-HMF in the third cycle and the crystallinity was maintained well after cycling performance, indicating the robust nature of MIL-53.

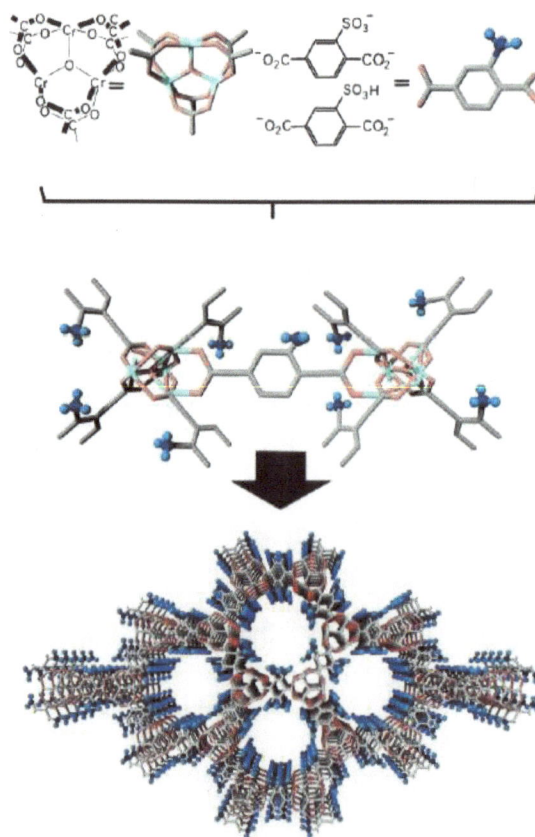

Fig. (5). The topology structures of MIL-101-SO$_3$H. Reprinted from ref. 81 with permission from John Wiley and Sons, Copyright 2013.

Table 1. The summary of catalytic performance for CMC to 5-HMF with different MOF catalysts.

Catalyst	5-HMF Yield (mol%)	TRS Yield (mol%)
UiO-66(Zr)	24.5	34.2
MOF-5(Zn)	-	29.9

(Table 1) cont.....

Catalyst	5-HMF Yield (mol%)	TRS Yield (mol%)
ZIF-8(Zn)	-	36.7
MIL-100(Cr)	25.8	38.9
HKUST-1(Cu)	31.8	42.1
MIL-53(Al)	40.3	54.2
MIL-101(Cr)	25.3	40.3
MIL-125(Ti)	30.9	32.4

In addition to MOFs themselves with active sites for biomass conversion, MOFs can also be host or support for guest component as catalysts employed in this field. Chen group designed a water-resistant catalyst, Ru-PTA/MIL-100(Cr), for the transformation of cellulose and cellobiose in water (Fig. **6**) [83]. The obtained bifunctional catalyst with appropriate Ru dispersion as well as acid site density, gave sorbitol selectivity of 57.9% and 95.1%, from cellulose and cellobiose, respectively. However, the recycling capacity was very poor, and just afforded 8.5% yield for sorbitol in the second cycle, which may ascribe to the insoluble substance blocked the pores of MOFs. In addition, the same group reported another multifunctional catalyst of Ru/NENU-3 [84], which was produced from the impregnation of Ru on NENU-3 (HKUST-1 and PTA composite) for ethylene glycol (EG) production from cellulose. As a result, the resulting functional catalyst Ru/NENU-3 showed a 50.2% yield of EG directly from cellulose.

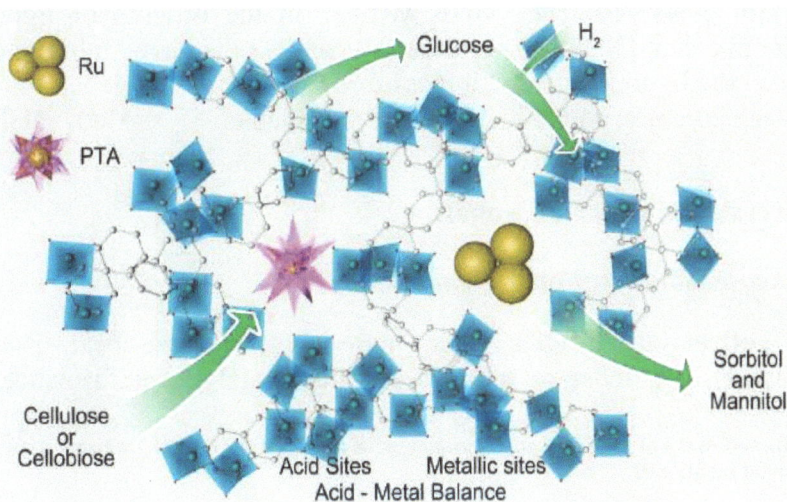

Fig. (6). Ru-PTA/MIL-100(Cr) catalyst for cellulose or cellobiose conversion. Reprinted from ref. 83 with permission from John Wiley and Sons, Copyright 2013.

Apart from metal or heteropoly acid, bio-enzyme was also introduced into the pore structures of MOFs for biorefinery. Xie *et al* reported a cellulase@UiO-6--NH$_2$ composite fabricated by a physical adsorption strategy for the hydrolysis of carboxymethyl cellulose (CMC) [85]. Cellulase@UiO-66-NH$_2$ catalyst displayed high PH tolerance, thermostability, reusability and lifetime. Moreover, the conversion of CMC could be kept for 72% after ten cycling tests. The excellent catalytic performance was due to the abundant -NH$_2$ and -COOH groups in UiO-66-NH$_2$ that could adsorb cellulase and improve its stability. Even so, the authors just presented the activity of catalyst, the information regarding the selectivity or yield of target products in this reaction were not mentioned. Similar to the case of cellulase@UiO-66-NH$_2$, another enzyme β-glucosidase was embedded into MOF Cu-PABA with high acidic stability by coprecipitation method [86]. The obtained β-G@Cu(PABA) hybrid combined with the cellulase showed 98% glucose productivity from CMC, which was much higher in comparison to co-immobilized cellulose alone. Furthermore, 90% activity could be retained after 10 cycles, suggesting the excellent reusability of β-G@Cu (PABA) bio-composites. Accordingly, applying MOF as a shielding to guard the enzymes under acidic conditions afforded a potential approach for further exploitation of bio-composites. Although the catalyst displayed excellent catalytic performance for CMC upgrading, the reason for deactivation was not discussed.

Except for MOFs and MOF composites, MOFs-derived materials are also applied in the application of biomass conversion. For example, metallic Ni nanoparticles were *in situ* generated from MOF MIL-77 in the process of lignocellulosic biomass conversion [87]. The resultant Ni catalyst displayed a high activity and transformed the lignocellulosic biomass through hydrodeoxygenation, which was superior than commercially Ni nanoparticles loaded on SiO$_2$-Al$_2$O$_3$ catalyst.

2.2. Conversion of C5 or C6 Sugars

2.2.1. Glucose Isomerization to Fructose

Table **2** collects the MOFs based materials that have been used for the transformation of glucose to fructose and their catalytic performance.

Table 2. The summary of catalytic performance with different MOFs for isomerization of glucose to fructose. N: not mentioned.

Catalyst	Reaction Conditions	Yield (%)	Cycle Number	Ref.
MIL-101-SO$_3$H	Water, 373 K, 24 h	21.6	3	[88]
MIL-101-GLY	Ethanol, 373 K, 24 h	59.3	4	[89]

(Table 2) cont.....

Catalyst	Reaction Conditions	Yield (%)	Cycle Number	Ref.
UiO-66-Fm	1-propanol, 363K, 24 h	56	N	[90]
UiO-66-MSBDC(20)	Water, 413 K, 3 h	25	4	[91]
Cr-MIL-101	Water, 373K, 24 h	12	2	[92]
Zr_6@SiO_2	Ethanol, 363 K, 70 h	23	5	[93]

The glucose isomerization to fructose was first reported by Kitagawa and co-authors [88]. The employed MIL-100 (Cr) and MIL-101(Cr) displayed better catalytic performance for glucose isomerization compared with Cr_2O_3 and amberlyst-15 under the same reaction conditions, which was due to the presence of large pore window sizes as well as open metal sites (OMS). In addition, MIL-101 exhibited higher yield of 12.6% fructose than that of 3.8% for MIL-100, the difference in performance was resulted from the pore windows of MIL-101 (1.6 and 1.2 nm) was much larger than those of MIL-100 (0.55 and 0.86 nm). The large pore windows were beneficial for the diffusion and transfer of glucose (0.8 nm in diameter). Furthermore, the electronic effect was also studied by changing the ligands with different groups (-NO_2, -NH_2, -SO_3H, and -$(CH_3)_2$,). The recovery experiment showed good result, however, the byproducts like 5-HMF and LA were observed.

Another example of MIL-101-based catalyst, $Cr(OH)_3$/MIL-101-GLY [89], was synthesized by $Cr(OH)_3$ nanoparticles precipitation on the surface of amino acid glycine modified MIL-101. This material showed high stability and the wrapped $Cr(OH)_3$ nanoparticles could be maintained on the surface of MIL-101 structures even after repeated sonication and washing. MIL-101-(GLY) displayed 59.3% yield of fructose from glucose, which was similar to Sn-containing zeolites that are benchmark in this field. The proximity of $Cr(OH)_3$ and Lewis acidic MIL-101 played an essential role in the glucose conversion. Besides, MIL-101-(GLY) exhibited a decay in conversion for the second run followed by a slight decrease up to four runs with stable selectivity to fructose, the deactivation might ascribe to the partially blocking of pores by the adsorbed byproducts.

Except for MIL-101(Cr) and its derived materials, Zr-based MOF UiO-66 was also used to investigate the catalytic isomerization of glucose owing to its high thermal and chemical stability. Tsapatsis reported a defected UiO-66 (UiO-6--Fm) with missing clusters in the framework by acid modulation [90]. The UiO-66-Fm catalytic performance was largely dependent on the solvent, 82% conversion of glucose and 56% yield for fructose were achieved in propanol. While in methanol or ethanol, the glucose showed a strong adsorption with the Brønsted acidic bridged OH group as well as the high polarizability of methanol and ethanol, leading to the formation of alky-glucosides as byproducts. While in propanol, the interaction was weaker, resulting in higher yield to fructose. Although the high yield of fructose was obtained, the two-step method involved multi-process, complex operation, and high cost, while a much more efficient, simple and environment approach should be developed for the glucose isomerization. In the same year, Degirmenci *et al* [91] designed a multivariate (MTV) UiO-66 structure by replacing different percentages of terephthalic acid (BDC) with sulfonic acid functionalized BDC linkers, the obtained catalysts were donated as UiO-66-MSBDC(y) (Fig. **7**).

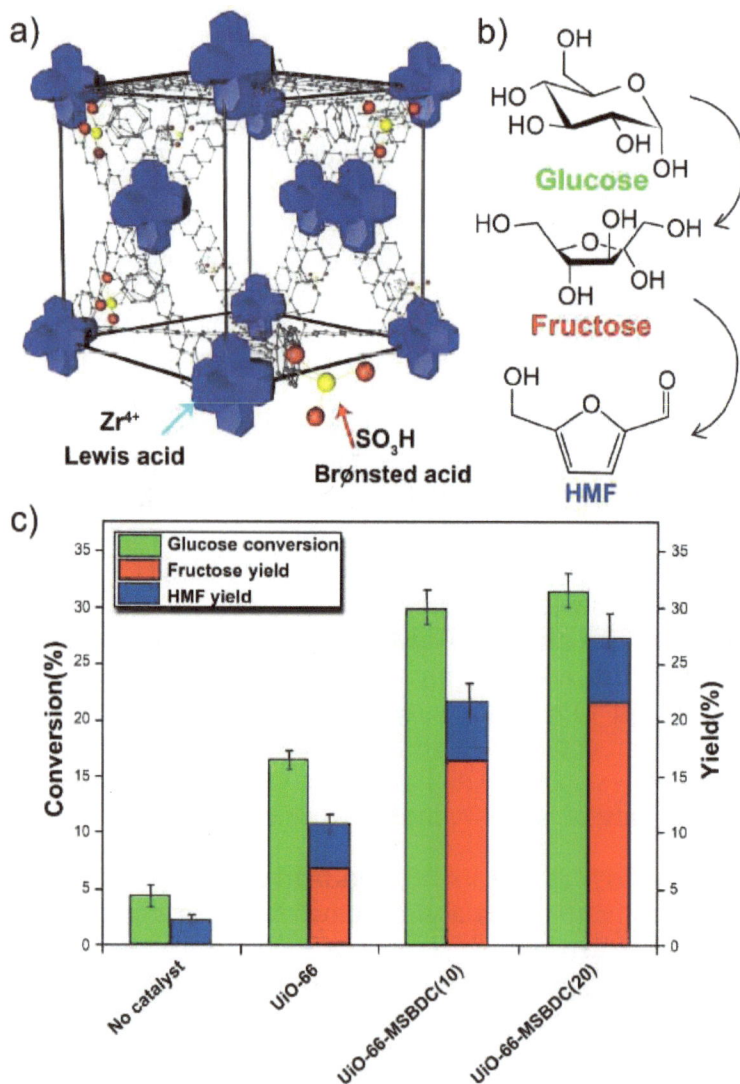

Fig. (7). Structures of UiO-66 and its performance for the conversion of glucose. Reprinted from ref. 91 with permission from John Wiley and Sons, Copyright 2018.

The resulting UiO-66-MSBDC(20) exhibited a 25% yield of fructose at glucose conversion of 48%, which was close to the performance of commercial Sn-beta zeolite. The existence of both Brønsted and Lewis acidity was reasonable for the high catalytic performance. Some undesired byproducts like 5-HMF (~10% yield) was formed during the reaction, so the aim of isomerization for glucose to fructose should be the development of no by-product generation in the future.

Adjusting the reaction pathway of glucose epimerization and isomerization is of significant importance for the generation of high yield fructose. The acid sites and local environment are essential to the activity and product selectivity in glucose isomerization. In this regard,MIL-101 with Lewis acidic sites originated from single-site active centers (SSACs) of Cr (III) cluster as a heterogeneous catalyst for glucose isomerization [92]. In contrast, UiO-66 with Lewis acidic Zr (IV) sites displayed both glucose epimerization and isomerization in water, and the glucose epimerization took place by an intramolecular carbon skeleton rearrangement (Fig. **8**). It was speculated that Zr atom has a larger atomic radius than Cr atom, resulting in higher polarizability of Zr nodes and strong electronegativity for neighboring oxygen atoms. As a result, the UiO-66 tends to cause epimerization, the cleavage of C-1 and C-2 bond, and the generation of C-3 and C-1 bond, while Cr sites tend to result in the formation of hexa-coordinate binuclear nodes with glucose, which facilitate the isomerization of glucose without the rearrangement of a C-C bond.

Besides the formerly mentioned strategies involving the pristine or modified MOFs as well as MOF composites as catalysts for the glucose isomerization, metal or metal oxide nanoparticles dispersed on porous carbon or other supports have also been developed employing MOFs as precursors. Stein reported a facile SiO_2 nano-casting method followed by thermal calcination in the air to fabricate stable single Zr site from NU-1000 (Fig. **9**) [93]. The obtained $Zr_6@SiO_2$ materials owned Lewis acid sites from the oxozirconium clusters that enabled to catalyze the glucose isomerization reaction in two steps, which indicated that the Zr clusters had access to the reagents. In contrast, the cal-NU-1000 sample without coated SiO_2 showed very low activity, resulting from the aggregation of active Zr sites. The recycling test displayed a decreasing trend for the overall activity within the five cycles, while the loss of fructose yield was less pronounced, which may ascribe to the reduction of Brønsted/Lewis ratio during the recycling test. The power to maintain high catalytic activity *via* templated porosity and site isolation makes nano-casting a versatile technique for employing MOFs-based materials for catalytic processes that require high temperatures.

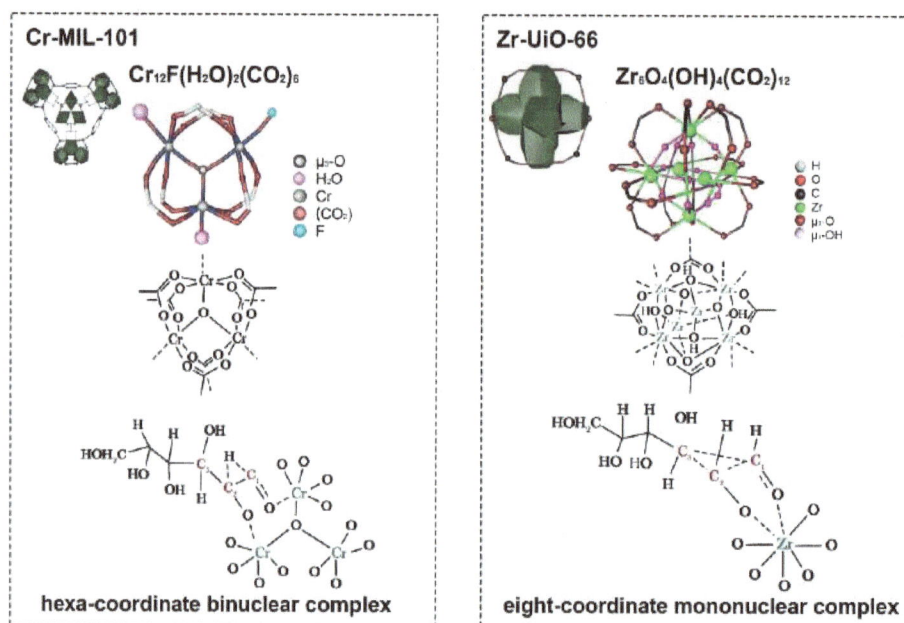

Fig. (8). Basic building units and local coordination environment of Cr-MIL-101 and Zr-UiO-66, and intermediate formed by glucose interacting with open SSACs. Reprinted from ref. 92 with permission from John Wiley and Sons, Copyright 2019.

Fig. (9). The preparation method for the Zr_6@SiO_2 from NU-1000. Reprinted from ref. 93 with permission from American Chemical Society, Copyright 2016.

2.2.2. Conversion of C5/C6 Sugars to 5-Hydroxymethylfurfural (HMF)

5-hydroxymethylfurfural (HMF) is considered as a most promising biomass-derived platform molecules, which could be transformed into key building blocks or significant intermediates, including 2,5-dimethylfuran, levulinic acid, 2,5-dihydroxymethyl tetrahydrofuran, 2,5-furan dicarboxylic acid, 2,5-diformylfuran, and 1,6-hexanediol. Table **3** summarizes the MOFs-based materials that have been applied for C5/C6 sugars to HMF combined with the catalytic performance.

Table 3. Comparison of the MOFs based materials that have been used for glucose isomerization. N: not mentioned.

Catalyst	Substrate	Reaction Conditions	Yield (%)	Cycle Number	Ref
PTA/MIL-101	fructose	DMSO, 403 K, 0.5 h	63	3	[94]
MOF-PMAi-Br	fructose	DMSO, 373 K, 1 h	80	4	[95]
MIL-101-SO$_3$H	fructose	DMSO, 393 K, 1 h	90	5	[96]
NUS-6(Hf)	fructose	DMSO, 373 K, 1 h	98	4	[97]
MIL-101-SO$_3$H	glucose	GVL/H$_2$O(9:1),425K, 2 h	45.8	5	[98]
MIL-101-SO$_3$H	glucose	THF/H$_2$O(39:1),403 K, 24 h	29	N	[99]
PO$_4$/NU	glucose	H$_2$O/2-propanol(1:9), 413 K, 7h	64	N	[100]
UiO-66	glucose	DMSO/water (39:1), 433 K, 0.5h	37	5	[101]
MIL-88(Fe)	glucose	DMSO, 413 K, 3 h	24.9	3	[102]
Yb-MOF	glucose	H$_2$O, 413 K, 24 h	15	4	[103]
UiO-66-SO$_3$H-NH$_2$ /PDA@PU	glucose	DMSO, 393 K, 2 h	70.3	5	[104]
UiO-66-NH$_2$-SO$_3$H-2/C$_3$N$_4$ @PDA	glucose	Isopropanol/ DMSO (9:1), 393 K, 6 h	54.9	5	[105]

The first example of selective upgrading of fructose to HMF using MOF-based materials was conducted by the Hensen group [94]. A bifunctional catalyst of PTA/MIL-101 with different PTA loadings were employed for HMF production from fructose. About 63% yield of HMF at 84% fructose conversion was obtained for PTA(3.0)/MIL-101 in EMIMCl, and the maximum HMF yield could reach 79%. Ascribing to the loss of PTA to the ionic liquid, DMSO was used to replace the EMIMCl solvent ascribing to the superior yield to produce HMF over various catalysts. The PTA(3.0)/MIL-101 displayed a yield of 63% after 0.5 h, which decreased to 47% in the third cycle because of the formation of undesired byproducts. The characterization of the reused catalyst regarding the stability and crystallinity as well as the BET surface areas was missing.

Hatton *et al* designed a porous polymer/MIL-101 composite through polymerization of maleimide and impregnated in the pores of MOFs, followed by functionalization with a polymetric analogue of *N*-bromosuccinimide (Fig. **10**) [95]. The resulting MOF-PMAi-Br displayed superior performance in the transformation of fructose to HMF, giving 85% HMF yield with full fructose conversion in DMSO. As the reaction time went on, HMF yield displayed a decreasing tendency due to the transformation of HMF to undesired byproducts. Meanwhile, the performance for the conversion of glucose was also tested, MOF-

PMAi-Br just gave 16% yield in 6 h, such a poor catalytic performance was ascribed to the absence of dissociative Cr in the solution as well as the halamine groups in framework were not active for glucose isomerization. After four cycles, the yield of HMF declined about 6% compared with the initial cycle with mass loss of 3-4% for the catalyst.

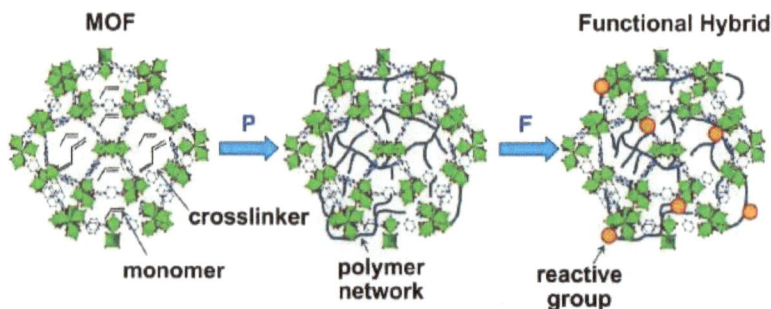

Fig. (10). Scheme for the fabrication process of functional hybrids of MOFs and polymer networks. Reprinted from ref. 95 with permission from American Chemical Society, Copyright 2014.

MOF functionalized with the sulfonic acid group usually displayed excellent activity in biomass conversion. Here, different MOFs (UiO-66, MIL-101, MIL53) functionalized with sulfonic acid were synthesized by post-synthetic modification method involving the sulfation of MOF with chlorosulfonic acid in dichloromethane [96]. The yield of HMF displayed an increasing trend with the increase of -SO$_3$H for fructose conversion, and the highest yield of 90% was gained for MIL-101(Cr)-SO$_3$H (15%). For comparison, -SO$_3$H functionalized MIL-53(Al) and UiO-66(Zr) showed HMF yield of 80% and 86%, respectively, which was consistent with their corresponding sulfonic acid density. In the recycling test, the HMF yield showed a decreasing trend, which may ascribe to the inevitable leaching of protons and the adsorption of insoluble products on active sites or pores of MOFs. The authors should exclude the effect of different acid densities with the same -SO$_3$H content in different MOFs.

It is still challenging to synthesize MOFs with strong Brønsted acidity in one-pot strategy due to the groups (like -SO$_3$H) on the organic linkers that would destroy the frameworks of MOF during the synthesis process. Zhao *et al* synthesized sulfonated Zr or Hf based MOF (NUS-6) directly from sulfonic acid functionalized organic linker *via* modulated hydrothermal method [97]. Ascribing to the impact between strong Brønsted acidity and suitable pore sizes in MOFs, the obtained NUS-6 displayed an outstanding performance for the production of HMF from fructose dehydration (Fig. **11**), whereas NUS-6 (Hf) gave a 98% yield of HMF at fructose conversion of 99%, which was superior to that of NUS-6(Zr).

Besides, NUS-6(Hf) can be reused for several times, about 90% yield of HMF was obtained in the fifth cycling test.

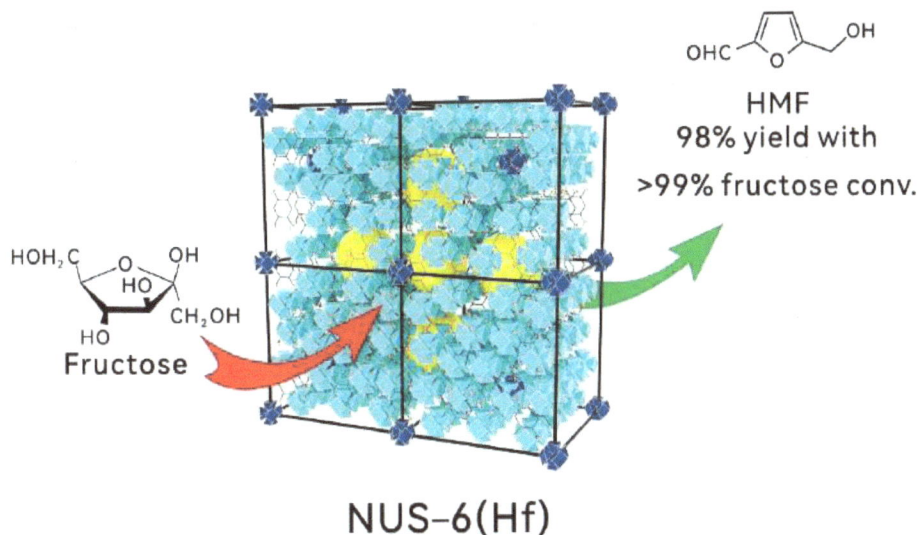

Fig. (11). The structures of Hf-based MOF NUS-6(Hf) and it as a catalyst for the conversion of fructose to HMF. Reprinted from ref. 97 with permission from American Chemical Society, Copyright 2014.

It is a great challenge to produce HMF with high yield from glucose due to the complex reactions occurred during the process. Functionalized MIL-101(Cr) with Brønsted and Lewis acid sites originated from -SO$_3$H functional groups and Cr SBU [98], was employed for the direct production of HMF from glucose, giving 44.9% HMF yield at glucose conversion of 45.8% in a mixed solvent of GVL and H$_2$O. The Cr^{3+} centers in MIL-101(Cr)-SO$_3$H played a role of Lewis acid that catalyzed the glucose isomerization, while the -SO$_3$H groups catalyzed the transformation of fructose to HMF. The resultant catalyst displayed excellent recycling performance with no obvious decay in the HMF yield but a decline in the glucose conversion in batch reaction. Furthermore, the catalyst was subjected to the fixed-bed reaction, and displayed relatively stable performance during the 50 h time on stream. A similar work was done by Janiak group [99], they test the glucose conversion over MIL-101(Cr)-SO$_3$H in the mixed solvent (THF/H$_2$O, 39:1 vol./vol.), and it displayed an HMF yield of 29%, which was lower than Bao's work [98]. In addition, the MIL-101-NO$_2$ catalyst without obvious Brønsted acid displayed a much higher yield of HMF compared with pure MIL-101, which may result from the stronger polarized aqua ligands from the electron-withdrawing nitro groups.

Apart from the sulfonic acid group functionalized MOF for glucose conversion to HMF, phosphate modified MOF could also be employed in the glucose transformation. Phosphate-modified NU-1000 was synthesized by simple impregnation of NU-1000 in the phosphoric acid aqueous solution and employed for glucose upgrading [100]. The obtained PO_4/NU (half) catalyst displayed a high yield of 64% for HMF at the full conversion of glucose into water/2-propanol (1:9, vol./vol.). In contrast, although the bulk ZrO_2 based catalyst displayed similar Brønsted acidity to NU-1000, it showed no selectivity to HMF. The difference in performance for NU-1000 and bulk ZrO_2 was probably because of the hydrophobic nature of NU-1000.

MOFs themselves can be employed as catalysts for the production of HMF from glucose because of the Lewis acidity of metal centers in frameworks. Kerton *et al* applied iso-reticular topology MOFs of UiO-66 with different functional groups in glucose conversion [101]. UiO-66 exhibited a high yield of 20% for HMF in DMSO, while UiO-66 with -SO_3H and -NH_2 gave HMF yield of 5% and 16%, respectively. This observation was much opposite to the anticipation and general rules which were associated with the acidity density of MOFs. The surface area may be one of the most significant impacts in determining the catalytic performance (515, 1045, and 1650 m^2 g^{-1} for UiO-66-SO_3H, UiO-66-NH_2, and UiO-66, respectively). Under the optimized reaction conditions, HMF yield reached 28%, and could be further increased to 37% for UiO-66 in the mixed solvent of DMSO/water (39:1, v/v).

The mixed-metal MIL-88B(Fe, Sc) was employed for the conversion of glucose to HMF, and displayed a 24.9% yield of HMF with 70.7% conversion of glucose [102]. Under the same reaction parameters, however, a single metal of MIL-88B(Fe) gave HMF yield of only 3.8%. The remarkable differences between the two MOFs may result from the existence of more Lewis acid sites in mixed-metal MOF, caused by structure disorder and defects (coordinatively unsaturated metal sites). Although the activity of glucose did not drop in the third cycles, the yield of HMF declined to a large extent from the second cycle. Later, a Yb based MOF, $Yb_6(BDC)_7(OH)_4(H_2O)_4$ constructed from Yb SBU and BDC linkers with coordinated H_2O was prepared, and applied for the glucose transformation to HMF [103]. Benefiting from the presence of Brønsted and Lewis acids sites, the obtained Yb-MOF exhibited HMF yield of about 15% at glucose conversion of 28%. The conversion of glucose displayed an increasing trend in the recycling test, in contrast, the yield of HMF was declined slightly. However, the reason was not clear.

Besides MOFs and MOF-guest materials for the dehydration of glucose, MOF composites (MOF with polymer or carbon-based materials) are also promising candidates for glucose conversion to HMF. Base on this concept, multifunctional UiO-66-SO$_3$H-NH$_2$/PDA@PU material was prepared *via* single-step method (Fig. 12) [104]. Benefiting from the existence of both acidity and basicity in UiO-6--SO$_3$H-NH$_2$/PDA@PU composite, this catalyst provided 70.3% yield of HMF, which was better than UiO-66-SO$_3$H/PDA@PU (60%). A slight decrease in the HMF yield appeared during the cycling test, which may result from the loss of active sites during the recycle process. In addition, the same group reported another UiO-66 based bi-functionalized catalyst, UiO-66-NH$_2$-SO$_3$H-2/C$_3$N$_4$@ PDA for the dehydration of glucose [105]. This catalyst displayed HMF yield of 54.9% in the mixed solvent of DMSO and isopropanol. The decline in the HMF yield inevitably occurred in the recycling run due to the blocking of insoluble byproducts on the surface of MOFs, which was consistence with the previous report [104].

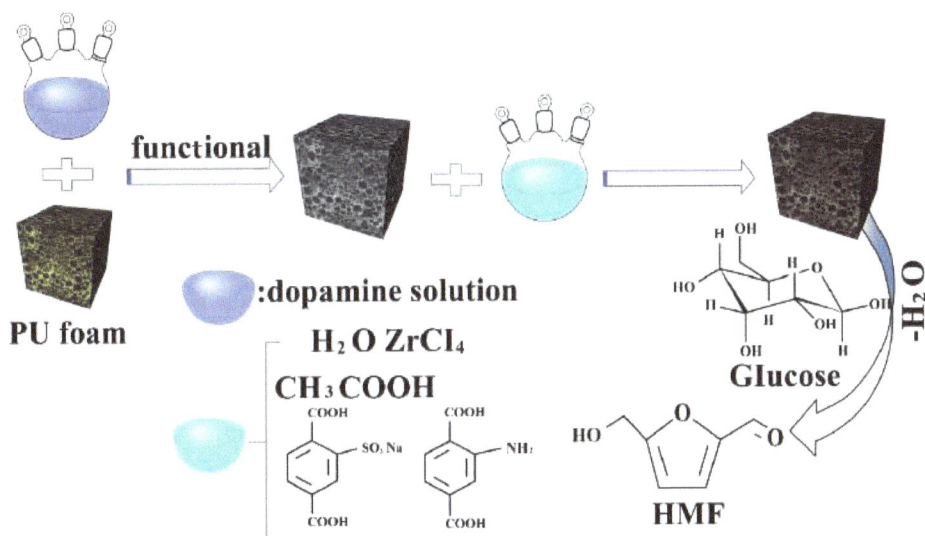

Fig. (12). The versatile method for preparing UiO-66-SO$_3$H-NH$_2$/PDA@PU and its catalytic behavior. Reprinted from ref. 104 with permission from Wiley-VCH Verlag GmbH & Co. KGaA, Weinheim, Copyright 2018.

2.2.3. Transformation of C5/C6 Sugars to other Feedstocks

2,5-diformylfuran (DFF), an important feedstock, can be used as a promising intermediate for heterocyclic ligands, fungicides, resins, and pharmaceuticals. DFF could be obtained from fructose and the reaction usually contains two separate processes, one is the fructose dehydration to HMF, followed by HMF oxidation to DFF. Recently, a potential catalyst with both redox potential and

Brønsted acidity, was employed for the direct production of DFF from fructose [106]. The obtained PMA-MIL-101 catalyst displayed a DFF yield of 75.1% at almost 100% fructose conversion, where the formation of DFF was accompanied by the sacrifice of HMF. Although the catalyst exhibited constant activity toward fructose in the recycling test, the yield of DFF displayed a decreasing trend along with the increasing yield of HMF, which was due to the loss of Mo species into the solution. Moreover, the characterizations of reused catalyst regarding the BET surface areas, morphology, and crystallinity were missing.

In addition, MOFs can also act as precursors for the preparation of a porous carbon-supported metal oxide. Li group designed a S doped Fe_3O_4 derived from Fe-based MOF (MIL-88B) and displayed good performance for the synthesis of DFF from fructose efficiently [107]. The resulting Fe-C/S catalyst exhibited an excellent DFF yield of 99% at a full conversion of fructose. In contrast, the undoped Fe/C catalyst just gave DFF yield of 10% with 73% fructose conversion, indicating the inevitable role of S in the catalytic reaction. In addition, the Fe/C-S(H^+) catalyst without Fe showed no activity for HMF oxidation, suggesting Fe species was crucial to the reaction. The superior performance of Fe-C/S was attributed to the fully exposed Fe_3O_4 (111) facets together with the non-oxidized elemental sulfur in the carbonaceous matrix.

2,5-dimethylfuran (2,5-DMF) is a very useful chemical and renewable fuel, the direct synthesis of 2,5-DMF from fructose with high productivity is attractive but challenging. Kim *et al* designed a bifunctional material synthesized by loading Pd nanoparticles on UiO-66, followed by depositing on sulfonated graphene oxide (GO) [108]. The resultant Pd/UiO-66@SGO composite with both Lewis and Brønsted acidity displayed good performance for the fructose transformation into 2,5-DMF in a one-pot reaction with 70.5% yield of 2,5-DMF. In addition, high 2,5-DMTHF selectivity with low selectivity of 2,5-DMF was observed for SiO_2 supported catalyst, suggesting that excess hydrogenation occurred caused by the high adsorption of 2,5-DMF in 4.8Pd/SiO_2. This phenomenon was due to the π–π interaction between C=C sp^2 of the graphitic carbon and the furan unsaturated ring (Fig. **13**). The yield of 2,5-DMF dropped sharply to 58.3% in the second run, and then displayed a slightly decreasing trend in the following cycles.

Transformation of sugars into methyl lactate *via* one-pot reaction process may hold many promising advantages over a fermentation derived procedure and has drawn considerable concern. Based on this, iso-reticular MOF-74 (Ni, Zn, Co, Mg) were used for the production of methyl lactate (ML) from glucose [109]. Ascribing to the larger surface area and smaller crystal size as well as the strong Lewis acidity, Mg-MOF-74 displayed the highest ML yield of 27% with a fructose yield of 1.7% in methanol, while MOF-74 with Ni, Zn and Co nodes only afforded ML yield of 16%, 19% and 20%, respectively. The catalyst displayed a slightly decreasing trend in the thrice cycling performance.

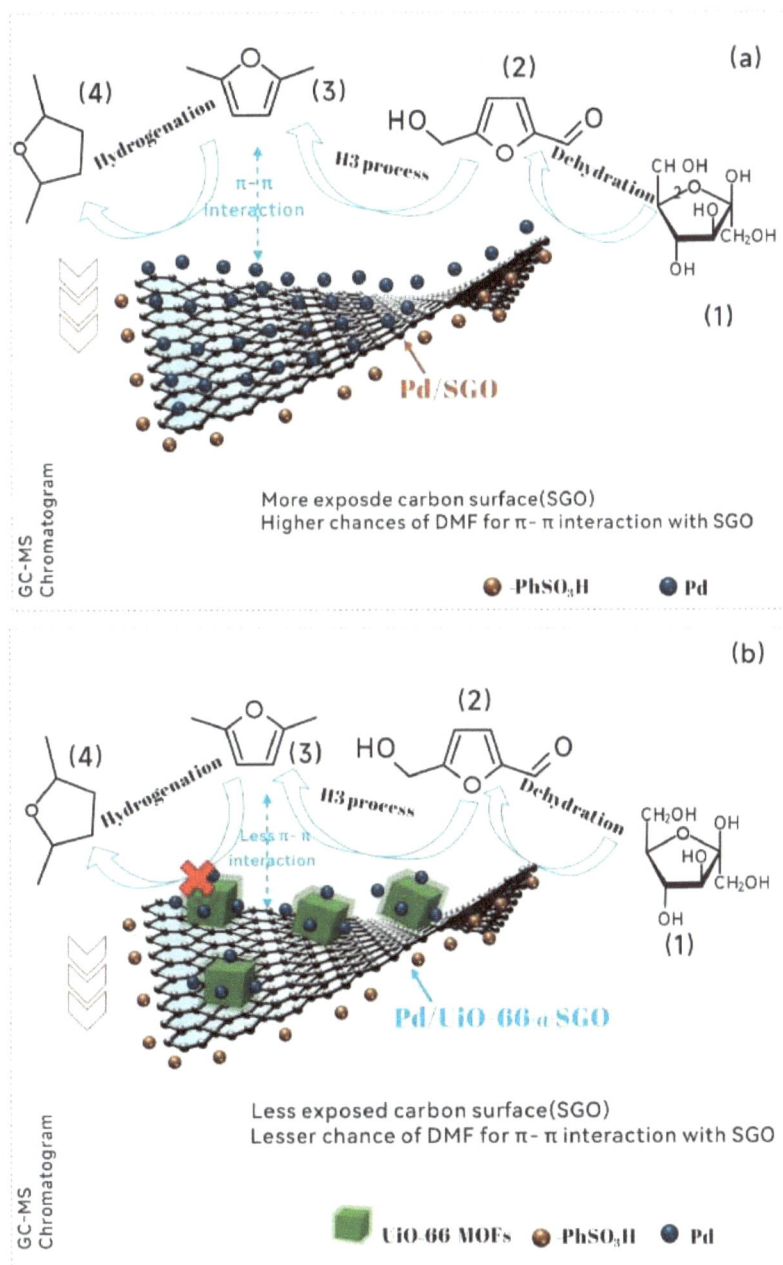

Fig. (13). The possible reaction route for the Pd-supported catalyst. Reprinted from ref. 108 with permission from The Royal Society of Chemistry, Copyright 2017.

Glucose can be transformed to levulinic acid (LA) through fructose as an intermediate by isomerization and dehydration process. Zhou *et al*. designed a bifunctional catalyst by loading lysine functionalized phosphotungstic acid ($H_3PW_{12}O_{40}$) (Lys-PTA) in MIL-100(Fe) for the upgrading of glucose to LA directly (Fig. **14**) [110]. The resultant Lys-PM$_2$ displayed the highest LA yield of 57.9% in comparison to other catalysts including Lys, PTA, Lys-PTA, PM$_2$ and MIL-100(Fe). The hydrogen bonding between the hydroxyl group and carboxyl group combined with the Lewis acid and Brøndsted acid as well as the mesoporous structures and large surface were responsible to the good catalytic performance. The LA yield decreased to 49% in the fourth cycling test, ascribing to the mass loss of catalysts during the washing procedure.

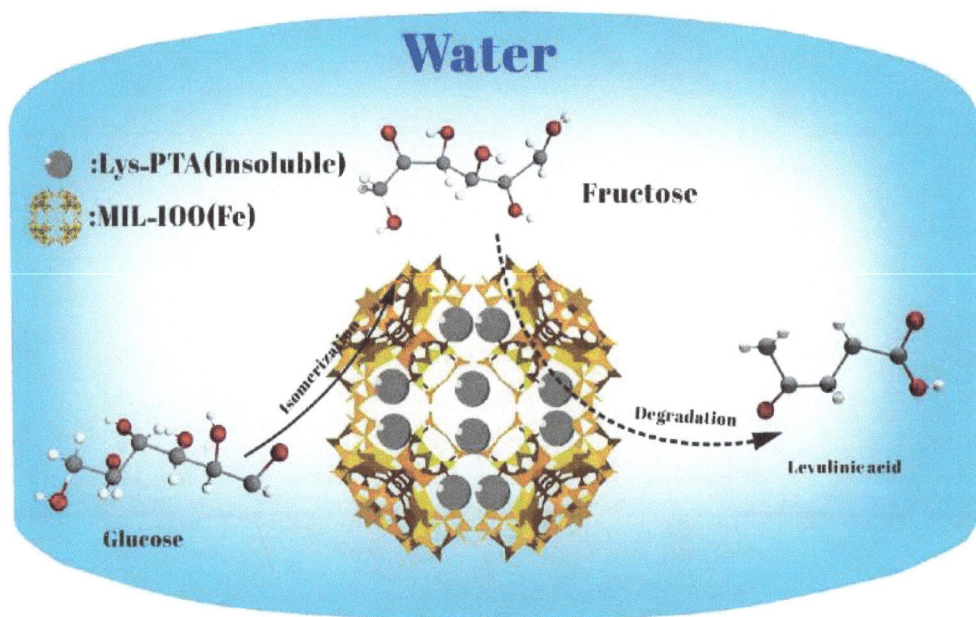

Fig. (14). The reaction process for the synthesis of LA from glucose over MIL-100(Fe) loaded Lys-PTA. Reprinted from ref. 110 with permission from Elsevier B.V, Copyright 2019.

MOF-derived materials have shown potential applications in dehydration of xylose owing to the presence of Lewis and Brønsted acid sites. Lam *et al* prepared a composite constructed from MIL-101 and mesoporous phosphate nanoparticles [111]. The obtained catalyst MF-2 with moderate SnP displayed the highest furfural (FF) yield of 86.7% under the optimized reaction conditions. MF-1 with less SnP content showed low activity and FF yield ascribed to the lack of active acid sites, while excess SnP in MF-3 would induce side reactions and decrease the yield of FF. Furthermore, the FF yield could be improved to 92.3% when 35 ppt

NaCl was added to the reaction solvent. No apparent decay in activity was observed in the initial seven runs and then a decrease was observed in the following cycles. Similar work was investigated by the same group, where a MOF composite constructed from MIL-101 and fly ash was synthesized and used for the xylose transformation [112]. Ascribing to the existence of Lewis acid sites from Cr center and Brøndsted acid sites from hydroxyl groups in fly ash, the obtained composite displayed a furfural yield of 71%, which was higher than bare MIL-101.

Besides, sulfonic acid-functionalized MOF was developed for the transformation of xylose to furfural in mixed solvents (Fig. **15**) [113]. The MIL-101(Cr) with SO$_3$H groups displayed 70.8% furfural yield at xylose conversion of 97.8%. Apparently, the good performance resulted from the synergistic effect between Cr acid center in the MIL-101 framework and the Brøndsted acid sites of sulfonic acid functional groups, which was also clarified by previous studies. As for the recycling test, MIL-101-SO$_3$H catalyst displayed a decreasing trend, and 63.2% furfural yield was achieved after five runs.

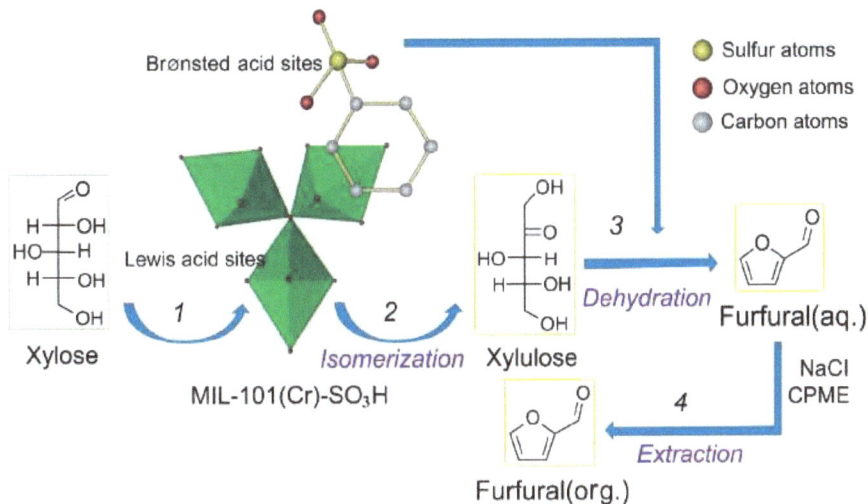

Fig. (15). Schematic illustration of MIL-101(Cr)-SO$_3$H catalyst for the conversion of xylose to furfural. Reprinted from ref. 113 with permission from Elsevier B.V, Copyright 2018.

2.3. Conversion of Platform Chemicals

2.3.1. Conversion of Levulinic Acid and its Derivates

Levulinic acid (LA) can be hydrogenated to synthesize γ-valerolactone (GVL), which is widely used in the field of solvent, fuel, fuel additive, and intermediates for other crucial fine chemicals. The hydrogenation of LA can be performed either with molecular H_2 as hydrogen sources, or using alcohols as a hydrogen donor through the "catalytic transfer hydrogenation" (CTH) process. Table **4** summarizes the MOFs based materials that have been applied in LA hydrogenation as well as the catalytic performance.

Table 4. Comparison of MOFs based materials for the synthesis of GVL from LA and the catalytic performance of yields. a: the solvent was not mentioned.

Catalyst	Reaction Conditions	Yield (%)	Cycle Number	Ref.
F-ZrF	2-PrOH, 473 K, 0.5 h	98	6	[114]
MIL-88(Fe)	2-PrOH, NaOH, 373 K, 10 h	99	5	[115]
HPW@MOF-808	2-PrOH, 433 K, 6 h	86	5	[116]
Ru/MIL-101(Cr)	H_2O, 343 K, 1MPa H_2, 5 h	99	4	[117]
Ru/Zr-BDC[a] nanosheets	363 K, 3MPa H_2, 5 h	99	6	[118]
Ru/MOF-808	H_2O, 363 K, 2MPa H_2, 24 h	99	5	[119]
Ru-CNF[a]	423 K, 4.5MPa H_2, 8 h	95	7	[120]
Ru/ZrO$_2$@C	H_2O, 413 K, 1MPa, 2 h	96.4	6	[121]
Ru/HfO$_2$@CN	H_2O, 353 K, 1MPa, 1.5 h	92.3	10	[122]
Ru@C-Al$_2$O$_3$	H_2O, 298 K, 0.1MPa, 2 h	99.9	7	[123]
Pd/UiO-66-NH$_2$	H_2O, 413 K, 2MPa, 2 h	98.2	5	[124]
Ni/C-500	2-PrOH, 473 K, 1MPa H_2, 5 h	98.2	8	[125]
Co^{R}NC/SMCNF	H_2O, 453 K, 4.5MPa H_2, 4 h	99	3	[126]

Until now, non-noble catalysts including Ni, Zr, and Fe have been used as catalysts for LA hydrogenation through CTH method. As a result, a series of Zr-based MOFs were synthesized from zircon salt and fumaric acid with different monocarboxylic acids as modulators in water and used for the synthesis of GVL from LA through CTH process [114]. Specifically, the formic acid modulated ZrF MOFs (F-ZrF) catalyst exhibited the highest yield of 98% at LA conversion of 96%, which was slightly higher than acetic acid and propanoic acid modulated ZrF MOFs. Moreover, the D-ZrF MOFs synthesized in DMF solvent gave only 78% yield of GVL. This may result from the lower surface areas and lower acidity of D-ZrF MOFs. The information about the spent catalyst regarding the crystallinity and surface areas seems missing.

Besides Zr-based MOFs, Fe-based MOFs were also applied in the transformation of LA to GVL. Fe-based MIL-88B displayed 99% GVL selectivity at full LA conversion in isopropanol with NaOH as a base additive [115]. In addition, MIL-88B(Cr) and NH$_2$-MIL-88B(Fe) also exhibited excellent catalytic performance in

LA conversion to GVL with 99% selectivity. However, MIL-53(Fe) showed a lower GVL yield of just 52%. The difference was owing to the higher magnitude of swelling and specific surface area in comparison to MIL-53(Fe). The conversion of LA to GVL was conducted to test the reusability of MIL-88(Fe), and it could maintain stable activity during the five runs. Polyoxometalates (POMs) and MOF-808 were integrated to fabricate HPW@MOF-808 composites for the production of GVL from biomass-based LA *via* the CTH approach [116]. The HPW@MOF-808 catalysts with different HPW loading were tested. It was found that their catalytic activity decreased in the order of 14%-HPW@MOF-808 > 20%-HPW@MOF-808 ≈ 9%-HPW@MOF-808 > 27%-HPW@MOF-808 > MOF-808, the best performance was achieved for 14%-HPW@MOF-808 with GVL yield of 86% at 99% LA conversion. The excellent performance was due to the presence of Lewis acidic Zr^{4+} sites and Brønsted acidic HPW sites (Fig. **16**).

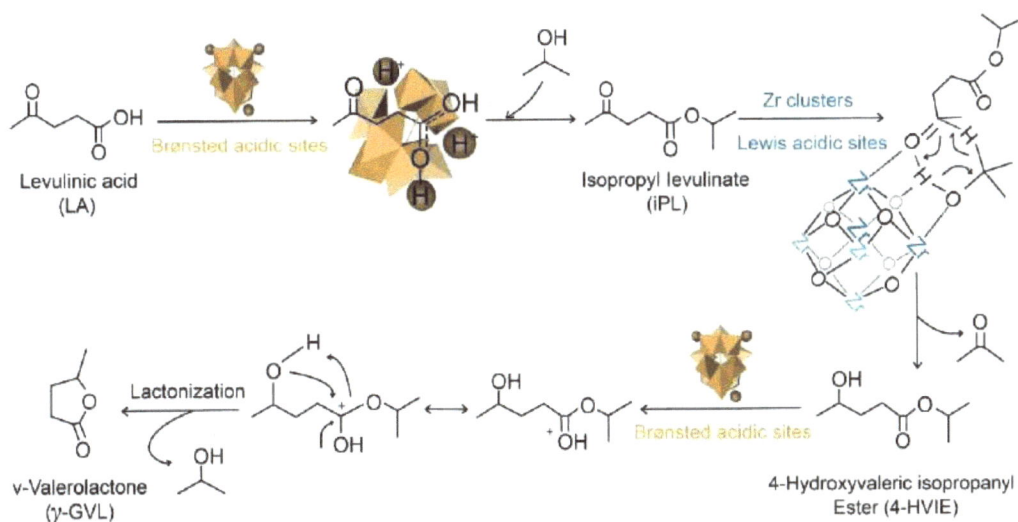

Fig. (16). Possible reaction pathway of LA to GVL over HPW@MOF-808. Reprinted from ref. 116 with permission from American Chemical Society, Copyright 2021.

MOFs could be used as catalysts in the hydrogenation of LA owing to their Lewis acid sites from metal centers. MOFs supported catalysts have also attracted considerable attention ascribing to their superior hydrogenation ability. Specially, numerous studies have been focusing on noble metals (Ru or Pd) supported on various MOFs for hydrogenation of LA to GVL. Chen *et al.* reported noble metals loaded on MOFs synthesized through impregnation for LA conversion into GVL [117]. Ru/MIL-101 could afford a 99% yield of GVL at the full conversion of LA in water. In contrast, Rh/MIL-101(Cr) gave a GVL yield of only 18% at LA conversion of 20%. While MIL-101 supported Ir, Pt and Pd catalysts displayed no activity to this reaction. Meanwhile, Ru supported on MIL-100(Cr) (composed of Cr SBU and trimesic acid) only gave a TOF of 18 mol_{LA} $mol_{Ru}^{-1} \cdot h^{-1}$, which was much lower than that of Ru/MIL-101(135 mol_{LA} $mol_{Ru}^{-1} \cdot h^{-1}$), attributing to the nature of MOF structures. The excellent performance was ascribed to the presence of Lewis acid sites and evenly metal species dispersion. The activity showed a sharp drop in the fourth run over Ru/MIL-101(Cr) catalyst, ascribing to the leaching of Ru metals (26%).

Ultrathin 2D MOF nanosheets were prepared by the surfactant assistance method through the self-assembly process. Ru nanoparticles (< 2 nm) were supported on the 2D MOF nanosheets for GVL production from LA [118]. As a result, the obtained Ru/Zr-BDC nanosheets exhibited an excellent 99% yield of GVL with complete LA conversion, which was higher than bulk Ru/Zr-BDC catalyst (28% yield of GVL). The remarkable difference between the 2D Zr-BDC and bulk sample resulted from the abundant exposed active sites and avoidable geometric constraints in 2D Zr-BDC nanosheets. A stable Zr-based MOF, MOF-808, was employed as a support for Ru metal and used as a catalyst for LA conversion (Fig. 17) [119]. The Ru nanoparticles existed both on the surface and in the pores of MOF-808. The obtained Ru/MOF-808 catalyst displayed 99% selectivity at the full conversion of LA in the aqueous solution. It can be used repeatedly for five runs without an obvious decline in LA conversion and GVL yield, suggesting the high stability of the Ru/MOF-808 catalyst.

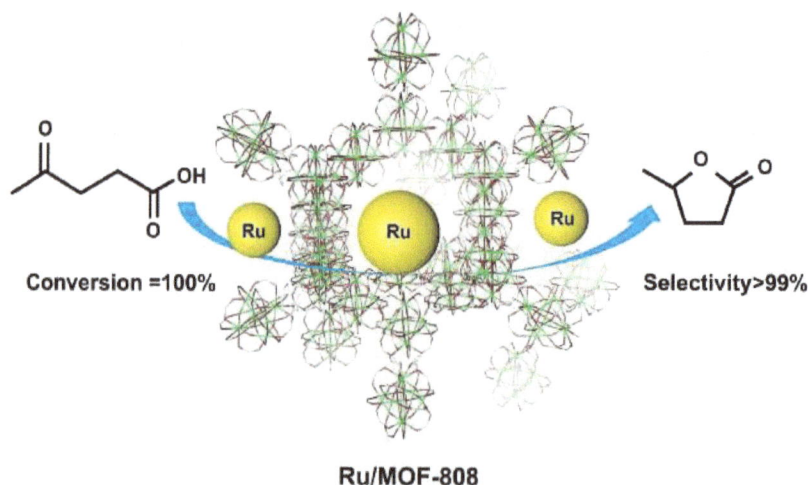

Fig. (17). Schematic illustration of Ru/MOF-808 for LA hydrogenation to GVL. Reprinted from ref. 119 with permission from Springer Science+Business Media, LLC, part of Springer Nature, Copyright 2021.

Ru nanoparticles supported on a porous carbon or metal oxide derived from MOFs have shown promising application in the hydrogenation of biomass-based LA to GVL. Yang *et al.* [120] reported a Ru catalyst encapsulated in porous carbon nanofibers (CNF) by doping Ru ions in the precursor solution of Zn-BTC MOF, followed by *in situ* thermal annealing in N_2 atmosphere. This Ru-CNF catalyst showed an increasing GVL yield with the increase of Ru doping, and the best performance was achieved for 0.300Ru-CNF with 95% yield of GVL at 96% LA conversion. In the recycling performance, the yield of GVL declined from 96 to 63% in the seven consecutive runs.

Ru/ZrO_2@C, synthesized by calcination of Zr-based MOF UiO-66 under N_2 atmosphere to obtain ZrO_2@C support, followed by wet impregnation of Ru precursor and reduced by aqueous $NaBH_4$ solution, was employed to the LA hydrogenation reaction [121]. The resulting Ru/ZrO_2@C catalyst with 0.85 wt% Ru displayed 96.4% GVL yield at LA conversion of 97.5%, affording a TOF of 0.17 s^{-1}. Ru/ZrO_2@C exhibited no obvious decline in the conversion of LA as well as the yield of GVL in the six runs. The even dispersion of Ru clusters as well as the strong interactions between Ru species and ZrO_2@C may account for the excellent performance. Moreover, Ru/ZrO_2@C catalyst displayed superior resistance to acid in the reaction system. A similar work was reported by Zhu group, they reported a Ru nanocluster catalyst loaded on HfO_2@CN (nitrogen dopped carbon) obtained from Hf-based MOF, and used for GVL production from LA [122]. This catalyst also displayed excellent GVL yield at full LA conversion as well as stable cycling tests compared with other catalysts, ascribing to the uniform dispersion of Ru clusters as well as the acidic sites. Soon afterwards, Wu *et al.* designed a Ru nano-catalyst embedded in carbon supported Al_2O_3 derived from MIL-53-NH_2 [123]. Different from previous method, the Ru ions were *in situ* encapsulated in the frameworks of MOF by adding the Ru salts in the precursor solution of MIL-53-NH_2 (Fig. **18**).

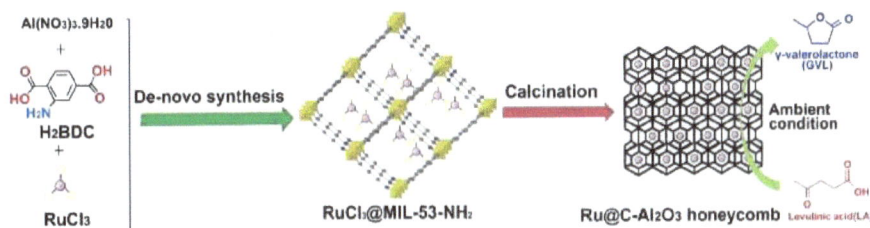

Fig. (18). Illustration of the Ru@C-Al_2O_3 catalyst synthesis and its application for LA conversion to GVL. Reprinted from ref. 123 with permission from Elsevier B.V, Copyright 2019.

The obtained $Ru@C-Al_2O_3$ catalyst displayed 99.9% yield of GVL at room temperature. Meanwhile, the $Ru@C-Al_2O_3$ catalyst could be reused for many runs without an obvious decay for LA conversion and GVL yield, the superior catalytic performance was resulted from the ultra-small Ru nanoparticles and high electron density around Ru nanoparticles.

Pd nanoparticles encapsulated in $UiO-66-NH_2$ were synthesized and investigated as catalysts for GVL production from LA (Fig. **19**) [124]. The resulting catalyst represented excellent GVL yield at almost complete LA conversion, and gave a high GVL productivity of 1741.1 mmol $g^{-1} \cdot h^{-1}$. On the contrary, $Pd/UiO-66-NH_2$ (through common impregnation way) exhibited a 94.5% yield of GVL for a much longer time. This may be due to the larger Pd nanoparticles and poor metal dispersion compared with a $Pd@UiO-66-NH_2$ catalyst. This catalyst exhibited no decay for activity and yield of GVL in the consecutive cycling and no leaching of Pd was observed. The excellent catalytic performance was due to the even Pd nanoparticles distribution as well as the encapsulation effect of $UiO-66-NH_2$.

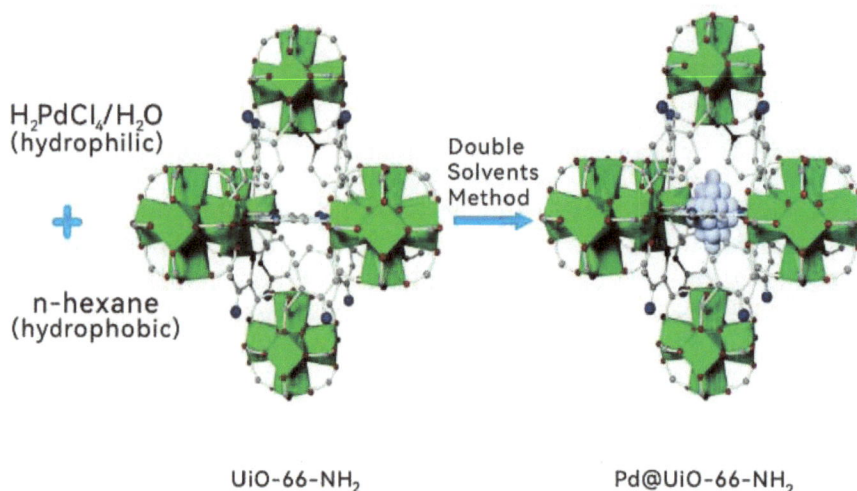

Fig. (19). The preparation method of $Pd@UiO-66-NH_2$ composite. Reprinted from ref. 124 with permission from Elsevier Inc, Copyright 2020.

A Ni/C catalyst derived from Ni-BTC MOF was reported by Yan group [125], and employed for LA hydrogenation into biofuels. The solvent used in this system had a significant impact on the type of product. The Ni/C-500 catalyst exhibited the highest GVL yield of 98.2% and 86.1% in 1,4-dioxane and IPA, while 86.5% methyl levulinate (ML) and 45.6% ethyl levulinate (EL) were obtained in MeOH and EtOH solvent, respectively. The Ni/C-500 catalyst displayed good recycling performance about six times without a marked decay, which may attribute to the protection from porous carbon, restraining the aggregation and leaching of Ni particles during the reaction process. Besides Ni catalyst, Co nanoparticles supported on nanofibers from the composite of ZIF-67 and sodium dodecyl sulfate modified wipe fiber, were reported and used for the transformation of LA to GVL [126]. The resulting Co^{R}NC/SMCNF catalyst displayed a 99% yield of GVL with almost 100% LA conversion, yielding a TOF of 206 h^{-1}, which was much better than other comparison samples. The excellent performance resulted from the strong interactions between the metal and the support, preventing Co nanoparticles from aggregation and leaching. This catalyst displayed a gradual decline in the GVL yield from 99% to 77% in the last cycle. On the contrary, the unmodified samples of Co^{R}NC/MCNF just can be reused for three runs with a sharp reduction in LA conversion and GVL yield.

2.3.2. Transformation of Alkyl Levulinates

In addition to LA, the upgrading of alkyl levulinates including methyl levulinate (ML) or ethyl levulinate (EL) to GVL is a promising candidate for biofuel production. An acid modulated UiO-66 was synthesized and employed for ML conversion to GVL [127]. The resultant UiO-66 displayed 96% GVL yield at 99% ML conversion with continuous flow in a fix-bed reactor. This catalyst exhibited excellent stability in the initial 9 h with constant ML conversion and GVL yield. As the time went on, the conversion of ML decreased gradually to 56% after 30 h time on stream, affording a GVL productivity of 58.94 mmol GVL g^{-1}·h^{-1}. The deactivation was ascribed to the loss of organic linker from the framework of UiO-66 and the partial collapse of structures.

UiO-66 with different loading of -SO$_3$H groups was prepared for the hydrogenation of ML to GVL (Fig. 20) [128]. Pure UiO-66 just afforded a 36% yield of GVL at ML conversion of 70%. In contrast, sulfonic acid functionalized UiO-66 displayed improved catalytic performance owing to the Brønsted and Lewis acid sites, and the highest yield of 85% GVL was achieved for UiO-66-S$_{60}$ with 99% ML conversion. This UiO-66-SO$_3$H displayed an excellent catalyst for the production of GVL from ML *via* CTH reaction, mainly because of the acid properties as well as its microporous structure combined with even dispersion of active sites. In addition, the UiO-S$_{60}$ catalyst displayed a decreasing activity of ML and yield of GVL during the recycling performance.

Zhang *et al.* prepared a Ru catalyst on SO$_3$H-UiO-66 and used it for the production of GVL from ML in an aqueous solution [129]. The Ru/SO$_3$H-UiO-66 catalyst with 5 wt.% Ru loading displayed 74.5% selectivity for GVL at 99.9% ML conversion, which was higher than Ru/C, Ru/NH$_2$-UiO-66, and Ru/UiO-66 catalyst under similar parameters. The superior performance of Ru/SO$_3$H-UiO-66 was ascribed to the uniform metal dispersions, indicating the significant role of support. Furthermore, this catalyst could be run for several cycles without distinct deactivation. Later, the Ru catalyst loaded on MIL-101-SO$_3$H was prepared for ML conversion to GVL [130]. Unlike the result of the Ru/SO$_3$H-UiO-66 catalyst, the activity of ML and selectivity of GVL increased with the amount of -SO$_3$H groups, and the highest performance was obtained for Ru/MIL-101-S$_{100}$ catalyst, corresponding to 86.6% GVL selectivity and 98.8% ML conversion, which was ascribed to the highest acid density in Ru/MIL-101-S$_{100}$. In contrast, the physical mixture of Ru/MIL-101 and MIL-101-SO$_3$H gave a lower GVL yield of 70.5%, indicating the importance of the intimate contact between metal and acid active species. Recently, the Zhu group prepared a hybrid composite composed of Ru nanoparticles on a UiO-66 packed heteropolyacid H$_4$SiW$_{12}$O$_{40}$ (SiW) as a solid catalyst for the production of GVL from ML [131]. The obtained catalyst displayed an exclusive yield of GVL with full ML conversion, which was better than Ru/UiO-66 catalyst. This result suggested the Ru species and SiW combined with UiO-66 support contributed to the excellent catalytic performance. The catalyst displayed a stable durability in the five cycles, resulting from the local defects and confinement effect of UiO-66 frameworks.

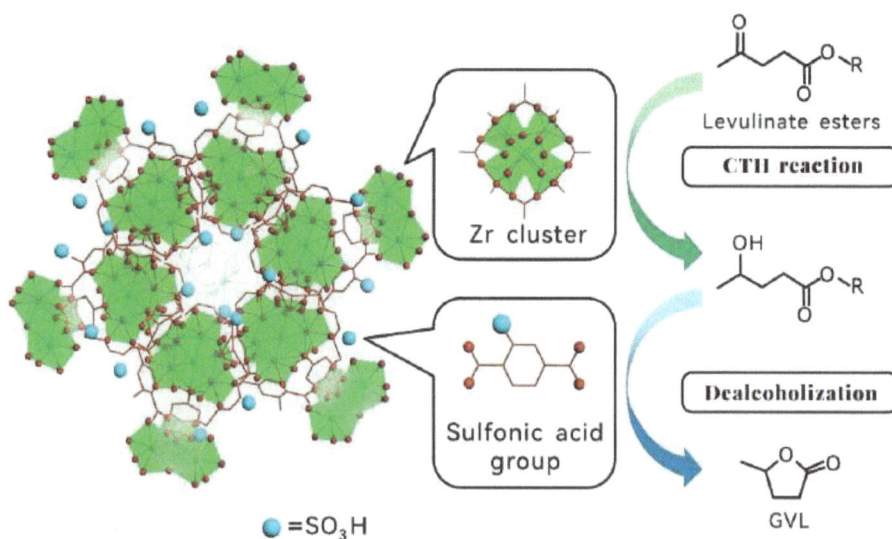

Fig. (20). The production of GVL from levulinate esters *via* CTH route. Reprinted from ref. 128 with permission from American Chemical Society, Copyright 2017.

In addition to ML, ethyl levulinate (EL) was also studied for the formation of GVL. Hwang and co-authors reported a Zr-based MOF UiO-66 with different ligand functionalities (H, COOH, and NH_2) for GVL production from EL *via* CTH route [132]. UiO-66(Zr) displayed the highest yield of GVL (53.5%) with complete EL conversion, while -COOH and -NH_2 functionalized with UiO-66 just gave a GVL yield of 13.9% and 27.3% at EL conversion of 28.5% and 97.6%, respectively. The low activity of UiO-66(Zr)-COOH may result from the limited access to reactive sites with narrow pore size in UiO-66(Zr)-COOH framework. While the transesterification reaction was predominant in UiO-66(Zr)-NH_2 catalyst. As a result, the pore size, surface area as well as an acid-base feature were crucial to the selective hydrogenation of EL to GVL. After five cycles, no visible reduction in EL conversion and GVL yield was detected, suggesting the high stability of the catalyst.

DUT-67 framework is built by 8-connected metal nodes, while UiO-66 are composed of saturated 12-connected metal nodes. The performance of these heterogeneous catalysts was tested for EL conversion in alcohols [133]. The highest yield of GVL (90.5%) was obtained for DUT-67(Hf), which was higher than DUT-67(Zr) (45.3%). In addition, the performance of DUT6-7(Zr or Hf) was better than UiO-66(Zr or Hf), which can be attributed to the unsaturated metal sites that existed in DUT-67, thus facilitating the reactants' access to the active sites. A very similar work was done by the Park group [134], they reported the DUT series MOFs as catalysts for the transformation of EL to GVL. Interestingly, some opposite conclusion was achieved compared with Zhang's work [133]. The best performance was achieved for DUT-52(Zr) with 94% GVL yield at 99% EL conversion. This result was entirely in contrast with a previous work. The superior catalytic performance of DUT-52(Zr) might result from the larger surface area, moderately acidic and basic sites along with porous framework.

The impact of various functionalized groups (-SO_3H, -NH_2 and -NO_2) in Ru/UiO-66 was systematically studied for EL conversion to GVL [135]. Ru/UiO-66 without functionalized groups displayed the highest EL conversion of 100% in water at 80 °C for 1 h, which was higher than –SO_3H, –NO_2 and –NH_2 functionalized UiO-66. The -SO_3H group had a little effect on the activity, while -NH_2 groups might have a strong interaction with Ru nanoparticles, inhibiting the H_2 activation and EL adsorption. When the dosage of Ru/UiO-66 decreased to half, the conversion of EL was still nearly 100%, indicating the fast reaction process. It is a pity that the durability and stability of the catalyst were not discussed.

2.3.3. Transformation of 5-Hydroxymethylfurfural (HMF)

HMF is generally regarded as one of the key platform chemicals for upgrading of lignocellulosic to high valuable feedstocks or fuels. Zeng *et al.* prepared CuNPs@ZIF-8 catalyst by simple impregnation of Cu^{2+} ions in ZIF-8 dispersion, followed by *in situ* reduction (Fig. **21**) [136]. The resultant catalyst displayed a 99% yield of 2,5-dihydroxymethylfuran (DHMF) with the completed conversion of HMF. The selectivity to DHMF remained constant with a gradually decrease in the HMF conversion in the four times cycling tests. The loss of activity may due to the aggregation of Cu nanoparticles. Meanwhile, this CuNPs@ZIF-8 can be used for hydrogenation of various unsaturated aldehydes to their corresponding alcohols with good to excellent yield.

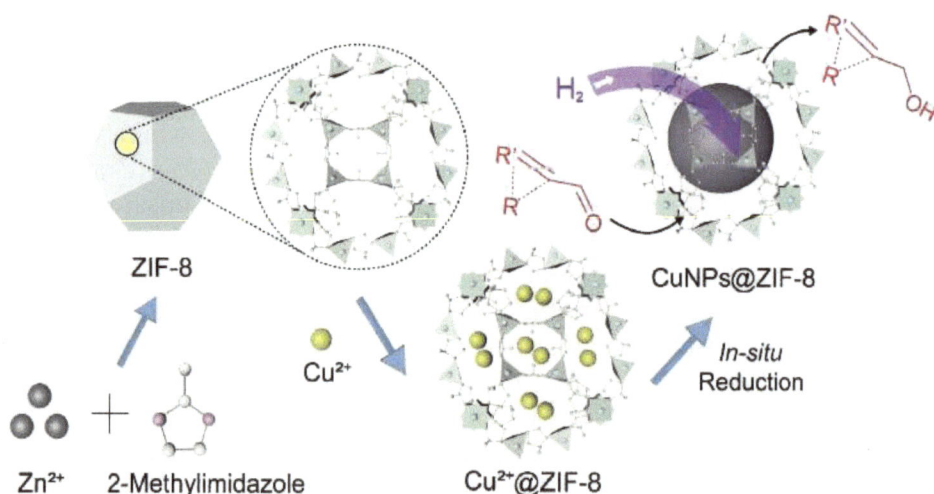

Fig. (21). The synthetic route for Cu@ZIF-8 and catalysis procedure. Reprinted from ref. 136 with permission from The Royal Society of Chemistry, Copyright 2019.

Ethyl levulinate (EL) and 5-ethoxymethylfurfural (EMF) can also be produced from HMF, and they are promising candidates in energy and chemical feedstocks. Chen *et al.* reported a MOF-based phosphomolybdic acid (Cu-BTC/HPM) composite by one pot *in situ* encapsulating method (Fig. **22**) [137]. The highest yield of 55% for EMF and 11% for EL was achieved at the optimized reaction conditions for Cu-BTC/HPM. This catalyst could be recycled for five times, and the EMF yield displayed a decreasing trend along with an increasing yield for EL. This phenomenon may result from the leaching of Cu species from the framework of Cu-BTC. A similar work was done by the same group of Chen. Sulfonic-aci-

-functionalized porous carbon (C-SO$_3$H) was synthesized by calcination of the HKUST-1 precursor [138]. About 71% yield of EMF with a 22% yield of EL was obtained at a low temperature of 100 °C over C-SO$_3$H catalyst, while 73% yield of EL with about 1% EMF was achieved at a temperature of 140 °C.

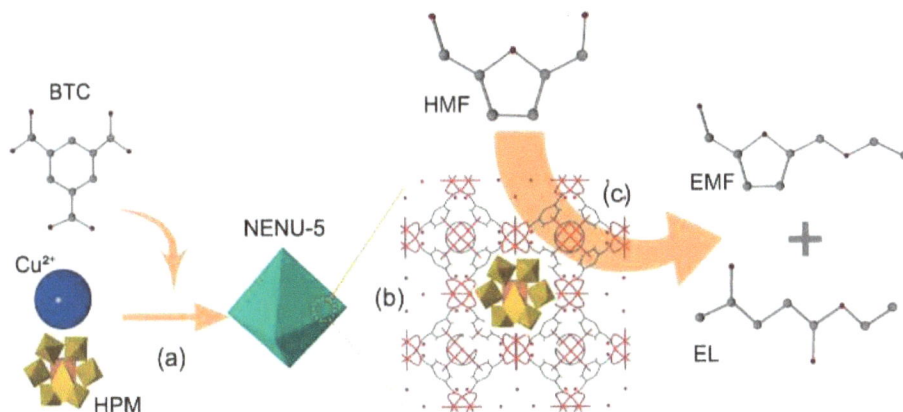

Fig. (22). The fabrication route of catalyst and catalysis procedure. Reprinted from ref. 137 with permission from The Royal Society of Chemistry, Copyright 2016.

HMF can be hydrogenated to DHMTHF over metal supported catalyst, and DHMTHF is an important solvent and intermediate for the synthesis of 1,6-hexanediol. Gao group reported a Pd catalyst supported on amine-functionalized MOF for DHMTHF production from HMF (Fig. **23**) [139]. The resultant Pd/MIL-101-NH$_2$ catalyst displayed 96% DHMTHF yield. On the contrary, MIL-101(Cr), MIL-53(Al)-NH$_2$, and MIL-53(Al) supported Pd catalyst gave 76%, 88%, and 61% selectivity for DHMTHF. Notably, the amine-functionalized MOFs exhibited superior catalytic performance than the unfunctionalized ones, attributing to the anchoring of Pd on the NH$_2$ groups. Of course, the performance was also consistence with the particle size of Pd and surface areas of supports as well as the absorption ability to reactants. The selectivity to DHMTHF reduced to 80% in the fifth run over the Pd/MIL-101-NH$_2$ catalyst, which may ascribe to the decrease in crystallinity and surface area caused by the blocking of the insoluble polymer.

The transformation of HMF to DMF has drawn a considerable concern because of its high energy density, octane number, stability and so on. Pt nanoparticles embedded in UiO-67 were prepared by replacing some BPDC linkers with BPYDC linkers and employed for the upgrading of HMF to DMF [140]. HMF conversion decreased in the order: PtNPs@UiO-67 (31%) > Pt@UiO-67 (11%) >

UiO-67 (3%). Soon after, they synthesized a 10%Pt@MOFs-T3 composite through *in situ* one-step route [141]. The resultant 10%Pt@MOFs-T3 catalyst displayed 86.1% yield of 2,5-DMF with complete conversion of HMF, which was much higher than Pt@MOFs-T2 and Pt@MOFs-T1 with a 2.5-DMF yield of 65.3% and 25.6%, respectively. The superior performance was resulted from the unique structures as well as much more exposed active sites. The post-modified sample (10% Pt/UiO-67) also showed poor yield for target product, ascribing to the larger particle size of Pt and aggregation of nanoparticles.

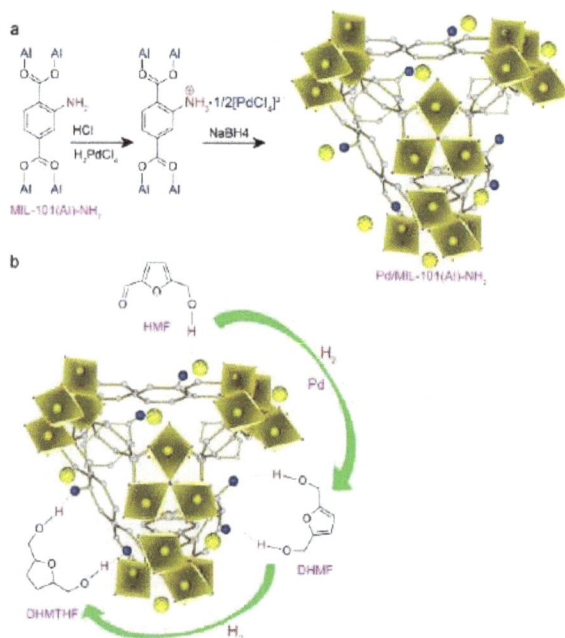

Fig. (23). The synthesis of Pd/MIL-101-NH$_2$ catalyst and catalysis procedure for Pd/MIL-101-NH$_2$ catalyst. Reprinted from ref. 139 with permission from American Chemical Society, Copyright 2015.

Besides the guest-MOF system, metal or metal oxides from MOFs were also reported as heterogeneous catalysts for HMF transformation to 2,5-DMF. The Liu group prepared a CuO$_x$@C catalyst by using HKUST-1 as a precursor through two step annealing process (Fig. **24**) [142]. The highest performance was achieved for the Cu-N450 catalyst with 2,5-DMF yield and HMF conversion of 56% and 98% respectively. After treatment in 1% O$_2$ atmosphere, the 2,5-DMF yield was further increased to 91.2% with complete conversion of HMF over Cu-O250. Recycling performance was conducted for Cu-O250, however, the yield of 2,5-

DMF and conversion of HMF decreased sharply from 91.2% to 32.4% and 100% to 85.6%, respectively. Such poor durability was attributed to the disappearance of Cu^{1+} and formation of humins, thus weakening the synergistic effect between Cu^{1+} and Cu^0 as well as covering the active sites.

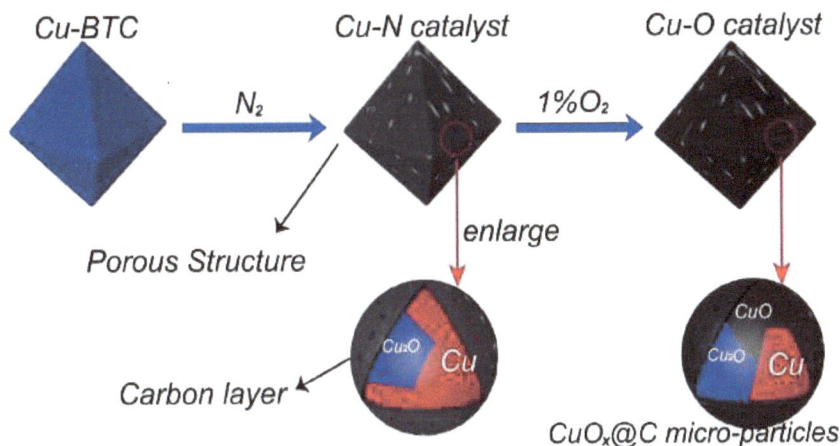

Fig. (24). Schematic illustration of structural changes of Cu-BTC calcined. Reprinted from ref. 142 with permission from Wiley-VCH Verlag GmbH & Co. KGaA, Weinheim, Copyright 2019.

A single metal displayed excellent catalytic performance in HMF hydrogenolysis, and dual metals also exhibited potential application in this field. The Cu-Pd catalyst encapsulated in porous carbon has been synthesized through thermal treatment Cu-BTC supported Pd composite [143]. The obtained Cu-Pd@C-B catalyst displayed 97% yield of 2,5-DMF with complete HMF conversion in THF. In contrast, the Cu-Pd@C-A just showed 36.6% 2,5-DMF at the poor conversion of HMF (48.8%). In addition, the performance of the Cu-Pd@C catalyst was better than other dual metal-based catalysts such as Cu-Co@C, Cu-Fe@C, and Cu-Ni@C. The superior catalytic performance of Cu-Pd@C-B was resulted from the generation of Cu-Pd alloy rendering reduced Cu-Cu distance and increased Pd-Pd distance. Last, the Cu-Pd@C-B showed no obvious decline in the 2,5-DMF yield in the consecutive five cycles. The Cu-Pd alloy effect and the protection of metal sites by wrapped carbon contributed to the improved performance of catalyst. Gui *et al*. designed a multifunctional Pd immobilized carbon catalyst by using ZIF-67 as a template for HMF hydrogenation [144]. All the catalysts showed a 100% HMF conversion, and 2,5-DMF yield displayed a volcanic type with the increasing roasting temperature, and the highest yield of 97.8% 2,5-DMF was obtained for PCNC-700. This PCNC-700 catalyst displayed excellent stability during the six runs without an apparent loss for HMF activity, and 97% yield of 2,5-DMF could be retained after reuse. The good catalytic performance was associated with the well Pd stabilization on the carbon-based support.

DFF is very crucial in industry due to its wide application in the production of fungicides, polymers and pharmaceuticals. No noble Fe-Co catalyst derived from the bimetal MOF of MIL-45b was reported by Li group [145]. The resultant FeCo/C(500) catalyst displayed the highest catalytic performance of 99% DFF selectivity and full conversion of HMF. Among these FeCo/C catalysts, the highest HMF adsorption was observed over FeCo/C(500) with no obvious difference in the DFF adsorption, which was in accordance with the catalytic results. The magnetic FerCo/C(500) catalyst could be recycled for six times with no noticeable decrease in the DFF yield, further indicating the stability of the catalyst. Lin *et al.* employed the Cu-based HKUST-1 for the first time as a solid catalyst for HMF oxidation to DFF [146]. The HKUST-1 combined with TEMPO exhibited a 96% DFF yield at 96% HMF conversion. In contrast, HKUST-1 or TEMPO alone with or without O_2 could not catalyze this reaction, however, HKUST-1 plus TEMPO without O_2 showed 65% DFF yield, suggesting the indispensable role of HKUST-1, TEMPO and O_2 in this reaction. The HKUST-1 catalyst displayed a gradual decrease in the DFF yield and HMF conversion in the cycling test. However, the reason for deactivation was not discussed. A similar work was reported by the same Lin group [147]. At this time, they combined the HKUST-1 with TEMPO for the conversion of HMF to DFF under microwave (MW) instead of O_2. And the yield of DFF could be further improved to 99% using MW in a shorter time of 1 h. The HMF was first adsorbed on a Cu-O cluster of HKUST-1 to form intermediates, followed by reaction with TEMPO to produce TEMPOH, which further yielded DFF under MW (Fig. **25**).

The transformation of HMF into dimethylfuran dicarboxylate (DMFDCA) has attracted increasing attention due to its promising application of DMFDCA in polyester synthesis. A Co@C-N material derived from ZIF-67 was employed in selective conversion of HMF to DMFDCA [148]. Co@C-N(800) displayed the highest yield of 91% DMFDCA at HMF conversion of 99%, which was ascribed to the suitable nitrogen content and relative high proportion of graphitic-N/C. Thus, this balance between the amounts and species of nitrogen led to higher electronic mobility and more active sites. This Co@C-N(800) catalyst displayed no appreciable change in the DMFDCA yield and HMF conversion after reduction in H_2 atmosphere.

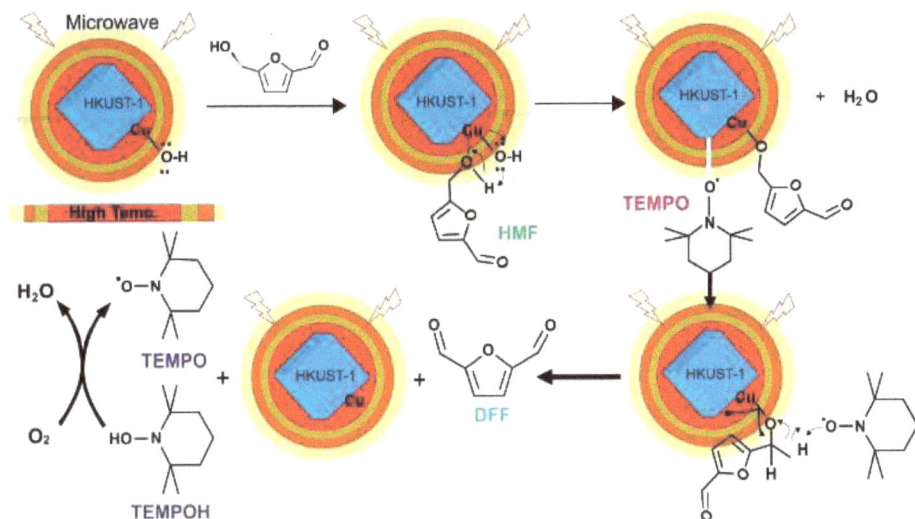

Fig. (25). A proposed mechanism for enhanced HMF oxidation to DFF using HKUST-1/TEMPO by MW. Reprinted from ref. 147 with permission from Springer-Verlag GmbH Germany, part of Springer Nature, Copyright 2020.

The oxidation of HMF to 2,5-furandicarboxylic acid (FDCA) has been considered as an important monomer for producing valuable polymers. Wu group prepared a N-doped nano-porous carbon (NNC) material by using ZIF-8 as a sacrifice template for HMF oxidation to FDCA [149]. The FDCA yield increased with the calcination temperature of catalyst from 11% for NNC-600 to 80% for NNC-900. The superior catalytic performance was associated with a higher graphitic nitrogen (N-Q) content in NNC-900. The presence of N-Q sites could facilitate the generation of oxygen radicals for oxidation. This metal free NNC-900 catalyst showed a gradual decrease in the FDCA yield from 80% to 70% in the last cycle, while the conversion of HMF remained constant. The loss of FDCA yield was due to the slight reduction in the N-Q amount. Li *et al.* prepared a metal oxide nanoparticle confined in the mesoporous KIT-6 derived from the pyrolysis of self-assembly MOFs in silica mesopores and applied in the application of HMF oxidation (Fig. **26**) [150]. Among the metal oxide@KIT-6 including Co, Ni, Cu and Fe catalysts, Co@KIT-6 displayed the highest catalytic performance of 99% yield of FDCA. In addition, the Co/KIT-6 synthesized by impregnation method just showed 9.5% FDCA yield with 10% HMF conversion. Furthermore, the Co@KIT-6 catalyst showed no noticeable loss in the activity and yield in cycling performance. The superior performance of Co@KIT-6 resulted from the presence of ultrafine Co_3O_4 nanoparticles and the confinement effect of KIT-6.

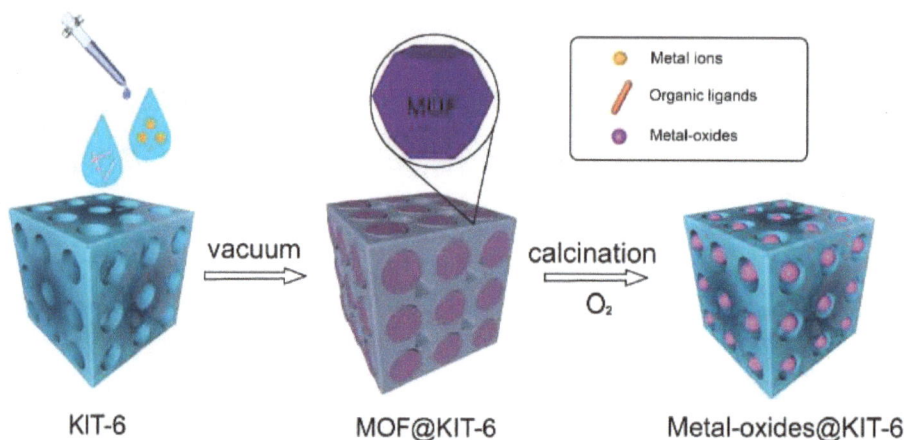

Fig. (26). Synthesis route of metal-oxides@KIT-6 from ZIF-67-KIT composites. Reprinted from ref. 150 with permission from The Royal Society of Chemistry, Copyright 2018.

Besides Co_3O_4 metal oxide, Mn-MOF derived Mn_2O_3 nanoflakes were also subjected to the oxidation of HMF to FDCA [151]. The pyrolysis temperature had a crucial impact on the catalytic performance, M400 sample showed 99.5% yield of FDCA with complete HMF conversion in the presence of $NaHCO_3$. While M350 displayed a lower yield of 76.1% FDCA, which may ascribe to the lower reactivity of Mn_3O_4 in the M350 catalyst. For comparison, Mn_2O_3 and MnO_2 exhibited FDCA yields of 5% and 12%, respectively. The high performance for M400 was due to the rich surface pores, which can facilitate the diffusion of reactants and enrich the active sites.

The alloy effect has a great impact on various catalytic reactions. The Co_3O_4 loaded Au-Pd alloy *via* pyrolysis of Au-Pd encapsulated in ZIF-67 composite was prepared and employed for HMF oxidation (Fig. **27**) [152]. The yield of FDCA was largely dependent on the ratio of the Au/Pd in $AuPd@Co_3O_4$ catalyst, and the highest yield of 87% was achieved over $Au_{0.5}Pd@Co_3O_4$. Furthermore, the performance of $Au_{0.5}Pd@Co_3O_4$ was much better than the individual $Au@Co_3O_4$ and $Pd@Co_3O_4$ with FDCA yields of 22% and 0%, respectively. The yield of FDCA could be further improved to 95% after adding a certain amount of sodium carbonate under similar reaction conditions. This $Au_{0.5}Pd@Co_3O_4$ catalyst displayed stable cycling performance during the 10 runs without a marked loss in HMF conversion and FDCA yield, which was due to the strong metal-support interactions combined with the higher percentage of Au^+ as well as the hydroperoxyl radicals produced from H_2O_2.

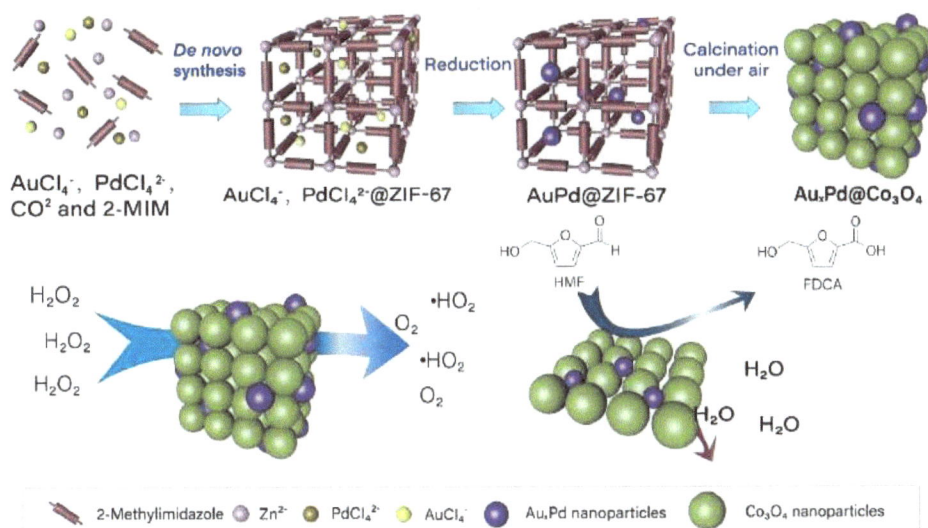

Fig. (27). *De Novo* synthesis of zeolitic imidazolate framework-67-derived Au-Pd embedded cobalt oxide cages applied for HMF oxidation into 2,5-furandicar-boxylic acid. Reprinted from ref. 152 with permission from Elsevier B.V, Copyright 2020.

2.3.4. Conversion of Furfural (FF)

Furfural (FF) is also a very important platform chemical. About 60-70% FF is converted into furfuryl alcohol (FFA) in the world by selective hydrogenation. FFA compound has wide potential applications in the polymer industry, fine chemicals and so on. Table **5** summarizes the MOFs-based materials that have been used for FF hydrogenation to FFA.

Table 5. Comparison of MOFs-based materials for FF hydrogenation to FFA and the catalytic performance of yields. a: the value meant selectivity to FFA.

Catalyst	Reaction Conditions	Yield (%)	Cycle Number	Ref.
Hf-MOF-808	2-PrOH, 373 K, 2 h	97	5	[153]
M-MOF-808	2-PrOH, 355 K, 2 h (reflux)	94.1	5	[155]
UiO-66	2-PrOH, 413 K, 5 h	97	5	[156]
Ru/UiO-66	H_2O, 298 K, 0.1MPa H_2, 4 h	94.9	5	[157]
Ru/Al-MIL-53-BDC	H_2O, 293 K, 0.5MPa H_2, 2 h	100[a]	N	[158]
CeO_2/Pd@MIL-53	H_2O, 353 K, 6MPa H_2, 2 h	85.3[a]	10	[159]
Pt-CeO_2@UIO-66-NH_2	2-PrOH, 353 K, 1MPa H_2, 30 h	99	5	[160]
Pd/H-UiO-66	H_2O, 333 K, 0.5MPa H_2, 3 h	99	10	[161]

(Table 5) cont.....

Catalyst	Reaction Conditions	Yield (%)	Cycle Number	Ref.
Pt-Sn@UiO-66-NH$_2$	433 K, 11.4 mL/min, 0.1MPa (fixed-bed)	93[a]	--	[162]
PtSn$_{0.1}$@UiO-66-NH$_2$	433 K, 11.4 mL/min, 0.1MPa (fixed-bed)	98[a]	--	[163]
P/UiO-66	2-PrOH, 453 K, 2 h	99.4[a]	5	[164]
Ni/MFC-500	MeOH, 433 K, 2MPa H$_2$, 4 h	59.5[a]	6	[165]
Ni/Al$_2$O$_3$-C-500	EtOH, 433K, 4MPa H$_2$, 6 h	93.2[a]	4	[166]
Cu@C-600	2-PrOH, 373 K, 1MPa H$_2$, 24 h	99.4	--	[167]
ZJU-199-350	2-PrOH, 403 K, 1MPa H$_2$, 3 h	97	5	[168]
Fe$_3$O$_4$/C	2-PrOH, 473 K, 2MPa N$_2$, 4 h	98.5[a]	4	[169]
CuCo/C-873	EtOH, 413K, 3MPa H$_2$, 1 h	97.7[a]	4	[170]
CoNi@NC-800	MeOH, 393 K, 2 MPa H$_2$, 1 h	99	5	[171]

Corma *et al.* applied a series of UiO-66 (12 connected) and MOF-808 (6-connected) for the transformation of FF to FFA in isopropanol *via* the CTH route (Fig. **28**) [153]. Among which, Hf-MOF-808 with unsaturated metal sites displayed 97% yield of FFA, and the reaction rate was two times higher than the Zr-MOF-808. Of course, the catalytic performance of MOF-808 was higher than those of UiO-66 series, attributing to the high density of Brønsted acid sites in MOF-808. Furthermore, MOF-808-Hf could be extended to other α,β-unsaturated carbonyl compounds and converted to their corresponding allylic alcohols with high selectivity.

Besides, Corma group employed these series of Hf and Zr-based MOF-808 or UiO-66 in FF condensation with acetone [154]. In this case, UiO-66(Hf) displayed the highest activity with a TOF value of 9.7 h^{-1}, while UiO-66(Zr) afforded a lower reaction rate (TOF of 1.7 h^{-1}). As for MOF-808, both Zr and Hf-based MOF showed lower activity (TOF=5 h^{-1}) than UiO-66(Hf) catalyst. And no obvious difference was observed for MOF-808(Zr) and MOF-808(Hf). In addition, the UiO-66-NH$_2$(Hf) catalyst displayed a lower TOF of 1.5 h^{-1} compared with UiO-66(Hf), which may ascribe to the sufficient space of acid and basic sites interacting with the carbonyl group of FF that increases the polarization and accelerates the attack of acetone. MOF-808 is really attractive due to its unique chemical and physical properties, high surface areas, six coordination features with unsaturated metal sites. The Zr centers in SBU can act as Lewis acid sites and the coordinated OH group or formic acid can be regarded as Brønsted acid sites. Hwang *et al.* prepared a methanol modified Zr-MOF-808 (M-MOF-808) as a solid catalyst for FF hydrogenation to FFA [155]. At high reaction temperature of 82 °C, no significant difference was observed between the M-MOF-808 and unmodified one, which may ascribe to reaching the equilibrium point. The huge difference was observed when the temperature decreased to 40 °C, a high FFA yield of 85.5% with 96.5% FF conversion was displayed for M-MOF-808, whereas MOF-808 just gave the value of 25.4 and 27.5% under the same reaction conditions, respectively. This catalyst displayed a FFA yield drop in the second cycle and then kept constant in the following runs.

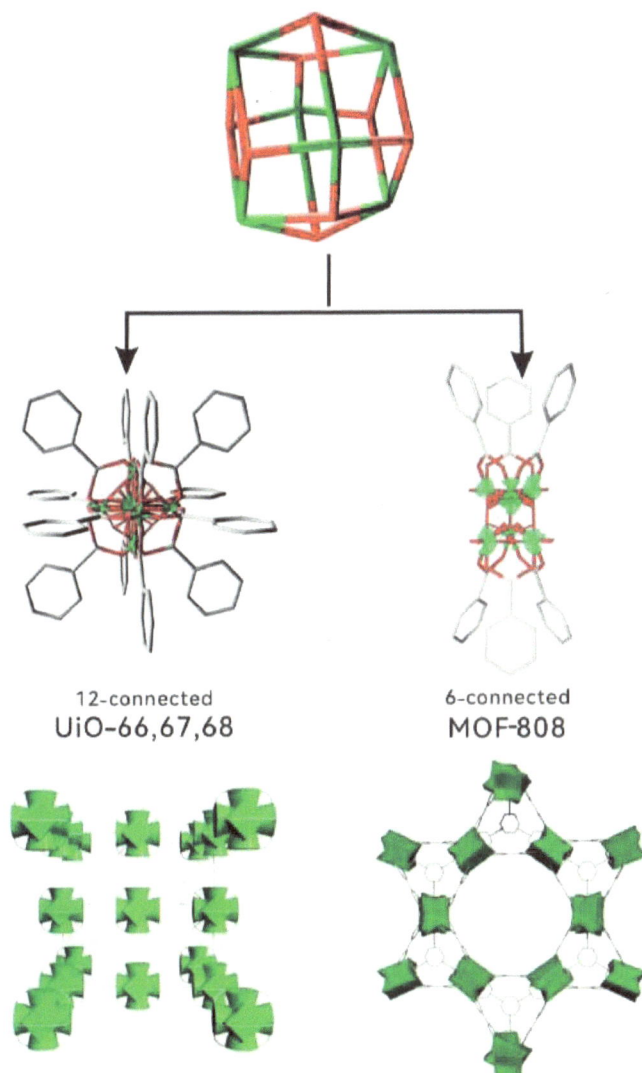

12-connected
UiO-66,67,68

6-connected
MOF-808

Fig. (28). The topology of different Zr-based MOFs. Reprinted from ref. 153 with permission from Wiley-VCH Verlag GmbH & Co. KGaA, Weinheim, Copyright 2018.

In addition, other Zr-based MOFs were also employed for the direct conversion of FF into FFA in isopropanol *via* the CTH route [156]. Amongst, UiO-66 (12 connected) displayed the best catalytic performance with 97% yield of FFA at full conversion of FF. MOF-808 with 6 connected structures showed lower FFA yield of 89%, while UiO-66-NH$_2$ exhibited a poor yield of just 42%. The discrepancy for these Zr-MOFs may result from the Lewis' acidity and structural characteristics. The yield of FFA over UiO-66 catalyst displayed a gradual decrease from 97% for the first cycle to 80% for the fifth run, which is probably due to the deposition of polymers on the MOF surface. Meanwhile, this UiO-66 catalyst was capable of catalyzing other aldehydes to their corresponding alcohols. Additionally, MOFs-guest compartments like MOF-supported metals (Ru, Pd, Pt, and Cu), and MOF-encapsulated objects were also investigated for FF hydrogenation. The first report of metal supported on the MOF for the conversion of FF to FFA was conducted by the Xue group [157]. The highest yield for FFA was achieved for Ru/UiO-66(94.9%), while MIL-140C, MIL-140B, MIL140A, Zr$_6$-NDC, and UiO-67 gave the FFA yield of 20.4%, 46.7%, 50.1%, 63.3%, and 88.4%, respectively, which was associated with the particle size, redox properties of RuO$_x$ and the interaction between Ru nanoparticles and MOF support as well as the structures of MOFs. The same group reported another work regarding Ru nanoparticles supported on Al-based MOF for the conversion of FF to FFA [158]. 2.9Ru/Al-MIL-53-BDC displayed 99.9% selectivity to FFA with full conversion of FF, while 3.0Ru/Al-MIL-53-ADP showed lower FF conversion of 44%. The higher activity for Ru/Al-MIL-BDC may attribute to its higher surface area combined with the lower reduction temperature of Ru.

The MOF shell was grown in situ on CeO$_2$/Pd by using Al$_2$O$_3$ as a sacrificial template *via* atomic layer deposition (ALD) method (Fig. **29**). The obtained CeO$_2$/Pd@MIL-53(Al) core-shell composites, on the one hand, can stabilize the Pd nanoparticles, and can also improve the surface area of CeO$_2$/Pd [159]. The selectivity for FFA was 85.3% with full conversion of FF over CeO$_2$/Pd@MIL-53, while CeO$_2$/Pd gave only 0.2% selectivity to FFA with tetrahydrofurfuryl alcohol as the main byproduct due to excessive hydrogenation. The recycling performance for cinnamaldehyde hydrogenation kept a stable activity and selectivity without obvious deactivation during the ten cycles. The superior catalytic performance for CeO$_2$/Pd@MIL-53(Al) was ascribed to the unique sandwich-like structures that the MIL-53 shell can prevent the Pd nanoparticles from leaching aggregation as well as restrain the excess hydrogenation reaction. Another similar work was done by Zhang group [160], they reported that Pt-CeO$_2$@UiO-66-NH$_2$ core-shell structures fabricated through modification of sodium polystyrene sulfonated (PSS) on Pt-CeO$_2$ followed by microwave-assisted growth of UiO-66-NH$_2$. About 99% FFA yield at complete conversion of FF was gained over Pt-CeO$_2$@UiO-66-NH$_2$ within 30 h, while the Pt@UiO-66-NH$_2$

sample need 45 h to reach the similar level, indicating the faster reaction rate of this catalyst. The introduction of CeO_2 in this composite was beneficial to the H-H bond dissociation. In addition, the FFA was the only product over Pt-CeO_2@UiO-66-NH_2 catalyst even after 70 h, which was attributed to the size selectivity of UiO-66-NH_2 shell.

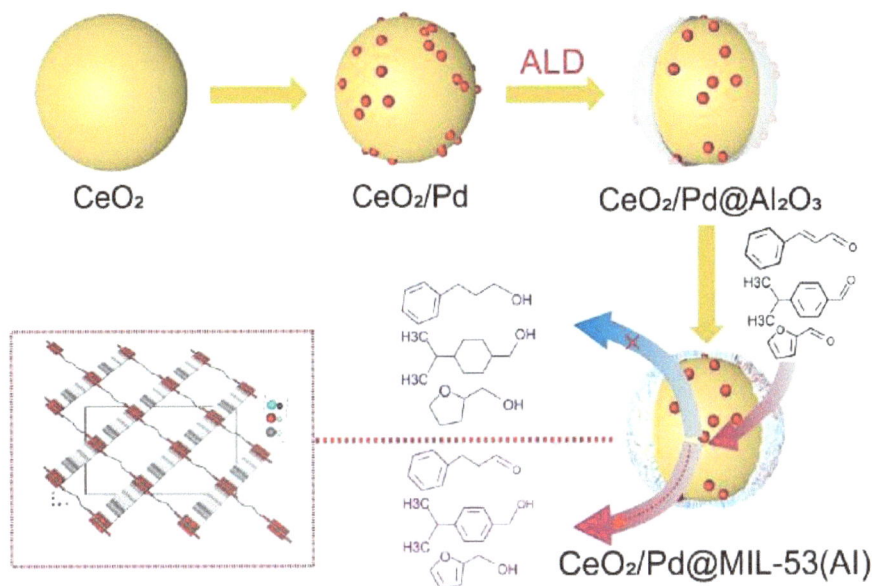

Fig. (29). The preparation route of MOF supported catalyst. Reprinted from ref. 159 with permission from The Royal Society of Chemistry, Copyright 2016.

Hierarchical porous UiO-66 (H-UiO-66) was synthesized by benzoic acid modification and used as a porous support for Pd nanoparticles in the application of FF conversion to FFA [161]. The Pd/H-UiO-66 displayed a high yield of FFA over 99% with FF conversion of nearly 99%, affording a TOF of 66.7 h^{-1}, which was better than the Pd/UiO-66 catalyst (63%). Besides, other H-UiO-66 supported metals (Au, Pt and Cu) were also examined for comparison with FFA yield of 75%, 49% and 11%, respectively. The Pd/H-UiO-66 showed no obvious change in FF conversion and FFA selectivity, indicating the excellent durability of catalyst. The abundant pores produced *in situ* during the growing process of MOFs facilitated the mass diffusion, thus leading to the high catalytic performance. Huang *et al.* synthesized Pt-Sn bimetals in UiO-66-NH_2 by two-step incipient wetness impregnation method and used for FF hydrogenation to FFA [162]. The resultant Pt-Sn@UiO-66-NH_2 catalyst displayed high selectivity to

FFA (93%) in fix-bed reaction, which was much higher than the single the Pt metal on UiO-66-NH$_2$ as well as the physical mixture of Pt and Sn supported catalyst with 51% and 60% FFA selectivity, respectively. The addition of Sn into the composites changed the electronic surface property of Pt and formed the Pt-Sn active interface. However, the reusability and stability of this Pt-Sn@UiO-66-NH$_2$ catalyst were not performed. A similar work was reported by the same group for FF conversion in a continuous flow [163]. The prepared PtSn$_{0.1}$@UiO-66-NH$_2$ catalyst displayed the highest selectivity of FFA (98%) with FF conversion of 60%, which was higher than other PtSn$_x$@UiO-66-NH$_2$ samples with a different Pt/Sn ratio. This catalyst could be run for 10 days continuously without a distinct decay in FF conversion and FFA selectivity.

Phosphorus-modified UiO-66 (P/UiO-66) was synthesized by sublimation of $(NH_4)_2HPO_4$ in the presence of UiO-66 powder, where some O atoms would be replaced by P atoms during the pyrolysis process [164]. This P/UiO-66 was applied for FF hydrogenation into FFA *via* the CTH route. The selectivity to FFA increased with the increasing P/Zr molar ratio, and the highest FFA selectivity of 99.4% was obtained over P/UiO-66 (P/Zr=2:1). The best performance was obtained with 98.9% conversion of FF and 99.4% selectivity to FFA. The superior performance for P/UiO-66 may ascribe to its high acidity and basicity resulting from the O-Zr-P and P-Zr-P species, which was in favor of hexatomic ring-catalytic reactions.

Besides the former mentioned methods by using pristine or guest-MOFs as heterogeneous catalysts, a new class of catalyst composed of metal nanoparticles encapsulated in porous carbon applying MOF as a sacrificed template has also been developed. The processes usually involve the calcination of MOF where the metal ions were converted into their metallic or metal oxide phase, while the organic ligands would be transformed into porous carbon. The morphology of the pristine MOFs usually could be remained during the high temperature thermal treatment, which was a benefit to the catalytic performance to some extent. In addition, this method could be widely used in the preparation of bimetal/metal oxide through the doping of the second metal precursor. For instance, a series of Ni-based catalysts implanted in carbon matrix synthesized by the calcination of a 2D MOF precursor and used a catalyst for the hydrogenation of FF [165]. With the increase of calcination temperature, the selectivity to FFA increased from 21.4% for Ni-MFC-300 to 59.5% for Ni-MFC-500, while the conversion of FF displayed no significant difference. It further elevated the roasting temperature, the conversion for FF and selectivity to FFA would decrease gradually along with the increase of tetrahydrofurfuryl alcohol (THFA). The highest selectivity of 51% for THFA with 91.8% FF conversion was obtained for the Ni-MFC-700 catalyst. However, these non-noble Ni catalysts did not present excellent high selectivity to one target product. Another Ni catalyst supported on alumina-porous carbon matrix derived from the Ni loaded Al-based MOF (MIL-96) was prepared by thermal treatment and employed in the selective hydrogenation of FF to FFA (Fig. 30) [166]. The catalytic performance was largely dependent on the calcination temperature, the highest FFA selectivity of 93.2% and FF conversion of 98.7% was achieved for $Ni_{0.15}/Al_2O_3$-C-500 sample. At low calcination temperature, the organic linkers decomposed incompletely, while at a higher temperature, larger Ni nanoparticles formed, which were not beneficial to the FF hydrogenation.

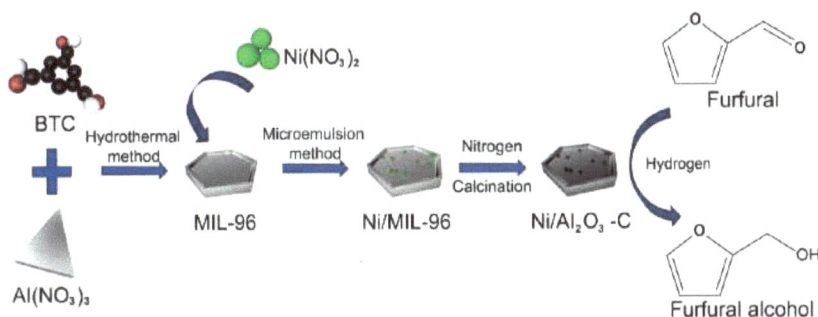

Fig. (30). Preparation of Ni/Al_2O_3-C catalyst from Ni/MIL-96 for FF conversion to FFA. Reprinted from ref. 166 with permission from Springer Science+Business Media, LLC, part of Springer Nature, Copyright 2019.

Cu nanoparticles embedded in a few layers of graphitic carbon (Cu@C) were prepared by direct high temperature pyrolysis of HKUST-1 [167]. The resultant Cu@C catalyst was applied for FF hydrogenation to FFA under visible light irradiation. Cu@C-600 exhibited 99.4% FFA yield, which was much higher than that samples pyrolysis at 400 °C (33.7%) and 600 °C (58.3%). The lower yield for Cu@C-400 and Cu@C-600 was due to the existence of a partial amount of CuO that was not active for the hydrogenation and a thicker carbon layer that would hinder the adsorption of visible light. The control experiments combined with characterization techniques suggested the significant role of metallic Cu and suitable carbon layer that were crucial to the plasmon-enhanced furfural hydrogenation. Meanwhile, this Cu@C-600 catalyst displayed no obvious deactivation, indicating the relative stability during the reaction process. A Cu/Cu_2O nanojunction with rich discrete domains encapsulated in porous organic matrices was synthesized from the Cu-based MOF ZJU-199 (Fig. 31) [168]. The acrylate species in the organic linkers could significantly inhibit the deep decomposition of organic moieties under thermal treatment. The obtained ZJU-199-350 consisted of abundant Cu_2O species with a small amount of metallic Cu species. This catalyst was employed for FF hydrogenation to FFA, and ZJU-19-

-350 displayed the highest performance with 97% FFA yield at 97.1% FF conversion. The higher yield for the ZJU-199-350 catalyst was due to the small Cu/Cu_2O nanoparticles with abundant $Cu-Cu_2O$ interfaces in the porous organic materials. For comparison, other Cu-based MOFs (HKUST-1, ZJU-35 and ZJU-36) were also examined at the same reaction conditions, they all showed inferior catalytic performance compared with ZJU-199, which was ascribed to the inaccessible active metallic Cu species covered by Cu_2O species.

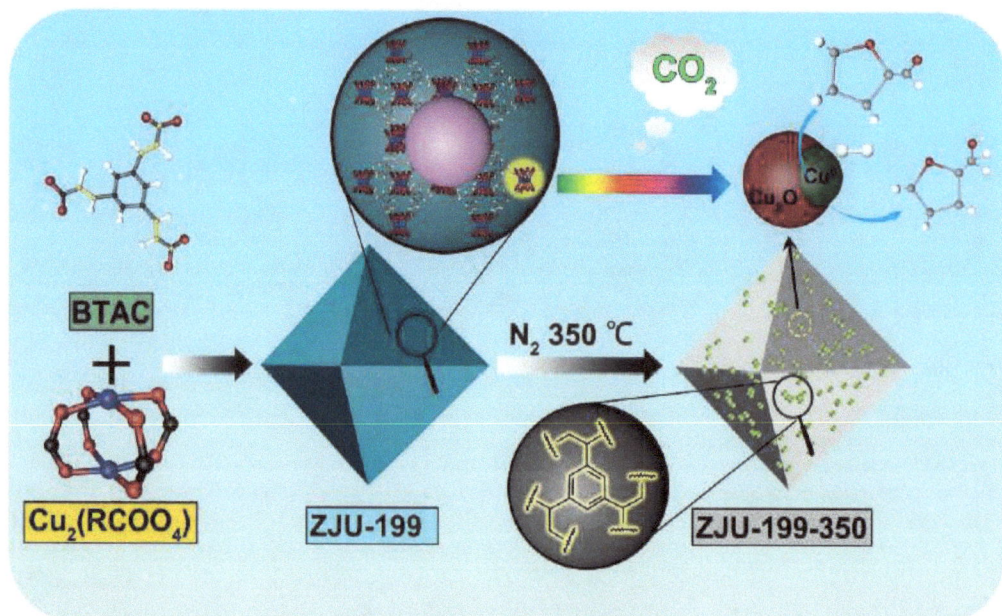

Fig. (31). Illustration of the preparation of porous organic frameworks, consisting of hetero-structural Cu/CuO_x nanoparticles generated *in situ*, for liquid-phase hydrogenation of FF into FFA. Reprinted from ref. 168 with permission from Wiley-VCH Verlag GmbH & Co. KGaA, Weinheim, Copyright 2020.

In addition to Ni and Cu catalyst, Fe catalyst was also active in FF hydrogenation to FFA. Li group reported Fe_3O_4 nanoparticles supported on porous carbon (Fe_3O_4/C) from Fe-based MOF and employed as a solid catalyst for the hydrogenation of FF [169]. The catalytic performance was largely dependent on the structures, synthetic solvent and reaction time of MOFs. The best performance was achieved for using Fe-BDC-DMF as a template with FFA selectivity of 98.5% with 76.4% FF conversion. The Fe_3O_4/C catalyst showed a slight decline in the activity and selectivity in the recycling test.

Bimetals or multiple metals usually inherent unique advantages from their parent metals and exhibit superior catalytic performance than the individual metal. A CuCo bimetallic catalyst encapsulated in porous carbon (CuCo-C) was synthesized from Co-doped Cu-BTC MOF *via* thermal pyrolysis [170]. The resultant CuCo-C catalyst was applied for FF hydrogenation to FFA, FFA conversion increased from 55.9% to 98.7% with the increase of the Co/Cu ratio from 0.1 to 0.4, the selectivity would decline slightly with a further increase in the Co/Cu ratio. The best performance was gained for $CuCo_{0.4}/C$-873 sample, the activity displayed a decreasing trend with the increasing temperature, which may ascribe to the larger particles and aggregation of CuCo nanoparticles at a higher temperature. Xiong *et al.* prepared Co-Ni alloy nanoparticles encapsulated in a porous nitrogen-dopped carbon (Co-Ni@NC) derived from Co-Ni bimetal MOFs and used for FF hydrogenation [171]. 2Co-1Ni@NC-800 displayed the best catalytic performance with 92.7% selectivity to FFA at 99.0% conversion of FF. With the increase of Ni content in the precursor, the yield for FFA decreased accompanied by the increased selectivity to tetrahydrofurfuryl alcohol (THFA), which may attribute to the excess hydrogenation of FFA. Furthermore, the performance of 2Co-1Ni@NC-800 catalyst was better than the individual Co@NC-800, Ni@NC-800 and Co@C-800 with FFA yield of 68.1%, 42.8% and 38.1% respectively, indicating the important role of Co, Ni metal sites and doped N. Interesting, the best performance was obtained in methanol other than secondary alcohols like isopropanol, which was probably due to the stronger methanol adsorption than FFA but weaker than FF.

Nowadays, the hydrogenation of furfural has been considerably investigated, which not only can be converted into FF, but also to the tetrahydrofurfuryl alcohol (THFA). THFA is widely used in the field of green solvent, intermediates for fine chemicals, and liquid transportation fuel or additive. Liang *et al.* prepared a Pd catalyst supported on amino functionalized MOF and employed as an efficient catalyst for FF hydrogenation to THFA [172]. The free amino groups on MOFs can help anchor the Pd nanoparticles and facilitate the dispersion of metal sites. The resultant Pd@MIL-101(Cr)-NH_2 catalyst displayed an increasing selectivity to THFA and activity with the increasing Pd loading, and the best performance was achieved for 3.0 wt% Pd loading with 99.9% selectivity to THFA at nearly complete conversion of FF. However, 10% drop in THFA selectivity in the second cycle was observed for 3.0wt% Pd@MIL-101(Cr)-NH_2, the loss of THFA selectivity was resulted from the aggregation of Pd nanoparticles as well as the partial collapse of MOF frameworks.

In addition, another Pd catalyst supported on UiO-66 was reported recently and employed for the complete FF hydrogenation to THFA [173]. The resultant Pd/UiO-66 displayed the best performance of 99.9% selectivity to THFA at full conversion of FA. Meanwhile, the performance was superior to the conventional support like SiO_2 and γ-Al_2O_3 with low THFA selectivity of 14% and 36%, respectively. The higher performance of Pd/UiO-66 might ascribe to the higher

FF adsorption capacity than γ-Al_2O_3 and SiO_2. Besides, the uniform dispersion of Pd nanoparticles on UiO-66 may also contribute to the high catalytic performance. Wang *et al.* reported a Ni catalyst embedded in carbon matrix (Ni/C) from Ni-based MOF through direct pyrolysis under N_2 atmosphere [174]. The catalytic performance showed a volcano type *versus* the calcination temperature over Ni/C catalyst. The best performance was achieved for Ni/C-500 with 99.9% selectivity to THFA at the complete conversion of FF, the improved catalytic performance might result from the uniform distribution and smaller Ni particle size as well as the abundant active sites. Besides, this catalyst displayed relatively stable recycling performance in the five runs, suggesting the high stability of catalyst.

The combination of transition-noble metal catalysts was developed from the Pd-doped Co-based MOF and employed as solid catalysts for FF hydrogenation to THFA [175]. The resultant $PdCo_3O_4$@NC exhibited the high yield of 95% THFA with the full conversion of FF. In contrast, the $PdFe_3O_4$@NC catalyst gave a lower yield of 70% at the same conditions. All the control experiments indicated an important role of Pd and Co_3O_4 on the improved performance. In addition, the carbon layers were beneficial to the dispersion of metal sites as well as preventing the aggregating of nanoparticles during the reaction process.

Furfuryl alcohol (FFA) can also be transferred to tetrahydrofurfuryl alcohol (THFA). Wang *et al.* prepared ultra-small NiCo bimetallic alloy nanoparticles by using NiCo-MOF-SiO_2 composite as a template and applied for the synthesis of THFA from FFA [176]. The resultant NiCo/SiO_2-MOF catalyst displayed the best catalytic performance with 99.1% selectivity to THFA at 99.8% conversion of FA, which was better than the individual Ni/SiO_2-MOF or Co/SiO_2-MOF with selectivity of 95.3% and 50.1%, and conversion of 96.9% and 28.1%, respectively. The ultrafine NiCo metal alloy combined with the synergistic effect between Ni and Co nanoparticles together with the large pore volume may account for the excellent catalytic performance over NiCo/SiO_2-MOF catalyst. The NiCo/SiO_2-MOF catalyst displayed a slight gradual decrease in the activity during the consecutive cycling experiment, which may ascribe to the leaching of Ni and Co. Another example of the FFA hydrogenation to THFA was reported by Duan group. They synthesized a Pd/UiO-66-v catalyst by *in situ* reduction of Pd ions on UiO-66-v by vinyl as the reductant [177]. The resultant Pd/UiO-66-v catalyst displayed high catalytic performance with 91% selectivity to THFA at 92% conversion of FFA. The conversion of FFA could be further improved to 99% with 90% selectivity to THFA after reaction of 12 h. Besides, Pd/UiO-66-v was capable of catalyzing other different furan compounds including 2-methylfuran, 2,5-di(hydroxymethyl)furan and 2,5-dimethylfuran, and showed good to excellent selectivity to their corresponding alcohols.

The conversion of FF to cyclopentene (CPO) has gained considerable concern, CPO could be widely applied in the fields of rubber chemicals, insecticides, and pharmaceuticals. Li *et al.* prepared a Ru/MIL-101 composite for selective transformation of FF to CPO [178]. The resultant catalyst displayed an increased selectivity to CPO and conversion of FF with the increased Ru loading, and the best performance was achieved for 3 wt%Ru/MIL-101 with 96% CPO selectivity at 99% FF conversion. While the catalytic performance would decline with a further increase of Ru loading, which was attributed to the aggregation of Ru nanoparticles. Furthermore, the Ru/MIL-101 catalyst was able to catalyze other α,β-unsaturated aldehydes with moderate to good yields.

Zhang group prepared a series of MOF supported Pd nanoparticles for the production of cyclopentanone compounds [179]. The Pd/Cu-BTC catalyst displayed higher selectivity (96.5%) to CPO than Pd/MIL-101 (71.2%) and Pd/FeCu-DMC (2.3%). The unique structure of Cu-BTC was beneficial to the dispersion of Pd nanoparticles as well as the formation of small metal sites, thus improving the catalytic performance. Meanwhile, the strong Lewis acidity in Cu-BTC MOF may also account for the high performance. Furthermore, this Pd/Cu-BTC catalyst was able to convert the HMF into 3-hydroxymethyl cyclopentanone (HCPN) with 90.9% selectivity at 99.5% conversion. Besides the guest-MOF composites, MOFs derived metal/metal oxides were also used for furfural rearrangement. Xiao *et al.* prepared a CuNi nanoparticle encapsulated in porous carbon by using Ni-dopped Cu-BTC MOF as a sacrificial template [180]. For Cu@C catalyst without Ni doping, it just displayed a 51.8% yield of CPO at 79.1% FF conversion. The catalytic performance increased with the Ni loading, and the best performance was obtained for $CuNi_{0.5}$@C with 96.9% CPO yield and 99.3% FF conversion. However, the activity and yield declined sharply by further increasing the Ni loading. The lower specific surface areas may attribute to the inferior catalytic performance for these catalysts.

Li *et al.* designed a multi-shell hollow nitrogen doped porous carbon supported Co catalyst (Co@NC) with fine and hierarchical structures derived from the direct pyrolysis of multi-layer ZIF-67 synthesized by step-by-step growth approach and was applied to the selective hydrogenation of FF to cyclopentanol (CPL) [181]. Interesting, the catalytic performance increased gradually with the increase of MOF shell number, and the highest performance was achieved for 4LH-Co@NC with 97% yield of CPL, which was superior to S-Co@NC and single shell sample (1LH-Co@NC). The corresponding turnover frequencies (TOF) for one to four layered Co@NC catalysts were 0.17, 0.34, 0.57, and 0.62 h^{-1}, respectively, consistent with the result of the activity. The superior performance for 4LH-Co@NC might result from multi-shell hollow structures that could stabilize the Co nanoparticles and facilitate the uniform dispersion of Co sites. Besides, the hierarchical pores were in favor of mass transfer and nitrogen-doped porous carbon could help modify the chemical environment of Co nanoparticles.

A Co@Cu core-shell nano-catalyst loaded on porous carbon was synthesized by calcinating multivariate (MTV) MOF-74 at high temperature and used as a solid catalyst for the hydrogenation of FF in a fix-bed reactor [182]. The Co/Cu ratio could be modulated by changing the feed ratio of Cu and Co in the synthetic precursor, and the formed CoCu bimetallic materials were composed of Co core and Cu shell. The monometallic Cu catalyst displayed a high selectivity to furfuryl alcohol, and the selectivity to 2-methylfuran (2-MF) would increase with the Co content in the MOF precursor. The obtained $Co_{0.5}Cu_{0.5}$ could be stable for 48 h time on stream with constant selectivity to 2-MF and FFA while gradually decreasing the FF conversion. The deactivation of this catalyst was due to the loss of active metals from the carbon support and the carbon deposition on active sites. $Cu/CuFe_2O_4$ supported on carbon matrix was prepared through calcinating Cu ions doped Fe-MIL-88B under mixture H_2/Ar atmosphere and employed as a heterogeneous catalyst for HDO reaction of FF to 2-MF [183]. The resultant catalyst displayed excellent catalytic performance with nearly 100% selectivity to 2-MF at full conversion of FF, which was much higher than the common CuFe catalyst supported on graphite (about 35.5% 2-MF selectivity with 52% FF conversion). Furthermore, this $Cu/CuFe_2O_4$ bimetallic catalyst displayed excellent cycling performance during five cycles without a significant drop in selectivity and activity. The superior performance of MOF-derived catalyst might result from the uniform distribution of metals and highly active metal surface areas along with the strong metal support interaction.

3. TRANSFORMATION OF VANILLIN

Vanillin (4-hydroxy-3methoxybenzaldehyde), is a very important content of pyrolysis oil obtained from lignin, which can be transformed to various useful chemicals like 2-methoxy-4-methylphenol or vanillin alcohol. Xu *et al*. prepared a Pd/MIL-101 catalyst by double solvent method and used it for vanillin hydrodeoxygenation (Fig. **32**) [184]. Profit from the ultrafine nanoparticles (1.0-2.2 nm) encapsulated in MIL-101 pores and the steric hindrance as well as the strong interactions between reactant and intermediate, the resultant Pd@MIL-101 displayed excellent catalytic performance with 100% selectivity to 2-methox--4methylphenol at 100% vanillin conversion. While the Pd/MIL-101 samples with larger Pd nanoparticles on the surface of MOF exhibited much lower performance than Pd@MIL-101. Additionally, no significant loss in activity and selectivity was observed for Pd@MIL-101 in the consecutive runs.

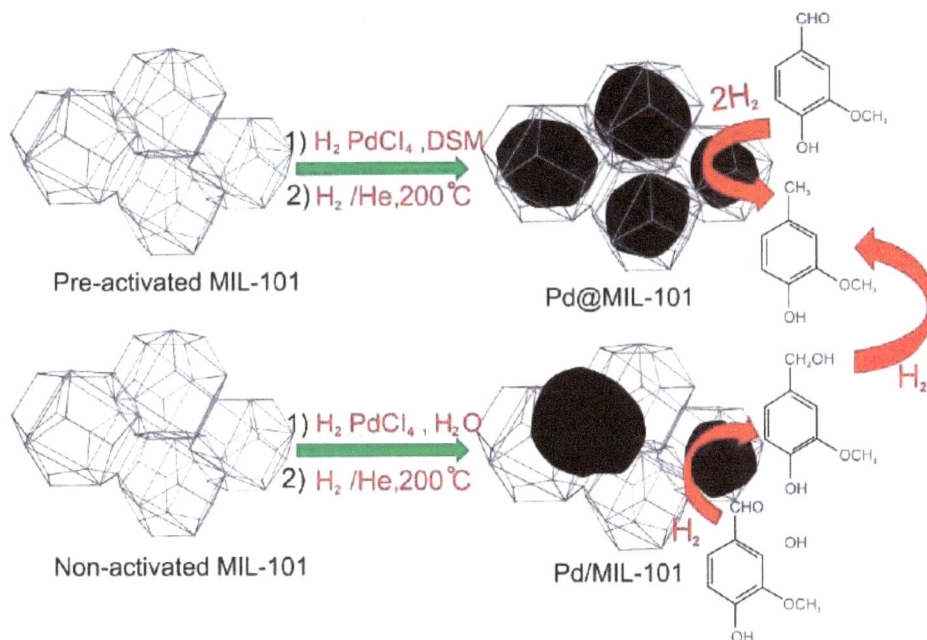

Fig. (32). Schematic representation of Pd nanoparticles into/onto MIL-101, and their use for vanillin conversion. Reprinted from ref. 184 with permission from The Royal Society of Chemistry, Copyright 2015.

Zhang group anchored Pd nanoparticles on sulfonic acid-functionalized MIL-101(Cr) for the transformation of vanillin [185]. Benefitting from the uniform distribution of Pd nanoparticles and enough Bronsted acid sites, the resultant Pd/SO$_3$H-MIL-101(Cr) displayed excellent catalytic performance with exclusive 2-methoxy-4-methylphenol selectivity at full vanillin conversion, which was remarkably higher than Pd/MIL-101(Cr) (17.9% selectivity and 86.4% conversion). Before long, Zhang *et al.* prepared Pd nanoparticles with a small size embedded in NH$_2$-UiO-66 by post-synthetic approach [186]. The obtained Pd@NH$_2$-UiO-66 exhibited exclusive selectivity to 2-methoxy-4-methylphenol at complete conversion of vanillin within 75 min, while the unfunctionalized Pd@UiO-66 just gave 43.1% 2-methoxy-4-methylphenol selectivity at 49.3% conversion of vanillin. This catalyst showed good durability in the cycling performance. The superior performance might result from the existence of free amine groups in MOF frameworks that facilitated the Pd nanoparticles' distribution and prevented their leaching. Soon afterwards, Ibrahim *et al.* reported a Pd nano-catalyst embedded in a 3D hierarchical composite assembled from reduced graphene oxide and Ce-MOF composite (Fig. **33**) [187]. Benefitting from the even dispersion of Pd nanoparticles and the existence of acidic active sites, the resultant 5wt%Pd/12.5wt%PRGO/Ce-MOF displayed the best performance with nearly 100% selectivity to 2-methoxy-4-methylphenol at 100% conversion of vanillin. However, the recycling performance was not ideal due to the blocking of residuals on the surface or in the pores of the catalyst.

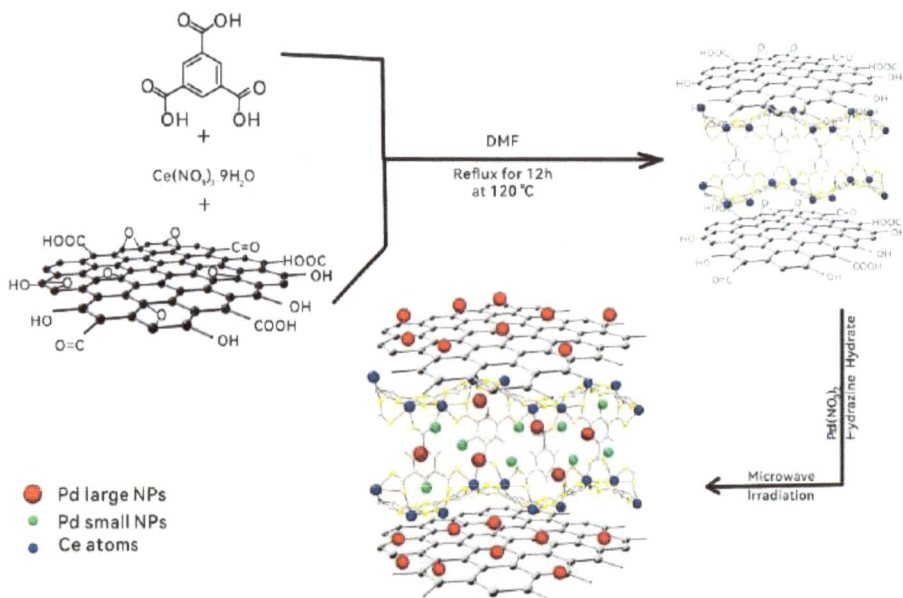

Fig. (33). Synthesis design approach for the hierarchical Pd/PRGO/Ce-MOF composite. Reprinted from ref. 187 with permission from Wiley-VCH Verlag GmbH & Co. KGaA, Weinheim, Copyright 2017.

Jiang group prepared a series of Pd nanoparticles in a carbon matrix from MOFs (including ZIF-8, ZIF-67, and MOF-5) by high temperature pyrolysis [188]. The resultant Pd/NPC-ZIF-8 catalyst exhibited the highest catalytic performance for vanillin hydrodeoxygenation with 100% selectivity to 2-methoxy-4-methylphenol at complete conversion of vanillin, and affording a TOF of 100 h^{-1}, which was much better than other catalysts like Pd/ZIF-8-urea (93 h^{-1}), Pd/NPC-ZIF-67 (54 h^{-1}), Pd/C-MOF-5 (36 h^{-1}) and Pd/C-MOF-5 urea (78 h^{-1}). The superior catalytic performance was due to the even dispersion of Pd nanoparticles, rich pores and high hydrophilicity of NPC-ZIF-8 support.

Garcia *et al.* synthesized a Fe-based MOF and used it for oxidation reactions [189]. MIL-100(Fe) displayed similar performance for acetal formation of benzaldehyde and cinnamaldehyde compared with the commercial Fe-BTC, which was also composed of Fe ions and BTC linkers with lower surface area.

Ascribing to the ring-opening reaction of epoxy ethylbenzene, Fe-BTC showed higher conversion (99%) than MIL-100(Fe) (62%), which was attributed to the higher Brønsted acid density. In contrast, the MIL-100(Fe) displayed superior performance for aerobic oxidation of thiophenol than Fe-BTC, ascribing to the existence of Fe^{3+}/Fe^{2+} pairs and structural stability.

Huang *et al.* prepared monodisperse Pt nanoparticles embedded in the cavities of UiO-66-NH$_2$ through simple impregnation method [190]. The loading content of Pt nanoparticles could be tuned from 0.97 wt% to 10.7 wt% without an obvious change in the particle size (1.16 ± 0.16 nm), which was consistent with the diameters of pore cages in MOF. The resultant Pt@UiO-66-NH$_2$ displayed 91.7% cinnamyl alcohol selectivity with 98.7% cinnamaldehyde conversion at the tenth cycle, affording a TON of 10900, which was much better than Pt/UiO-66-NH$_2$ catalyst (52.2% conversion and 71.6% selectivity). Such excellent performance was ascribed to the ultra small monodispersed Pt nanoparticles encapsulated in UiO-66-NH$_2$ cavities as well as steric effect caused by the pore windows of MOFs. Li group synthesized a Pd nano-catalyst anchored on nitrogen containing organic linkers of UiO-67 *via* a facile one-pot approach [191]. The organic linker in UiO-67 had a strong coordinating ability to Pd metal sites, which could anchor the Pd nanoparticles effectively during the self-assembling process of MOFs, therefore leading to the outstanding performance for cinnamaldehyde (CAL) hydrogenation. Pd@UiO-67 exhibited 100% selectivity for hydrocinnamaldehyde (HCAL) with full conversion of CAL. Meanwhile, this catalyst was capable of catalyzing various cinnamaldehydes with good yield.

Allendorf group demonstrated Ti and Ni based IRMOF-74 *via* the atomic layer deposition of Cp$_2$Ni and TiCl$_4$ precursor in the pores of MOFs [192]. The resultant catalyst displayed moderate to excellent catalytic performance for diphenyl ether benzylphenyl ether, and phenylethylphenyl ether, that containing the 4-O-5, α--4, and β-O-4 linkages, respectively. The highest PPE conversion of 82% with 98% selectivity to ethylbenzene and phenol was obtained over Ni@IRMOF-74 without the addition of base. Pd-Ni bimetallic nanoparticles supported on MIL-100(Fe) were fabricated and used as heterogeneous catalysts for the hydrogenolysis of β-O-4 linkages [193]. Ascribing to the confinement effect of MIL-100 that facilitated the stabilization and dispersion of bimetals, as well as the synergistic effect between Pd and Ni that created much more active sites, the obtained PdNi$_4$/MIL-100(Fe) exhibited high selectivity for benzaldehyde (95%) with 99% conversion of 2-phenoxyl-1-phenethanol, which was much better than the single Pd catalyst (Pd/MIL-100) with 71% selectivity and 77% conversion. This catalyst exhibited about 20% drop in selectivity after four recycling tests, which resulted from the loss of catalyst as well as the leaching of Pd and Ni during the cycling process. Furthermore, PdNi$_4$/MIL-100(Fe) catalyst was able to catalyze other substrates with good to excellent yield.

4. CONCLUSIONS AND FUTURE PROSPECT

The flexibility in which the metal ions/clusters and organic linkers could be varied has resulted in thousands of MOFs being synthesized and studied every year. More than 20,000 kinds of MOFs have been reported so far, and they are widely used in the field of gas storage, separation, and catalysis. In particular, numerous studies have been devoted to the investigations of photocatalysis, electrocatalysis, and energy storage. The applications in biomass conversion were developed in recent years and need to be further investigated. Pristine MOFs with Lewis acidity or Brønsted acidity, functionalized MOFs with adjustable acid sites by post-modification, guest-MOF compartment with polyoxometalates or polymers, MOFs supported metal nanoparticles, and metal/metal oxide supported on porous carbon derived from MOFs, in this context, have been reviewed for the biomass conversion. And the reactions include the hydrolysis of lignin and cellulose model compounds, along with the isomerization of glucose, dehydration of saccharides, and hydrogenation or oxidations for platform chemicals. Especially, the transformation of LA, FF, and 5-HMF from biomass was systematically summarized. The results summarized in literature have demonstrated outstanding performance of MOFs than conventional zeolites, molecular sieve, or other heterogeneous catalysts. The considerable increase in the studies, which applied MOFs or MOF derived materials in this new domain, suggests their promising capacity for further research. In addition to the unique strength of MOFs or MOF based constructions for the transformation of biomass, indeed, some key issues should be addressed in the future.

Most of the studies in the literatures focus on the star MOFs like UiO-66(Zr), MIL-101(Cr), and HKUST-1(Cu), while the Cr clusters in MOF which are unamiable to the environment, and the involved MOFs for biomass conversion are just a tip of the iceberg in MOFs family. In the future, the metal ions/clusters in MOF should be no-toxic and friendly to the environment. Meanwhile, the attempts and efforts for the possible and potential applications should be made for other promising MOFs.

As for the upstream reactions like cellulose upgrading, MOFs seem to be less efficient and effective compared with other solid catalysts, attributing to the lack of acid sites like Lewis or Brønsted acids in MOFs that were insufficient to catalyze these reactions as well as the harsh reaction conditions. It is very important and urgent to develop the application of MOFs in this field. There is plenty of room and market for the conversion of lignocellulose by MOFs in the future if one or more MOFs could undergo hydrolysis with high efficiency and selectivity.

Although many studies showed that MOFs-based catalysts could be recycled for five or more times during the cycling performance, majority of them were performed in organic solvents. Furthermore, the crystallinity usually weakened along with the structural damage in the consecutive process. H_2O is the most environmental starting solvent for various catalytic reactions, however, the MOFs based catalyst exhibited low activity in water and they are not stable in water in general. Therefore, it is necessary to enhance the water stability of MOFs and improve their catalytic efficiency in water.

Last but not the least, metal-supported catalysts have displayed excellent performance in photocatalysis and electrocatalysis, especially the single atom or noble metals. However, in our knowledge, the single atom-based catalyst on MOFs has not been reported in biomass conversion at present. Meanwhile, more research should be investigated on the non-noble metal-based catalysts for biomass conversion owing to their low cost and practical industry application.

In addition, the cost for the production of MOFs is high due to the high price of raw materials including the organic linkers, organic solvent, as well as the activation process compared with the conventional zeolites or metal oxides. Besides, the yield of MOFs is usually very low and limited to gram scale in lab. Hence, it is quite essential to synthesize MOFs on a large scale to satisfy the demand for numerous studies and reduce the synthetic cost of MOFs.

LIST OF ABBREVIATIONS

MOFs	metal-organic frameworks
POMs	polyoxometalates
HMF	5-hydroxymethylfurfural
DFF	furan-2,5-diformylfuran
DHMTHF	2,5-dihydroxymethyl-tetrahydrofuran
DHMF	di(hydroxymethyl) furfural
LA	levulinic acid
GVL	γ-valerolactone
FA	furfuryl alcohol
MTHF	2-methyl tetrahydrofuran
MF	2-methyl furan
THFOL	tetrahydrofurfuryl alcohol

CPO	cyclopentanone
CMC	carboxymethyl cellulose
PTA	phosphotungstic acid
EG	ethylene glycol
PABA	paminobenzoic acid
OMS	open metal sites
GLY	amino acid glycine
Fm	modified frameworks
MTV	multivariate
BDC	terephthalic acid
MSBDC	sulfonic acid functionalized BDC
SSACs	single-site active centers
EMIMCl	1-ethyl-3-methylimidazolium chloride
DMSO	dimethyl sulfoxide
PMAi	polymetric analogue of N-bromosuccinimide
THF	tetrahydrofuran
SBU	secondary building units
PDA	polydopamine
PU	polyurethane
PMA	phosphomolybdic acid
DMF	2,5-dimethylfuran
GO	graphene oxide
DMTHF	2,5-dimethyltetrafuran
ML	methyl lactate
MF	fly-ash-MOF-composite
FF	furfural
CTH	catalytic transfer hydrogenation
F-ZrF	Formic acid modulated Zr-based MOF
HPW	$H_3PW_{12}O_{40}$
BTC	benzene tricarboxylic acid
CNF	carbon nanofibers
CN	nitrogen dopped carbon
EL	ethyl levulinate
SMCNF	sulfate modified carbon nanofibers
SiW	silicotungstic acid

EMF	5-ethoxymethylfurfural
HPM	phosphomolybdic acid hydrate
BPDC	biphenyldicarboxylate
NP	nanoparticles
N450	annealing in N_2 atmosphere at 450 °C
O-250	annealing in O_2 atmosphere at 450 °C
PCNC	Pd loaded carbon nanocomposites
TEMPO	2,2,6,6-tetramethylpiperidine-1-oxyl
MW	microwave
DMFDCA	dimethylfuran dicarboxylate
FDCA	2,5-furandicarboxylic acid
NNC	nano-porous carbon
M400	Mn_2O_3 nanoflakes obtained by calcinating at 400 °C
FFA	furfuryl alcohol
TOF	turnover frequency
ALD	atomic layer deposition
PSS	polystyrene sulfonated
THFA	tetrahydrofurfuryl alcohol
Cu@C	Cu nanoparticles embedded graphitic carbon
CPO	cyclopentene
HCPN	3-hydroxymethyl cyclopentanone
CPL	cyclopentanol
2-MF	2-methylfuran
HDO	hydrodeoxygenation
CAL	cinnamaldehyde
HCAL	hydrocinnamaldehyde

CONSENT FOR PUBLICATION

Not applicable.

CONFLICT OF INTEREST

The author declares no conflict of interest, financial or otherwise.

ACKNOWLEDGEMENTS

Declared none.

REFERENCES

[1]   Hoel, M.; Kverndokk, S. Depletion of fossil fuels and the impacts of global warming. *Resour. Energy Econ.,* **1996**, *18*(2), 115-136.
        [http://dx.doi.org/10.1016/0928-7655(96)00005-X]

[2]   Day, C.; Day, G. Climate change, fossil fuel prices and depletion: The rationale for a falling export tax. *Econ. Model.,* **2017**, *63*, 153-160.
        [http://dx.doi.org/10.1016/j.econmod.2017.01.006]

[3]   Corma, A.; Iborra, S.; Velty, A. Chemical routes for the transformation of biomass into chemicals. *Chem. Rev.,* **2007**, *107*(6), 2411-2502.
        [http://dx.doi.org/10.1021/cr050989d] [PMID: 17535020]

[4]   van Putten, R.J.; van der Waal, J.C.; de Jong, E.; Rasrendra, C.B.; Heeres, H.J.; de Vries, J.G. Hydroxymethylfurfural, a versatile platform chemical made from renewable resources. *Chem. Rev.,* **2013**, *113*(3), 1499-1597.
        [http://dx.doi.org/10.1021/cr300182k] [PMID: 23394139]

[5]   Zhou, C.H.; Xia, X.; Lin, C.X.; Tong, D.S.; Beltramini, J. Catalytic conversion of lignocellulosic biomass to fine chemicals and fuels. *Chem. Soc. Rev.,* **2011**, *40*(11), 5588-5617.
        [http://dx.doi.org/10.1039/c1cs15124j] [PMID: 21863197]

[6]   Zhang, Z.; Huber, G.W. Catalytic oxidation of carbohydrates into organic acids and furan chemicals. *Chem. Soc. Rev.,* **2018**, *47*(4), 1351-1390.
        [http://dx.doi.org/10.1039/C7CS00213K] [PMID: 29297525]

[7]   Xu, C.; Arancon, R.A.D.; Labidi, J.; Luque, R. Lignin depolymerisation strategies: towards valuable chemicals and fuels. *Chem. Soc. Rev.,* **2014**, *43*(22), 7485-7500.
        [http://dx.doi.org/10.1039/C4CS00235K] [PMID: 25287249]

[8]   Long, H.; Li, X.; Wang, H.; Jia, J. Biomass resources and their bioenergy potential estimation: A review. *Renew. Sustain. Energy Rev.,* **2013**, *26*, 344-352.
        [http://dx.doi.org/10.1016/j.rser.2013.05.035]

[9]   Tong, X.; Ma, Y.; Li, Y. Biomass into chemicals: Conversion of sugars to furan derivatives by catalytic processes. *Appl. Catal. A Gen.,* **2010**, *385*(1-2), 1-13.
        [http://dx.doi.org/10.1016/j.apcata.2010.06.049]

[10]   Liu, Y.; Nie, Y.; Lu, X.; Zhang, X.; He, H.; Pan, F.; Zhou, L.; Liu, X.; Ji, X.; Zhang, S. Cascade utilization of lignocellulosic biomass to high-value products. *Green Chem.,* **2019**, *21*(13), 3499-3535.
        [http://dx.doi.org/10.1039/C9GC00473D]

[11]   Viggiano, R.P., III; Schiraldi, D.A. Fabrication and mechanical characterization of lignin-based aerogels. *Green Mater.,* **2014**, *2*(3), 153-158.
        [http://dx.doi.org/10.1680/gmat.14.00004]

[12]   Besson, M.; Gallezot, P.; Pinel, C. Conversion of biomass into chemicals over metal catalysts. *Chem. Rev.,* **2014**, *114*(3), 1827-1870.
        [http://dx.doi.org/10.1021/cr4002269] [PMID: 24083630]

[13]   Yang, H.; Yan, R.; Chen, H.; Lee, D.H.; Zheng, C. Characteristics of hemicellulose, cellulose and lignin pyrolysis. *Fuel,* **2007**, *86*(12-13), 1781-1788.
        [http://dx.doi.org/10.1016/j.fuel.2006.12.013]

[14]   Mika, L.T.; Cséfalvay, E.; Németh, Á. Catalytic Conversion of Carbohydrates to Initial Platform Chemicals: Chemistry and Sustainability. *Chem. Rev.,* **2018**, *118*(2), 505-613.
        [http://dx.doi.org/10.1021/acs.chemrev.7b00395] [PMID: 29155579]

[15]   Dhepe, P.L.; Sahu, R. A solid-acid-based process for the conversion of hemicellulose. *Green Chem.,* **2010**, *12*(12), 2153-2156.
        [http://dx.doi.org/10.1039/c004128a]

[16]   Huang, Y.B.; Fu, Y. Hydrolysis of cellulose to glucose by solid acid catalysts. *Green Chem.*, **2013**, *15*(5), 1095-1111.
[http://dx.doi.org/10.1039/c3gc40136g]

[17]   Shrotri, A.; Kobayashi, H.; Fukuoka, A. Cellulose Depolymerization over Heterogeneous Catalysts. *Acc. Chem. Res.*, **2018**, *51*(3), 761-768.
[http://dx.doi.org/10.1021/acs.accounts.7b00614] [PMID: 29443505]

[18]   Hu, S.; Zhang, Z.; Song, J.; Zhou, Y.; Han, B. Efficient conversion of glucose into 5-hydroxymethylfurfural catalyzed by a common Lewis acid $SnCl_4$ in an ionic liquid. *Green Chem.*, **2009**, *11*(11), 1746-1749.
[http://dx.doi.org/10.1039/b914601f]

[19]   McKenna, S.M.; Mines, P.; Law, P.; Kovacs-Schreiner, K.; Birmingham, W.R.; Turner, N.J.; Leimkühler, S.; Carnell, A.J. The continuous oxidation of HMF to FDCA and the immobilisation and stabilisation of periplasmic aldehyde oxidase (PaoABC). *Green Chem.*, **2017**, *19*(19), 4660-4665.
[http://dx.doi.org/10.1039/C7GC01696D]

[20]   Alamillo, R.; Tucker, M.; Chia, M.; Pagán-Torres, Y.; Dumesic, J. The selective hydrogenation of biomass-derived 5-hydroxymethylfurfural using heterogeneous catalysts. *Green Chem.*, **2012**, *14*(5), 1413-1419.
[http://dx.doi.org/10.1039/c2gc35039d]

[21]   Xiao, B.; Zheng, M.; Li, X.; Pang, J.; Sun, R.; Wang, H.; Pang, X.; Wang, A.; Wang, X.; Zhang, T. Synthesis of 1,6-hexanediol from HMF over double-layered catalysts of Pd/SiO_2 + $Ir–ReO_x/SiO_2$ in a fixed-bed reactor. *Green Chem.*, **2016**, *18*(7), 2175-2184.
[http://dx.doi.org/10.1039/C5GC02228B]

[22]   Qi, L.; Horváth, I.T. Catalytic Conversion of Fructose to γ-Valerolactone in γ-Valerolactone. *ACS Catal.*, **2012**, *2*(11), 2247-2249.
[http://dx.doi.org/10.1021/cs300428f]

[23]   Liao, Y.T.; Matsagar, B.M.; Wu, K.C.W. Metal–Organic Framework (MOF)-Derived Effective Solid Catalysts for Valorization of Lignocellulosic Biomass. *ACS Sustain. Chem.& Eng.*, **2018**, *6*(11), 13628-13643.
[http://dx.doi.org/10.1021/acssuschemeng.8b03683]

[24]   Saha, B.C. Hemicellulose bioconversion. *J. Ind. Microbiol. Biotechnol.*, **2003**, *30*(5), 279-291.
[http://dx.doi.org/10.1007/s10295-003-0049-x] [PMID: 12698321]

[25]   Patwardhan, P.R.; Brown, R.C.; Shanks, B.H. Product distribution from the fast pyrolysis of hemicellulose. *ChemSusChem*, **2011**, *4*(5), 636-643.
[http://dx.doi.org/10.1002/cssc.201000425] [PMID: 21548106]

[26]   Zhou, X.; Li, W.; Mabon, R.; Broadbelt, L.J. A Critical Review on Hemicellulose Pyrolysis. *Energy Technol. (Weinheim)*, **2017**, *5*(1), 52-79.
[http://dx.doi.org/10.1002/ente.201600327]

[27]   Lange, J.P.; van der Heide, E.; van Buijtenen, J.; Price, R. Furfural--a promising platform for lignocellulosic biofuels. *ChemSusChem*, **2012**, *5*(1), 150-166.
[http://dx.doi.org/10.1002/cssc.201100648] [PMID: 22213717]

[28]   Mariscal, R.; Maireles-Torres, P.; Ojeda, M.; Sádaba, I.; López Granados, M. Furfural: a renewable and versatile platform molecule for the synthesis of chemicals and fuels. *Energy Environ. Sci.*, **2016**, *9*(4), 1144-1189.
[http://dx.doi.org/10.1039/C5EE02666K]

[29]   Yan, K.; Wu, G.; Lafleur, T.; Jarvis, C. Production, properties and catalytic hydrogenation of furfural to fuel additives and value-added chemicals. *Renew. Sustain. Energy Rev.*, **2014**, *38*, 663-676.
[http://dx.doi.org/10.1016/j.rser.2014.07.003]

[30]   Li, X.; Jia, P.; Wang, T. Furfural: A Promising Platform Compound for Sustainable Production of C_4

and C $_5$ Chemicals. *ACS Catal.,* **2016**, *6*(11), 7621-7640.
[http://dx.doi.org/10.1021/acscatal.6b01838]

[31] Ragauskas, A.J.; Beckham, G.T.; Biddy, M.J.; Chandra, R.; Chen, F.; Davis, M.F.; Davison, B.H.; Dixon, R.A.; Gilna, P.; Keller, M.; Langan, P.; Naskar, A.K.; Saddler, J.N.; Tschaplinski, T.J.; Tuskan, G.A.; Wyman, C.E. Lignin valorization: improving lignin processing in the biorefinery. *Science,* **2014**, *344*(6185), 1246843.
[http://dx.doi.org/10.1126/science.1246843] [PMID: 24833396]

[32] Calvo-Flores, F.G.; Dobado, J.A. Lignin as renewable raw material. *ChemSusChem,* **2010**, *3*(11), 1227-1235.
[http://dx.doi.org/10.1002/cssc.201000157] [PMID: 20839280]

[33] Wang, H.; Pu, Y.; Ragauskas, A.; Yang, B. From lignin to valuable products–strategies, challenges, and prospects. *Bioresour. Technol.,* **2019**, *271*, 449-461.
[http://dx.doi.org/10.1016/j.biortech.2018.09.072] [PMID: 30266464]

[34] Kleinert, M.; Barth, T. Phenols from Lignin. *Chem. Eng. Technol.,* **2008**, *31*(5), 736-745.
[http://dx.doi.org/10.1002/ceat.200800073]

[35] Hatakeyama, H.; Hatakeyama, T. Lignin Structure, Properties, and Applications. *Biopolymers,* **2009**, 1-63.

[36] Chui, S.S.Y.; Lo, S.M.F.; Charmant, J.P.H.; Orpen, A.G.; Williams, I.D. A chemically functionalizable nanoporous material. *Science,* **1999**, *283*(5405), 1148-1150.
[http://dx.doi.org/10.1126/science.283.5405.1148] [PMID: 10024237]

[37] Yaghi, O.M.; O'Keeffe, M.; Ockwig, N.W.; Chae, H.K.; Eddaoudi, M.; Kim, J. Reticular synthesis and the design of new materials. *Nature,* **2003**, *423*(6941), 705-714.
[http://dx.doi.org/10.1038/nature01650] [PMID: 12802325]

[38] Kitagawa, S.; Kitaura, R.; Noro, S. Functional porous coordination polymers. *Angew. Chem. Int. Ed.,* **2004**, *43*(18), 2334-2375.
[http://dx.doi.org/10.1002/anie.200300610] [PMID: 15114565]

[39] Férey, G.; Mellot-Draznieks, C.; Serre, C.; Millange, F.; Dutour, J.; Surblé, S.; Margiolaki, I. A chromium terephthalate-based solid with unusually large pore volumes and surface area. *Science,* **2005**, *309*(5743), 2040-2042.
[http://dx.doi.org/10.1126/science.1116275] [PMID: 16179475]

[40] Ma, S.; Zhou, H.C. A metal-organic framework with entatic metal centers exhibiting high gas adsorption affinity. *J. Am. Chem. Soc.,* **2006**, *128*(36), 11734-11735.
[http://dx.doi.org/10.1021/ja063538z] [PMID: 16953594]

[41] Li, H.; Eddaoudi, M.; O'Keeffe, M.; Yaghi, O.M. Design and synthesis of an exceptionally stable and highly porous metal-organic framework. *Nature,* **1999**, *402*(6759), 276-279.
[http://dx.doi.org/10.1038/46248]

[42] Ghosh, S.K.; Bharadwaj, P.K. Puckered-boat conformation hexameric water clusters stabilized in a 2D metal-organic framework structure built from CuII and 1,2,4,5-benzenetetracarboxylic acid. *Inorg. Chem.,* **2004**, *43*(17), 5180-5182.
[http://dx.doi.org/10.1021/ic049739q] [PMID: 15310189]

[43] Park, K.S.; Ni, Z.; Côté, A.P.; Choi, J.Y.; Huang, R.; Uribe-Romo, F.J.; Chae, H.K.; O'Keeffe, M.; Yaghi, O.M. Exceptional chemical and thermal stability of zeolitic imidazolate frameworks. *Proc. Natl. Acad. Sci. USA,* **2006**, *103*(27), 10186-10191.
[http://dx.doi.org/10.1073/pnas.0602439103] [PMID: 16798880]

[44] Huang, X.C.; Lin, Y.Y.; Zhang, J.P.; Chen, X.M. Ligand-directed strategy for zeolite-type metal-organic frameworks: zinc(II) imidazolates with unusual zeolitic topologies. *Angew. Chem. Int. Ed.,* **2006**, *45*(10), 1557-1559.
[http://dx.doi.org/10.1002/anie.200503778] [PMID: 16440383]

[45]   Furukawa, H.; Cordova, K.E.; O'Keeffe, M.; Yaghi, O.M. The chemistry and applications of metal-organic frameworks. *Science,* **2013**, *341*(6149), 1230444-1230455.
[http://dx.doi.org/10.1126/science.1230444] [PMID: 23990564]

[46]   He, C.; Liu, D.; Lin, W. Nanomedicine Applications of Hybrid Nanomaterials Built from Metal–Ligand Coordination Bonds: Nanoscale Metal–Organic Frameworks and Nanoscale Coordination Polymers. *Chem. Rev.,* **2015**, *115*(19), 11079-11108.
[http://dx.doi.org/10.1021/acs.chemrev.5b00125] [PMID: 26312730]

[47]   Farha, O.K.; Eryazici, I.; Jeong, N.C.; Hauser, B.G.; Wilmer, C.E.; Sarjeant, A.A.; Snurr, R.Q.; Nguyen, S.T.; Yazaydın, A.Ö.; Hupp, J.T. Metal-organic framework materials with ultrahigh surface areas: is the sky the limit? *J. Am. Chem. Soc.,* **2012**, *134*(36), 15016-15021.
[http://dx.doi.org/10.1021/ja3055639] [PMID: 22906112]

[48]   Li, G.; Kobayashi, H.; Taylor, J.M.; Ikeda, R.; Kubota, Y.; Kato, K.; Takata, M.; Yamamoto, T.; Toh, S.; Matsumura, S.; Kitagawa, H. Hydrogen storage in Pd nanocrystals covered with a metal–organic framework. *Nat. Mater.,* **2014**, *13*(8), 802-806.
[http://dx.doi.org/10.1038/nmat4030] [PMID: 25017188]

[49]   Sheberla, D.; Sun, L.; Blood-Forsythe, M.A.; Er, S.; Wade, C.R.; Brozek, C.K.; Aspuru-Guzik, A.; Dincă, M. High electrical conductivity in Ni$_3$(2,3,6,7,10,11-hexaiminotriphenylene)$_2$, a semiconducting metal-organic graphene analogue. *J. Am. Chem. Soc.,* **2014**, *136*(25), 8859-8862.
[http://dx.doi.org/10.1021/ja502765n] [PMID: 24750124]

[50]   Shekhah, O.; Liu, J.; Fischer, R.A.; Wöll, C. MOF thin films: existing and future applications. *Chem. Soc. Rev.,* **2011**, *40*(2), 1081-1106.
[http://dx.doi.org/10.1039/c0cs00147c] [PMID: 21225034]

[51]   Lee, J.; Farha, O.K.; Roberts, J.; Scheidt, K.A.; Nguyen, S.T.; Hupp, J.T. Metal–organic framework materials as catalysts. *Chem. Soc. Rev.,* **2009**, *38*(5), 1450-1459.
[http://dx.doi.org/10.1039/b807080f] [PMID: 19384447]

[52]   Fang, Z.; Bueken, B.; De Vos, D.E.; Fischer, R.A. Defect-Engineered Metal-Organic Frameworks. *Angew. Chem. Int. Ed.,* **2015**, *54*(25), 7234-7254.
[http://dx.doi.org/10.1002/anie.201411540] [PMID: 26036179]

[53]   Dissegna, S.; Epp, K.; Heinz, W.R.; Kieslich, G.; Fischer, R.A. Defective Metal-Organic Frameworks. *Adv. Mater.,* **2018**, *30*(37), 1704501.
[http://dx.doi.org/10.1002/adma.201704501]

[54]   Vermoortele, F.; Vandichel, M.; Van de Voorde, B.; Ameloot, R.; Waroquier, M.; Van Speybroeck, V.; De Vos, D.E. Electronic effects of linker substitution on Lewis acid catalysis with metal-organic frameworks. *Angew. Chem. Int. Ed.,* **2012**, *51*(20), 4887-4890.
[http://dx.doi.org/10.1002/anie.201108565] [PMID: 22488675]

[55]   Deng, H.; Doonan, C.J.; Furukawa, H.; Ferreira, R.B.; Towne, J.; Knobler, C.B.; Wang, B.; Yaghi, O.M. Multiple functional groups of varying ratios in metal-organic frameworks. *Science,* **2010**, *327*(5967), 846-850.
[http://dx.doi.org/10.1126/science.1181761] [PMID: 20150497]

[56]   Hu, Z.; Peng, Y.; Kang, Z.; Qian, Y.; Zhao, D. A Modulated Hydrothermal (MHT) Approach for the Facile Synthesis of UiO-66-Type MOFs. *Inorg. Chem.,* **2015**, *54*(10), 4862-4868.
[http://dx.doi.org/10.1021/acs.inorgchem.5b00435] [PMID: 25932655]

[57]   Kong, X.; Deng, H.; Yan, F.; Kim, J.; Swisher, J.A.; Smit, B.; Yaghi, O.M.; Reimer, J.A. Mapping of functional groups in metal-organic frameworks. *Science,* **2013**, *341*(6148), 882-885.
[http://dx.doi.org/10.1126/science.1238339] [PMID: 23887875]

[58]   Akiyama, G.; Matsuda, R.; Sato, H.; Takata, M.; Kitagawa, S. Cellulose hydrolysis by a new porous coordination polymer decorated with sulfonic acid functional groups. *Adv. Mater.,* **2011**, *23*(29), 3294-3297.

[http://dx.doi.org/10.1002/adma.201101356] [PMID: 21661069]

[59]   Shultz, A.M.; Farha, O.K.; Hupp, J.T.; Nguyen, S.T. A catalytically active, permanently microporous MOF with metalloporphyrin struts. *J. Am. Chem. Soc.,* **2009**, *131*(12), 4204-4205.
[http://dx.doi.org/10.1021/ja900203f] [PMID: 19271705]

[60]   Feng, D.; Gu, Z.Y.; Li, J.R.; Jiang, H.L.; Wei, Z.; Zhou, H.C. Zirconium-metalloporphyrin PCN-222: mesoporous metal-organic frameworks with ultrahigh stability as biomimetic catalysts. *Angew. Chem. Int. Ed.,* **2012**, *51*(41), 10307-10310.
[http://dx.doi.org/10.1002/anie.201204475] [PMID: 22907870]

[61]   Tu, B.; Pang, Q.; Xu, H.; Li, X.; Wang, Y.; Ma, Z.; Weng, L.; Li, Q. Reversible Redox Activity in Multicomponent Metal–Organic Frameworks Constructed from Trinuclear Copper Pyrazolate Building Blocks. *J. Am. Chem. Soc.,* **2017**, *139*(23), 7998-8007.
[http://dx.doi.org/10.1021/jacs.7b03578] [PMID: 28541696]

[62]   Hermes, S. Schröter. M.; Schmid, R.; Khodeir, L.; Muhler, M.; Tissler, A.; Fischer, R.W.; Fischer, R.A. Metal@MOF: Loading of Highly Porous Coordination Polymers Host Lattices by Metal Organic Chemical Vapor Deposition. *Angew. Chem. Int. Ed.,* **2005**, *44*, 6237-6241.
[http://dx.doi.org/10.1002/anie.200462515]

[63]   Müller, M.; Turner, S.; Lebedev, O.I.; Wang, Y.; van Tendeloo, G.; Fischer, R.A. Au@MOF-5 and Au/MO $_x$ @MOF-5 (M = Zn, Ti; x = 1, 2): Preparation and Microstructural Characterisation. *Eur. J. Inorg. Chem.,* **2011**, *2011*(12), 1876-1887.
[http://dx.doi.org/10.1002/ejic.201001297]

[64]   Kitao, T.; Zhang, Y.; Kitagawa, S.; Wang, B.; Uemura, T. Hybridization of MOFs and polymers. *Chem. Soc. Rev.,* **2017**, *46*(11), 3108-3133.
[http://dx.doi.org/10.1039/C7CS00041C] [PMID: 28368064]

[65]   Aguilera-Sigalat, J.; Bradshaw, D. Synthesis and applications of metal-organic framework–quantum dot (QD@MOF) composites. *Coord. Chem. Rev.,* **2016**, *307*, 267-291.
[http://dx.doi.org/10.1016/j.ccr.2015.08.004]

[66]   Lian, X.; Fang, Y.; Joseph, E.; Wang, Q.; Li, J.; Banerjee, S.; Lollar, C.; Wang, X.; Zhou, H.C. Enzyme–MOF (metal–organic framework) composites. *Chem. Soc. Rev.,* **2017**, *46*(11), 3386-3401.
[http://dx.doi.org/10.1039/C7CS00058H] [PMID: 28451673]

[67]   Juan-Alcañiz, J.; Ramos-Fernandez, E.V.; Lafont, U.; Gascon, J.; Kapteijn, F. Building MOF bottles around phosphotungstic acid ships: One-pot synthesis of bi-functional polyoxometalate-MIL-101 catalysts. *J. Catal.,* **2010**, *269*(1), 229-241.
[http://dx.doi.org/10.1016/j.jcat.2009.11.011]

[68]   Wang, Q.; Astruc, D. State of the Art and Prospects in Metal–Organic Framework (MOF)-Based and MOF-Derived Nanocatalysis. *Chem. Rev.,* **2020**, *120*(2), 1438-1511.
[http://dx.doi.org/10.1021/acs.chemrev.9b00223] [PMID: 31246430]

[69]   Shen, K.; Chen, X.; Chen, J.; Li, Y. Development of MOF-Derived Carbon-Based Nanomaterials for Efficient Catalysis. *ACS Catal.,* **2016**, *6*(9), 5887-5903.
[http://dx.doi.org/10.1021/acscatal.6b01222]

[70]   Wu, H.B.; Lou, X.W.D. Metal-organic frameworks and their derived materials for electrochemical energy storage and conversion: Promises and challenges. *Sci. Adv.,* **2017**, *3*(12), eaap9252.
[http://dx.doi.org/10.1126/sciadv.aap9252] [PMID: 29214220]

[71]   Chen, Y.Z.; Zhang, R.; Jiao, L.; Jiang, H.L. Metal–organic framework-derived porous materials for catalysis. *Coord. Chem. Rev.,* **2018**, *362*, 1-23.
[http://dx.doi.org/10.1016/j.ccr.2018.02.008]

[72]   Salunkhe, R.R.; Kaneti, Y.V.; Yamauchi, Y. Metal–Organic Framework-Derived Nanoporous Metal Oxides toward Supercapacitor Applications: Progress and Prospects. *ACS Nano,* **2017**, *11*(6), 5293-5308.

[http://dx.doi.org/10.1021/acsnano.7b02796] [PMID: 28613076]

[73]   Chaikittisilp, W.; Ariga, K.; Yamauchi, Y. A new family of carbon materials: synthesis of MOF-derived nanoporous carbons and their promising applications. *J. Mater. Chem. A Mater. Energy Sustain.,* **2013**, *1*(1), 14-19.
[http://dx.doi.org/10.1039/C2TA00278G]

[74]   Yang, S.J.; Kim, T.; Im, J.H.; Kim, Y.S.; Lee, K.; Jung, H.; Park, C.R. MOF-Derived Hierarchically Porous Carbon with Exceptional Porosity and Hydrogen Storage Capacity. *Chem. Mater.,* **2012**, *24*(3), 464-470.
[http://dx.doi.org/10.1021/cm202554j]

[75]   Jiang, H.; Jin, S.; Wang, C.; Ma, R.; Song, Y.; Gao, M.; Liu, X.; Shen, A.; Cheng, G.J.; Deng, H. Nanoscale Laser Metallurgy and Patterning in Air Using MOFs. *J. Am. Chem. Soc.,* **2019**, *141*(13), 5481-5489.
[http://dx.doi.org/10.1021/jacs.9b00355] [PMID: 30823704]

[76]   Bavykina, A.; Kolobov, N.; Khan, I.S.; Bau, J.A.; Ramirez, A.; Gascon, J. Metal–Organic Frameworks in Heterogeneous Catalysis: Recent Progress, New Trends, and Future Perspectives. *Chem. Rev.,* **2020**, *120*(16), 8468-8535.
[http://dx.doi.org/10.1021/acs.chemrev.9b00685] [PMID: 32223183]

[77]   Herbst, A.; Janiak, C. MOF catalysts in biomass upgrading towards value-added fine chemicals. *CrystEngComm,* **2017**, *19*(29), 4092-4117.
[http://dx.doi.org/10.1039/C6CE01782G]

[78]   Aljammal, N.; Jabbour, C.; Thybaut, J.W.; Demeestere, K.; Verpoort, F.; Heynderickx, P.M. Metal-organic frameworks as catalysts for sugar conversion into platform chemicals: State-of-the-art and prospects. *Coord. Chem. Rev.,* **2019**, *401*, 213064-213087.
[http://dx.doi.org/10.1016/j.ccr.2019.213064]

[79]   Sudarsanam, P.; Zhong, R.; Van den Bosch, S.; Coman, S.M.; Parvulescu, V.I.; Sels, B.F. Functionalised heterogeneous catalysts for sustainable biomass valorisation. *Chem. Soc. Rev.,* **2018**, *47*(22), 8349-8402.
[http://dx.doi.org/10.1039/C8CS00410B] [PMID: 30226518]

[80]   Fang, R.; Dhakshinamoorthy, A.; Li, Y.; Garcia, H. Metal organic frameworks for biomass conversion. *Chem. Soc. Rev.,* **2020**, *49*(11), 3638-3687.
[http://dx.doi.org/10.1039/D0CS00070A] [PMID: 32396593]

[81]   Akiyama, G.; Matsuda, R.; Sato, H.; Takata, M.; Kitagawa, S. Cellulose hydrolysis by a new porous coordination polymer decorated with sulfonic acid functional groups. *Adv. Mater.,* **2011**, *23*(29), 3294-3297.
[http://dx.doi.org/10.1002/adma.201101356] [PMID: 21661069]

[82]   Zi, G.; Yan, Z.; Wang, Y.; Chen, Y.; Guo, Y.; Yuan, F.; Gao, W.; Wang, Y.; Wang, J. Catalytic hydrothermal conversion of carboxymethyl cellulose to value-added chemicals over metal–organic framework MIL-53(Al). *Carbohydr. Polym.,* **2015**, *115*, 146-151.
[http://dx.doi.org/10.1016/j.carbpol.2014.08.065] [PMID: 25439879]

[83]   Chen, J.; Wang, S.; Huang, J.; Chen, L.; Ma, L.; Huang, X. Conversion of cellulose and cellobiose into sorbitol catalyzed by ruthenium supported on a polyoxometalate/metal-organic framework hybrid. *ChemSusChem,* **2013**, *6*(8), 1545-1555.
[http://dx.doi.org/10.1002/cssc.201200914] [PMID: 23619979]

[84]   Wang, S.; Chen, J.; Chen, L. Selective Conversion of Cellulose into Ethylene Glycol over Metal–Organic Framework-Derived Multifunctional Catalysts. *Catal. Lett.,* **2014**, *144*(10), 1728-1734.
[http://dx.doi.org/10.1007/s10562-014-1334-1]

[85]   Ahmed, I.N.; Yang, X.L.; Dubale, A.A.; Li, R.F.; Ma, Y.M.; Wang, L.M.; Hou, G.H.; Guan, R.F.; Xie, M.H. Hydrolysis of cellulose using cellulase physically immobilized on highly stable zirconium based metal-organic frameworks. *Bioresour. Technol.,* **2018**, *270*, 377-382.

[http://dx.doi.org/10.1016/j.biortech.2018.09.077] [PMID: 30243245]

[86]   Wang, L.; Zhi, W.; Wan, J.; Han, J.; Li, C.; Wang, Y. Recyclable β-Glucosidase by One-Pot Encapsulation with Cu-MOFs for Enhanced Hydrolysis of Cellulose to Glucose. *ACS Sustain. Chem.& Eng.,* **2019,** *7*(3), 3339-3348.
[http://dx.doi.org/10.1021/acssuschemeng.8b05489]

[87]   Čelič, T.B.; Grilc, M.; Likozar, B.; Tušar, N.N. In situ generation of Ni nanoparticles from metal-organic framework precursors and their use for biomass hydrodeoxygenation. *ChemSusChem,* **2015,** *8*(10), 1703-1710.
[http://dx.doi.org/10.1002/cssc.201403300] [PMID: 25755008]

[88]   Akiyama, G.; Matsuda, R.; Sato, H.; Kitagawa, S. Catalytic glucose isomerization by porous coordination polymers with open metal sites. *Chem. Asian J.,* **2014,** *9*(10), 2772-2777.
[http://dx.doi.org/10.1002/asia.201402119] [PMID: 25080129]

[89]   Guo, Q.; Ren, L.; Kumar, P.; Cybulskis, V.J.; Mkhoyan, K.A.; Davis, M.E.; Tsapatsis, M. A Chromium Hydroxide/MIL-101(Cr) MOF Composite Catalyst and Its Use for the Selective Isomerization of Glucose to Fructose. *Angew. Chem. Int. Ed.,* **2018,** *57*(18), 4926-4930.
[http://dx.doi.org/10.1002/anie.201712818] [PMID: 29490110]

[90]   de Mello, M.D.; Tsapatsis, M. Selective Glucose-to-Fructose Isomerization over Modified Zirconium UiO-66 in Alcohol Media. *ChemCatChem,* **2018,** *10*(11), 2417-2423.
[http://dx.doi.org/10.1002/cctc.201800371]

[91]   Oozeerally, R.; Burnett, D.L.; Chamberlain, T.W.; Walton, R.I.; Degirmenci, V. Exceptionally Efficient and Recyclable Heterogeneous Metal-Organic Framework Catalyst for Glucose Isomerization in Water. *ChemCatChem,* **2018,** *10*(4), 706-709.
[http://dx.doi.org/10.1002/cctc.201701825] [PMID: 29541254]

[92]   Luo, Q.X.; Zhang, Y.B.; Qi, L.; Scott, S.L. Glucose Isomerization and Epimerization over Metal-Organic Frameworks with Single-Site Active Centers. *ChemCatChem,* **2019,** *11*(7), 1903-1909.
[http://dx.doi.org/10.1002/cctc.201801889]

[93]   Malonzo, C.D.; Shaker, S.M.; Ren, L.; Prinslow, S.D.; Platero-Prats, A.E.; Gallington, L.C.; Borycz, J.; Thompson, A.B.; Wang, T.C.; Farha, O.K.; Hupp, J.T.; Lu, C.C.; Chapman, K.W.; Myers, J.C.; Penn, R.L.; Gagliardi, L.; Tsapatsis, M.; Stein, A. Thermal Stabilization of Metal–Organic Framework-Derived Single-Site Catalytic Clusters through Nanocasting. *J. Am. Chem. Soc.,* **2016,** *138*(8), 2739-2748.
[http://dx.doi.org/10.1021/jacs.5b12688] [PMID: 26848741]

[94]   Zhang, Y.; Degirmenci, V.; Li, C.; Hensen, E.J.M. Phosphotungstic acid encapsulated in metal-organic framework as catalysts for carbohydrate dehydration to 5-hydroxymethylfurfural. *ChemSusChem,* **2011,** *4*(1), 59-64.
[http://dx.doi.org/10.1002/cssc.201000284] [PMID: 21226212]

[95]   Bromberg, L.; Su, X.; Hatton, T.A. Functional Networks of Organic and Coordination Polymers: Catalysis of Fructose Conversion. *Chem. Mater.,* **2014,** *26*(21), 6257-6264.
[http://dx.doi.org/10.1021/cm503098p]

[96]   Chen, J.; Li, K.; Chen, L.; Liu, R.; Huang, X.; Ye, D. Conversion of fructose into 5-hydroxymethylfurfural catalyzed by recyclable sulfonic acid-functionalized metal–organic frameworks. *Green Chem.,* **2014,** *16*(5), 2490-2499.
[http://dx.doi.org/10.1039/C3GC42414F]

[97]   Hu, Z.; Peng, Y.; Gao, Y.; Qian, Y.; Ying, S.; Yuan, D.; Horike, S.; Ogiwara, N.; Babarao, R.; Wang, Y.; Yan, N.; Zhao, D. Direct Synthesis of Hierarchically Porous Metal–Organic Frameworks with High Stability and Strong Brønsted Acidity: The Decisive Role of Hafnium in Efficient and Selective Fructose Dehydration. *Chem. Mater.,* **2016,** *28*(8), 2659-2667.
[http://dx.doi.org/10.1021/acs.chemmater.6b00139]

[98]   Su, Y.; Chang, G.; Zhang, Z.; Xing, H.; Su, B.; Yang, Q.; Ren, Q.; Yang, Y.; Bao, Z. Catalytic

dehydration of glucose to 5-hydroxymethylfurfural with a bifunctional metal-organic framework. *AIChE J.,* **2016**, *62*(12), 4403-4417.
[http://dx.doi.org/10.1002/aic.15356]

[99] Herbst, A.; Janiak, C. Selective glucose conversion to 5-hydroxymethylfurfural (5-HMF) instead of levulinic acid with MIL-101Cr MOF-derivatives. *New J. Chem.,* **2016**, *40*(9), 7958-7967.
[http://dx.doi.org/10.1039/C6NJ01399F]

[100] Yabushita, M.; Li, P.; Islamoglu, T.; Kobayashi, H.; Fukuoka, A.; Farha, O.K.; Katz, A. Selective Metal–Organic Framework Catalysis of Glucose to 5-Hydroxymethylfurfural Using Phosphate-Modified NU-1000. *Ind. Eng. Chem. Res.,* **2017**, *56*(25), 7141-7148.
[http://dx.doi.org/10.1021/acs.iecr.7b01164]

[101] Gong, J.; Katz, M.J.; Kerton, F.M. Catalytic conversion of glucose to 5-hydroxymethylfurfural using zirconium-containing metal–organic frameworks using microwave heating. *RSC Advances,* **2018**, *8*(55), 31618-31627.
[http://dx.doi.org/10.1039/C8RA06021E] [PMID: 35548202]

[102] Pertiwi, R.; Oozeerally, R.; Burnett, D.L.; Chamberlain, T.W.; Cherkasov, N.; Walker, M.; Kashtiban, R.J.; Krisnandi, Y.K.; Degirmenci, V.; Walton, R.I. Replacement of Chromium by Non-Toxic Metals in Lewis-Acid MOFs: Assessment of Stability as Glucose Conversion Catalysts. *Catalysts,* **2019**, *9*(5), 437.
[http://dx.doi.org/10.3390/catal9050437]

[103] Burnett, D.L.; Oozeerally, R.; Pertiwi, R.; Chamberlain, T.W.; Cherkasov, N.; Clarkson, G.J.; Krisnandi, Y.K.; Degirmenci, V.; Walton, R.I. A hydrothermally stable ytterbium metal–organic framework as a bifunctional solid-acid catalyst for glucose conversion. *Chem. Commun. (Camb.),* **2019**, *55*(76), 11446-11449.
[http://dx.doi.org/10.1039/C9CC05364F] [PMID: 31486470]

[104] Zhang, Y.; Zhao, J.; Wang, K.; Gao, L.; Meng, M.; Yan, Y. Green Synthesis of Acid-Base Bi-functional UiO-66-Type Metal-Organic Frameworks Membranes Supported on Polyurethane Foam for Glucose Conversion. *ChemistrySelect,* **2018**, *3*(32), 9378-9387.
[http://dx.doi.org/10.1002/slct.201801893]

[105] Zhang, Y.; Guan, W.; Song, H.; Wei, Y.; Jin, P.; Li, B.; Yan, C.; Pan, J.; Yan, Y. Coupled acid and base UiO-66-type MOFs supported on g-C$_3$N$_4$ as a bi-functional catalyst for one-pot production of 5-HMF from glucose. *Microporous Mesoporous Mater.,* **2020**, *305*, 110328-110339.
[http://dx.doi.org/10.1016/j.micromeso.2020.110328]

[106] Zhao, J.; Anjali, J.; Yan, Y.; Lee, J.M. Cr-MIL-101-Encapsulated Keggin Phosphomolybdic Acid as a Catalyst for the One-Pot Synthesis of 2,5-Diformylfuran from Fructose. *ChemCatChem,* **2017**, *9*(7), 1187-1191.
[http://dx.doi.org/10.1002/cctc.201601546]

[107] Fang, R.; Luque, R.; Li, Y. Efficient one-pot fructose to DFF conversion using sulfonated magnetically separable MOF-derived Fe$_3$O$_4$ (111) catalysts. *Green Chem.,* **2017**, *19*(3), 647-655.
[http://dx.doi.org/10.1039/C6GC02018F]

[108] Insyani, R.; Verma, D.; Kim, S.M.; Kim, J. Direct one-pot conversion of monosaccharides into high-yield 2,5-dimethylfuran over a multifunctional Pd/Zr-based metal–organic framework@sulfonated graphene oxide catalyst. *Green Chem.,* **2017**, *19*(11), 2482-2490.
[http://dx.doi.org/10.1039/C7GC00269F]

[109] Lu, X.; Wang, L.; Lu, X. Catalytic conversion of sugars to methyl lactate over Mg-MOF-74 in near-critical methanol solutions. *Catal. Commun.,* **2018**, *110*, 23-27.
[http://dx.doi.org/10.1016/j.catcom.2018.02.027]

[110] Qu, H.; Liu, B.; Gao, G.; Ma, Y.; Zhou, Y.; Zhou, H.; Li, L.; Li, Y.; Liu, S. Metal-organic framework containing Brønsted acidity and Lewis acidity for efficient conversion glucose to levulinic acid. *Fuel Process. Technol.,* **2019**, *193*, 1-6.

[http://dx.doi.org/10.1016/j.fuproc.2019.04.035]

[111] Chatterjee, A.; Hu, X.; Lam, F.L.Y. A dual acidic hydrothermally stable MOF-composite for upgrading xylose to furfural. *Appl. Catal. A Gen.,* **2018**, *566*, 130-139.
[http://dx.doi.org/10.1016/j.apcata.2018.04.016]

[112] Chatterjee, A.; Hu, X.; Lam, F.L.Y. Catalytic activity of an economically sustainable fly-ash-met-l-organic- framework composite towards biomass valorization. *Catal. Today,* **2018**, *314*, 137-146.
[http://dx.doi.org/10.1016/j.cattod.2018.01.018]

[113] Liu, Y.; Ma, C.; Huang, C.; Fu, Y.; Chang, J. Efficient Conversion of Xylose into Furfural Using Sulfonic Acid-Functionalized Metal–Organic Frameworks in a Biphasic System. *Ind. Eng. Chem. Res.,* **2018**, *57*(49), 16628-16634.
[http://dx.doi.org/10.1021/acs.iecr.8b04070]

[114] Yun, W.C.; Yang, M.T.; Lin, K.Y.A. Water-born zirconium-based metal organic frameworks as green and effective catalysts for catalytic transfer hydrogenation of levulinic acid to γ-valerolactone: Critical roles of modulators. *J. Colloid Interface Sci.,* **2019**, *543*, 52-63.
[http://dx.doi.org/10.1016/j.jcis.2019.02.036] [PMID: 30779993]

[115] Vasanthakumar, P.; Sindhuja, D.; Senthil Raja, D.; Lin, C.H.; Karvembu, R. Iron and chromium MOFs as sustainable catalysts for transfer hydrogenation of carbonyl compounds and biomass conversions. *New J. Chem.,* **2020**, *44*(20), 8223-8231.
[http://dx.doi.org/10.1039/D0NJ00552E]

[116] Li, J.; Zhao, S.; Li, Z.; Liu, D.; Chi, Y.; Hu, C. Efficient Conversion of Biomass-Derived Levulinic Acid to γ-Valerolactone over Polyoxometalate@Zr-Based Metal–Organic Frameworks: The Synergistic Effect of Brønsted and Lewis Acidic Sites. *Inorg. Chem.,* **2021**, *60*(11), 7785-7793.
[http://dx.doi.org/10.1021/acs.inorgchem.1c00185] [PMID: 33755456]

[117] Guo, Y.; Li, Y.; Chen, J.; Chen, L. Hydrogenation of Levulinic Acid into γ-Valerolactone Over Ruthenium Catalysts Supported on Metal–Organic Frameworks in Aqueous Medium. *Catal. Lett.,* **2016**, *146*(10), 2041-2052.
[http://dx.doi.org/10.1007/s10562-016-1819-1]

[118] Zhang, X.; Zhang, P.; Chen, C.; Zhang, J.; Yang, G.; Zheng, L.; Zhang, J.; Han, B. Fabrication of 2D metal–organic framework nanosheets with tailorable thickness using bio-based surfactants and their application in catalysis. *Green Chem.,* **2019**, *21*(1), 54-58.
[http://dx.doi.org/10.1039/C8GC02835D]

[119] Feng, J.; Zhong, Y.; Xie, M.; Li, M.; Jiang, S. Using MOF-808 as a Promising Support to Immobilize Ru for Selective Hydrogenation of Levulinic Acid to γ-Valerolactone. *Catal. Lett.,* **2021**, *151*(1), 86-94.
[http://dx.doi.org/10.1007/s10562-020-03277-x]

[120] Yang, Y.; Sun, C.J.; Brown, D.E.; Zhang, L.; Yang, F.; Zhao, H.; Wang, Y.; Ma, X.; Zhang, X.; Ren, Y. A smart strategy to fabricate Ru nanoparticle inserted porous carbon nanofibers as highly efficient levulinic acid hydrogenation catalysts. *Green Chem.,* **2016**, *18*(12), 3558-3566.
[http://dx.doi.org/10.1039/C5GC02802G]

[121] Cao, W.; Luo, W.; Ge, H.; Su, Y.; Wang, A.; Zhang, T. UiO-66 derived Ru/ZrO$_2$@C as a highly stable catalyst for hydrogenation of levulinic acid to γ-valerolactone. *Green Chem.,* **2017**, *19*(9), 2201-2211.
[http://dx.doi.org/10.1039/C7GC00512A]

[122] Pan, J.; Xu, Q.; Fang, L.; Tu, G.; Fu, Y.; Chen, G.; Zhang, F.; Zhu, W. Ru nanoclusters supported on HfO$_2$@CN derived from NH$_2$-UiO-66(Hf) as stable catalysts for the hydrogenation of levulinic acid to γ-valerolactone. *Catal. Commun.,* **2019**, *128*, 105710.
[http://dx.doi.org/10.1016/j.catcom.2019.105710]

[123] Van Nguyen, C.; Matsagar, B.M.; Yeh, J.Y.; Chiang, W.H.; Wu, K.C.W. MIL-53-NH$_2$-derived carbon-Al$_2$O$_3$ composites supported Ru catalyst for effective hydrogenation of levulinic acid to γ-

valerolactone under ambient conditions. *Molecular Catalysis,* **2019**, *475*, 110478.
[http://dx.doi.org/10.1016/j.mcat.2019.110478]

[124] Feng, J.; Li, M.; Zhong, Y.; Xu, Y.; Meng, X.; Zhao, Z.; Feng, C. Hydrogenation of levulinic acid to γ-valerolactone over Pd@UiO-66-NH₂ with high metal dispersion and excellent reusability. *Microporous Mesoporous Mater.,* **2020**, *294*, 109858.
[http://dx.doi.org/10.1016/j.micromeso.2019.109858]

[125] Xu, H.; Hu, D.; Yi, Z.; Wu, Z.; Zhang, M.; Yan, K. Solvent Tuning the Selective Hydrogenation of Levulinic Acid into Biofuels over Ni-Metal Organic Framework-Derived Catalyst. *ACS Appl. Energy Mater.,* **2019**, *2*(10), 6979-6983.
[http://dx.doi.org/10.1021/acsaem.9b01439]

[126] Shao, S.; Yang, Y.; Guo, S.; Hao, S.; Yang, F.; Zhang, S.; Ren, Y.; Ke, Y. Highly active and stable Co nanoparticles embedded in nitrogen-doped mesoporous carbon nanofibers for aqueous-phase levulinic acid hydrogenation. *Green Energy & Environment,* **2021**, *6*(4), 567-577.
[http://dx.doi.org/10.1016/j.gee.2020.11.005]

[127] Ouyang, W.; Zhao, D.; Wang, Y.; Balu, A.M.; Len, C.; Luque, R. Continuous Flow Conversion of Biomass-Derived Methyl Levulinate into γ-Valerolactone Using Functional Metal Organic Frameworks. *ACS Sustain. Chem.& Eng.,* **2018**, *6*(5), 6746-6752.
[http://dx.doi.org/10.1021/acssuschemeng.8b00549]

[128] Kuwahara, Y.; Kango, H.; Yamashita, H. Catalytic Transfer Hydrogenation of Biomass-Derived Levulinic Acid and Its Esters to γ-Valerolactone over Sulfonic Acid-Functionalized UiO-66. *ACS Sustain. Chem.& Eng.,* **2017**, *5*(1), 1141-1152.
[http://dx.doi.org/10.1021/acssuschemeng.6b02464]

[129] Lin, Z.; Cai, X.; Fu, Y.; Zhu, W.; Zhang, F. Cascade catalytic hydrogenation–cyclization of methyl levulinate to form γ-valerolactone over Ru nanoparticles supported on a sulfonic acid-functionalized UiO-66 catalyst. *RSC Advances,* **2017**, *7*(70), 44082-44088.
[http://dx.doi.org/10.1039/C7RA06293A]

[130] Lin, Z.; Luo, M.; Zhang, Y.; Wu, X.; Fu, Y.; Zhang, F.; Zhu, W. Coupling Ru nanoparticles and sulfonic acid moieties on single MIL-101 microcrystals for upgrading methyl levulinate into γ-valerolactone. *Appl. Catal. A Gen.,* **2018**, *563*, 54-63.
[http://dx.doi.org/10.1016/j.apcata.2018.06.027]

[131] Cai, X.; Xu, Q.; Tu, G.; Fu, Y.; Zhang, F.; Zhu, W. Synergistic Catalysis of Ruthenium Nanoparticles and Polyoxometalate Integrated Within Single UiO-66 Microcrystals for Boosting the Efficiency of Methyl Levulinate to γ-Valerolactone. *Front Chem.,* **2019**, *7*, 42.
[http://dx.doi.org/10.3389/fchem.2019.00042] [PMID: 30775365]

[132] Valekar, A.H.; Cho, K.H.; Chitale, S.K.; Hong, D.Y.; Cha, G.Y.; Lee, U.H.; Hwang, D.W.; Serre, C.; Chang, J.S.; Hwang, Y.K. Catalytic transfer hydrogenation of ethyl levulinate to γ-valerolactone over zirconium-based metal–organic frameworks. *Green Chem.,* **2016**, *18*(16), 4542-4552.
[http://dx.doi.org/10.1039/C6GC00524A]

[133] Cai, Z.; Li, W.; Wang, F.; Zhang, X. Zirconium/hafnium-DUT67 for catalytic transfer hydrogenation of ethyl levulinate to γ-valerolactone. *J. Taiwan Inst. Chem. Eng.,* **2018**, *93*, 374-378.
[http://dx.doi.org/10.1016/j.jtice.2018.08.002]

[134] Kurisingal, J.F.; Rachuri, Y.; Palakkal, A.S.; Pillai, R.S.; Gu, Y.; Choe, Y.; Park, D.W. Water-Tolerant DUT-Series Metal–Organic Frameworks: A Theoretical–Experimental Study for the Chemical Fixation of CO₂ and Catalytic Transfer Hydrogenation of Ethyl Levulinate to γ-Valerolactone. *ACS Appl. Mater. Interfaces,* **2019**, *11*(44), 41458-41471.
[http://dx.doi.org/10.1021/acsami.9b16834] [PMID: 31613085]

[135] Yang, J.; Huang, W.; Liu, Y.; Zhou, T. Enhancing the conversion of ethyl levulinate to γ-valerolactone over Ru/UiO-66 by introducing sulfonic groups into the framework. *RSC Advances,* **2018**, *8*(30), 16611-16618.

[http://dx.doi.org/10.1039/C8RA01314D] [PMID: 35540507]

[136] Feng, Y.; Yan, G.; Wang, T.; Jia, W.; Zeng, X.; Sperry, J.; Sun, Y.; Tang, X.; Lei, T.; Lin, L. Cu ^{1}–Cu ^{0} bicomponent CuNPs@ZIF-8 for highly selective hydrogenation of biomass derived 5-hydroxymethylfurfural. *Green Chem.,* **2019**, *21*(16), 4319-4323.
[http://dx.doi.org/10.1039/C9GC01331H]

[137] Wang, Z.; Chen, Q. Conversion of 5-hydroxymethylfurfural into 5-ethoxymethylfurfural and ethyl levulinate catalyzed by MOF-based heteropolyacid materials. *Green Chem.,* **2016**, *18*(21), 5884-5889.
[http://dx.doi.org/10.1039/C6GC01206J]

[138] Wang, Z.; Chen, Q. Variations of Major Product Derived from Conversion of 5-Hydroxymethylfurfural over a Modified MOFs-Derived Carbon Material in Response to Reaction Conditions. *Nanomaterials (Basel),* **2018**, *8*(7), 492-510.
[http://dx.doi.org/10.3390/nano8070492] [PMID: 29976847]

[139] Chen, J.; Liu, R.; Guo, Y.; Chen, L.; Gao, H. Selective Hydrogenation of Biomass-Based 5-Hydroxymethylfurfural over Catalyst of Palladium Immobilized on Amine-Functionalized Metal–Organic Frameworks. *ACS Catal.,* **2015**, *5*(2), 722-733.
[http://dx.doi.org/10.1021/cs5012926]

[140] Hester, P.; Xu, S.; Liang, W.; Al-Janabi, N.; Vakili, R.; Hill, P.; Muryn, C.A.; Chen, X.; Martin, P.A.; Fan, X. On thermal stability and catalytic reactivity of Zr-based metal–organic framework (UiO-67) encapsulated Pt catalysts. *J. Catal.,* **2016**, *340*, 85-94.
[http://dx.doi.org/10.1016/j.jcat.2016.05.003]

[141] Wang, K.; Zhao, W.; Zhang, Q.; Li, H.; Zhang, F. *in situ* One-Step Synthesis of Platinum Nanoparticles Supported on Metal–Organic Frameworks as an Effective and Stable Catalyst for Selective Hydrogenation of 5-Hydroxymethylfurfural. *ACS Omega,* **2020**, *5*(26), 16183-16188.
[http://dx.doi.org/10.1021/acsomega.0c01759] [PMID: 32656440]

[142] Zhang, Q.; Zuo, J.; Peng, F.; Chen, S.; Wang, Q.; Liu, Z. A Non-Noble Monometallic Catalyst Derived from Cu–MOFs for Highly Selective Hydrogenation of 5-Hydroxymethylfurfural to 2,5-Dimethylfuran. *ChemistrySelect,* **2019**, *4*(46), 13517-13524.
[http://dx.doi.org/10.1002/slct.201903256]

[143] Sarkar, C.; Koley, P.; Shown, I.; Lee, J.; Liao, Y.F.; An, K.; Tardio, J.; Nakka, L.; Chen, K.H.; Mondal, J. Integration of Interfacial and Alloy Effects to Modulate Catalytic Performance of Metal–Organic-Framework-Derived Cu–Pd Nanocrystals toward Hydrogenolysis of 5-Hydroxymethylfurfural. *ACS Sustain. Chem.& Eng.,* **2019**, *7*(12), 10349-10362.
[http://dx.doi.org/10.1021/acssuschemeng.9b00350]

[144] Shang, Y.; Liu, C.; Zhang, Z.; Wang, S.; Zhao, C.; Yin, X.; Zhang, P.; Liu, D.; Gui, J. Insights into the Synergistic Effect in Pd Immobilized to MOF-Derived Co-CoO $_x$@N-Doped Carbon for Efficient Selective Hydrogenolysis of 5-Hydroxylmethylfurfural. *Ind. Eng. Chem. Res.,* **2020**, *59*(14), 6532-6542.
[http://dx.doi.org/10.1021/acs.iecr.9b07099]

[145] Fang, R.; Luque, R.; Li, Y. Selective aerobic oxidation of biomass-derived HMF to 2,5-diformylfuran using a MOF-derived magnetic hollow Fe–Co nanocatalyst. *Green Chem.,* **2016**, *18*(10), 3152-3157.
[http://dx.doi.org/10.1039/C5GC03051J]

[146] Lin, J.Y.; Yuan, M.H.; Lin, K.Y.A.; Lin, C.H. Selective aerobic oxidation of 5-hydroxymethylfurfural to 2,5-diformylfuran catalyzed by Cu-based metal organic frameworks with 2,2,6,6-tetramethylpiperidin-oxyl. *J. Taiwan Inst. Chem. Eng.,* **2019**, *102*, 242-249.
[http://dx.doi.org/10.1016/j.jtice.2019.06.008]

[147] Lin, J.; Thanh, B.X.; Kwon, E.; Lin, K. *A. Enhanced Catalytic Conversion of 5-Hydroxymethylfurfural to 2,5-diformylfuran by HKUST-1/TEMPO under Microwave Irradiation*; Biomass Conv. Bioref, **2020**.
[http://dx.doi.org/10.1007/s13399-020-00648-7]

[148] Feng, Y.; Jia, W.; Yan, G.; Zeng, X.; Sperry, J.; Xu, B.; Sun, Y.; Tang, X.; Lei, T.; Lin, L. Insights into the active sites and catalytic mechanism of oxidative esterification of 5-hydroxymethylfurfural by metal-organic frameworks-derived N-doped carbon. *J. Catal.,* **2020**, *381*, 570-578.
[http://dx.doi.org/10.1016/j.jcat.2019.11.029]

[149] Nguyen, C.V.; Liao, Y.T.; Kang, T.C.; Chen, J.E.; Yoshikawa, T.; Nakasaka, Y.; Masuda, T.; Wu, K.C.W. A metal-free, high nitrogen-doped nanoporous graphitic carbon catalyst for an effective aerobic HMF-to-FDCA conversion. *Green Chem.,* **2016**, *18*(22), 5957-5961.
[http://dx.doi.org/10.1039/C6GC02118B]

[150] Fang, R.; Tian, P.; Yang, X.; Luque, R.; Li, Y. Encapsulation of ultrafine metal-oxide nanoparticles within mesopores for biomass-derived catalytic applications. *Chem. Sci. (Camb.),* **2018**, *9*(7), 1854-1859.
[http://dx.doi.org/10.1039/C7SC04724J] [PMID: 29675231]

[151] Bao, L.; Sun, F.Z.; Zhang, G.Y.; Hu, T.L. Aerobic Oxidation of 5-Hydroxymethylfurfural to 2,5-Furandicarboxylic Acid over Holey 2D Mn_2O_3 Nanoflakes from a Mn-based MOF. *ChemSusChem,* **2020**, *13*(3), 548-555.
[http://dx.doi.org/10.1002/cssc.201903018] [PMID: 31714031]

[152] Liao, Y.T.; Nguyen, V.C.; Ishiguro, N.; Young, A.P.; Tsung, C.K.; Wu, K.C.W. Engineering a homogeneous alloy-oxide interface derived from metal-organic frameworks for selective oxidation of 5-hydroxymethylfurfural to 2,5-furandicarboxylic acid. *Appl. Catal. B,* **2020**, *270*, 118805-118816.
[http://dx.doi.org/10.1016/j.apcatb.2020.118805]

[153] Rojas-Buzo, S.; García-García, P.; Corma, A. Catalytic Transfer Hydrogenation of Biomass-Derived Carbonyls over Hafnium-Based Metal-Organic Frameworks. *ChemSusChem,* **2018**, *11*(2), 432-438.
[http://dx.doi.org/10.1002/cssc.201701708] [PMID: 29139603]

[154] Rojas-Buzo, S.; García-García, P.; Corma, A. Hf-based metal–organic frameworks as acid–base catalysts for the transformation of biomass-derived furanic compounds into chemicals. *Green Chem.,* **2018**, *20*(13), 3081-3091.
[http://dx.doi.org/10.1039/C8GC00806J]

[155] Valekar, A.H.; Lee, M.; Yoon, J.W.; Kwak, J.; Hong, D.Y.; Oh, K.R.; Cha, G.Y.; Kwon, Y.U.; Jung, J.; Chang, J.S.; Hwang, Y.K. Catalytic Transfer Hydrogenation of Furfural to Furfuryl Alcohol under Mild Conditions over Zr-MOFs: Exploring the Role of Metal Node Coordination and Modification. *ACS Catal.,* **2020**, *10*(6), 3720-3732.
[http://dx.doi.org/10.1021/acscatal.9b05085]

[156] Qiu, M.; Guo, T.; Xi, R.; Li, D.; Qi, X. Highly efficient catalytic transfer hydrogenation of biomass-derived furfural to furfuryl alcohol using UiO-66 without metal catalysts. *Appl. Catal. A Gen.,* **2020**, *602*, 117719-117725.
[http://dx.doi.org/10.1016/j.apcata.2020.117719]

[157] Yuan, Q.; Zhang, D.; Haandel, L.; Ye, F.; Xue, T.; Hensen, E.J.M.; Guan, Y. Selective liquid phase hydrogenation of furfural to furfuryl alcohol by Ru/Zr-MOFs. *J. Mol. Catal. Chem.,* **2015**, *406*, 58-64.
[http://dx.doi.org/10.1016/j.molcata.2015.05.015]

[158] Yang, J.; Ma, J.; Yuan, Q.; Zhang, P.; Guan, Y. Selective hydrogenation of furfural on Ru/Al-MIL-53: a comparative study on the effect of aromatic and aliphatic organic linkers. *RSC Advances,* **2016**, *6*(95), 92299-92304.
[http://dx.doi.org/10.1039/C6RA21701J]

[159] Xu, T.; Sun, K.; Gao, D.; Li, C.; Hu, X.; Chen, G. Atomic-layer-deposition-formed sacrificial template for the construction of an MIL-53 shell to increase selectivity of hydrogenation reactions. *Chem. Commun. (Camb.),* **2019**, *55*(53), 7651-7654.
[http://dx.doi.org/10.1039/C9CC02727K] [PMID: 31198911]

[160] Long, Y.; Song, S.; Li, J.; Wu, L.; Wang, Q.; Liu, Y.; Jin, R.; Zhang, H. Pt/CeO_2@MOF Core@Shell Nanoreactor for Selective Hydrogenation of Furfural *via* the Channel Screening Effect. *ACS Catal.,*

2018, *8*(9), 8506-8512.
[http://dx.doi.org/10.1021/acscatal.8b01851]

[161]   Fang, R.; Chen, L.; Shen, Z.; Li, Y. Efficient hydrogenation of furfural to fufuryl alcohol over hierarchical MOF immobilized metal catalysts. *Catal. Today,* **2021**, *368*, 217-223.
[http://dx.doi.org/10.1016/j.cattod.2020.03.019]

[162]   Goh, T.W.; Tsung, C.K.; Huang, W. Spectroscopy Identification of the Bimetallic Surface of Metal–Organic Framework-Confined Pt–Sn Nanoclusters with Enhanced Chemoselectivity in Furfural Hydrogenation. *ACS Appl. Mater. Interfaces,* **2019**, *11*(26), 23254-23260.
[http://dx.doi.org/10.1021/acsami.9b06229] [PMID: 31252478]

[163]   Zhang, B.; Pei, Y.; Maligal-Ganesh, R.V.; Li, X.; Cruz, A.; Spurling, R.J.; Chen, M.; Yu, J.; Wu, X.; Huang, W. Influence of Sn on Stability and Selectivity of Pt–Sn@UiO-66-NH$_2$ in Furfural Hydrogenation. *Ind. Eng. Chem. Res.,* **2020**, *59*(39), 17495-17501.
[http://dx.doi.org/10.1021/acs.iecr.0c01336]

[164]   Wang, Y.; Huang, J.; Lu, S.; Li, P.; Xia, X.; Li, C.; Li, F. Phosphorus-modified zirconium metal organic frameworks for catalytic transfer hydrogenation of furfural. *New J. Chem.,* **2020**, *44*(46), 20308-20315.
[http://dx.doi.org/10.1039/D0NJ04285D]

[165]   Guo, P.; Liao, S.; Tong, X. Heterogeneous Nickel Catalysts Derived from 2D Metal–Organic Frameworks for Regulating the Selectivity of Furfural Hydrogenation. *ACS Omega,* **2019**, *4*(26), 21724-21731.
[http://dx.doi.org/10.1021/acsomega.9b02443] [PMID: 31891051]

[166]   Hu, F.; Wang, Y.; Xu, S.; Zhang, Z.; Chen, Y.; Fan, J.; Yuan, H.; Gao, L.; Xiao, G. Efficient and Selective Ni/Al$_2$O$_3$–C Catalyst Derived from Metal–Organic Frameworks for the Hydrogenation of Furfural to Furfuryl Alcohol. *Catal. Lett.,* **2019**, *149*(8), 2158-2168.
[http://dx.doi.org/10.1007/s10562-019-02766-y]

[167]   Wang, R.; Liu, H.; Wang, X.; Li, X.; Gu, X.; Zheng, Z. Plasmon-enhanced furfural hydrogenation catalyzed by stable carbon-coated copper nanoparticles driven from metal–organic frameworks. *Catal. Sci. Technol.,* **2020**, *10*(19), 6483-6494.
[http://dx.doi.org/10.1039/D0CY01162B]

[168]   Chen, K.; Ling, J.L.; Wu, C.D. *in situ* Generation and Stabilization of Accessible Cu/Cu$_2$O Heterojunctions inside Organic Frameworks for Highly Efficient Catalysis. *Angew. Chem. Int. Ed.,* **2020**, *59*(5), 1925-1931.
[http://dx.doi.org/10.1002/anie.201913811] [PMID: 31755200]

[169]   Jiang, S.; Huang, J.; Wang, Y.; Lu, S.; Li, P.; Li, C.; Li, F. Metal–organic frameworks derived magnetic FE$_3$O$_4$ /C for catalytic transfer hydrogenation of furfural to furfuryl alcohol. *J. Chem. Technol. Biotechnol.,* **2021**, *96*(3), 639-649.
[http://dx.doi.org/10.1002/jctb.6577]

[170]   Wang, Y.; Miao, Y.; Li, S.; Gao, L.; Xiao, G. Metal-organic frameworks derived bimetallic Cu-Co catalyst for efficient and selective hydrogenation of biomass-derived furfural to furfuryl alcohol. *Molecular Catalysis,* **2017**, *436*, 128-137.
[http://dx.doi.org/10.1016/j.mcat.2017.04.018]

[171]   Huang, L.; Hao, F.; Lv, Y.; Liu, Y.; Liu, P.; Xiong, W.; Luo, H. MOF-derived well-structured bimetallic catalyst for highly selective conversion of furfural. *Fuel,* **2021**, *289*, 119910-119922.
[http://dx.doi.org/10.1016/j.fuel.2020.119910]

[172]   Yin, D.; Ren, H.; Li, C.; Liu, J.; Liang, C. Highly selective hydrogenation of furfural to tetrahydrofurfuryl alcohol over MIL-101(Cr)-NH$_2$ supported Pd catalyst at low temperature. *Chin. J. Catal.,* **2018**, *39*(2), 319-326.
[http://dx.doi.org/10.1016/S1872-2067(18)63009-8]

[173]   Wang, C.; Wang, A.; Yu, Z.; Wang, Y.; Sun, Z.; Kogan, V.M.; Liu, Y.Y. Aqueous phase

hydrogenation of furfural to tetrahydrofurfuryl alcohol over Pd/UiO-66. *Catal. Commun.,* **2021**, *148*, 106178-106183.
[http://dx.doi.org/10.1016/j.catcom.2020.106178]

[174] Su, Y.; Chen, C.; Zhu, X.; Zhang, Y.; Gong, W.; Zhang, H.; Zhao, H.; Wang, G. Carbon-embedded Ni nanocatalysts derived from MOFs by a sacrificial template method for efficient hydrogenation of furfural to tetrahydrofurfuryl alcohol. *Dalton Trans.,* **2017**, *46*(19), 6358-6365.
[http://dx.doi.org/10.1039/C7DT00628D] [PMID: 28463366]

[175] Pendem, S.; Bolla, S.R.; Morgan, D.J.; Shinde, D.B.; Lai, Z.; Nakka, L.; Mondal, J. Metal–organic-framework derived Co–Pd bond is preferred over Fe–Pd for reductive upgrading of furfural to tetrahydrofurfuryl alcohol. *Dalton Trans.,* **2019**, *48*(24), 8791-8802.
[http://dx.doi.org/10.1039/C9DT01190K] [PMID: 31124551]

[176] Wang, H.; Li, X.; Lan, X.; Wang, T. Supported Ultrafine NiCo Bimetallic Alloy Nanoparticles Derived from Bimetal–Organic Frameworks: A Highly Active Catalyst for Furfuryl Alcohol Hydrogenation. *ACS Catal.,* **2018**, *8*(3), 2121-2128.
[http://dx.doi.org/10.1021/acscatal.7b03795]

[177] Yang, Y.; Deng, D.; Sui, D.; Xie, Y.; Li, D.; Duan, Y. Facile Preparation of Pd/UiO-66-v for the Conversion of Furfuryl Alcohol to Tetrahydrofurfuryl Alcohol under Mild Conditions in Water. *Nanomaterials (Basel),* **2019**, *9*(12), 1698-1710.
[http://dx.doi.org/10.3390/nano9121698] [PMID: 31795102]

[178] Fang, R.; Liu, H.; Luque, R.; Li, Y. Efficient and selective hydrogenation of biomass-derived furfural to cyclopentanone using Ru catalysts. *Green Chem.,* **2015**, *17*(8), 4183-4188.
[http://dx.doi.org/10.1039/C5GC01462J]

[179] Deng, Q.; Wen, X.; Zhang, P. Pd/Cu-MOF as a highly efficient catalyst for synthesis of cyclopentanone compounds from biomass-derived furanic aldehydes. *Catal. Commun.,* **2019**, *126*, 5-9.
[http://dx.doi.org/10.1016/j.catcom.2019.04.008]

[180] Wang, Y.; Sang, S.; Zhu, W.; Gao, L.; Xiao, G. CuNi@C catalysts with high activity derived from metal–organic frameworks precursor for conversion of furfural to cyclopentanone. *Chem. Eng. J.,* **2016**, *299*, 104-111.
[http://dx.doi.org/10.1016/j.cej.2016.04.068]

[181] Chen, H.; Shen, K.; Tan, Y.; Li, Y. Multishell Hollow Metal/Nitrogen/Carbon Dodecahedrons with Precisely Controlled Architectures and Synergistically Enhanced Catalytic Properties. *ACS Nano,* **2019**, *13*(7), 7800-7810.
[http://dx.doi.org/10.1021/acsnano.9b01953] [PMID: 31287293]

[182] Golub, K.W.; Sulmonetti, T.P.; Darunte, L.A.; Shealy, M.S.; Jones, C.W. Metal–Organic-Framewor--Derived Co/Cu–Carbon Nanoparticle Catalysts for Furfural Hydrogenation. *ACS Appl. Nano Mater.,* **2019**, *2*(9), 6040-6056.
[http://dx.doi.org/10.1021/acsanm.9b01555]

[183] Koley, P.; Chandra Shit, S.; Joseph, B.; Pollastri, S.; Sabri, Y.M.; Mayes, E.L.H.; Nakka, L.; Tardio, J.; Mondal, J. Leveraging Cu/CuFe$_2$O$_4$ -Catalyzed Biomass-Derived Furfural Hydrodeoxygenation: A Nanoscale Metal–Organic-Framework Template Is the Prime Key. *ACS Appl. Mater. Interfaces,* **2020**, *12*(19), 21682-21700.
[http://dx.doi.org/10.1021/acsami.0c03683] [PMID: 32314915]

[184] Aijaz, A.; Zhu, Q.L.; Tsumori, N.; Akita, T.; Xu, Q. Surfactant-free Pd nanoparticles immobilized to a metal–organic framework with size- and location-dependent catalytic selectivity. *Chem. Commun. (Camb.),* **2015**, *51*(13), 2577-2580.
[http://dx.doi.org/10.1039/C4CC09139F] [PMID: 25569372]

[185] Zhang, F.; Jin, Y.; Fu, Y.; Zhong, Y.; Zhu, W.; Ibrahim, A.A.; El-Shall, M.S. Palladium nanoparticles incorporated within sulfonic acid-functionalized MIL-101(Cr) for efficient catalytic conversion of vanillin. *J. Mater. Chem. A Mater. Energy Sustain.,* **2015**, *3*(33), 17008-17015.

[http://dx.doi.org/10.1039/C5TA03524D]

[186] Zhang, F.; Zheng, S.; Xiao, Q.; Zhong, Y.; Zhu, W.; Lin, A.; Samy El-Shall, M. Synergetic catalysis of palladium nanoparticles encaged within amine-functionalized UiO-66 in the hydrodeoxygenation of vanillin in water. *Green Chem.*, **2016**, *18*(9), 2900-2908.
[http://dx.doi.org/10.1039/C5GC02615F]

[187] Ibrahim, A.A.; Lin, A.; Zhang, F.; AbouZeid, K.M.; El-Shall, M.S. Palladium Nanoparticles Supported on a Metal-Organic Framework-Partially Reduced Graphene Oxide Hybrid for the Catalytic Hydrodeoxygenation of Vanillin as a Model for Biofuel Upgrade Reactions. *ChemCatChem*, **2017**, *9*(3), 469-480.
[http://dx.doi.org/10.1002/cctc.201600956]

[188] Chen, Y.Z.; Cai, G.; Wang, Y.; Xu, Q.; Yu, S.H.; Jiang, H.L. Palladium nanoparticles stabilized with N-doped porous carbons derived from metal–organic frameworks for selective catalysis in biofuel upgrade: the role of catalyst wettability. *Green Chem.*, **2016**, *18*(5), 1212-1217.
[http://dx.doi.org/10.1039/C5GC02530C]

[189] Dhakshinamoorthy, A.; Alvaro, M.; Horcajada, P.; Gibson, E.; Vishnuvarthan, M.; Vimont, A.; Grenèche, J.M.; Serre, C.; Daturi, M.; Garcia, H. Comparison of Porous Iron Trimesates Basolite F300 and MIL-100(Fe) As Heterogeneous Catalysts for Lewis Acid and Oxidation Reactions: Roles of Structural Defects and Stability. *ACS Catal.*, **2012**, *2*(10), 2060-2065.
[http://dx.doi.org/10.1021/cs300345b]

[190] Guo, Z.; Xiao, C.; Maligal-Ganesh, R.V.; Zhou, L.; Goh, T.W.; Li, X.; Tesfagaber, D.; Thiel, A.; Huang, W. Pt Nanoclusters Confined within Metal–Organic Framework Cavities for Chemoselective Cinnamaldehyde Hydrogenation. *ACS Catal.*, **2014**, *4*(5), 1340-1348.
[http://dx.doi.org/10.1021/cs400982n]

[191] Chen, L.; Chen, H.; Li, Y. One-pot synthesis of Pd@MOF composites without the addition of stabilizing agents. *Chem. Commun. (Camb.)*, **2014**, *50*(94), 14752-14755.
[http://dx.doi.org/10.1039/C4CC06568A] [PMID: 25318046]

[192] Stavila, V.; Parthasarathi, R.; Davis, R.W.; El Gabaly, F.; Sale, K.L.; Simmons, B.A.; Singh, S.; Allendorf, M.D. MOF-Based Catalysts for Selective Hydrogenolysis of Carbon–Oxygen Ether Bonds. *ACS Catal.*, **2016**, *6*(1), 55-59.
[http://dx.doi.org/10.1021/acscatal.5b02061]

[193] Zhang, J.; Lu, G.; Cai, C. Self-hydrogen transfer hydrogenolysis of β-O-4 linkages in lignin catalyzed by MIL-100(Fe) supported Pd–Ni BMNPs. *Green Chem.*, **2017**, *19*(19), 4538-4543.
[http://dx.doi.org/10.1039/C7GC02087B]

CHAPTER 5

MOF-Based Materials for CO$_2$ Conversion

Dinesh De[1,*], **Vivekanand Sharma**[2] and **Mayank Gupta**[3]

[1] *Department of Basic Science, Vishwavidyalaya Engineering College, Ambikapur, CSVTU-Bhilai, Chhatisgarh-497001, India*

[2] *Advanced Membranes and Porous Materials Center, Bldg. 4. Level 3, Office 3266-WS10, King Abdullah University of Science and Technology (KAUST), Thuwal 23955-6900, Kingdom of Saudi Arabia*

[3] *Department of Physics, National University of Singapore, 3 Science Drive 3, 117543, Singapore*

Abstract: Due to the rapid and continuous increase in CO$_2$ concentrations in the atmosphere by the massive combustion of fossil fuels, the global ecosystem is being affected severely. Therefore, balancing the CO$_2$ content in the atmosphere should be our main agenda nowadays. For minimization of CO$_2$ concentration, carbon capture and its conversion to valuable chemicals are being perused worldwide. Metal-organic framework (MOF)-based materials having a porous structure and tuneable structural features, are best candidates for the purpose. Herein, we provide a detailed discussion on the design, synthesis and catalytic applications of MOF-based materials for various CO$_2$ conversion reactions.

Keywords: Carbon dioxide, Catalysis, Cyclic carbonate, Metal-organic frameworks (MOFs), Methanol, MOF composites, Photocatalyst.

1. INTRODUCTION

Considering the global climate issue, the conversion of atmospheric carbon dioxide (CO$_2$) into energy and other useful chemicals becomes a burgeoning field of scientific research. It will not be early to say that the next quest for the human race is for sustainable growth. Increased CO$_2$ emissions from the burning of fossil fuel is a key factor for environmental concern as the increase in the transport of people and goods is set to continue over the coming years. According to a study carried out in 2019, the CO$_2$ emissions from human activity have reached 34 Gt [1]. Hence, the need to develop a safe, economic and clean methodology for CO$_2$ capture and its conversion into a valuable chemical could be a great milestone for

---

* **Corresponding author Dinesh De:** Department of Basic Science, Vishwavidyalaya Engineering College, Ambikapur, CSVTU-Bhilai, Chhatisgarh-497001, India; E-mail: d2chem@gmail.com

Junkuo Gao & Reza Abazari (Eds.)

the survival of the human race. Therefore, it is inevitable to examine and address the development of new and efficient catalysts for catalytic conversion of CO_2 into fuels and valuable chemicals. Significant efforts are given to develop a highly efficient catalyst bearing high surface area, high CO_2 uptake property, and stable catalytic activity to facilitate energy efficient capture and subsequent conversion [2]. However, the biggest obstacle in the process of its conversion is high thermodynamic stability and kinetic inertness as the carbon of CO_2 is in the most oxidized state keeping its energy level low. Hence a catalyst is needed to activate it and this will be going to possess long-term goals [3].

MOFs are an exciting class of new materials possessing very high surface areas, high porosity and myriads of chemical functionalities which are tunable [4]. They've been used with great success in many applications, including adsorption/separation [5], water splitting [6], gas separation [7], drug delivery [8], conversion of CO_2 to economically valuable products [9], photovoltaics [10], catalysis [11], batteries [12] and so on. In the forthcoming discussion of this chapter, the current state of research focused on the application of MOF-based materials as catalysts for CO_2 conversion into fuels and chemicals has been discussed. There are mainly three different routes available for CO_2 conversion, 1) thermal catalysis of the majority of reactions involves hydrogenation reaction at a relatively low temperature and produces carbon monoxide, methanol and methane, 2) photocatalytic conversion where the catalyst is exposed to solar light and generated photoelectrons induce a redox reaction involving CO_2, and 3) electrocatalytic conversion involves the reduction of CO_2 by the transfer of two, four, six, or eight number of electrons and controlling the formation of single and desired products is the grand challenge behind this pathways. This route of CO_2 conversion is gaining significant momentum as it offers higher cost-effectiveness because of milder operating conditions, high efficiency, controllable reaction conditions, and recyclability of the electrolyte and the catalyst. There are other methods too but fundamentally, all are just the same or a combination of the aforementioned three routes.

Finally, we end up this discussion pertaining to the challenges associated with MOF-based catalysts, their pertinent solutions, and some highlights on their future scenarios for the conversion of CO_2 to fuels and valuable chemicals.

2. SYNTHESIS OF VALUE-ADDED ORGANIC COMPOUNDS USING CO_2 AS PRECURSOR

MOFs have porous structures and high CO_2 uptake properties. Therefore, plenty of MOFs or MOF-based materials have been exploited for the conversion of CO_2

to several valuable organic compounds. Some of the representative CO_2 conversion reactions catalyzed by-MOF-materials are recorded in Table **1**.

Table 1. List of representative CO_2 conversion reactions catalyzed by MOF-based materials.

Type of Reaction	Equation	Active site(s)
Cycloaddition	R–(epoxide) $+ CO_2 \xrightarrow{\text{Catalyst}}$ cyclic carbonate (R)	Lewis acid, Lewis base, Brønsted acid, organic salt, organic base, etc.
	R_2-aziridine (R_1) $+ CO_2 \xrightarrow{\text{Catalyst}}$ oxazolidinone (R_2, N–R_1) $+$ oxazolidinone (N–R_1, R_2)	
Olefin oxidative carboxylation	R–(olefin) $+ [O] + CO_2 \xrightarrow{\text{Catalyst}}$ cyclic carbonate (R)	Oxidant, Lewis acid, Lewis base, halide
Terminal alkyne carboxylation	$R^1{-}\!\!\equiv\!\!{-}H + R^2X + CO_2 \xrightarrow{\text{Catalyst}} R^1{-}\!\!\equiv\!\!{-}\!\!\overset{O}{\underset{O-R^2}{}}$	Cu(I)
	$R{-}\!\!\equiv\!\!{-}H + CO_2 \xrightarrow{\text{Catalyst}} R{-}\!\!\equiv\!\!{-}\!\!\overset{O}{\underset{OH}{}}$	Ag, Pd–Cu NPs
Propargylic alcohol carboxylic cyclization	$R^3{-}\!\!\equiv\!\!{-}C(R^1)(R^2){-}OH + CO_2 \xrightarrow{\text{Catalyst}}$ cyclic carbonate (R^1, R^2, $=R^3$)	Ag(I), Cu(I)
Three-component carboxylic cyclization of propargyl alcohols, CO_2, and primary amines	$\equiv\!\!{-}C(R^1)(R^2){-}OH + R^3{-}NH_2 + CO_2 \xrightarrow{\text{Catalyst}}$ oxazolidinone (R^1, R^2, $N{-}R^3$)	Ag(I)
Propargyl amine carboxylic cyclization	$R^3{-}\!\!\equiv\!\!{-}C(R^2)(HN{-}R^1) + CO_2 \xrightarrow{\text{Catalyst}}$ oxazolidinone ($R^3=$, R^2, $N{-}R^1$)	Ag(I)

The transformation of CO_2 to the value-added compounds will be "killing two birds with one stone" *i.e.* the atmospheric CO_2 level can be reduced while producing essential chemicals. To do so, several MOFs-materials have proven to be excellent heterogeneous catalysts. Conversion of CO_2 mainly happens through the following processes:

a) cycloaddition of CO_2 to epoxides to form cyclic carbonates,

b) fixation of CO_2 through carboxylation of terminal alkynes

c) hydrogenation reactions of CO_2, and

d) photochemical/ electrochemical reductions of CO_2.

2.1. Conversion of CO_2 into Cyclic Carbonates (CC)

Cycloaddition reactions of CO_2 to epoxide to yield cyclic carbonates have been widely explored. These cyclic carbonates have wide applications in chemical and pharmaceutical industries [13]. The proposed mechanism of this reaction is given in Scheme 1. The epoxide ring is activated by the influence of a Lewis acid (LA) or a Brønsted acid (BA) site present in the MOF-based material. The halide ions produced from the co-catalyst, tetrabutylammonium bromide (TBAB), attack the less crowded carbon in an S_N2-type reaction to open the ring. Afterward, the cycloaddition of CO_2 with alkoxide intermediate forms the cyclic carbonate. Often CO_2 gets activated by Lewis base (LB) through dipolar interactions. The yield of cyclic carbonates could be encouraged by the collective effect of LA and LB sites present in a single MOF [14]. Remarkably, some MOFs show outstanding catalytic properties without a co-catalyst [15]. Sometimes the halide nucleophiles present in the MOFs cavity exclude the external use of the co-catalyst [16].

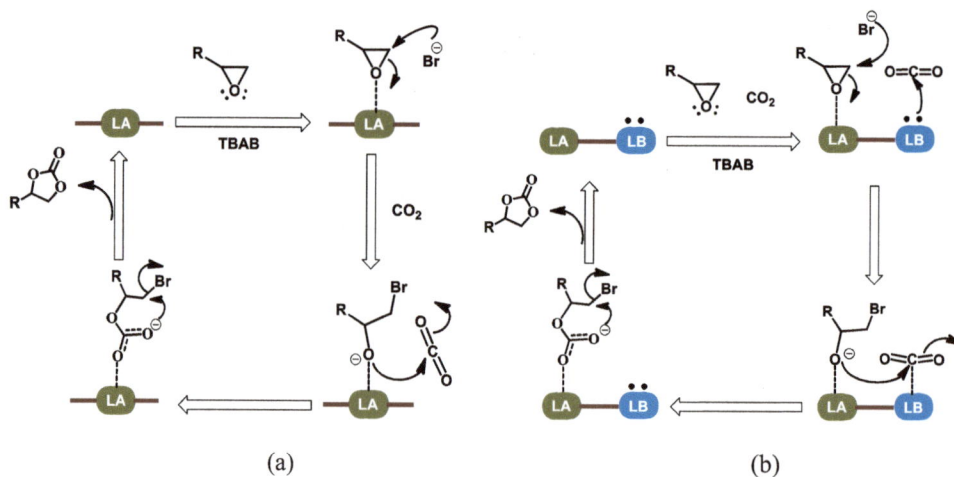

(a)   (b)

Scheme 1. Probable mechanism for the reaction of CO_2 with epoxide involving the catalytic sites: **(a)** LA (or BA), **(b)** LA (or BA) along with LB, TBAB is used as co-catalyst.

2.1.1. CO₂ Conversion into Cyclic Carbonates Using MOF Catalysts

Lots of MOFs have been implemented to catalyze the cycloaddition of CO_2 with epoxides. As for example, an aromatic tetracarboxylate linker was used in the presence of L-proline to isolate [17] two Ni(II) containing compounds: discrete tubular, Ni−TCPE1 and Ni−TCPE2 as a 3D MOF (Fig. **1**). The discrete Ni−TCPE1 containing a big aperture showed an admirable yield and enantioselectivity (ee~92%) of styrene carbonate in case of reaction with a pure enantiomer (R or S) of styrene oxide and CO_2. In contrast, Ni−TCPE2 displayed a comparatively lower yield possibly due to pore blockage that hindered the approach of an epoxide.

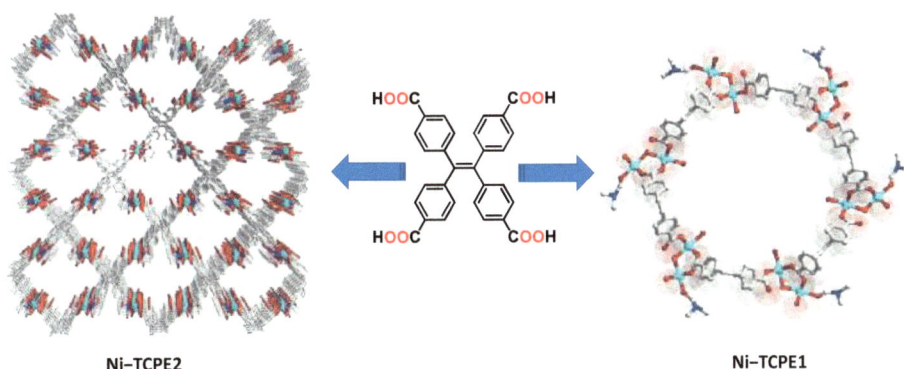

Fig. (1). Structure of **(a)** Ni−TCPE2, and **(b)** Ni−TCPE1 synthesized from a tetracarboxylate ligand. Reproduced with permission from [17] American Chemical Society. Copyright 2015.

Cycloaddition of CO_2 with epoxides could be carried out by several ZIFs [18]. Several Cu(II)- and Zn(II)-MOFs showed significant catalytic activity. A MOF is constructed as a porous structure with accessible nitrogen-rich groups (Fig. **2**) with Cu(II) and the amide functionalized tetracarboxylate ligand [19]. Only small size epoxides gave admirable yields of CCs at ambient conditions, as bulky epoxides could not easily reach the coordination space.

MOFs constructed from the ligands incorporating Lewis basic sites can exhibit a cooperative effect [20 - 22]. A Cu(II)-MOF built with a straight tetracarboxylate ligand having free amine group and paddlewheel SBU (Fig. **3**) catalyzed CO_2 cycloaddition in the presence of TBAB, at 20 bar pressure of CO_2 at 393 K [23]. Although behaved as a recoverable catalyst, only ~ 50% conversion was achieved with styrene epoxide.

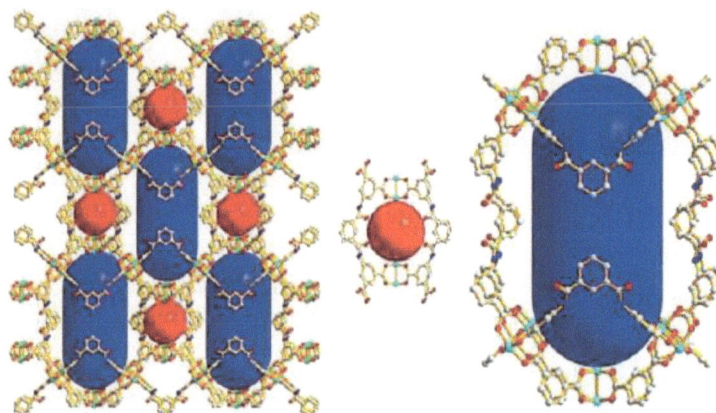

Fig. (2). A view of the Cu(II)-based MOF showing two different voids. Reproduced with permission from [19] American Chemical Society. Copyright 2017.

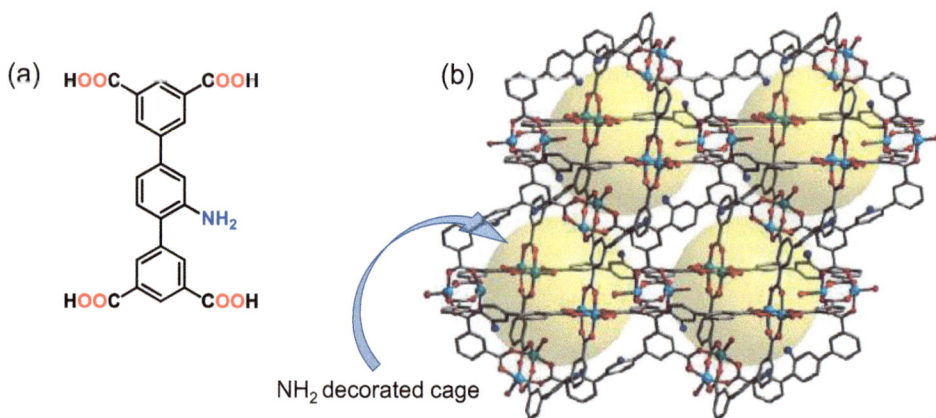

Fig. (3). (a) An amino group incorporated tetracarboxylic acid ligand, and **(b)** the Cu(II)-MOF.

Using a V-shaped tetracarboxylate linker (Fig. **4**) having an outward -NH$_2$ group, a porous MOF was synthesized where free amine groups were positioned near the cavity [24]. This showed remarkably high (60 wt. %) CO$_2$ adsorption at 32 bar pressure and 25 °C. The cyclic carbonates' formation with various epoxides could be achieved in an excellent yield at ambient conditions. Most significantly, when laboratory air was bubbled through the mixture of TBAB, epoxide and suspended MOF, significant conversion was observed. The catalyst was reused four times with no noticeable drop in the yield of CC.

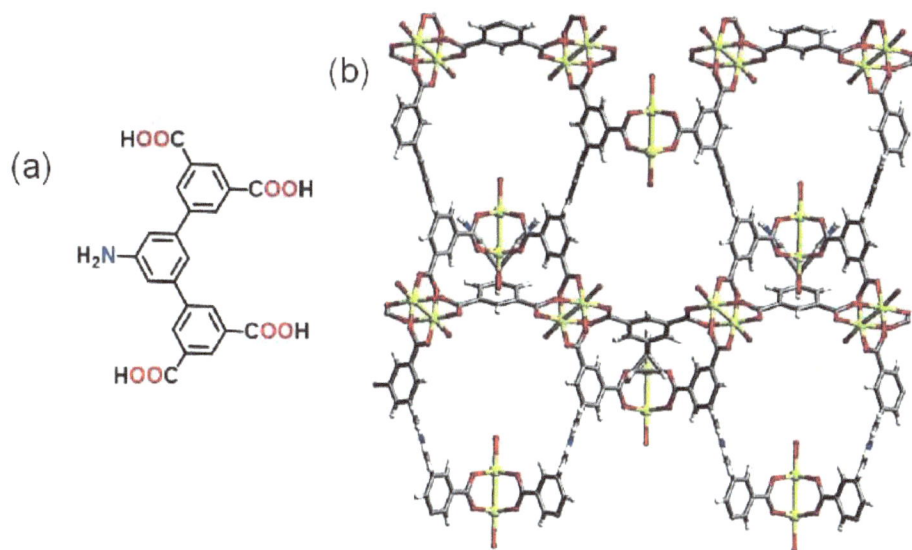

Fig. (4). (a) V-shaped tetracarboxylic acid ligand and (b) the pore of the Cu(II)-MOF.

However, with linkers having a -NH$_2$ group inward like in the following cases (Scheme **2**), Cu(II)-MOF with similar structures could be obtained although their catalytic performance was lower compared to the above system. This suggests that both coordinatively unsaturated metal centers and Lewis basic sites in the ligand are important for their enhanced catalytic performance of the CC formation from epoxide and CO$_2$. However, it is also important to note that the distance and relative orientation of the -NH$_2$ and carboxylate groups are important parameters that should be kept in mind during ligand design.

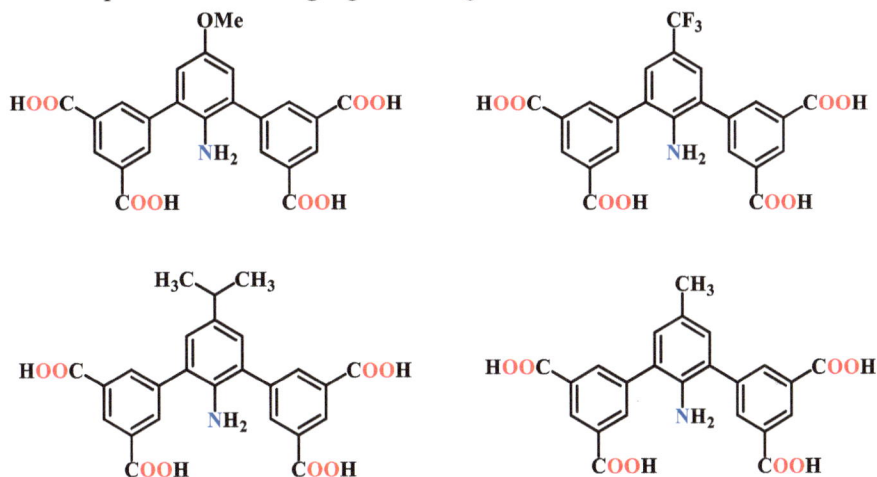

Scheme 2. The V-shaped tetracarboxylic acid ligand with inward amino group.

Two MOFs, Zn-NTTA and Cu-NTTA could be achieved with the linker, H_6NTTA (Fig. **5**) [25]. The MOF, Zn-NTTA contained three cages of different sizes all with diameters >15 Å. Both Zn-NTTA and Cu-NTTA showed high catalytic activity in cyclic carbonate formation with complete conversion in the presence of TBAB and pure CO_2 purged at 1.0 MPa and 373 K. Mixed gas as a CO_2 source afforded lower yields. However, no adverse effect of moisture could be observed.

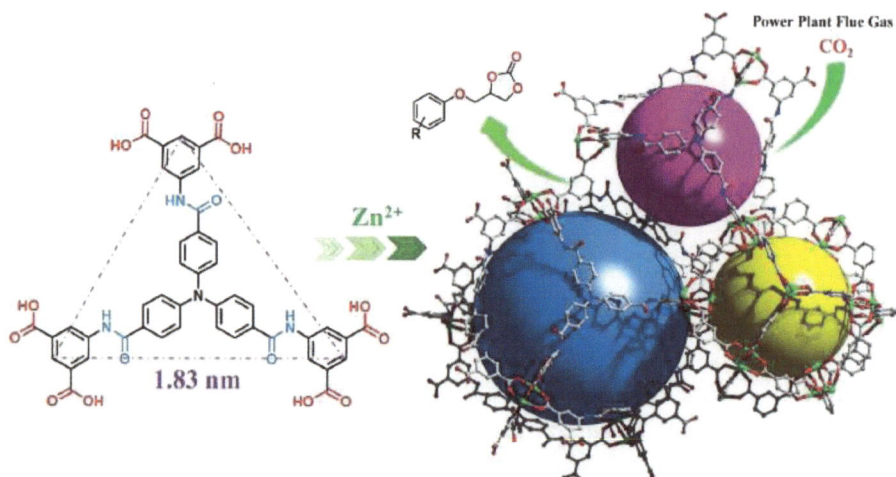

Fig. (5). The tripodal linker, H_6NTTA and three different voids in Zn-NTTA. Reproduced with permission from [25] American Chemical Society. Copyright 2016.

An anionic Zn- MOF, synthesized with an amine functionalized tricarboxylate linker (Fig. **6**) exhibited catalytic ability upon activation in the presence of TBAB at atmospheric pressure and 373 K temperature under solvent-free conditions [26].

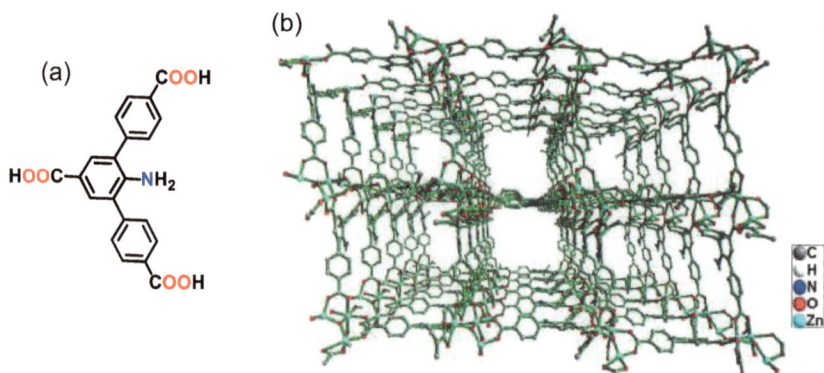

Fig. (6). (a) Structure of the amine functionalized tricarboxylate linker and **(b)** a perspective view of the framework.

In order to probe the effect of imidazole group, a dicarboxylate ligand was used to construct the anionic framework having double-walled (Fig. 7) hexagonal channels [27]. It exhibited catalytic activity under mild conditions with recyclability.

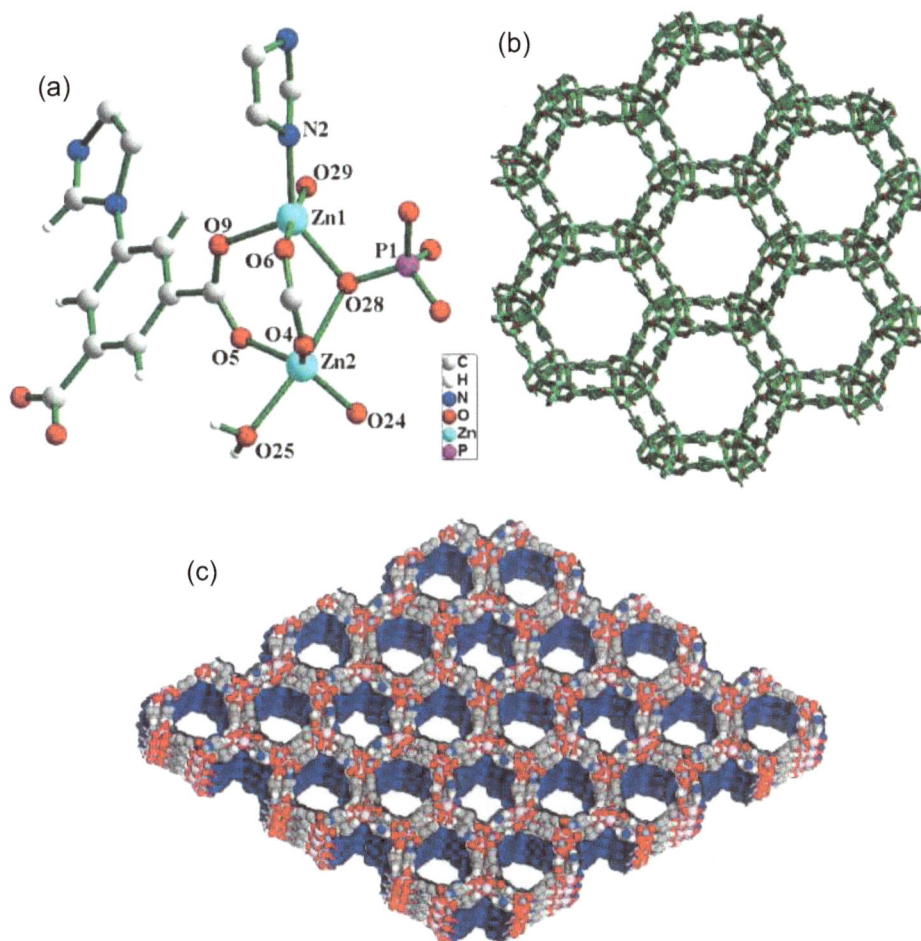

Fig. (7). (a) Coordination environment of Zn(II); **(b, c)** double-walled channels along *c*-axis. Reproduced with permission from [27] American Chemical Society. Copyright 2017.

Two MOFs Hf-NU-1000 and Zr-NU-1000 showed the catalytic activity in CO_2 fixation under mild conditions in the presence of TBAB. The activity Hf-N-1000 was superior to that of Zr-NU-1000 owing to larger no of LA/BA sites and wider channel aperture (13-29 Å) [28]. Hf-NU-1000 provided 100% yield of CC formation.

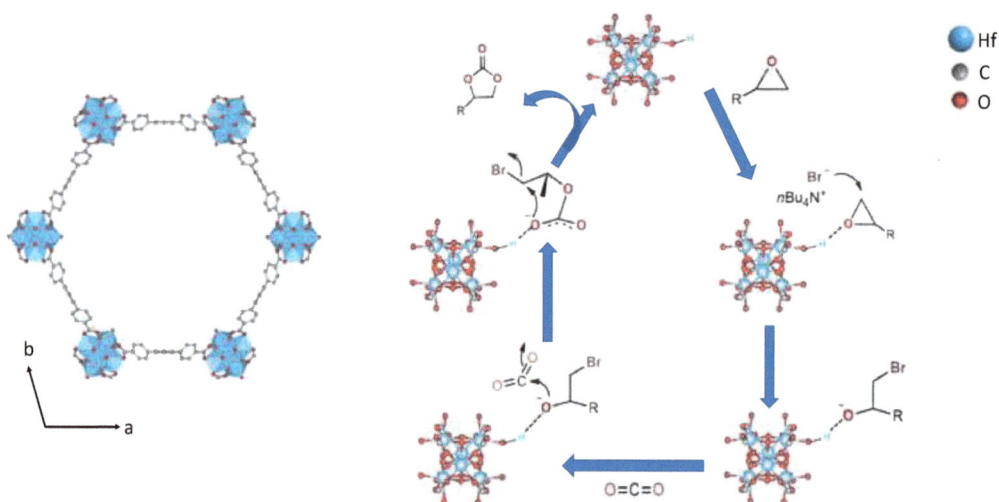

Fig. (8). (a) Porous framework of Hf-NU1000. **(b)** Probable mechanism for the Hf-NU1000 catalysed reaction of CO_2 and epoxide in presence of TBAB. Role of SBU is shown. Reproduced with permission from [28] American Chemical Society. Copyright 2014.

It is convenient to have the halide ion present closer to the catalytic site of the MOF. Thus, the process becomes simpler and cost-effective. Towards this end, a bifunctional imidazolium functionalized Zr-based MOF (I⁻)Meim-UiO-66 was synthesized [29] post-synthetically from the parent MOF, Im-UiO-66 (Scheme 3).

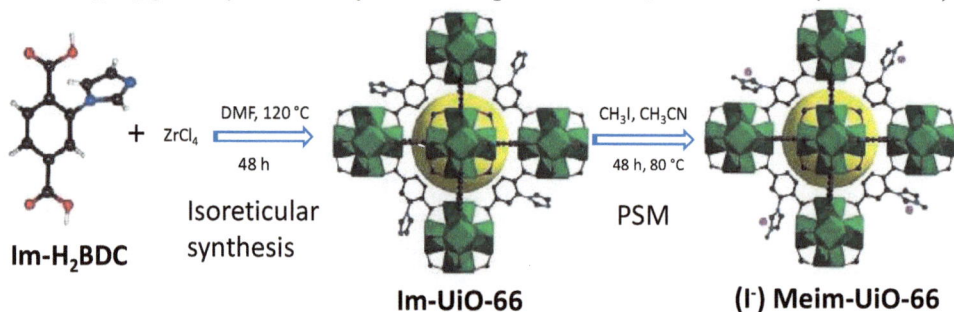

Scheme 3. Synthetic scheme for the synthesis of **(I⁻) Meim-UiO-66** from **Im-UiO-66**. (PSM: post synthetic modification) Green polyhedral represents Zr_6 SBUs and yellow ball represents cavity.

The bifunctional framework, (I⁻) Meim-UiO-66, containing both BA sites and I⁻, exhibited prominent catalytic activity in CO_2 cycloaddition reaction, in the absence of a co-catalyst. In this MOF, the I⁻ ions located in the cavity play the role of a co-catalyst. At 100 °C and 1 bar CO_2 pressure, 88.5% conversion of epichlorohydrin was found that was improved to 100% at 120 °C. The BA (Zr -OH or Zr-OH$_2$) and LB (I$^-$) cooperatively catalyzed the reaction.

2.1.2. CO$_2$ Conversion into Cyclic Carbonates Catalyzed by MOF Composites

MOF composites have also been utilized in cycloaddition reactions of CO$_2$ to epoxide [30]. MOFs encapsulated with metal oxides, NPs, and species that act as catalysts, have been studied as CO$_2$ conversion catalysts. As an illustrative example, an -NH$_2$ incorporated ligand was used to afford UiO-66-NH$_2$ (labeled as ZrMOF). This MOF was then loaded with Ni nanoparticles to have Ni@ZrOF (Fig. **9**) [31]. Ni@ZrOF in the presence of TBAB showed excellent catalytic activity for cycloaddition of CO$_2$ to styrene oxide (yield: 98%, selectivity: 99%). The Ni° and Zr^{4+} present in Ni@ZrOF collectively catalyzed the reaction. This catalyst showed a wide substrate scope and five times recycling property.

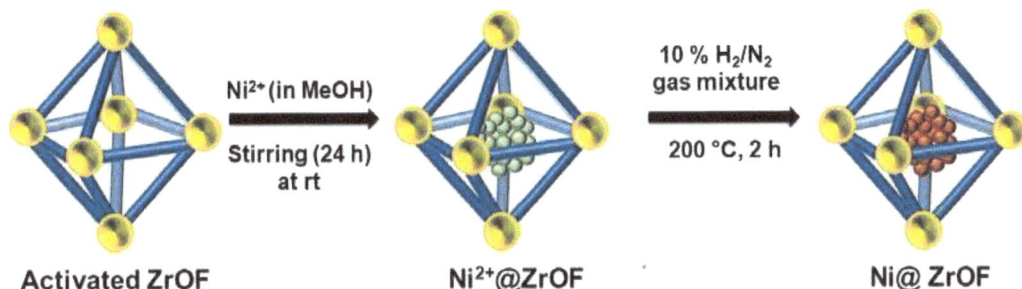

Fig. (9). Encapsulation of Ni NPs to obtain Ni@ZrOF composite. Reproduced with permission from [31] American Chemical Society. Copyright 2019.

In another example, optically active Co(III)-salen complex (R,R configuration) has been encapsulated within the pores of IRMOF-3 [32]. The optically active composite, Co-salen@IRMOF-3 exhibited 25% conversion and 21% ee in case of reaction of CO$_2$ with propylene oxide. Subsequently, two Cu(II)-salen-complexes were incorporated within the cage of MIL-101 for the reaction of CO$_2$ with epoxide [33]. 0.3 g Cu(II)-Salen-^{t}Bu loaded in 1 g MIL-101 provided the maximum 87.8% conversion for the above reaction at 25 °C under 1 atm CO$_2$.

It is more convenient to keep away the utilization of co-catalyst from an external source rather it should be present inside the MOF itself. Therefore, poly(ionic liquid)s (polyILs) based on imidazolium ions were encapsulated in the cages of MIL-101 applying the *in situ* polymerization method (Fig. **10**) [34]. The resultant composite material, polyILs@ MIL-101 in the absence of a co-catalyst, exhibited 94% conversion of epichlorohydrin with high selectivity (~99%) to corresponding cyclic carbonate at 343 K, 1 bar CO_2 pressure. After ten times recycling, the composite catalyst exhibited no significant decrease in the catalytic activity.

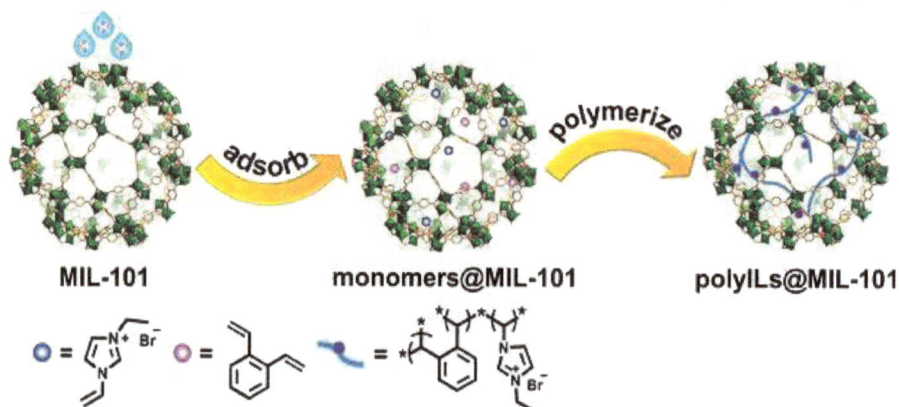

Fig. (10). Synthesis of polyILs@MIL-101 (right) by the *in-situ* polymerization technique. Reproduced with permission from [34] American Chemical Society. Copyright 2018.

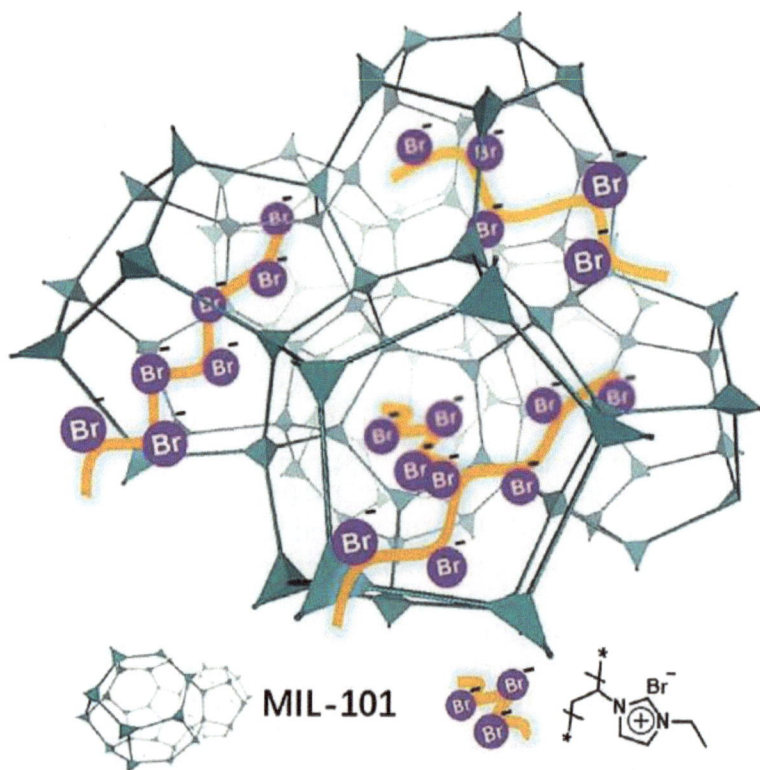

Fig. (11). The ionic polymer (IP) entrapped within the channels of the host (Cr)-MIL-101.

Similarly, in another report, an IP (ionic polymer) was incorporated within the pores of (Cr)-MIL-101 [35]. The Br⁻ ions (counter ions) of the ionic polymer were situated near the unsaturated Cr(III)-sites (Fig. **11**).

The catalytic performance of IP entrapped MIL-101 was superior (99%) to the individual IP (3%) and MIL-101 (32%), or MIL-101 and IP together (80%) under the identical conditions. Further, MIL-101 had been modified *via* a double-click synthetic route to a cationic MOF (Scheme **4**) that could include several Br⁻ anions in the cavity and exclude the need for the co-catalyst, TBAB [36]. This system gave good and rapid product formation with small epoxides under mild conditions although long and bulky epoxides required longer reaction times.

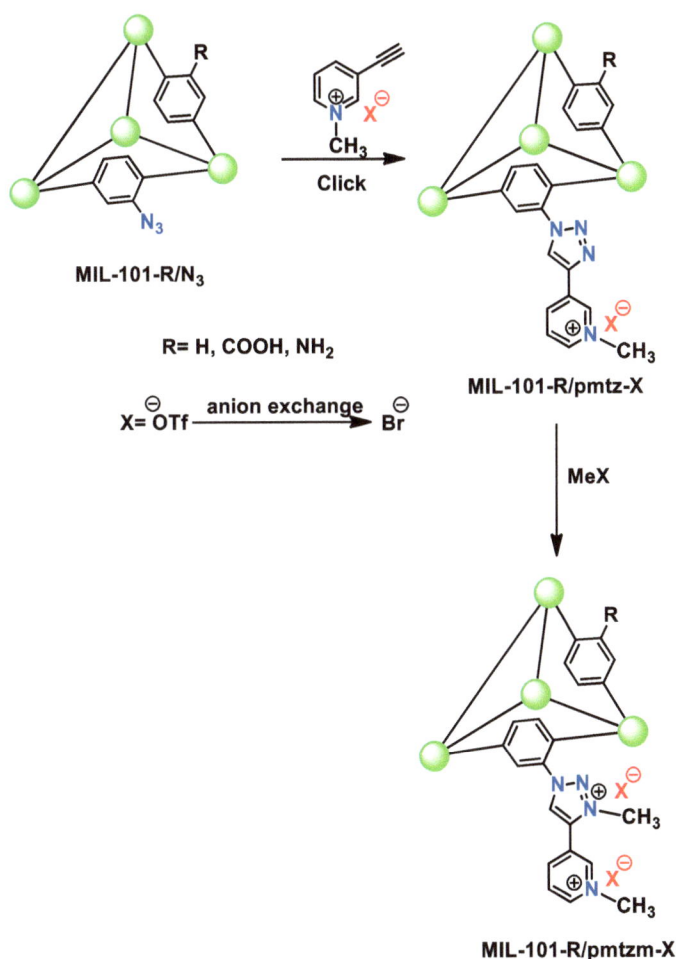

Scheme 4. Synthesis of high anionic environment in MIL-101 *via* double click post synthetic modification process.

2.1.3. CO₂ Conversion into Cyclic Carbonates Using MOF Derivatives

MOF derivatives are very fascinating materials produced by the pyrolysis of transition metals-based MOFs constructed from organic ligands preferentially having heteroatoms. In these materials, a plenty of newly developed active sites are observed which were absent in precursor MOFs [37]. For example, ZIF-7, ZIF-8, ZIF-9, and ZIF-67 were pyrolyzed to obtain nano-porous carbons doped with N and metal. These materials were used for CO_2 cycloaddition with epoxides [38]. When the pyrolysis was carried out at 600 °C under argon atmosphere, ZIF-9 gave the highest yield (90%) of CC formation compared to all other ZIFs. The high activity was due to the cooperative effect of partially oxidized LA Co-NPs and LB N-species. Later on, an oxidized porous carbon was prepared from the mixture of ZIF-8 (template) and NaOCl (oxidant) [39]. Few oxygenated functional moieties like carboxylic, alkoxy, and lactone moiety could be attached on the exterior of the N-doped porous carbon (NPC) without affecting the old active sites (ZnO and pyridine N). The catalyst, ZnO@NPC-Ox-700 along with TBAB displayed brilliant catalytic activity for reactions of CO_2 with epoxides under ambient conditions. So, MOFs derivative could be potential catalysts for various organic transformations.

A membrane-based MOF material exhibited fixation of CO_2 with organic epoxides efficiently [40]. The ionic liquid (IL)-based MOFs, UiO-66- IL-X (X = Br, PF_6, SO_3CF_3, and ClO_4) could be synthesized (Scheme **5**) by the exchange of Br⁻ ion in UiO-66-IL-Br. In case of X= ClO_4 the resultant MOF UiO-66-IL-ClO_4 reacted with polyurethane oligomer to produce a cross-linked membrane bound MOF. This confined MOF exhibited an outstanding reaction rate for selective CC formation from CO_2 and epoxide.

2.2. Fixation of CO₂ in Carboxylation of Terminal Alkynes by MOF-Based Materials

Reactions of CO_2 with terminal alkynes provide various useful alkynyl compounds. These products have several pharmaceutical applications [41]. As an example, for carboxylation of terminal alkynes, the Gd and Cu cluster-based MOFs showed admirable catalytic activities in the carboxylation reactions of CO_2 with acetylene moiety in the absence of any co-catalyst or additive under 1 atm and ambient conditions [42]. These two MOFs have catalytic sites [$Cu_{12}I_{12}$] and [Cu_3I_2].

The Ag(I)-containing MOF, Ag-TCPE [43] with porous structure showed excellent conversion for the [4+2] cycloisomerization reaction between the derivative of propargyl alcohols and CO_2 in the presence of PPh₃ to yield α-alkylidene cyclic carbonates. The Ag(I) ions in the framework had a specific alkynophilic character, so activated the alkyne moieties by both σ- and π-interactions with the C≡C group of substrate (Fig. **12**). As a result, the activated propargylic alcohol favourably cyclized with CO_2 to give α-alkylidene cyclic carbonates.

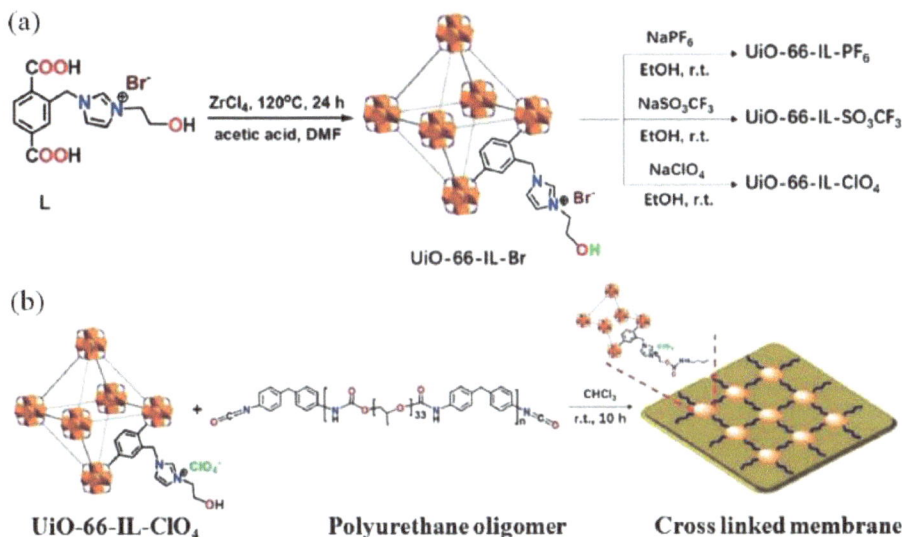

Scheme 5. Synthetic schemes: **(a)** UiO-66-IL-X, and **(b)** membrane bound UiO-66-IL-ClO₄. Reproduced with permission from [40] American Chemical Society. Copyright 2017.

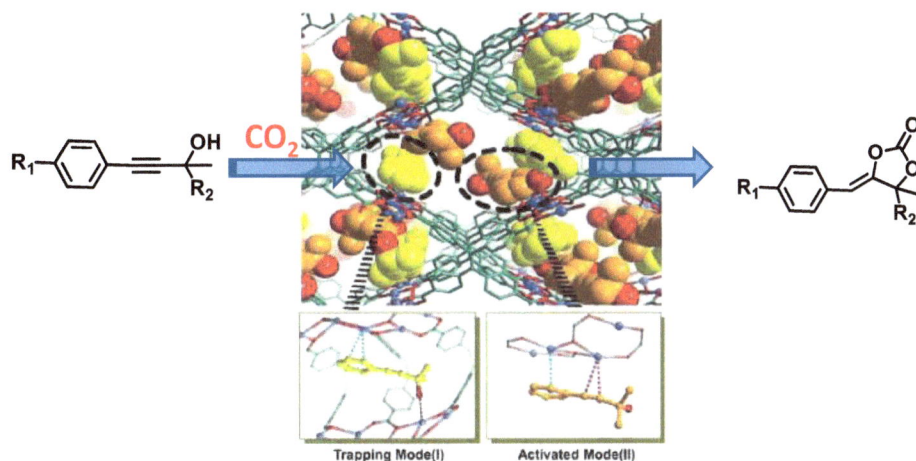

Fig. (12). Framework structure of Ag-TCPE impregnated with propargylic alcohol showing different interactions. Reproduced with permission from [43] American Chemical Society. Copyright 2017.

In a similar way, the organosulfonate-based MOF, TMOF-3-Ag exhibited the above CO_2 cyclization reactions with the aid of DBU in excellent yields (more than 86%). Additionally, TMOF-3-Ag showed efficient catalytic activity in the synthesis of oxazolidinones from CO_2 (yields >97%) by tandem reaction [44].

Interestingly, one Cd-based flexible MOF ornamented with NH_2 groups showed the cyclization reaction between CO_2 and propargylamines to afford five-membered urethane derivatives with high TON 9300. The NH_2 groups helped in the ring-closing step to the oxazolidinones derivative [45].

MOFs can be excellent heterogeneous catalysts for the carboxylation of several aromatic terminal acetylene moieties to the corresponding carboxylic acids. The ability of silver nanoparticles supported on microcrystalline cellulose catalyzed the reaction between a metal-bound terminal alkyne and CO_2 to carboxylic acids [46]. MIL-101 could support the growth of Ag NPs and the AgNPs@MIL-101 in DMF exhibited its catalytic ability in the presence of a base such as Cs_2CO_3 through C-H bond activation under 1 atm. CO_2 pressure at 323 K (Scheme **6**) [47].

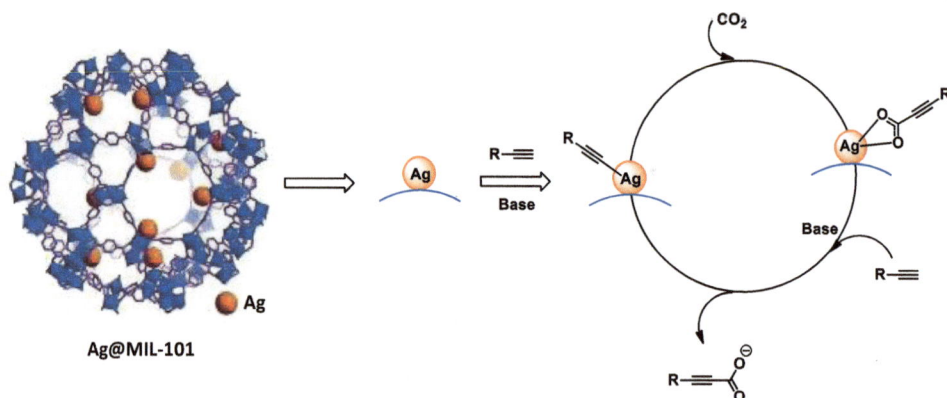

Scheme 6. The mechanism of carboxylation of alkyne with CO_2 by AgNPs@MIL-101.

The individually electron-accepting or electron-releasing group attached to the alkyne could be easily carboxylated to the corresponding carboxylic acids affording almost quantitative yields. Besides, the encapsulated Ag nanoparticles being stable, their catalytic ability remained intact at least up to five cycles. Since the host MIL-101 contains toxic Cr(III) ions, other MOFs such as (Zr)-UIO-66, (Fe)-MIL-100, and a porous Co(II)-MOF constructed from salicylate ligand had been used as hosts for Ag nanoparticles. All these systems gave excellent yields of the desired product under mild conditions [48, 49].

3. CONVERSION OF CO_2 INTO FUELS AND HYDROCARBONS

Hydrogenation reactions and photocatalytic reactions of CO_2 generate several valuable substances like formic acid, methanol, methane and so on. MOF catalysts have several benefits: (1) MOF structure can be designed and monitored by incorporating highly reactive catalytic sites *e.g.*, Lewis acids or Lewis bases (or both) in a single MOF [50], (2) Classical organo catalysts, metals, metal complexes, metal oxides, metal nanoparticles, and clusters of metals can be encapsulated and in MOFs to form MOF-based materials for highly efficient CO_2 conversion reactions [51].

3.1. MOF-Based Materials as Photocatalysts for CO_2 Conversion

The photocatalytic reduction of CO_2 to form valuable compounds is definitely a hopeful policy to diminish CO_2 emissions and to address the energy crisis. MOF-materials have been exploited as photocatalysts for CO_2 photoreduction owing to their affinity to CO_2, having satiable catalytic sites and light adsorption property. The CO_2 reduction is an endothermic one and can be started by the influence of UV/visible light in the presence of a catalyst with an appropriate band gap. The photoreduction reactions proceed through a sequence of multi-electron reactions assisted by protons. In most of the cases, it gives low CO_2-conversion, thus it is challenging to the synthesis of effective photocatalysts.

It is found that -NH_2 functional MOFs increase the CO_2 adsorption capacity and also the electron density in MOF, thereby reducing the HOMO-LUMO energy gap. This shifts the absorption maximum to the higher wavelengths. In some cases, wavelength shifts to the visible range which is effective for CO_2 photoreduction. Towards this end, Ti-based -NH_2 functionalized framework, (NH_2-MIL-125(Ti) has been synthesized [52]. Both CO_2 uptake capacity (132.2 cc/g) and the absorption maximum (550 nm) of NH_2-MIL-125(Ti) were enhanced than the non-amino framework MIL-125(Ti) (98.6 cc/g and 350 nm). Thus, NH_2-MIL-125(Ti) exhibited CO_2 photoreduction under visible light in acetonitrile solvent in the presence of a sacrificial reagent, triethanolamine (TEOA) to afford $HCOO^-$. The rate of formation of $HCOO^-$ was 16.28 μmol h^{-1} g^{-1} in the case of NH_2-MIL-125(Ti) catalyst which was significantly larger than that of MIL-125(Ti) catalyzed reaction. Upon irradiation, the photoinduced electron transfer from the ligand to Ti^{4+} ion in Ti-O cluster led to the reduction of Ti^{4+} to Ti^{3+} (Fig. **13**). A charge-separated species was generated in MOF that enhanced the performance for CO_2 reduction in NH_2-MIL-125(Ti).

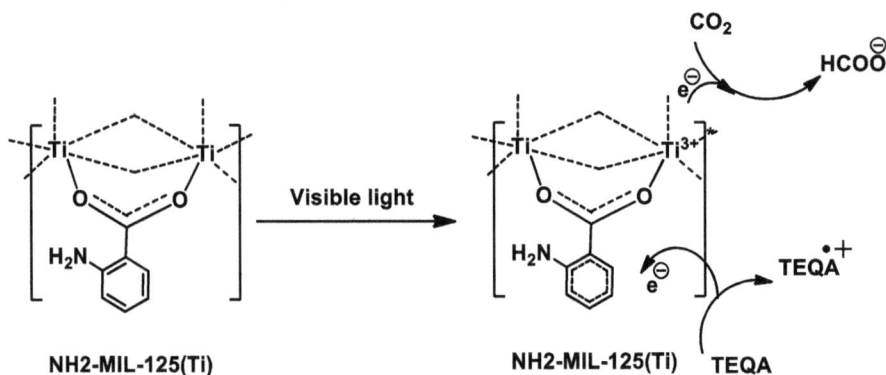

Fig. (13). Photoreduction of CO_2 catalyzed by NH_2-MIL-125(Ti).

The photocatalyst Cp*Rh@UiO-67 was synthesized [53] by partial incorporation of a light-harvesting ligand [Cp*Rh(5,5'-dcbpy)]$^+$Cl$^-$ into UiO-67 (Fig. **14**). This modified MOF exhibited photoreduction CO_2 to HCOO$^-$ with a TON of 47 under visible light. Similarly, the photosensitizer, H_2RuCO was introduced in UiO-67 to afford Zr-bpdc/RuCO MOF [54]. Under visible light, the resulting MOF showed a reduction of CO_2 to formic acid with the aid of TEOA as a sacrificial reagent.

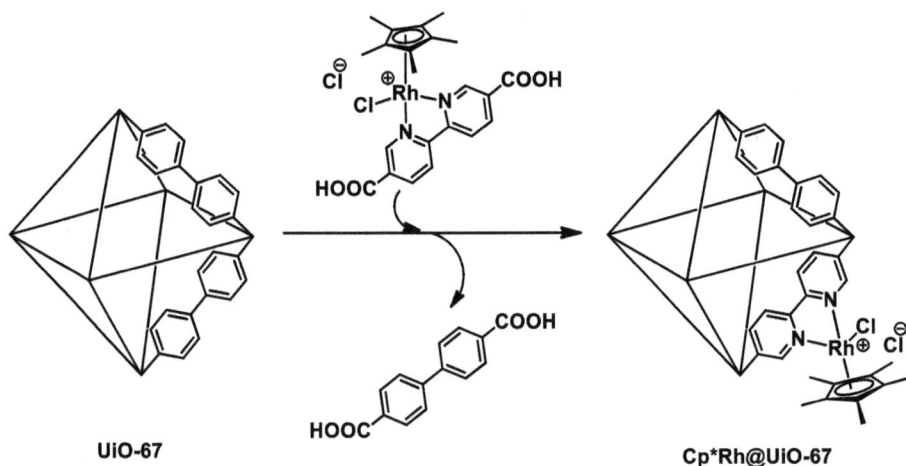

Fig. (14). Synthetic scheme of Cp*Rh@UiO-67.

MOFs constructed from highly conjugated porphyrin linkers could be potential for CO_2 photoreduction. A Zr-MOF (PCN-22) was constructed with a porphyrin ligand for the photoreduction of CO_2 to formate [55]. The MOF behaved like an *n*-type semiconductor with the HOMO-LUMO band energies 1.35 V and -0.4 V respectively. Under visible light, PCN-22 along with TEOA efficiently reduced CO_2 in the CH_3CN medium to form $HCOO^-$ with a rate of 60 µmol/h /g. Also, PCN-22 was reused three times with no obvious changes in the yield of $HCOO^-$ formation. The probable mechanism is shown in Fig. (**15B**). The Zn/PMOF which was porphyrin-based [56] exhibited photoreduction of CO_2 to CH_4 selectively under ultraviolet light in addition to water vapor. The rate of formation of methane was 8.7 µmol/ h/ g.

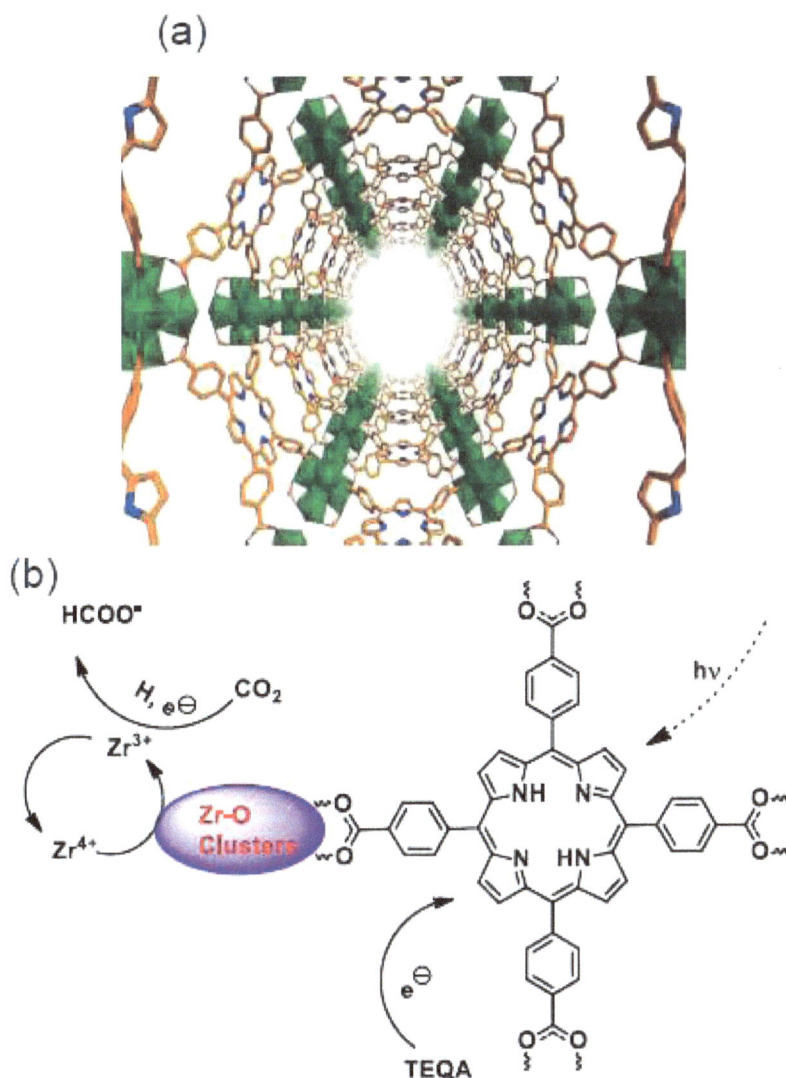

Fig. (**15**). (**a**) View of PCN-222, (**b**) probable mechanism for CO_2 reduction. Reproduced with permission from [55] American Chemical Society. Copyright 2015.

A zirconium-MOF, NNU-28 was synthesized with an electron rich anthracene-based linker to reduce CO_2 photocatalytically [57]. When exposed to visible light for 10 hours, NNU-28 showed a reduction of CO_2 to formate with a rate of 52.8 µmol/ h/ g. Two probable catalytic routes for the CO_2 reduction to formate have been proposed (Fig. **16**).

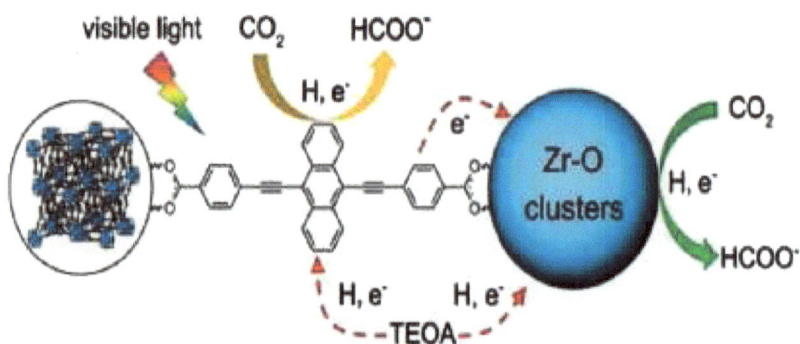

Fig. (16). CO_2 reduction to formate by NNU-28.

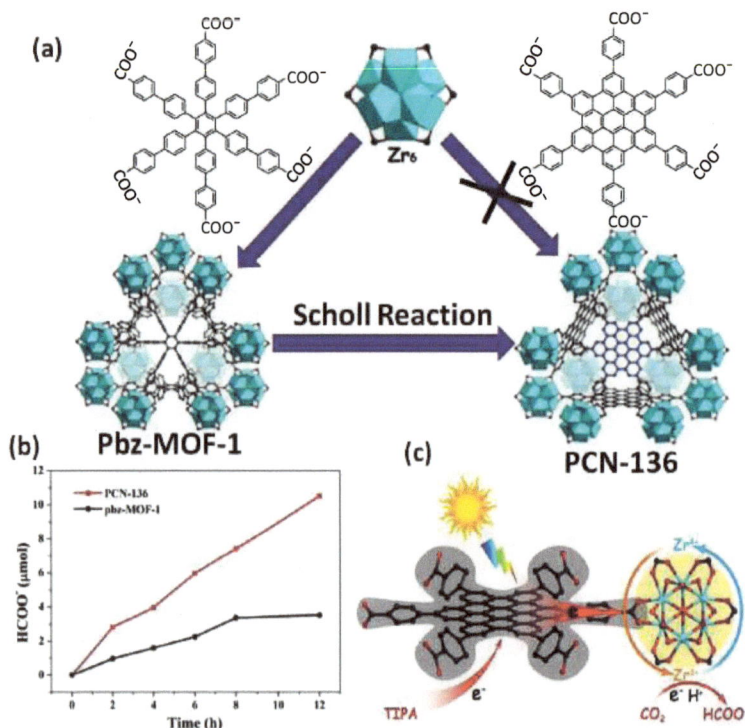

Fig. (17). (a) The synthetic scheme of PCN-136, **(b)** rate of photocatalytic reduction of CO_2 to HCOO, **(c)** PCN-136 catalyzed probable mechanism of CO_2 reduction. Reproduced with permission from [58] American Chemical Society. Copyright 2019.

Using a hexabenzocoronene-based ligand (Fig.17), a zirconium-based MOF (PCN-136) was synthesized by SC to SC conversion method [58]. In this MOF, the hexabenzocoronene moiety performed the role of a photosensitizer and Zr-oxo SBU performed the role of catalytic center. PCN-136 exhibited the conversion of CO_2 into formate in the presence of visible light more efficiently than that of parent Pbz-MOF-1.

The nanocomposite $Cd_{0.2}Zn_{0.8}S@UiO-66-NH_2$ was found to be a very good photocatalyst for the reduction of CO_2 to MeOH [59]. The rate of methanol production in the presence of this was 6.8 μmol/ h/ g.

3.2. Application of MOF in CO_2 Conversion to Methanol

The capture of atmospheric or dissolved CO_2 by the use of MOF and its conversion to fuels like methanol *via* hydrogenation or photo/electrocatalysis offers a significant way to decrease the atmospheric CO_2 content [60].

In one interesting example, photocatalytic conversion of CO_2 into methanol was achieved using Zn_2GeO_4/ZIF-8 hybrid nanorods [61]. These materials were procured *via* deposition of ZIF-8 NPs on the surfaces of Zn_2GeO_4 nanorods (Fig. **18a**). These hybrid materials exhibited 62% improvement in the formation of methanol from CO_2 by photoreduction with relative to only Zn_2GeO_4 nanorods (Fig. **18b**). This enhancement in the rate of methanol formation was due to ZIF-8 having high adsorption property of dissolved CO_2 which helped the composite in CO_2 reduction in water.

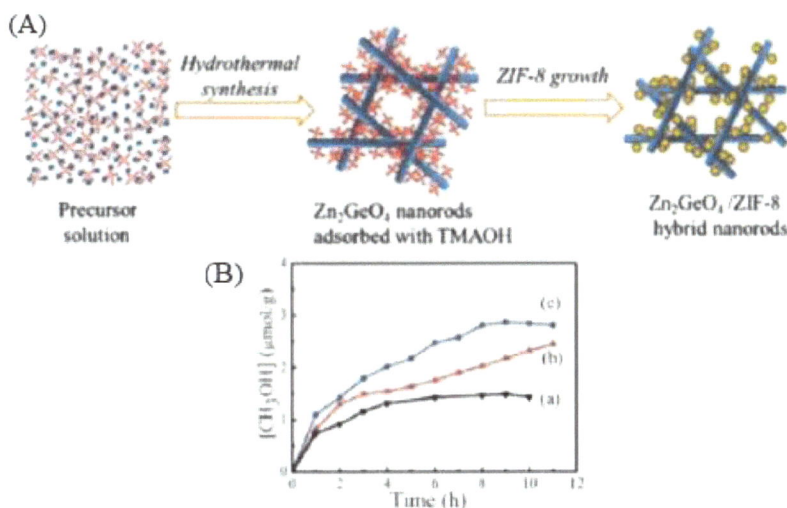

Fig. (18). (A) Preparation of composite catalyst. (B) Formation of methanol using the catalyst: (a) only Zn_2GeO_4 nanorods, (b) Zn_2GeO_4/ZIF-8 composite, and (c) 1 w t% Pt-doped Zn_2GeO_4/ZIF-8 composite.

Electrocatalytic reduction of CO_2 into CH_3OH was investigated [62] utilizing four porous Cu-based MOFs supported on gas diffusion electrodes, namely (a) HKUST-1, (b) CuAdeAce, (c) CuDTA, and (d) CuZnDTA. An enhanced cumulative Faradaic efficiency for the CO_2 conversion was achieved in case of HKUST-1 (15.9%) and CuZnDTA (9.9%). These results depicted that the unsaturated coordination sites located toward the channel enhance the CO_2 electrocatalytic reduction.

It was found that doping of Pd NPs in carbonized Cu(BTC) MOF selectively reduced CO_2 to CH_3OH [63]. Based on DFT calculations, it was found that the energy to produce methanol was lower (19.5 eV) than that of producing formic acid. Such enhanced catalytic activity was due to the many defects present on the carbon base. These defects facilitated the electron transfer in the reaction pathway and as well as offered more reactive centers to adsorb the intermediates (Fig. **19**).

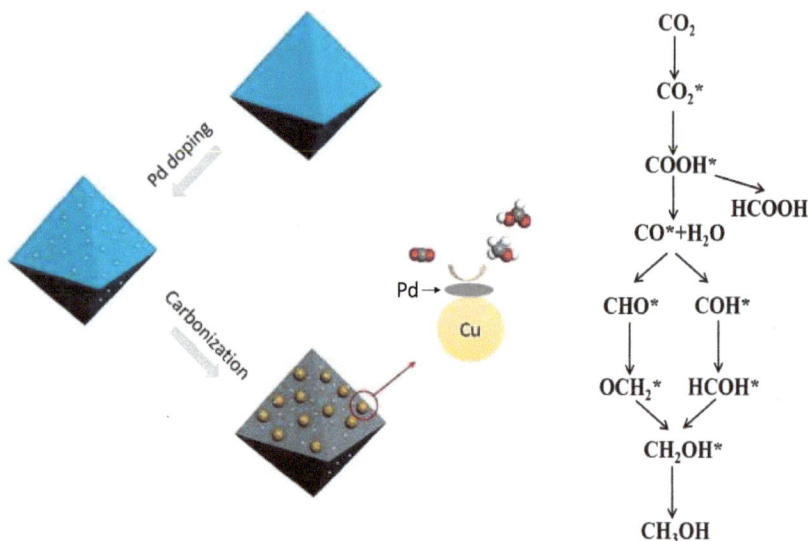

Fig. (19). Reaction scheme and pathways of CO_2 photoelectrochemical reduction to produce Methanol.

The conventional copper nanocrystal catalyst over Zirconium-based MOF UiO-66 showed efficient CO_2 hydrogenation to methanol [64]. The promoted catalyst exhibited 100% selectivity and 8-fold enhanced yield of CH_3OH formation relative to ordinary $Cu/ZnO/Al_2O_3$ catalyst. This higher performance of the catalyst was due the large area of interface between Zirconium oxide clusters of UiO-66 and copper nanocrystals along with the presence of multiple oxidation states of Cu (Fig. **20**).

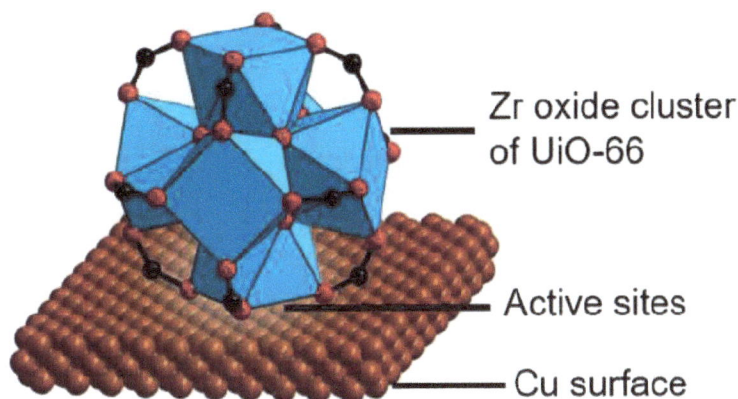

Fig. (20). Active site present in interface of copper nonocrystal and Zirconium oxide cluster.

The Cu/ZnO$_x$ NPs were effectively stabilized within the cage of UiO-bpy MOF [65]. As shown in Fig. (**21**), the interactions between Cu with bpy sites and Zn with Zr-SBU stabilized the ultrasmall Cu/ZnO$_x$ NPs. These interactions not only averted the agglomeration but also phase separation between Cu and ZnO$_x$ NPs in MOF cavity. The mixed nanoparticles confined-MOF exhibited excellent yields of methanol up to 2.59 gMeOH kgCu^{-1} h^{-1} with cent percent selectivity at 250 °C, 39.5 atm pressure (H$_2$: CO$_2$ = 3:1). The schematic diagram (Scheme **7**) illustrates how the H$_2$ molecules got dissociated and hydrogenated to activated CO$_2$ to ultimately form methanol.

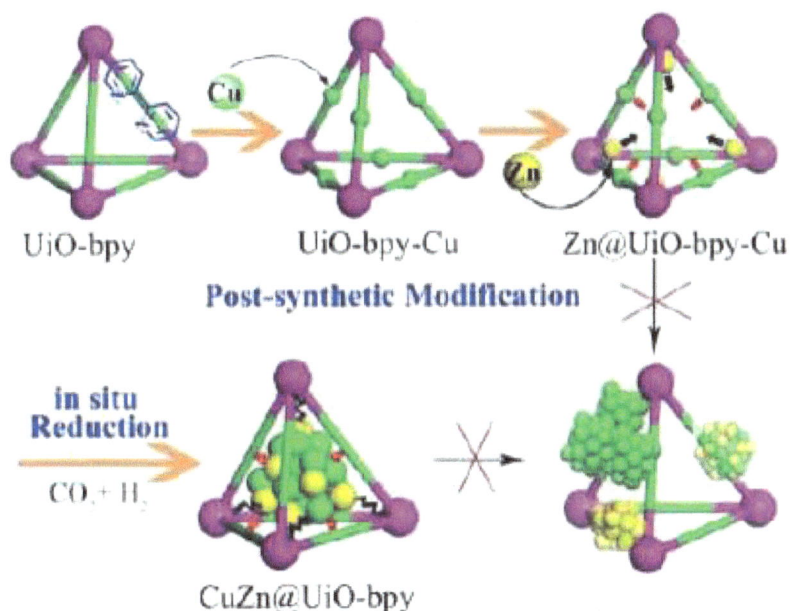

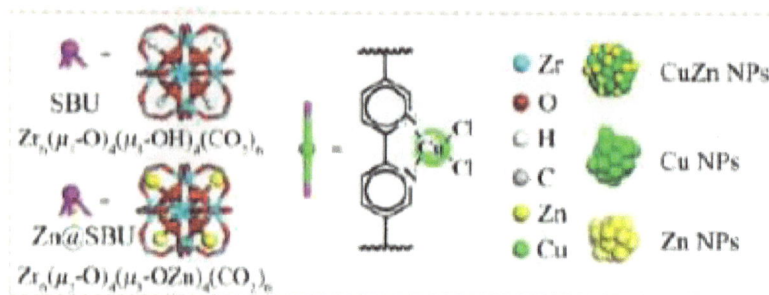

Fig. (21). Synthesis of CuZn@UiO-bpy. Reproduced with permission from [65] American Chemical Society. Copyright 2017.

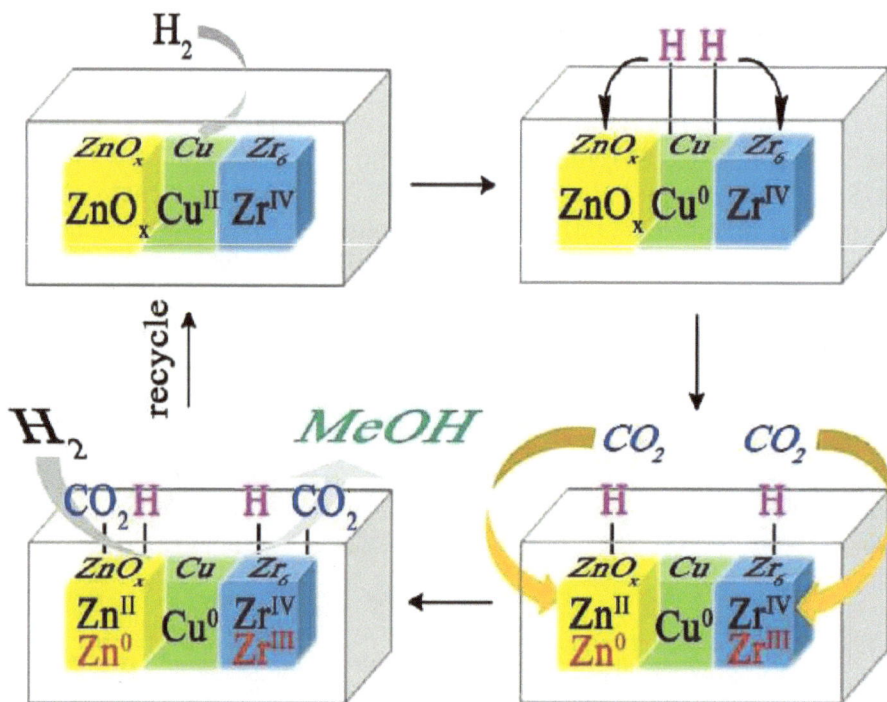

Scheme 7. Schematic diagram to show the mechanism of catalytic CO_2 hydrogenation to form methanol. Reproduced with permission from [65] American Chemical Society. Copyright 2017.

The Pd/ZnO alloy catalyst was found to be active toward the carbon dioxide hydrogenation to CH_3OH (maximum TOF 972 h^{-1}) [66]. Firstly, Pd NPs were encapsulated in a highly porous structure of ZIF-8 *via* the reduction of Pd^{2+} with NaBH$_4$ to form Pd@ZIF-8 precursor. Direct pyrolysis of Pd@ZIF-8 under air at different temperatures afforded Pd/Zn catalyst. Embedded Pd accelerates the decomposition of ZIF-8 to ZnO with a large number of surface oxygen defects. Upon pre-reduction by H_2, nearly half of the surface Pd atoms were replaced by

Zn atoms after the formation of PdZn alloy, which exhibited high methanol yield (Fig. **22**).

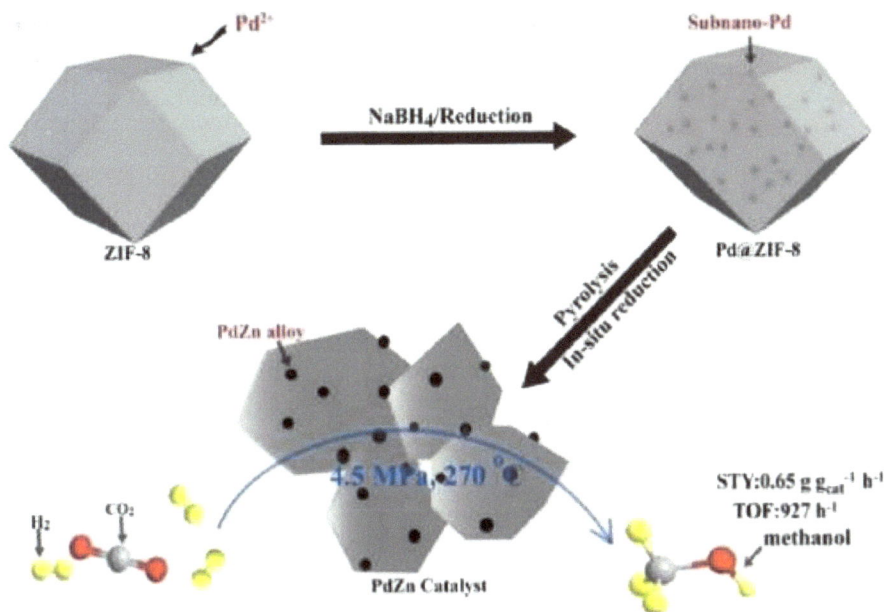

Fig. (22). The preparation process of the Pd/Zn hybrid catalysts for CO_2 hydrogenation to CH_3OH.

When CO_2 hydrogenation was investigated for Pt NPs loaded in UiO-67 at 170 °C, and 1-8 atm [67], it was found that CH_3OH formation occurred at the interface of linker-deficient Zr_6O_8 sites and Pt NPs. The reaction proceeds *via* hydrogen activation at Pt NPs while $HCOO^-$ ions are attached to the adjacent Zr sites (Fig. **23**). These results depict the strong interaction formed between Pt NPs ad defects created at Zr sites. DFT study revealed that during prior-treatment, Pt NPs' growth facilitates the linker's displacement and formation of vacant Zr sites.

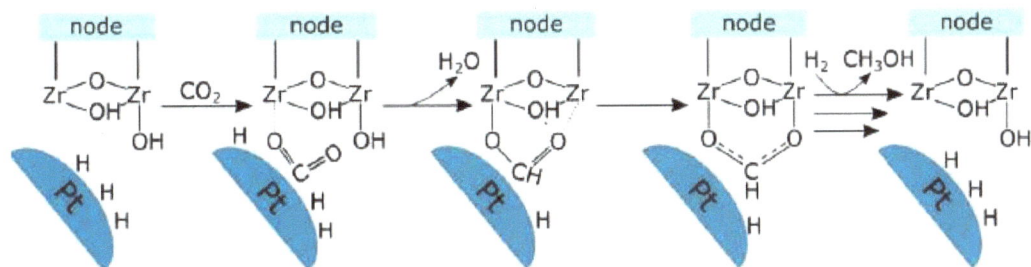

Fig. (23). Probable reaction pathways of CO_2 conversion to methanol. Reproduced with permission from [67] American Chemical Society. Copyright 2019.

Direct formation of MeOH by carbon dioxide hydrogenation was found using Cu loaded (Zr)-UiO-66 and (Ce/Zr)-UiO-66 MOF [68]. When 50% Zr(IV) ions in UiO-66(Zr) were replaced by Ce(IV) ions, the resulting mixed metal (Ce/Zr)-UiO-66 showed an improvement of selectivity towards MeOH formation without affecting the rate. Further, higher MeOH production was found for Cu@(Zr)-Ui--66 as compared to Cu@(Ce/Zr)-UiO-66 with very similar MeOH selectivity.

3.3. Conversion of CO_2 into Formic Acid/ Formate

Hydrogenation of CO_2 to produce formic acid (HCOOH) is another fascinating way of conversion of CO_2. Several MOF-materials have been utilized as catalysts for this reaction. Molecular Ir-catalysts were attached to the linkers strut of two UiO-type MOFs (Fig. **24**) [69]. To maximize the contact between the reacting species, the solvent (H_2O), gases (CO_2 and H_2), and the MOF catalyst were put in a condensing compartment of a Soxhlet-type extractor. The solution was neutralized using sodium bicarbonate ($NaHCO_3$) for progress the following reaction in the forward direction.

$$H_2(g) + CO_2(g) \rightleftharpoons HCOOH(aq)$$

Furthermore, the hydroxyl group present in the ligand functions as an electron and a hydrogen-bond donor, which again accelerates the reaction rate. An appreciable TOF value (410 ± 3 h^{-1}) was obtained at 358 K and 1 MPa H_2/CO_2 (taken equal volume). To understand the mechanism of the hydrogenation process, a DFT study was performed on UiO-66 attached with frustrated Lewis pairs (FLPs) [70]. It showed that initially cleavage of H_2 molecules happened heterolytically by the help of FLPs. Subsequently, H^- and H^+ were added to CO_2 simultaneously to produce formic acid (HCOOH) (Fig. **25**). Later, using the framework UiO-67, the mechanism was further studied [71].

The ruthenium complex (^{tBu}PNP)Ru(CO)HCl could be encapsulated within the cage of UiO-66 [72] by an aperture-opening process ensuing from dissociative ligand replacement (Fig. **26**). The final encapsulated complex [Ru]@UiO-66 catalyzed the hydrogenation of CO_2 to formate in a heterogeneous way. The catalyst was recycled for five times with no decomposition of the encapsulated complex.

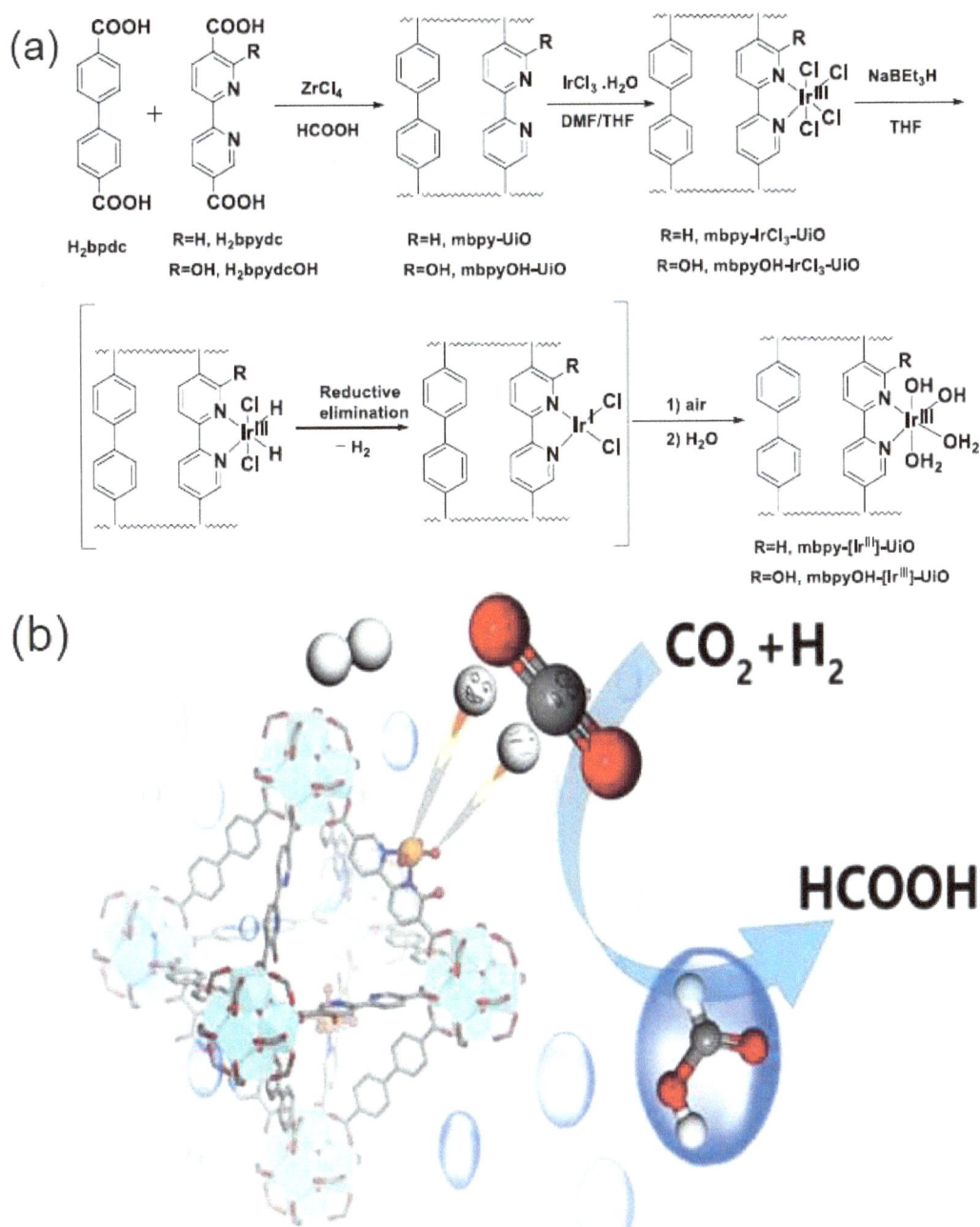

Fig. (24). (a) synthesis of iridium catalysts containing MOFs. **(b)** a portion of mbpyOH-[Ir^{III}]-UiO structure showing CO$_2$ hydrogenation. Reproduced with permission from [69] American Chemical Society. Copyright 2017.

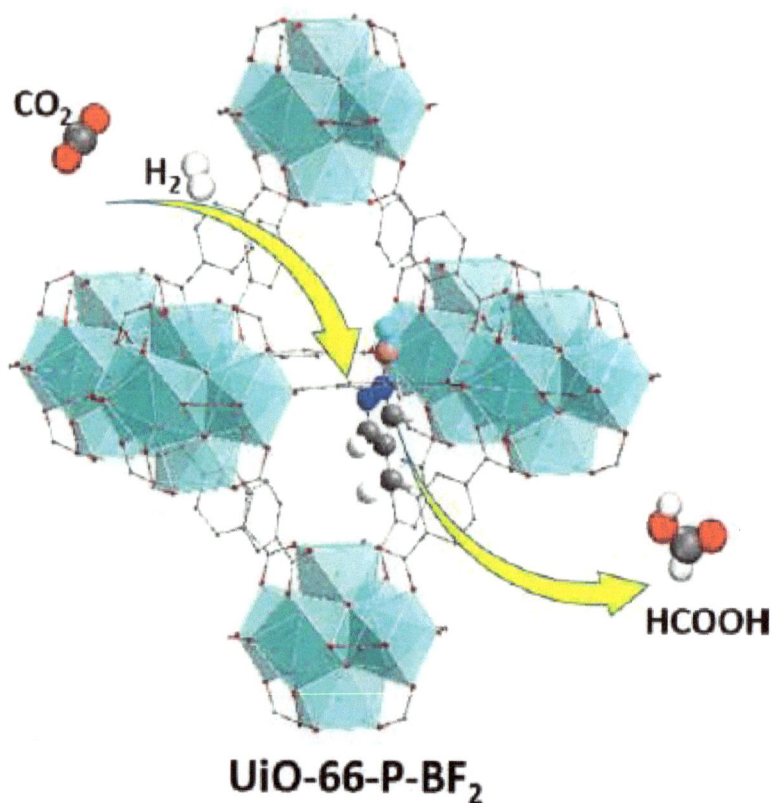

Fig. (25). The frustrated Lewis pairs functionalize MOF catalyzed the hydrogenation of CO_2 to formic acid. Reproduced with permission from [70] American Chemical Society. Copyright 2015.

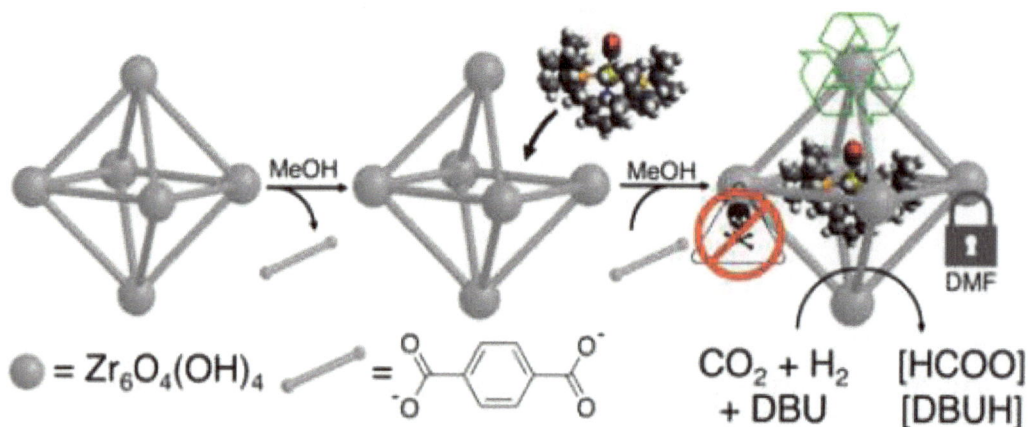

Fig. (26). Aperture-opening process to encapsulate ruthenium complex in UiO-66 for CO_2 hydrogenation to $HCOO^-$. Reproduced with permission from [72] American Chemical Society. Copyright 2018.

3.4. Conversion of CO_2 to Methane

Reaction of H_2 and CO_2 to form CH_4 is called the methanation reaction. This reaction is typically carried out by Ru, Rh, Ni and other transition metal-based catalysts [73]. For example, a highly efficient CO_2 methanation catalyst xRu/UiO-66 (where x is wt % of Ru) was developed [74] by doping Ru nanoparticles into UiO-66, The final catalyst contained a mixture of Ru-nanoparticles (size: 2-5 nm) supported on ZrO_2 nanoparticles (size:10-20 nm). This composite catalyst exhibited 96 - 98% CO_2 conversions with 99% CH_4 selectivity under 5 bar pressure of the gas mixture (H_2:CO_2 = 4:1) (Fig. 27).

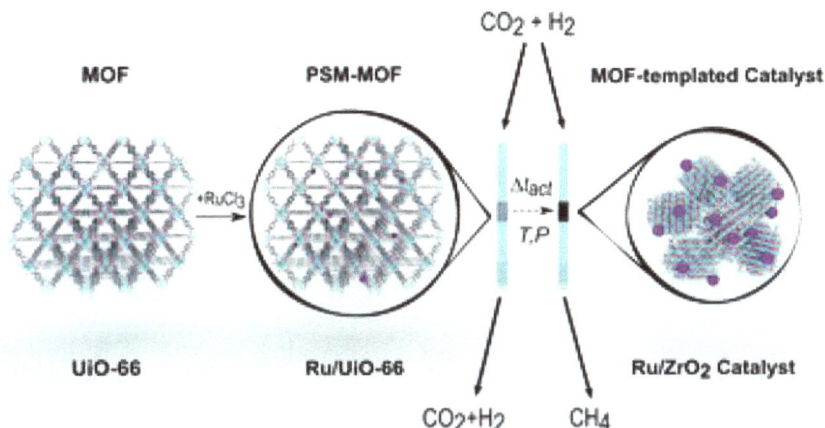

Fig. (27). Synthesis of Ru/UiO 66 followed by *in situ* formation of MOF derived catalyst. Here Δt_{act} is period of activation.

Ni@MOF-5 was found to be useful for CO_2 methanation [75]. Different Ni loading (x = 5, 7.5, 10, 12.5) produced xNi@MOF-5 catalysts. Among these catalysts, 10Ni@MOF-5 showed the highest surface area (2961 m^2 g^{-1}) and large pore volume (1.037cm^3 g^{-1}) and as a result, a large amount of Ni (41.8%) could be dispersed in a highly uniform way. 10Ni@MOF-5 exhibited 75.09% CO_2 conversion to CH_4 with 100% selectivity at 320 °C. Additionally, this catalyst showed high stability up to 100 h without any significant deactivation.

Co-nanoparticle could be incorporated into porous carbon by the pyrolysis of Co-based ZIF-67 at 873 K [76]. The particles of size 7 to 20 nm were encapsulated by graphite-like carbon. The resultant catalysts showed excellent catalytic performance in the methanation of CO_2 at a lower temperature. Besides, in the presence of 0.01% CTAB, the catalyst showed a maximum of 52.5% CO_2 conversion and 99.2% CH_4 selectivity at 270 °C.

CONCLUDING REMARKS

The rise in the number of MOF based catalytic materials with improved performance has opened a new avenue for CO_2 capture and conversion. One of the most important attributes an MOF has is its chemical tunability along with its interactions with other substrates. We have discussed that the MOFs can be combined with other mechanically strong substrates like graphene, carbon nanotubes, *etc.* to deal with their thermal/chemical instability keeping their own properties quite intact. Broadly speaking, the MOF-based materials offer favorable catalytic performance owing to their unique structural attributes and subsequent modulation. Their range of chemical functionalities and porosities facilitate to adsorb/activate other substrates/CO_2 leading to facile CO_2 conversion. MOFs based materials have been reported for the utilization of CO_2 by converting them into cyclic carbonates under mild reaction conditions using epoxides. We have seen the examples where MOFs have been combined with electrically active component to make them electroactive, thereby further increasing their scope and potential application in CO_2 conversion by electroactive means into methanol, carbon monoxide, formic acid *etc.* Many photoactive ligands/components have been used to make photoactive MOFs to impart visible-light absorbance capabilities. These features have further opened a new photocatalytic route for CO_2 conversion. Thus, so far, MOF based materials have shown promising results in the mitigation of CO_2 and have converted them into many industrially useful chemicals. In light of the above discussion, we should not ignore the CO_2 adsorption capacities of MOFs. The MOFs based catalyst would not be more effective unless or until the pristine MOFs lack the CO_2 adsorption capacities. Thus, the CO_2 capture adsorption capabilities of pristine MOFs are crucial and we should focus on enhancing the CO_2 uptake capacities along with its catalytic abilities. The MOFs having a very high CO_2 adsorption value can make these materials superior in terms of high catalytic activities and selectivities in the presence of many gases even at a low concentration and pressure of CO_2. Though many advancements happened in the MOF based materials, many challenges are still existing. From the industrial point of view, more research and development are needed to employ them at the point sources for CO_2 capture and conversion. We can expect more interesting work in the coming future on MOFs-based catalysts that could be able to address a broad range of CO_2 conversion reactions.

LIST OF ABBREVIATIONS

MOF Metal organic framework

LA   Lewis acid

BA   Brønsted acid

TBAB	Tetrabutylammonium bromide
3D	Three dimensional
SBU	secondary building unit
NPs	Nanoparticles
IP	Ionic polymer
FLPs	Frustrated Lewis pairs
CC	Cyclic Carbonate

CONSENT FOR PUBLICATION

Not applicable.

CONFLICT OF INTEREST

The author declares no conflict of interest, financial or otherwise.

ACKNOWLEDGEMENTS

We gratefully acknowledge the financial support received from the "TEQIP Collaborative Research Scheme (CRS)", NPIU, New Delhi, India (to D.D.)

REFERENCES

[1]   Global CO_2 emissions in 2019, IEA, Paris. **2019**.https:// www.iea.org/articles/global-CO

[2]   a. Zhong, H.; Ghorbani-Asl, M.; Ly, K.H.; Zhang, J.; Ge, J.; Wang, M.; Liao, Z.; Makarov, D.; Zschech, E.; Brunner, E.; Weidinger, I.M.; Zhang, J.; Krasheninnikov, A.V.; Kaskel, S.; Dong, R.; Feng, X. Synergistic electroreduction of carbon dioxide to carbon monoxide on bimetallic layered conjugated metal-organic frameworks. *Nat. Commun.,* **2020,** *11*(1), 1409.
[http://dx.doi.org/10.1038/s41467-020-15141-y] [PMID: 32179738] b. Pander, M.; Janeta, M.; Bury, W. Quest for an Efficient 2-in-1 MOF-Based Catalytic System for Cycloaddition of CO_2 to Epoxides under Mild Conditions. *ACS Appl. Mater. Interfaces,* **2021,** *13*(7), 8344-8352.
[http://dx.doi.org/10.1021/acsami.0c20437] [PMID: 33560110] c. Xu, L.; Xiu, Y.; Liu, F.; Liang, Y.; Wang, S. Research Progress in Conversion of CO_2 to Valuable Fuels. *Molecules,* **2020,** *25*(16), 3653.
[http://dx.doi.org/10.3390/molecules25163653] [PMID: 32796612] d. Liu, A.; Gao, M.; Ren, X.; Meng, F.; Yang, Y.; Gao, L.; Yang, Q.; Ma, T. Current progress in electrocatalytic carbon dioxide reduction to fuels on heterogeneous catalysts. *J. Mater. Chem. A Mater. Energy Sustain.,* **2020,** *8*(7), 3541-3562.
[http://dx.doi.org/10.1039/C9TA11966C]

[3]   Álvarez, A.; Borges, M.; Corral-Pérez, J.J.; Olcina, J.G.; Hu, L.; Cornu, D.; Huang, R.; Stoian, D.; Urakawa, A. CO_2 Activation over Catalytic Surfaces. *ChemPhysChem,* **2017,** *18*(22), 3135-3141.
[http://dx.doi.org/10.1002/cphc.201700782] [PMID: 28851111]

[4]   a. Kitagawa, S.; Kitaura, R.; Noro, S. Functional porous coordination polymers. *Angew. Chem. Int. Ed.,* **2004,** *43*(18), 2334-2375.
[http://dx.doi.org/10.1002/anie.200300610] [PMID: 15114565] b. Furukawa, H.; Cordova, K.E.; O'Keeffe, M.; Yaghi, O.M. The chemistry and applications of metal-organic frameworks. *Science,* **2013,** *341*(6149), 1230444.
[http://dx.doi.org/10.1126/science.1230444] [PMID: 23990564]

[5]   Russo, V.; Hmoudah, M.; Broccoli, F.; Iesce, M.R.; Jung, O-S.; Di Serio, M. Applications of Metal Organic Frameworks in Wastewater Treatment: A Review on Adsorption and Photodegradation. *Frontiers in Chemical Engineering,* **2020**, *2*, 581487.
[http://dx.doi.org/10.3389/fceng.2020.581487]

[6]   Li, X.; Wang, Z.; Wang, L. Metal-Organic Framework-Based Materials for Solar Water Splitting. *Small Sci.,* **2021**, *1*(5), 2000074.
[http://dx.doi.org/10.1002/smsc.202000074]

[7]   aLin, R.B.; Xiang, S.; Zhou, W.; Chen, B. Microporous Metal-Organic Framework Materials for Gas Separation. *Chem,* **2020**, *6*(2), 337-363.
[http://dx.doi.org/10.1016/j.chempr.2019.10.012] bHiraide, S.; Sakanaka, Y.; Kajiro, H.; Kawaguchi, S.; Miyahara, M.T.; Tanaka, H. High-throughput gas separation by flexible metal-organic frameworks with fast gating and thermal management capabilities. *Nat. Commun.,* **2020**, *11*(1), 3867.
[http://dx.doi.org/10.1038/s41467-020-17625-3] [PMID: 32747638]

[8]   Lawson, H.D.; Walton, S.P.; Chan, C. Metal-Organic Frameworks for Drug Delivery: A Design Perspective. *ACS Appl. Mater. Interfaces,* **2021**, *13*(6), 7004-7020.
[http://dx.doi.org/10.1021/acsami.1c01089] [PMID: 33554591]

[9]   Ding, M.; Flaig, R.W.; Jiang, H.L.; Yaghi, O.M. Carbon capture and conversion using metal-organic frameworks and MOF-based materials. *Chem. Soc. Rev.,* **2019**, *48*(10), 2783-2828.
[http://dx.doi.org/10.1039/C8CS00829A] [PMID: 31032507]

[10]   Kaur, R.; Kim, K.H.; Paul, A.K.; Deep, A. Recent advances in the photovoltaic applications of coordination polymers and metal organic frameworks. *J. Mater. Chem. A Mater. Energy Sustain.,* **2016**, *4*(11), 3991-4002.
[http://dx.doi.org/10.1039/C5TA09668E]

[11]   Lee, J.; Farha, O.K.; Roberts, J.; Scheidt, K.A.; Nguyen, S.T.; Hupp, J.T. Metal-organic framework materials as catalysts. *Chem. Soc. Rev.,* **2009**, *38*(5), 1450-1459.
[http://dx.doi.org/10.1039/b807080f] [PMID: 19384447]

[12]   Zhu, J.P.; Wang, X.H.; Zuo, X.X. The application of metal-organic frameworks in electrode materials for lithium-ion and lithium-sulfur batteries. *R. Soc. Open Sci.,* **2019**, *6*(7), 190634.
[http://dx.doi.org/10.1098/rsos.190634] [PMID: 31417758]

[13]   aSchäffner, B.; Schäffner, F.; Verevkin, S.P.; Börner, A. Organic carbonates as solvents in synthesis and catalysis. *Chem. Rev.,* **2010**, *110*(8), 4554-4581.
[http://dx.doi.org/10.1021/cr900393d] [PMID: 20345182] bShaikh, A.A.G.; Sivaram, S. Organic Carbonates. *Chem. Rev.,* **1996**, *96*(3), 951-976.
[http://dx.doi.org/10.1021/cr950067i] [PMID: 11848777]

[14]   Beyzavi, M.H.; Stephenson, C.J.; Liu, Y.; Karagiaridi, O.; Hupp, J.T.; Farha, O.K. Metal-organic framework-based catalysts: Chemical fixation of CO_2 with epoxides leading to cyclic organic carbonates. *Front. Energy Res.,* **2015**, *2*(63), 1-10.

[15]   Tharun, J.; Bhin, K.M.; Roshan, R.; Kim, D.W.; Kathalikkattil, A.C.; Babu, R.; Ahn, H.Y.; Won, Y.S.; Park, D.W. Ionic liquid tethered post functionalized ZIF-90 framework for the cycloaddition of propylene oxide and CO_2. *Green Chem.,* **2016**, *18*(8), 2479-2487.
[http://dx.doi.org/10.1039/C5GC02153G]

[16]   a. Liang, J.; Xie, Y.Q.; Wang, X.S.; Wang, Q.; Liu, T.T.; Huang, Y.B.; Cao, R. An imidazolium-functionalized mesoporous cationic metal-organic framework for cooperative CO_2 fixation into cyclic carbonate. *Chem. Commun. (Camb.),* **2018**, *54*(4), 342-345.
[http://dx.doi.org/10.1039/C7CC08630J] [PMID: 29182177] b. Liang, J.; Xie, Y.Q.; Wu, Q.; Wang, X.Y.; Liu, T.T.; Li, H.F.; Huang, Y.B.; Cao, R. Zinc Porphyrin/Imidazolium Integrated Multivariate Zirconium Metal-Organic Frameworks for Transformation of CO_2 into Cyclic Carbonates. *Inorg. Chem.,* **2018**, *57*(5), 2584-2593.
[http://dx.doi.org/10.1021/acs.inorgchem.7b02983] [PMID: 29430915]

[17]   Zhou, Z.; He, C.; Xiu, J.; Yang, L.; Duan, C. Metal-Organic Polymers Containing Discrete Single-Walled Nanotube as a Heterogeneous Catalyst for the Cycloaddition of Carbon Dioxide to Epoxides. *J. Am. Chem. Soc.,* **2015,** *137*(48), 15066-15069.
[http://dx.doi.org/10.1021/jacs.5b07925] [PMID: 26584402]

[18]   a. He, H.; Perman, J.A.; Zhu, G.; Ma, S. Metal-Organic Frameworks for CO_2 Chemical Transformations. *Small,* **2016,** *12*(46), 6309-6324.
[http://dx.doi.org/10.1002/smll.201602711] [PMID: 27762496] b. Timofeeva, M.N.; Lukoyanov, I.A.; Panchenko, V.N.; Bhadra, B.N.; Gerasimov, E.Y.; Jhung, S.H. Effect of MAF-6 crystal size on its physicochemical and catalytic properties in the cycloaddition of CO_2 to propylene oxide. *Catalysts,* **2021,** *11*(9), 1061-1077.
[http://dx.doi.org/10.3390/catal11091061]

[19]   Li, P.Z.; Wang, X.J.; Liu, J.; Phang, H.S.; Li, Y.; Zhao, Y. Highly Effective Carbon Fixation *via* Catalytic Conversion of CO_2 by an Acylamide-Containing Metal-Organic Framework. *Chem. Mater.,* **2017,** *29*(21), 9256-9261.
[http://dx.doi.org/10.1021/acs.chemmater.7b03183]

[20]   Kathalikkattil, A.C.; Roshan, R.; Tharun, J.; Babu, R.; Jeong, G.S.; Kim, D.W.; Cho, S.J.; Park, D.W. A sustainable protocol for the facile synthesis of zinc-glutamate MOF: an efficient catalyst for room temperature CO_2 fixation reactions under wet conditions. *Chem. Commun. (Camb.),* **2016,** *52*(2), 280-283.
[http://dx.doi.org/10.1039/C5CC07781H] [PMID: 26515327]

[21]   Liang, L.; Liu, C.; Jiang, F.; Chen, Q.; Zhang, L.; Xue, H.; Jiang, H.L.; Qian, J.; Yuan, D.; Hong, M. Carbon dioxide capture and conversion by an acid-base resistant metal-organic framework. *Nat. Commun.,* **2017,** *8*(1), 1233.
[http://dx.doi.org/10.1038/s41467-017-01166-3] [PMID: 29089480]

[22]   Zhao, D.; Liu, X.H.; Guo, J.H.; Xu, H.J.; Zhao, Y.; Lu, Y.; Sun, W.Y. Porous Metal-Organic Frameworks with Chelating Multiamine Sites for Selective Adsorption and Chemical Conversion of Carbon Dioxide. *Inorg. Chem.,* **2018,** *57*(5), 2695-2704.
[http://dx.doi.org/10.1021/acs.inorgchem.7b03099] [PMID: 29446625]

[23]   De, D.; Pal, T.K.; Neogi, S.; Senthilkumar, S.; Das, D.; Gupta, S.S.; Bharadwaj, P.K. A Versatile Cu[II] Metal-Organic Framework Exhibiting High Gas Storage Capacity with Selectivity for CO_2 : Conversion of CO_2 to Cyclic Carbonate and Other Catalytic Abilities. *Chemistry,* **2016,** *22*(10), 3387-3396.
[http://dx.doi.org/10.1002/chem.201504747] [PMID: 26833880]

[24]   Sharma, V.; De, D.; Saha, R.; Das, R.; Chattaraj, P.K.; Bharadwaj, P.K. A Cu(II)-MOF capable of fixing CO_2 from air and showing high capacity H_2 and CO_2 adsorption. *Chem. Commun. (Camb.),* **2017,** *53*(100), 13371-13374.
[http://dx.doi.org/10.1039/C7CC08315G] [PMID: 29199739]

[25]   Guo, X.; Zhou, Z.; Chen, C.; Bai, J.; He, C.; Duan, C. New *rht* -Type Metal-Organic Frameworks Decorated with Acylamide Groups for Efficient Carbon Dioxide Capture and Chemical Fixation from Raw Power Plant Flue Gas. *ACS Appl. Mater. Interfaces,* **2016,** *8*(46), 31746-31756.
[http://dx.doi.org/10.1021/acsami.6b13928] [PMID: 27933976]

[26]   Verma, A.; De, D.; Tomar, K.; Bharadwaj, P.K. An Amine Functionalized Metal-Organic Framework as an Effective Catalyst for Conversion of CO_2 and Biginelli Reactions. *Inorg. Chem.,* **2017,** *56*(16), 9765-9771.
[http://dx.doi.org/10.1021/acs.inorgchem.7b01286] [PMID: 28771347]

[27]   Gupta, M.; De, D.; Tomar, K.; Bharadwaj, P.K. From Zn(II)-Carboxylate to Double-Walled Zn(II)-Carboxylato Phosphate MOF: Change in the Framework Topology, Capture and Conversion of CO_2, and Catalysis of Strecker Reaction. *Inorg. Chem.,* **2017,** *56*(23), 14605-14611.
[http://dx.doi.org/10.1021/acs.inorgchem.7b02443] [PMID: 29131604]

[28]　Beyzavi, H.; Klet, R.C.; Tussupbayev, S.; Borycz, J.; Vermeulen, N.A.; Cramer, C.J.; Stoddart, J.F.; Hupp, J.T.; Farha, O.K. A hafnium-based metal-organic framework as an efficient and multifunctional catalyst for facile CO_2 fixation and regioselective and enantioretentive epoxide activation. *J. Am. Chem. Soc.,* **2014**, *136*(45), 15861-15864.
　　　　[http://dx.doi.org/10.1021/ja508626n] [PMID: 25357020]

[29]　Liang, J.; Chen, R.P.; Wang, X.Y.; Liu, T.T.; Wang, X.S.; Huang, Y.B.; Cao, R. Postsynthetic ionization of an imidazole-containing metal-organic framework for the cycloaddition of carbon dioxide and epoxides. *Chem. Sci. (Camb.),* **2017**, *8*(2), 1570-1575.
　　　　[http://dx.doi.org/10.1039/C6SC04357G] [PMID: 28451286]

[30]　a. Chen, D.; Luo, R.; Li, M.; Wen, M.; Li, Y.; Chen, C.; Zhang, N. Salen(Co(III)) imprisoned within pores of a metal-organic framework by post-synthetic modification and its asymmetric catalysis for CO_2 fixation at room temperature. *Chem. Commun. (Camb.),* **2017**, *53*(79), 10930-10933.
　　　　[http://dx.doi.org/10.1039/C7CC06522A] [PMID: 28932854] b. Aguila, B.; Sun, Q.; Wang, X.; O'Rourke, E.; Al-Enizi, A.M.; Nafady, A.; Ma, S. Lower Activation Energy for Catalytic Reactions through Host-Guest Cooperation within Metal-Organic Frameworks. *Angew. Chem. Int. Ed.,* **2018**, *57*(32), 10107-10111.
　　　　[http://dx.doi.org/10.1002/anie.201803081] [PMID: 29766629]

[31]　Singh, M.; Solanki, P.; Patel, P.; Mondal, A.; Neogi, S. Highly Active Ultrasmall Ni Nanoparticle Embedded Inside a Robust Metal-Organic Framework: Remarkably Improved Adsorption, Selectivity, and Solvent-Free Efficient Fixation of CO_2. *Inorg. Chem.,* **2019**, *58*(12), 8100-8110.
　　　　[http://dx.doi.org/10.1021/acs.inorgchem.9b00833] [PMID: 31144809]

[32]　Chen, D.; Luo, R.; Li, M.; Wen, M.; Li, Y.; Chen, C.; Zhang, N. Salen(Co(III)) imprisoned within pores of a metal-organic framework by post-synthetic modification and its asymmetric catalysis for CO_2 fixation at room temperature. *Chem. Commun. (Camb.),* **2017**, *53*(79), 10930-10933.
　　　　[http://dx.doi.org/10.1039/C7CC06522A] [PMID: 28932854]

[33]　Liu, C.; Liu, X.H.; Li, B.; Zhang, L.; Ma, J.G.; Cheng, P. Salen-Cu(II)@MIL-101(Cr) as an efficient heterogeneous catalyst for cycloaddition of CO_2 to epoxides under mild conditions. *Journal of Energy Chemistry,* **2017**, *26*(5), 821-824.
　　　　[http://dx.doi.org/10.1016/j.jechem.2017.07.022]

[34]　Ding, M.; Jiang, H.L. Incorporation of Imidazolium-Based Poly(ionic liquid)s into a Metal-Organic Framework for CO_2 Capture and Conversion. *ACS Catal.,* **2018**, *8*(4), 3194-3201.
　　　　[http://dx.doi.org/10.1021/acscatal.7b03404]

[35]　Aguila, B.; Sun, Q.; Wang, X.; O'Rourke, E.; Al-Enizi, A.M.; Nafady, A.; Ma, S. Lower Activation Energy for Catalytic Reactions through Host-Guest Cooperation within Metal-Organic Frameworks. *Angew. Chem. Int. Ed.,* **2018**, *57*(32), 10107-10111.
　　　　[http://dx.doi.org/10.1002/anie.201803081] [PMID: 29766629]

[36]　Liu, W.S.; Zhou, L.J.; Li, G.; Yang, S.L.; Gao, E.Q. Double Cationization Approach toward Ionic Metal-Organic Frameworks with a High Bromide Content for CO_2 Cycloaddition to Epoxides. *ACS Sustain. Chem.& Eng.,* **2021**, *9*(4), 1880-1890.
　　　　[http://dx.doi.org/10.1021/acssuschemeng.0c08349]

[37]　Chen, Y.Z.; Zhang, R.; Jiao, L.; Jiang, H.L. Metal-organic framework-derived porous materials for catalysis. *Coord. Chem. Rev.,* **2018**, *362*, 1-23.
　　　　[http://dx.doi.org/10.1016/j.ccr.2018.02.008]

[38]　Toyao, T.; Fujiwaki, M.; Miyahara, K.; Kim, T.H.; Horiuchi, Y.; Matsuoka, M. Design of Zeolitic Imidazolate Framework Derived Nitrogen-Doped Nanoporous Carbons Containing Metal Species for Carbon Dioxide Fixation Reactions. *ChemSusChem,* **2015**, *8*(22), 3905-3912.
　　　　[http://dx.doi.org/10.1002/cssc.201500780] [PMID: 26395673]

[39]　Ding, M.; Chen, S.; Liu, X.Q.; Sun, L.B.; Lu, J.; Jiang, H.L. Metal-Organic Framework-Templated Catalyst: Synergy in Multiple Sites for Catalytic CO_2 Fixation. *ChemSusChem,* **2017**, *10*(9), 1898-

1903.
[http://dx.doi.org/10.1002/cssc.201700245] [PMID: 28322516]

[40]   Yao, B.J.; Ding, L.G.; Li, F.; Li, J.T.; Fu, Q.J.; Ban, Y.; Guo, A.; Dong, Y.B. Chemically cross-linked MOF membrane generated from imidazolium-based ionic liquiddecorated UiO-66 type NMOF and its application toward CO_2 separation and conversion. *ACS Appl. Mater. Interfaces,* **2017**, *9*(44), 38919-38930.
[http://dx.doi.org/10.1021/acsami.7b12697] [PMID: 29027785]

[41]   Manjolinho, F.; Arndt, M.; Gooßen, K.; Gooßen, L.J. Catalytic C-H carboxylation of terminal alkynes with carbon dioxide. *ACS Catal.,* **2012**, *2*(9), 2014-2021.
[http://dx.doi.org/10.1021/cs300448v]

[42]   Xiong, G.; Yu, B.; Dong, J.; Shi, Y.; Zhao, B.; He, L.N. Cluster-based MOFs with accelerated chemical conversion of CO_2 through C-C bond formation. *Chem. Commun. (Camb.),* **2017**, *53*(44), 6013-6016.
[http://dx.doi.org/10.1039/C7CC01136A] [PMID: 28518199]

[43]   Zhou, Z.; He, C.; Yang, L.; Wang, Y.; Liu, T.; Duan, C. Alkyne activation by a porous silver coordination polymer for heterogeneous catalysis of carbon dioxide cycloaddition. *ACS Catal.,* **2017**, *7*(3), 2248-2256.
[http://dx.doi.org/10.1021/acscatal.6b03404]

[44]   Zhang, G.; Yang, H.; Fei, H. Unusual missing linkers in an organosulfonate-based primitive-cubic (pcu)-type metal-organic framework for CO_2 capture and conversion under ambient conditions. *ACS Catal.,* **2018**, *8*(3), 2519-2525.
[http://dx.doi.org/10.1021/acscatal.7b04189]

[45]   Zhao, D.; Liu, X.H.; Zhu, C.; Kang, Y.S.; Wang, P.; Shi, Z.; Lu, Y.; Sun, W.Y. Efficient and reusable metal-organic framework catalysts for carboxylative cyclization of propargylamines with carbon dioxide. *ChemCatChem,* **2017**, *9*(24), 4598-4606.
[http://dx.doi.org/10.1002/cctc.201701190]

[46]   Shah, D.J.; Sharma, A.S.; Shah, A.P.; Sharma, V.S.; Athar, M.; Soni, J.Y. Fixation of CO_2 as a carboxylic acid precursor by microcrystalline cellulose (MCC) supported Ag NPs: a more efficient, sustainable, biodegradable and eco-friendly catalyst. *New J. Chem.,* **2019**, *43*(22), 8669-8676.
[http://dx.doi.org/10.1039/C8NJ06373G]

[47]   Liu, X.H.; Ma, J.G.; Niu, Z.; Yang, G.M.; Cheng, P. An efficient nanoscale heterogeneous catalyst for the capture and conversion of carbon dioxide at ambient pressure. *Angew. Chem. Int. Ed.,* **2015**, *54*(3), 988-991.
[http://dx.doi.org/10.1002/anie.201409103] [PMID: 25385217]

[48]   Zhu, N.N.; Liu, X.H.; Li, T.; Ma, J.G.; Cheng, P.; Yang, G.M. Composite System of Ag Nanoparticles and Metal-Organic Frameworks for the Capture and Conversion of Carbon Dioxide under Mild Conditions. *Inorg. Chem.,* **2017**, *56*(6), 3414-3420.
[http://dx.doi.org/10.1021/acs.inorgchem.6b02855] [PMID: 28263612]

[49]   Molla, R.A.; Ghosh, K.; Banerjee, B.; Iqubal, M.A.; Kundu, S.K.; Islam, S.M.; Bhaumik, A. Silver nanoparticles embedded over porous MOF for CO_2 fixation *via* carboxylation of terminal alkynes at ambient pressure. *J. Colloid Interface Sci.,* **2016**, *477*, 220-229.
[http://dx.doi.org/10.1016/j.jcis.2016.05.037] [PMID: 27309859]

[50]   Ye, J.; Johnson, J.K. Design of Lewis Pair-Functionalized Metal Organic Frameworks for CO_2 Hydrogenation. *ACS Catal.,* **2015**, *5*(5), 2921-2928.
[http://dx.doi.org/10.1021/acscatal.5b00396]

[51]   a. Lippi, R.; Howard, S.C.; Barron, H.; Easton, C.D.; Madsen, I.C.; Waddington, L.J.; Vogt, C.; Hill, M.R.; Sumby, C.J.; Doonan, C.J.; Kennedy, D.F. Highly active catalyst for CO_2 methanation derived from a metal organic framework template. *J. Mater. Chem. A Mater. Energy Sustain.,* **2017**, *5*(25), 12990-12997.

[http://dx.doi.org/10.1039/C7TA00958E] b. Yin, Y.; Hu, B.; Li, X.; Zhou, X.; Hong, X.; Liu, G. Pd@zeolitic imidazolate framework-8 derived PdZn alloy catalysts for efficient hydrogenation of CO_2 to methanol. *Appl. Catal. B,* **2018**, *234*, 143-152.
[http://dx.doi.org/10.1016/j.apcatb.2018.04.024] c. Zhang, H.; Wang, T.; Wang, J.; Liu, H.; Dao, T.D.; Li, M.; Liu, G.; Meng, X.; Chang, K.; Shi, L.; Nagao, T.; Ye, J. Surface-Plasmon-Enhanced Photodriven CO_2 Reduction Catalyzed by Metal-Organic-Framework-Derived Iron Nanoparticles Encapsulated by Ultrathin Carbon Layers. *Adv. Mater.,* **2016**, *28*(19), 3703-3710.
[http://dx.doi.org/10.1002/adma.201505187] [PMID: 27001900] d. Ramirez, A.; Gevers, L.; Bavykina, A.; Ould-Chikh, S.; Gascon, J. Metal Organic Framework-Derived Iron Catalysts for the Direct Hydrogenation of CO_2 to Short Chain Olefins. *ACS Catal.,* **2018**, *8*(10), 9174-9182.
[http://dx.doi.org/10.1021/acscatal.8b02892]

[52]   Fu, Y.; Sun, D.; Chen, Y.; Huang, R.; Ding, Z.; Fu, X.; Li, Z. An amine-functionalized titanium metal-organic framework photocatalyst with visible-light-induced activity for CO_2 reduction. *Angew. Chem. Int. Ed.,* **2012**, *51*(14), 3364-3367.
[http://dx.doi.org/10.1002/anie.201108357] [PMID: 22359408]

[53]   Chambers, M.B.; Wang, X.; Elgrishi, N.; Hendon, C.H.; Walsh, A.; Bonnefoy, J.; Canivet, J.; Quadrelli, E.A.; Farrusseng, D.; Mellot-Draznieks, C.; Fontecave, M. Photocatalytic carbon dioxide reduction with rhodium-based catalysts in solution and heterogenized within metal-organic frameworks. *ChemSusChem,* **2015**, *8*(4), 603-608.
[http://dx.doi.org/10.1002/cssc.201403345] [PMID: 25613479]

[54]   Kajiwara, T.; Fujii, M.; Tsujimoto, M.; Kobayashi, K.; Higuchi, M.; Tanaka, K.; Kitagawa, S. Photochemical Reduction of Low Concentrations of CO_2 in a Porous Coordination Polymer with a Ruthenium(II)-CO Complex. *Angew. Chem. Int. Ed.,* **2016**, *55*(8), 2697-2700.
[http://dx.doi.org/10.1002/anie.201508941] [PMID: 26800222]

[55]   Xu, H.Q.; Hu, J.; Wang, D.; Li, Z.; Zhang, Q.; Luo, Y.; Yu, S.H.; Jiang, H.L. Visible-Light Photoreduction of CO_2 in a Metal-Organic Framework: Boosting Electron-Hole Separation *via* Electron Trap States. *J. Am. Chem. Soc.,* **2015**, *137*(42), 13440-13443.
[http://dx.doi.org/10.1021/jacs.5b08773] [PMID: 26434687]

[56]   Sadeghi, N.; Sharifnia, S.; Arabi, M. S. A porphyrin-based metal organic framework for high rate photoreduction of CO_2 to CH_4 in gas phase. *J. CO_2 Util,* **2016**, *16*, 450-457.

[57]   Chen, D.; Xing, H.; Wang, C.; Su, Z. Highly efficient visible-light-driven CO_2 reduction to formate by a new anthracene-based zirconium MOF *via* dual catalytic routes. *J. Mater. Chem. A Mater. Energy Sustain.,* **2016**, *4*(7), 2657-2662.
[http://dx.doi.org/10.1039/C6TA00429F]

[58]   Qin, J.S.; Yuan, S.; Zhang, L.; Li, B.; Du, D.Y.; Huang, N.; Guan, W.; Drake, H.F.; Pang, J.; Lan, Y.Q.; Alsalme, A.; Zhou, H.C. Creating Well-Defined Hexabenzocoronene in Zirconium Metal-Organic Framework by Postsynthetic Annulation. *J. Am. Chem. Soc.,* **2019**, *141*(5), 2054-2060.
[http://dx.doi.org/10.1021/jacs.8b11042] [PMID: 30621391]

[59]   Su, Y.; Zhang, Z.; Liu, H.; Wang, Y. $Cd_{0.2}Zn_{0.8}S$@UiO-66-NH_2 nanocomposites as efficient and stable visible-light-driven photocatalyst for H_2 evolution and CO_2 reduction. *Appl. Catal. B,* **2017**, *200*, 448-457.
[http://dx.doi.org/10.1016/j.apcatb.2016.07.032]

[60]   Liu, W.C.; Baek, J.; Somorjai, G.A. The methanol economy: methane and carbon dioxide conversion. *Top. Catal.,* **2018**, *61*(7-8), 530-541.
[http://dx.doi.org/10.1007/s11244-018-0907-4]

[61]   Liu, Q.; Low, Z.X.; Li, L.; Razmjou, A.; Wang, K.; Yao, J.; Wang, H. ZIF-8/Zn_2GeO_4 nanorods with an enhanced CO_2 adsorption property in an aqueous medium for photocatalytic synthesis of liquid fuel. *J. Mater. Chem. A Mater. Energy Sustain.,* **2013**, *1*(38), 11563-11569.
[http://dx.doi.org/10.1039/c3ta12433a]

[62]   Albo, J.; Vallejo, D.; Beobide, G.; Castillo, O.; Castaño, P.; Irabien, A. Copper-based metal-organic

porous materials for CO_2 Electrocatalytic reduction to alcohols. *ChemSusChem,* **2017**, *10*(6), 1100-1109.
[http://dx.doi.org/10.1002/cssc.201600693] [PMID: 27557788]

[63] Cheng, J.; Xuan, X.; Yang, X.; Zhou, J.; Cen, K. Selective reduction of CO_2 to alcohol products on octahedral catalyst of carbonized Cu(BTC) doped with Pd nanoparticles in a photoelectrochemical cell. *Chem. Eng. J.,* **2019**, *358*, 860-868.
[http://dx.doi.org/10.1016/j.cej.2018.10.091]

[64] Rungtaweevoranit, B.; Baek, J.; Araujo, J.R.; Archanjo, B.S.; Choi, K.M.; Yaghi, O.M.; Somorjai, G.A. Copper nanocrystals encapsulated in Zr-based metal-Organic frameworks for highly selective CO_2 hydrogenation to methanol. *Nano Lett.,* **2016**, *16*(12), 7645-7649.
[http://dx.doi.org/10.1021/acs.nanolett.6b03637] [PMID: 27960445]

[65] An, B.; Zhang, J.; Cheng, K.; Ji, P.; Wang, C.; Lin, W. Confinement of ultrasmall Cu/ZnOx nanoparticles in metal-Organic frameworks for selective methanol synthesis from catalytic hydrogenation of CO_2. *J. Am. Chem. Soc.,* **2017**, *139*(10), 3834-3840.
[http://dx.doi.org/10.1021/jacs.7b00058] [PMID: 28209054]

[66] Yin, Y.; Hu, B.; Li, X.; Zhou, X.; Hong, X.; Liu, G. Pd@zeolitic imidazolate framework-8 derived PdZn alloy catalysts for efficient hydrogenation of CO_2 to methanol. *Appl. Catal. B,* **2018**, *234*, 143-152.
[http://dx.doi.org/10.1016/j.apcatb.2018.04.024]

[67] Gutterød, E.S.; Lazzarini, A.; Fjermestad, T.; Kaur, G.; Manzoli, M.; Bordiga, S.; Svelle, S.; Lillerud, K.P.; Skúlason, E.; Øien-Ødegaard, S.; Nova, A.; Olsbye, U. Hydrogenation of CO_2 to methanol by Pt nanoparticles encapsulated in UiO-67: deciphering the role of the metal-organic framework. *J. Am. Chem. Soc.,* **2020**, *142*(2), 999-1009.
[http://dx.doi.org/10.1021/jacs.9b10873] [PMID: 31794194]

[68] Stawowy, M.; Ciesielski, R.; Maniecki, T.; Matus, K.; Łużny, R.; Trawczynski, J.; Silvestre-Albero, J.; Łamacz, A. CO_2 hydrogenation to methanol over Ce and Zr containing UiO-66 and Cu/UiO-66. *Catalysts,* **2019**, *10*(1), 39.
[http://dx.doi.org/10.3390/catal10010039]

[69] An, B.; Zeng, L.; Jia, M.; Li, Z.; Lin, Z.; Song, Y.; Zhou, Y.; Cheng, J.; Wang, C.; Lin, W. Molecular Iridium Complexes in Metal-Organic Frameworks Catalyze CO_2 Hydrogenation *via* Concerted Proton and Hydride Transfer. *J. Am. Chem. Soc.,* **2017**, *139*(49), 17747-17750.
[http://dx.doi.org/10.1021/jacs.7b10922] [PMID: 29179548]

[70] Ye, J.; Johnson, J.K. Design of Lewis Pair-Functionalized Metal Organic Frameworks for CO_2 Hydrogenation. *ACS Catal.,* **2015**, *5*(5), 2921-2928.
[http://dx.doi.org/10.1021/acscatal.5b00396]

[71] Ye, J.; Johnson, J.K. Catalytic hydrogenation of CO_2 to methanol in a Lewis pair functionalized MOF. *Catal. Sci. Technol.,* **2016**, *6*(24), 8392-8405.
[http://dx.doi.org/10.1039/C6CY01245K]

[72] Li, Z.; Rayder, T.M.; Luo, L.; Byers, J.A.; Tsung, C.K. Aperture-opening encapsulation of a transition metal catalyst in a metal-organic framework for CO_2 hydrogenation. *J. Am. Chem. Soc.,* **2018**, *140*(26), 8082-8085.
[http://dx.doi.org/10.1021/jacs.8b04047] [PMID: 29909631]

[73] Younas, M.; Loong Kong, L.; Bashir, M.J.K.; Nadeem, H.; Shehzad, A.; Sethupathi, S. Recent advancements, fundamental challenges, and opportunities in catalytic methanation of CO_2. *Energy Fuels,* **2016**, *30*(11), 8815-8831.
[http://dx.doi.org/10.1021/acs.energyfuels.6b01723]

[74] Lippi, R.; Howard, S.C.; Barron, H.; Easton, C.D.; Madsen, I.C.; Waddington, L.J.; Vogt, C.; Hill, M.R.; Sumby, C.J.; Doonan, C.J.; Kennedy, D.F. Highly active catalyst for CO_2 methanation derived from a metal organic framework template. *J. Mater. Chem. A Mater. Energy Sustain.,* **2017**, *5*(25),

12990-12997.
[http://dx.doi.org/10.1039/C7TA00958E]

[75]   Zhen, W.; Li, B.; Lu, G.; Ma, J. Enhancing catalytic activity and stability for CO_2 methanation on Ni@MOF-5 *via* control of active species dispersion. *Chem. Commun. (Camb.),* **2015**, *51*(9), 1728-1731.
        [http://dx.doi.org/10.1039/C4CC08733J] [PMID: 25518948]

[76]   Li, W.; Zhang, A.; Jiang, X.; Chen, C.; Liu, Z.; Song, C.; Guo, X. Low Temperature CO_2 Methanation: ZIF-67-Derived Co-Based Porous Carbon Catalysts with Controlled Crystal Morphology and Size. *ACS Sustain. Chem.& Eng.,* **2017**, *5*(9), 7824-7831.
        [http://dx.doi.org/10.1021/acssuschemeng.7b01306]

CHAPTER 6

Metal-Organic Framework Composites for Photocatalytic Water Purification

Ning Yuan[1,*] and **Xinling Zhang**[1]

[1] *Department of Chemical Engineering, School of Chemical and Environmental Engineering, China University of Mining and Technology, Beijing 100083, China*

Abstract: The rapid rise in photocatalytic technology with efficient removal capabilities has attracted wide attention. Recently, metal-organic frameworks (MOFs), a kind of coordination polymers, have also been applied in the field of photocatalytic water purification due to their characteristics such as high specific surface area and adjustable pore structure. However, the weak water stability, low reutilization rate, and poor photocatalytic ability of the constructed MOFs restrict their application in environmental remediation. To tackle these problems, many researchers have devoted themselves to designing highly efficient MOF-based composites by adding other substances. This chapter mainly focuses on the research status of MOF-based composites in the photocatalytic elimination of various pollutants from water. Additionally, the synthetic strategies for MOFs and their composite materials as well as for photodegradation of pollutants in water are reviewed and exemplified. The possible removal mechanisms of some MOF-based composites have also been briefly analyzed. Finally, the achievements and prospects on future research of MOFs and their composite materials have been described in detail.

Keywords: Adsorption, Metal-organic framework composites, Photocatalysis, Water purification, Water treatment.

1. INTRODUCTION

To meet the needs of the alarming growth of the world population, the rapid development of the industrial sector has produced the required products and a lot of polluted organic wastes [1]. The generated organic pollutants can cause serious environmental pollution, which will threaten the existence and development of human beings [2]. Industrial wastewater discharged into lakes, rivers, and drinking water environment often contains heavy metal ions or common organic contaminants. They are highly likely to menace the lives of aquatic creatures and

* **Corresponding author Ning Yuan:** Department of Chemical Engineering, School of Chemical and Environmental Engineering, China University of Mining and Technology, Beijing 100083, China; E-mail: ning.yuan@hotmail.com

Junkuo Gao & Reza Abazari (Eds.)

the health of human beings [3]. Some common organic pollutants in rivers are dyestuffs, phenols, pharmaceuticals and personal care products (PPCPs), herbicides, pesticides, as well as other organics [4]. In addition, some inorganic pollutants are also included and mixed pollutants may coexist in real water [5]. These pollutants are highly stable, remain very long-term bioactive, cause a big effect on the type of microbial species, and they may also produce certain viruses [6]. Due to their high solubility, the decontamination process of inorganic pollutants is more durable and extremely resistant. The predominant concentrations of these pollutants in the water are virulent or deadly, even at ppm or ppb levels. These pollutants are of high risk because they can easily be amplified by organisms and cause harm to organisms higher up the tropical food chain [7]. As a consequence, it is urgent to search for a valuable water treatment technology to remove the pollutants. So far, numerous approaches for the elimination of these pollutants have been utilized, for example, membrane separation, ion exchange, photocatalysis, adsorption, and so on. Among these water treatment techniques, photocatalysis is a burgeoning method and is identified as the most valuable technique on account of its affordable cost, easy operation, and good practicability [8]. As a consequence, photocatalysis is deemed to be the most effective method for removing these pollutants [9]. Hence, fabricating an efficient photocatalyst with enhanced photocatalytic performance is of great practical importance.

MOFs can be self-assembled from organic ligands and metal salts or metal clusters. Up to now, these MOF materials have been considered as a new kind of crystalline porous material [10]. Their applications in the field of photocatalysis have gradually emerged and been systematically studied based on the large surface area and adjustable pore sizes [11 - 13]. Initially, they are mainly used for photocatalytic H_2 generation [14], photocatalytic degradation, and metal ion reduction. However, due to the high e^--h^+ recombination efficiency, low solar energy conversion efficiency, and poor electrical conductivity, photocatalytic efficiency is not ideal. Hence, many research groups have established MOF-based composites to boost the photodegradation capacity of MOFs. The constructed composites have been reported to be used in photocatalytic water purification. At the same time, they are expected to achieve better photocatalytic performance.

In this chapter, the applications of MOF-based composites in photocatalytic water purification in recent years are reviewed (Fig. **1**). Firstly, the present chapter provides a brief review of their preparation strategy and photocatalytic substrates of MOF-based composites. Then, we study MOF-based composite materials from the aspect of photocatalytic degradation of organic pollutants, inorganic pollutants, and mixed system pollutants in water in the following context. Subsequently, we introduce the application of adsorption-photocatalysis synergism in water purification. Finally, we summarize the photocatalytic water purification of MOF-based composites and give a prospect of the development of the composites in the future. We hope that this chapter will provide a better guide to the future challenges of MOF-based composites in photocatalytic water purification.

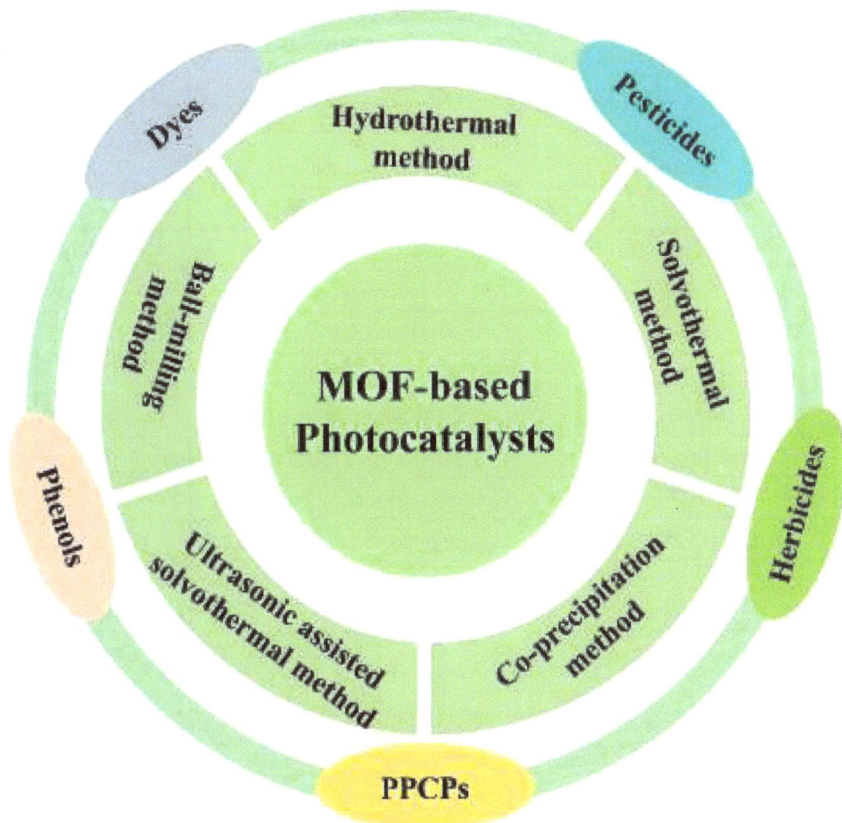

Fig. (1). Schematic diagram of the preparation method and photocatalytic water purification onto MOF-based composites.

2. SYNTHESIS OF MOF-BASED COMPOSITES

Recently, more and more researchers set about constructing MOF-based composites. The MOF-based composites with diverse synthetic methods can be generated by regulating and controlling the reaction conditions. Various synthesis strategies to prepare MOF-based composites have been exploited, leading to diverse morphologies and properties, so they can be applied to the photodegradation of contaminants in wastewater. In the current section, we delineate the preparation strategy of different MOF-based composites. Their use to treat the different organic pollutants in wastewater is also summarized in the following table and will be discussed successively in the next part. These reported studies on water purification using different MOF-based composites are provided in Tables **1** - **4**.

Table 1. Construction strategies of composites based on MILs and their photocatalytic properties in water purification.

Composite	Synthesis Strategy	Pollutant	Ref.
PHIK/MIL-125-NH$_2$	*in-situ* growth method	RhB	[15]
BiOBr/NH$_2$-MIL-125(Ti)	Co-precipitation method	RhB	[16]
NH$_2$-MIL-125(Ti)/BiOCl	One-pot hydrothermal strategy	TC; BPA	[17]
NH$_2$-MIL-125(Ti)/TTB-TTA	Solvothermal method	MO	[18]
RhB-sensitized MIL-125(Ti)	Post-impregnation strategy	MO	[19]
MIL-125(Ti)/MnO$_2$	Solvothermal method	RhB	[20]
NH$_2$-MIL-125(Ti)/TiO$_2$	One-pot solvothermal method	cyclohexane	[21]
NH$_2$-MIL-125(Ti)(TiO$_2$)/Ti$_3$C$_2$	One-step solvothermal method	TC	[22]
Pt/MIL-125(Ti)/Ag	Solvothermal method	ketoprofen	[23]
MIL-125(Ti)/ZnIn$_2$S$_4$	Solvothermal method	RhB; Cr(VI)	[24]
Fe/Ti-MOF-NH$_2$/PS	One-pot solvothermal method	orange II	[25]
CoS$_x$/NH$_2$-MIL-125	*in-situ* hydrothermal method	RhB	[26]
NH$_2$-MIL-125(Ti)@Bi$_2$MoO$_6$	Solvothermal method	dichlorophen; trichorophenol	[27]
Ag$_3$PO$_4$@NH$_2$-MIL-125	One-pot method	RhB; MB	[28]
Ag$_3$VO$_4$@NH$_2$-MIL-125 Ag$_2$WO$_4$@NH$_2$-MIL-125	*in-situ* synthesis method	MB&RhB	[29]
CdS@NH$_2$-MIL-125(Ti)	Solvothermal method	RhB; OTC phenol	[30]
Zr-NH$_2$-MIL-125(Ti)	Solvothermal method	acetaminophen	[31]
Ag/rGO/MIL-125(Ti)	One-pot self-assembly method	RhB	[32]
Ag/AgBr/NH$_2$-MIL-125(Ti)	*in-situ* growth method	MO	[33]
MIL-125-NH$_2$@Ag/AgCl	Deposition-photoreduction method	Cr(VI); RhB; MG	[34]
CdSe QDs sensitized MIL-125/TiO$_2$@SiO$_2$	Hydrothermal method	RhB	[35]
g-C$_3$N$_4$/MIL-53(Fe)	Solvothermal method	Cr(VI)	[36]
MIL-53(Fe)/Fe$_3$O$_4$	Solvothermal method	RhB; PNP	[37]
MIL-125(Ti)-ZnO MIL-53(Al)-ZnO	Hydrothermal/Calcination method	MO	[38]
AgI/MIL-53(Fe)	Solution method	RhB	[39]

(Table 1) cont.....

Composite	Synthesis Strategy	Pollutant	Ref.
Ag_3PO_4/MIL-53(Fe)	*in situ* precipitation strategy	TC; OTC; CTC; deoxytetracycline	[40]
Ag_2S/MIL-53(Fe)	Solvothermal method	TC; RhB	[41]
BiOBr/MIL-53(Fe)	Co-precipitation method	RhB; CBZ	[2]
Resin/MIL-53(Fe)	Immobilization process	RhB; MB; org II; sulphorhodamine	[42]
1T-MoS_2@MIL-53(Fe)	One-pot solvothermal approach	ibuprofen	[43]
$Ti_3C_2T_x$/MIL-53(Fe)	*in-situ* synthesis method	TC	[44]
CdS/MIL-53(Fe)	Solvothermal method	RhB	[45]
MIL-53(Fe)/CQDs	*in-situ* generation process	Cr(VI)	[46]
Co-doped MIL-53-NH_2	Step-by-step assembly strategy	BPA; ofloxacin	[47]
MIL-53(Al)@SiO_2	Solvothermal method	BPA	[48]
Bi_2WO_6/MIL-53(Al)/PVDF	Hydrothermal process combined with immersion phase inversion method	RhB	[49]
$BiVO_4$/MIL-53(Fe)/GO	Self-assembly method	RhB	[50]
In_2S_3/MIL-100(Fe)	Solvothermal method	TC	[51]
GO@MIL-101(Fe)	Hydrothermal method	tris(2-chloroethyl) phosphate	[52]
CdS-MIL-100(Fe)	Solvothermal method	benzyl alcohol	[53]
PCN/MIL-100(Fe)	Coating strategy	RhB; MB	[54]
CNNSs-MIL-100(Fe)	*in-situ* synthesis method	RhB	[55]
MIL-100(Fe)/g-C_3N_4	Ball-milling and annealing method	Cr(VI)	[56]
M.MIL-100(Fe)@ZnO	*in-situ* self-assembly method	phenol; BPA; atrazine	[57]
Fe_3O_4@MIL-100(Fe)	Assembly method	MB	[58]
α-Fe_2O_3/MIL-101(Cr)	Hydrothermal method	CBZ	[59]
TiO_2/MIL-100(Fe)	Two-pot hydrothermal method	MB	[60]
MIL-100(Fe)/TiO_2	Hydrothermal method	MB	[61]
TiO_2/MIL-101	Solvothermal method	MB; RhB; CV	[62]
TiO_2/mag-MIL-101(Cr)	Reduction-precipitation route	BPF; acid red 1	[63]
Pd/TiO_2@MIL-101(Cr)	MOCVD and thermally decomposition	RhB	[64]
TiO_2@NH_2-MIL-88B(Fe)	Solvothermal method	MB	[65]
$Bi_{12}O_{17}Cl_2$/MIL-100(Fe)	Ball-milling method	BPA	[66]
Ag_3PO_4/MIL-101/$NiFe_2O_4$	*in situ* precipitation method	RhB	[67]

(Table 1) cont.....

Composite	Synthesis Strategy	Pollutant	Ref.
Ce-doped MIL-101-NH$_2$/Ag$_3$PO$_4$	*in-situ* ion-exchange deposition method	MB; MO; RhB; TC	[68]
Ag@AgCl@MIL-100(Fe)/CCF	Layer-by-layer loading strategy	MB; RhB	[69]
Ag$_3$PO$_4$/Fe-MIL-88-NH$_2$	*in situ* solution method	RhB	[70]
MIL-88A/g-C$_3$N$_4$	Hydrothermal method	RhB; TC; phenol	[71]
BiOBr/MIL-88B(Fe)	Simple method	RhB	[72]
MIL-88A(Fe)@ZnIn$_2$S$_4$	Low-temperature solvothermal method	Cr(VI); sulfamethoxazole	[73]
GO/MIL-88A(Fe)	Vacuum filtration method	MB; BPA	[74]
MIL-68(In)-NH$_2$/GrO	Solvothermal method	amoxicillin	[75]
g-C$_3$N$_4$/MIL-68(In)-NH$_2$	Solvothermal synthesis assisted with ultrasonication	ibuprofen	[76]
Cu$_2$O/MIL(Fe/Cu)	*in-situ* Cu-bridging strategy	thiacloprid	[77]

Table 2. Construction strategies of composites based on UiO-66 and their photocatalytic properties in water purification.

Composite	Synthesis Strategy	Pollutant	Ref.
BiOBr/UiO-66	*in-situ* growth method	atrazine	[95]
BiOCl/UiO-66	Hydrothermal method	RhB	[96]
BiOBr/UiO-66-NH$_2$	Simple synthetic route in the ambient air	norfloxacin	[97]
NH$_2$-UiO-66/BiOBr	Solvothermal method	TC; Cr(VI)	[98]
UiO-66/GO	Hydrothermal method	DCF; CBZ	[99]
UiO-67/CdS/rGO	Solvothermal method	ofloxacin	[100]
AgI/UiO-66(NH$_2$)	Ion exchange precipitation method	TC	[101]
Ag$_2$CO$_3$/UiO-66(Zr)	Convenient solution method	RhB	[8]
UiO-66-NH$_2$/Ag$_2$CO$_3$	Ion-exchange-solution method	Cr(VI)	[102]
Ag$_2$CO$_3$@UiO-66-NH$_2$/GO	Vacuum filtration self-assembly	MB; RB; MR; Cr(VI)	[103]
Ag/Ag$_3$PO$_4$/UiO-66	Chemical bath deposition method	RhB	[104]
UiO-66-NH$_2$/Ag/AgCl	Precipitation-photoreduction method	RhB	[105]
UiO-66-NH$_2$/Ag/AgCl	UV irradiation method at room temperature	RhB; *p*-chlorophenol	[106]
UiO-66@ZnO/GO UiO-66@TiO$_2$/GO	Ultrasonic-solvothermal technique	TC; MA	[107]
UiO-66@WO$_3$/GO	Solvothermal method	TC; MA	[108]
TiO$_2$@NH$_2$-UiO-66	One-step solvothermal technique	TC	[109]

(Table 2) cont.....

Composite	Synthesis Strategy	Pollutant	Ref.
ZnO@NH$_2$-UiO-66	Post-synthetic modification method	TC	[110]
UiO-66/BiFeO$_3$	One-pot solvothermal method	RhB; MO	[111]
BiOI/BiFeO$_3$/UiO-66(Zr/Ti)	Deposition technique	urea	[112]
UiO-66/Cu$_x$S	Surface functionalization method	RhB	[113]
g-C$_3$N$_4$/UiO-66	Solvothermal method	RhB	[114]
Cd$_{0.5}$Zn$_{0.5}$S@UiO-66@g-C$_3$N$_4$	*in situ* hydrothermal and precipitation approach	MO	[115]
Bi$_2$WO$_6$/UiO-66	Hydrothermal method	RhB	[116]
UiO-66/CdIn$_2$S$_4$	Hydrothermal method	triclosan	[117]
ZnIn$_2$S$_4$/UiO-66-(SH)$_2$	Solvothermal method	RhB	[118]
UiO-66-(COOH)$_2$/MoS$_2$/ZnIn$_2$S$_4$	One-step hydrothermal method	Cr(VI)	[119]
CoTiO$_3$/UiO-66-NH$_2$	Hydrothermal method	norfloxacin	[120]
2-MI/UiO-66	*in-situ* one-pot solvothermal method	RhB	[121]
CdS/UiO-66-NH$_2$	Water bath deposition	RhB	[122]
perylene imide-NH$_2$-UiO-66	Solvothermal method	TC	[123]
UiO-66-NH$_2$@TpMA	Step-by-step assembly method	BPA	[124]

Table 3. Construction strategies of composites based on ZIFs and their photocatalytic properties in water purification.

Composite	Synthesis Strategy	Pollutant	Ref.
ZIF-8/C$_3$N$_4$ aerogels	Sol−gel process	CR; MB	[78]
ZIF-8/g-C$_3$N$_4$	Thermodynamic method	U(VI)	[79]
WO$_3$@ZIF-8	*in-situ* encapsulation	MB	[80]
SnO$_2$@ZIF-8	*in-situ* encapsulation	MB	[81]
MAPbBr$_3$@ZIF-8	Pore-encapsulated solvent-directed (PSD) approach	MO	[82]
ZIF-8@BiVO$_4$	*in-situ* growth method	MB	[83]
Bi$_2$MoO$_6$/ZIF-8	Self-assembly process	MB	[84]
MoS$_2$/ZIF-8	Solvothermal method	ciprofloxacin; TC	[85]
Cd$_{0.5}$Zn$_{0.5}$S@ZIF-8	Self-assembly process	Cr(VI)	[86]
CuInS$_2$@ZIF-8	Encapsulation procedure	RhB	[87]
CQDs@ZIF-8	Impregnation approach	MB	[88]
Ag@ZIF-8	UV irradiation method	RhB; MO	[89]
Ag$_3$VO$_4$/ZIF-8	Two-step method	RhB	[90]

(Table 3) cont.....

Composite	Synthesis Strategy	Pollutant	Ref.
Ag/AgCl/ZIF-8	Two-step approach	RhB	[91]
Fe_3O_4-COOH@ZIF-8/Ag/Ag_3PO_4	Three-step approach	diazinon	[92]
ZIF-67/red phosphorus (RP)	Hydrothermal method	Cr(VI); RhB	[93]
ZIF-67/AgCl/Ag	Three-step approach	MO	[94]

Table 4. Construction strategies of composites based on other MOFs and their photocatalytic properties in water purification.

Composite	Synthesis Strategy	Pollutant	Ref.
BUC-21/TNTs	Ball-milling method	Cr(VI)	[125]
BUC-21/N-$K_2Ti_4O_9$	Ball-milling method	Cr(VI)	[126]
BiOBr/GO/MOF-5	Ultrasonic assisted solvothermal method	RhB	[127]
MOF-5@rGO	One-step hydrothermal method	MB; MO; RhB	[128]
TiO_2@HKUST-1	Single-step hydrothermal method	MB	[129]
HKUST-1-$BiVO_4$	Ultrasound-assisted hydrothermal process	disulfine blue; rose bengal	[130]
$Cu_2(OH)PO_4$-HKUST-1	Sonophotocatalytic technique	abamectin	[131]
Ag_3PO_4/Bi_2S_3-HKUST-1	Sonochemical assisted solvothermal method	TB; VS	[132]
Bi_2O_3/Cu-MOF/GO	Self-assembly method	RhB	[133]
CdTe/NTU-9	Solvothermal method	Rh 6G	[134]
Co/Ni-MOF/$BiFeO_3$	Immobilization method	MO; 4-NP	[5]
CNTs/MOF-808	Solvothermal method	CBZ; diazinon	[135]
PCN-222-PW_{12}/TiO_2	One step solvothermal method	RhB; ofloxacin	[136]
Nd-MOF/GO/Fe_3O_4	Ultrasonic vibration process	MB	[137]

2.1. The Solvothermal Method

More recently, the facile *in situ* solvothermal method has been employed to fabricate MOF-based composites. For instance, Rodríguez *et al.* synthesized a composite material containing potassium poly(heptazine imide) (PHIK) and MIL-125-NH_2. The obtained composites were prepared *in situ* by simply dispersing both single materials in water and showed good photocatalytic activity. The diffusion of K^+ from PHIK to MIL-125-NH_2 and the electrostatic interaction were responsible for the generation of the complex [15]. In addition, Xue *et al.* also adopted the *in-situ* growth method to obtain the BiOBr/UiO-66 composite. Firstly, a modified solvothermal approach was adopted to obtain pure MOF. The composites with different molar ratios of Bi: Zr (Bi: Zr = 2: 1; 2.5: 1; 3: 1; 3.5: 1) were obtained according to the as-prepared UiO-66. It was observed that UiO-66 particles were scattered in the BiOBr flakes. In addition, due to the presence of UiO-66 nanoparticles, the agglomeration of these BiOBr flakes was reduced, which can increase the surface area. Therefore, compared with pure BiOBr, the composites showed significantly excellent removal capacity for atrazine with the help of visible light irradiation [95]. Apart from the methods mentioned above, the one-step solvothermal process has also been devoted to forming MOF-based composites. Li *et al.* immobilized TiO_2 nanoparticles on NH_2-MIL-88B(Fe) by this method, which improved the adsorption ability and photocatalytic capacity of the composite material TiO_2@NH_2-MIL-88B(Fe) (abbreviated as SU-x). They used the ability of the composite to remove methylene blue (MB) as a criterion to evaluate its ability of adsorption and photocatalysis. As the adsorption performance was effectively improved, the photogenerated e^- was transferred from MOF to the TiO_2 conduction band, and the catalytic ability was greatly improved. The synergy between TiO_2 and MOF played a crucial role in removing MB [65].

2.2. The Hydrothermal Method

The hydrothermal method has also been reported to generate MOF-based composites. For example, the Z-scheme α-Fe_2O_3/MIL-101(Cr) hybrids were obtained *via* the hydrothermal method by the Huo research group. Firstly, the pre-synthetic MOF was decentralized into deionized water to form a mixture. Then $FeCl_3 \cdot 6H_2O$ was added to the above-mentioned mixture, and the hybrids were prepared by hydrothermal reactions. The final result showed that after irradiation, 100.0% of carbamazepine (CBZ) could be completely degraded by prepared hybrid materials [59].

2.3. The Ball-milling Method

Recently, the ball-milling method is also utilized to synthesize MOF-based composites. Wang *et al.* reported the fabrication of BUC-21/titanate nanotube composites through the ball-milling method. Firstly, pure BUC-21 and titanate nanotubes (TNTs) were prepared with the help of the hydrothermal method, respectively. Then the two materials were mixed to form a mixture. The composites were acquired *via* ball-milling of the mixture under certain conditions. Finally, the photodegradation capacity of composite materials for Cr(VI) was studied under UV light [125]. Later, his research group also used the ball-milling method to prepare BUC-21/N-$K_2Ti_4O_9$ composite materials. The main preparation strategy was similar to the previous system, and its photocatalytic ability for Cr(VI) was also explored under UV light and white light [126].

2.4. The Ultrasonic-assisted Solvothermal Method

In addition to the solvothermal method, MOF-based composites can also be obtained by means of the ultrasonic method. Chen *et al.* synthesized BiOBr/GO/MOF-5 photocatalyst by using the ultrasonic-assisted solvothermal method. Firstly, three pure materials were prepared. The three materials were then dissolved separately in *N, N*-dimethylformamide (DMF) by ultrasonication. Afterward, DMF solutions containing MOF-5 and GO were added to the BiOBr solution one by one. After mixing evenly, the mixture was moved to the reactor to react at 160 °C for 24 h. Finally, the BiOBr/GO/MOF-5 samples were obtained. In addition, the result turned out that the composite exhibited an outstanding capability of degradaing rhodamine B (RhB) [127].

2.5. The Co-precipitation Method

Apart from the preparation methods mentioned above, the co-precipitation pathway was also recently used to prepare MOF-based composites. Zhu *et al.* prepared BiOBr/NH$_2$-MIL-125(Ti) composite by this method to further improve the synergistic effect of BiOBr flakes and MOFs. Firstly, the pure MOF was obtained by the solvothermal method. Then it was added to an aqueous solution containing KBr to obtain a dispersion with ultrasonication. At the same time, the Bi(NO$_3$)$_3$·5H$_2$O was added to acetic acid to obtain a solution. Finally, the two solutions were mixed and transferred to the reactor for reaction to obtain the composites. The composites showed excellent photocatalytic ability for the removal of RhB [16]. Currently, the Tang research group also adopted this preparation strategy to obtain BiOBr/MIL-53(Fe) composites. It showed good photocatalytic ability for the degradation of two kinds of organic compounds (RhB and CBZ) [2].

3. PHOTOCATALYTIC WATER PURIFICATION

The rapid growth of industrialization has given rise to tons of industrial effluents, most of which contain highly stable carcinogens. Therefore, the effective disposal of these hazardous substances is very necessary. MOF-based composites with fine-tuning frames and photocatalytic properties are ideal materials for the elimination of these contaminants.

Organic Pollutants

3.1. Dyes

Organic dyes are the by-products of many industrial processes, such as printing, dyeing, textile, paper-making, and other processes. The organic dye-containing effluent discharged from these processes is toxic to human beings and organisms. The ability to degrade organic pollutants into less toxic or even harmless compounds is critical for photocatalytic water purification [138]. Recently, MOF-based composites have been applied to remove organic dyes from wastewater by photocatalytic technology (Table **5**). In addition, Fig. (**2**) shows the most common chemical formulas for the dye contaminants in the wastewater.

Table 5. Summary of photocatalytic properties of composites based on MOFs for organic dyes.

Photocatalyst	Reactant	S_{BET} (m^2 g^{-1})	Pore Size (nm)	Photocatalytic Efficiency (%)	Ref.
TiO$_2$/MIL-100(Fe)	MB	461	6.10	99.0	[60]
Nd-MOF/GO/Fe$_3$O$_4$	MB	49	/	95.0	[137]
PHIK/MIL-125-NH$_2$	RhB	185	/	/	[15]
BiOBr/GO/MOF-5	RhB	185	/	92.0	[127]
PCN-222-PW12/TiO$_2$	RhB	312	/	98.5	[136]
Bi$_2$O$_3$/Cu-MOF/GO	RhB	117	4.90	/	[133]
CdTe/NTU-9	Rh 6G	880	/	95.0	[134]
UiO-66/BiFeO$_3$	MO	42	/	88.7	[111]
Ag@ZIF-8	MO	1485	/	100.0	[89]

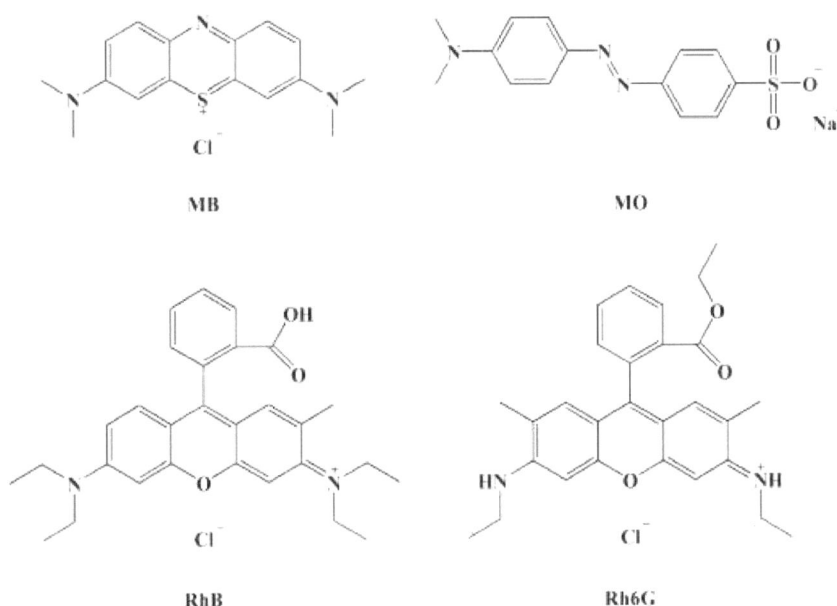

Fig. (2). The chemical structure formulas of common dyes in wastewater.

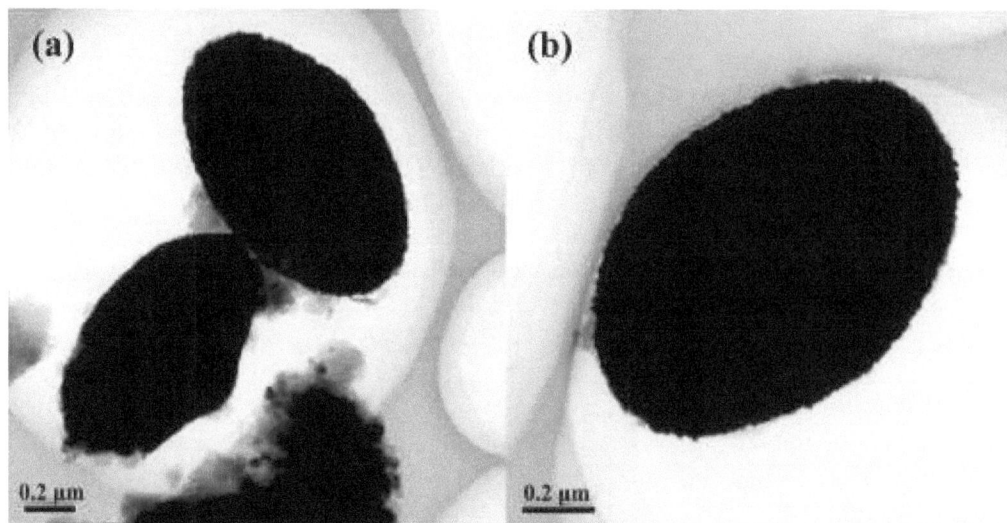

Fig. (3). (**a** and **b**) The TEM images of 30% TiO_2/MIL-100(Fe) under different perspectives. Reproduced with permission from Ref [60]. Copyright 2019, The Royal Society of Chemistry.

With the rapid development of industry, sewage has posed a serious threat to the public and the environment due to its toxic and teratogenic effects. Some researchers have designed and manufactured MOF-based composites to photodegrade organic dyes from wastewater [8]. Wu and co-workers firstly adopted a two-pot hydrothermal method to form an egg-like TiO_2/MIL-100(Fe) composite. The key strategy was adding {001} and {101} co-exposed TiO_2 to a solution of DMF containing Fe^{3+} and terephthalic acid. When the optimal molar ratio of two metal ions (Ti^{4+} and Fe^{3+}) was proved to be 0.3 to 1, the composite showed a complete egg-like (Figs. **3a** and **3b**) according to TEM results. The S_{BET} result of the composite was 461 m^2 g^{-1}. The 30% TiO_2/MIL-100(Fe) can achieve a better photocatalytic ability for MB (99.0%). The higher pore size not only supplied an efficient channel for the elimination of MB but also enhanced the interfacial effect between the two components. The transfer efficiency of the photogenerated e$^-$-h$^+$ was enhanced effectively. The results of these experiments were beneficial to the improvement of photocatalytic performance [60]. As demonstrated by some previously mentioned examples, the inclusion of other components into MOFs also provides a pathway for tuning their photochemical capacity. The ternary composites have recently been used in the study of photocatalytic water purification. Bai *et al.* used the Nd-MOF, GO, and Fe_3O_4 to fabricate a ternary nanocomposite photocatalyst (Nd-MOF/GO/Fe_3O_4). The characterization results showed that the Nd-MOF nanoparticles were integrated with GO flakes through a close interfacial contact. Moreover, many Fe_3O_4 microspheres were found to be uniformly dispersed on the GO surface. The Nd-

MOF/GO/Fe$_3$O$_4$ exhibited unexceptionable photocatalytic capacity in the degradation of MB (95.0%) in the sunlight in the aqueous solution. On the one hand, compared with Nd-MOF-1, the photocatalytic capacity of the composite can be increased by 90.0%, and the excitation wavelength of the composite changed from the UV to the visible-light region. On the other hand, the promotion of photocatalytic property was attributed to the existence of GO, which can promote e$^-$ transfer to some extent, reduce the recombination probability of photoinduced e$^-$-h$^+$ pairs, and improve the light absorption characteristics, thus generating synergistic promotion of the photocatalysis process [137].

Except for treating MB from wastewater, MOF-based composites can also be used to degrade other dyes in wastewater. Recently, for the sake of improving photocatalytic capacity, Rodríguez *et al.* reported PHIK/MIL-125-NH$_2$ composites by building a type I heterostructure. Photodegradation of RhB by the constructed type I heterostructure exhibited a synergy. It was mainly by efficient e$^-$ transfer between the two components. The surface potential analysis of the pure solid showed that the components can be close to each other *via* electrostatic force. The activity of COM50 (k = 0.020 min^{-1}) was twice that of MIL-125-NH$_2$ (k = 0.010 min^{-1}) and seven times that of PHIK (k = 0.003 min^{-1}). The wonderful photocatalytic performance of the composite was illustrated by degrading RhB. In general, the increase in the reaction rate was due to the effective charge separation within the composites [15]. At the same time, Chen *et al.* developed BiOBr/GO/MOF-5 with the help of an ultrasonic-assisted solvothermal process. Using MOF-5 as the substrate, they took advantage of a novel BiOBr catalyst and GO for a new composite. In order to estimate the photocatalytic ability of BiOBr/GO/MOF-5, the obtained samples were applied to remove RhB. The experiment results displayed that the generated composites possessed admirable photocatalytic degradation performance for RhB (92.0%) than the pure BiOBr. Due to the excellent e$^-$ mobility of the composite photocatalyst integrated graphene, it could efficiently accelerate the separation of photogenerated carriers to promote the photodegradation capacity. Besides, MOF-5 has a high S_{BET}, which can effectively disperse the obtained samples. In addition, due to the existence of the BiOBr/GO/MOF-5 sheet-like structure, this structure can supply an effective migration path for the reactants, thus providing more reactive active sites for the photocatalytic reaction. The presence of these results was helpful to improve the photocatalytic ability of the composites [127]. For the sake of further improving the photocatalytic degradation of RhB, Li *et al.* fabricated a novel PCN-22--PW12/TiO$_2$ composite *via* a one-step solvothermal method and explored its performance of photocatalytic degradation. The obtained composite material was a type II heterojunction photocatalyst. The result confirmed that 5 wt% PCN-22--PW12/TiO$_2$ showed an admirable ability of degradation, which was 10.69 times more than pure TiO$_2$. Meanwhile, it was also found that the composites had a certain photocatalytic performance to ofloxacin, which was 94.8%. The enhanced performance of obtained composites was attributable to the increased adsorption effect towards organic pollutants, the light absorption in the whole visible region, and the efficient separation of the light-induced carrier [136]. In addition, the Chen research group fabricated Bi_2O_3/Cu-MOF/GO (BCG) composites and evaluated their photocatalytic effect on RhB by using visible light. As expected, the prepared Bi_2O_3/Cu-MOF/GO composites exhibited the distinct ability of photodegradation of RhB. The introduction of MOF and GO into semiconductor Bi_2O_3 significantly improved the photocatalytic activity of Bi_2O_3, and BCG-2 had the best photodegradation efficiency. The enhanced photodegradation efficiency was due to that GO-containing Bi_2O_3/Cu-MOF/GO composites had excellent conductivity and MOF exhibited certain adsorption during the photodegradation process. The emergence of such synergy was conducive to photoinduced carrier transfer and separation [133].

In addition to the common MB and RhB molecules in wastewater, the MOF-based composites can also treat other dye molecules in wastewater. Kaur *et al.* mixed CdTe quantum dots (QDs) into the precursors (titanium isopropoxide $(Ti(i-OPr)_4)$ and 2,5-dihydroxyterephthalic acid (H_4DOBDC)) of NTU-9, to establish CdTe/NTU-9 nanocomposite and assessed its photocatalytic capacity by photodegradation rhodamine 6G (Rh 6G). The dye can be almost completely degraded (95.0%). This was mainly attributed to the large S_{BET} (880 m^2 g^{-1}) and porous properties of the CdTe/NTU-9 prepared. As a photocatalyst driven by visible light, CdTe/NTU-9 composites had better photodegradation kinetics than traditional P-25 TiO_2 [134].

With the help of a one-pot solvothermal method, Bargozideh *et al.* attempted to incorporate UiO-66 with $BiFeO_3$ to develop novel visible-light photocatalysts. The prepared samples were examined for degrading methyl orange (MO) to discuss their photocatalytic capacity. The results revealed that the obtained UiO-66/$BiFeO_3$ nanocomposite showed better photodegradation performance (88.7%) of MO by using visible light when the addition amount of pure MOF was 30 wt%. The improved photodegradation efficiency was attributable not only to the high S_{BET} of UiO-66, which can enlarge the contact area with MO molecule but also to the building of a heterojunction between the two components, so the photogeneration carriers can be used more efficiently [111]. In order to achieve a more complete photocatalytic degradation of MO, some researchers began to study the combination of precious metals and MOF. For instance, Abdi *et al.* adopted the UV irradiation method to synthesize Ag-X@ZIF-8 catalysts. The removal of RhB and MO was used to judge the photocatalytic ability of the prepared catalysts. The result turned out that the Ag-15%@ZIF-8 showed optimal photodegradation efficiency by using visible light. The photodegradation removal rate of MO by Ag-15%@ZIF-8 reached 100.0%. They also found that the obtained Ag-15%@ZIF-8 manifested high stability and reproducibility. The photodegradation ability of the Ag-X@ZIF-8 was superior to the physical mixture, which was mainly due to the heterojunction between the two components in the composite material [89].

3.2. Phenols

Phenols are one class of 129 pollutants listed by the U.S. Environmental Protection Agency (EPA) as priority pollutants that can harm aquatic life reproduction and survival, and they have very harmful effects on health. For example, chronic phenol poisoning in humans can cause symptoms such as headache, vomiting, difficulty swallowing, liver damage, and fainting. At the same time, phenols as environmental pollutants are difficult to remove due to good stability in the aqueous solutions [139]. Recently, the MOF-based composites have been reported for the photodegradation of phenols in aqueous solutions (Table **6**). Besides, Fig. (**4**) shows the chemical formulas for the common phenols in the wastewater.

Table 6. Summary of photocatalytic properties of composites based on MOFs for phenols.

Photocatalyst	Reactant	S_{BET} (m^2 g^{-1})	Pore Size (nm)	Photocatalytic Efficiency (%)	Ref.
M.MIL-100(Fe)@ZnO	phenol	654	35.30	95.0	[57]
MIL-101-NH$_2$@TpMA	BPA	129	/	99.0	[124]
UiO-66-NH$_2$@TpMA		531	/	82.0	
GO/MIL-88A(Fe)	BPA	/	/	97.3	[74]
UiO-66-NH$_2$/Ag/AgCl	4-CP	979	/	50.0	[106]
NH$_2$-MIL-125(Ti)@Bi$_2$MoO$_6$	DCL	88	8.30	93.3	[27]
	TCL			92.2	

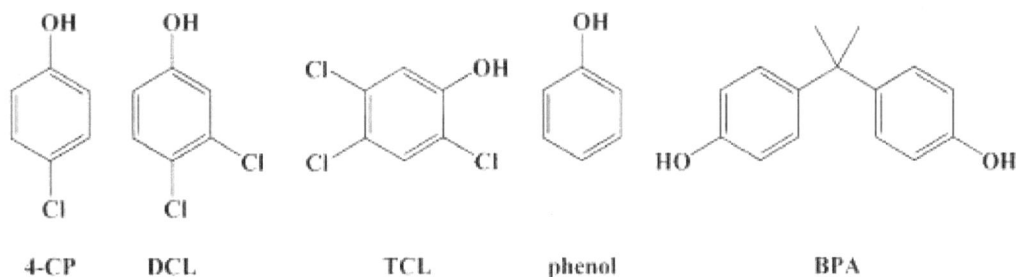

Fig. (4). The chemical structure formulas of common phenols in wastewater.

Phenols consist of aromatic compounds with one or more hydroxyl groups attached to the aromatic ring. Phenolic compounds and their derivatives have attracted more and more international attention [4]. Ahmad *et al.* adopted an *in-situ* self-assembly method to synthesize mesoporous MIL-100(Fe)@ZnO composites. The obtained composite presented a better removal effect for the photodegradation of phenol (95.0%). Additionally, the removal capacity of mesoporous MIL-100(Fe) was better compared with pure microporous MOF. This result may be due to the existence of the mesocellular structure, and the presence of this structure was conducive to improving molecular diffusion and accessibility in the MOF channels. The separation efficiency of photogenerated carriers and photo-Fenton property of mesoporous MIL-100(Fe) could further be improved *via* adding ZnO and H_2O_2, which was of great help to enhance the photocatalytic degradation ability of the MIL-100(Fe)@ZnO composites. In addition, the Fe in mesoporous MIL-100(Fe) can accelerate the decomposition of H_2O_2, resulting in the generation of more ·OH radicals, which contributes to the enhancement of photo-Fenton activity [57].

In a recent study, Lv *et al.* fabricated two novel MOFs@COFs hybrid materials (MIL-101-NH$_2$@TpMA and UiO-66-NH$_2$@TpMA) in virtue of step-by-step assembly method for the first time. And then the hybrids were employed for the removal of bisphenol A (BPA) by photodegradation with the help of visible light to appraise their photocatalytic capacity. The results of the detailed experimental analysis showed that the obtained composites can greatly enlarge the range of the visible-light absorption, and the heterojunction produced on the interface can effectively promote the migration rate of photo-induced carriers, and thus promote photocatalytic degradation performance for BPA. Importantly, the obtained solar/MOFs@COFs/PS system demonstrated a good removal capacity for BPA, eventually achieving unexceptionable photodegradation efficiencies of 99.0% over MIL-101-NH$_2$@TpMA and 82.0% over UiO-66-NH$_2$@TpMA, respectively [124].

To tackle the problem of low photocatalytic performance, some research groups began to prepare new MOF-based composites. Xie *et al.* firstly designed and prepared a GO/MIL-88A(Fe) membrane through the vacuum filtration method. The GO/M88A membrane was formed by inserting M88A into GO nanosheets. In addition, they used the prepared membrane for photodegradation of MB and BPA. The experimental results showed that the membrane exhibited high photo-Fenton catalytic degradation efficiency for both MB and BPA, reaching 98.8% and 97.3%, respectively. By inserting M88A with photo-Fenton catalysis into GO nanosheets, on the one hand, two-dimensional nano-channels can be regulated, and on the other hand, the prepared film can also have photo-Fenton catalytic activity. Importantly, the GO nanosheets could expose more catalytic active sites, thus boosting the photocatalytic ability. A possible photocatalytic mechanism of the GO/M88A membrane was also analyzed [74].

Other than the photocatalytic removal of BPA, MOF-based composites can also remove other phenolic organics. More recently, Jin *et al.* successfully formed UiO-66-NH$_2$/Ag/AgCl by deposition of Ag/AgCl onto defect-engineered MOF. The surface plasmon resonance (SPR) effect and heterostructure can appear in the composite system by using this method. And this ternary strategy was applied to get rid of RhB and *p*-chlorophenol (4-CP) to judge the photocatalytic activities. Experimental results confirmed that the appearance of structural connection defects on MOF was beneficial to the formation of metal Ag and enhanced the interaction between Ag and AgCl. Thus, the existence of the Ag SPR effect was confirmed. In addition, the experimental results also showed that the height of the Schottky barrier of Ag/AgCl was reduced. These results were helpful to improve the separation efficiency of photogenerated carriers. The photodegradation performance test verified that the UA-40 possessed a high photodegradation ability for RhB (99.0%) and degradation of 4-CP (50.0%). The high photodegradation rate might be generated by the Ag plasmons that extended the visible-light absorption and the heterostructure as well as the structural defects that promoted the separation of e$^-$-h$^+$ [106].

In addition, Zhang *et al.* recently utilized the solvothermal method to prepare mesoporous NH$_2$-MIL-125(Ti)@Bi$_2$MoO$_6$ core–shell heterojunctions. They demonstrated that the prepared heterojunction had surface defects. The MOF and Bi$_2$MoO$_6$ were served as core and shell in the composites. For purpose of assessing the photodegradation capacity of obtained samples, they selected two organic contaminants (dichlorophenol and trichlorophenols) in water as substrates. The 0.32-NH$_2$-MIL-125(Ti)@Bi$_2$MoO$_6$ manifested the optimal photodegradation ability. The efficiency of photocatalytic degradation of highly toxic dichlorophenol (DCL) and trichlorophenol (TCL) was 93.3% and 92.3%, respectively. The rate constants of the photocatalytic degradation of highly toxic DCL and TCL were 8 and 17 times higher than that of the original MOF, respectively. The high photodegradation removal ability can be attributed to pure BiMO and TiM lack of active points such as oxygen vacancy, h$^+$, and Ti^{3+}, and a higher content of BiMO can enlarge the pores of TiM, block photon transmission channels, reduce the lifetime of a photogenerated e$^-$, and facilitate the recombination of e$^-$-h$^+$ [27].

3.3. PPCPs

As a new type of pollutants, PPCPs including antibiotics, organic substances in cosmetics, food additives, hair dyes, and microbicides, have been paid more and more attention, because the overuse of PPCPs can cause serious water pollution problems. The PPCPs are ubiquitous in surface water, groundwater, and even some domestic water, causing a potential impact on human health [140]. Currently, the MOF-based composites have been employed for photocatalytic degradation of PPCPs in aqueous solutions (Table 7). The chemical formulas for the selected PPCPs contaminants in the wastewater are provided in Fig. (5).

Table 7. Summary of photocatalytic performance of composites based on MOFs for PPCPs.

Photocatalyst	Reactant	S_{BET} (m² g⁻¹)	Pore Size (nm)	Photocatalytic Efficiency (%)	Ref.
Zr-NH$_2$-MIL-125(Ti)	ACE	1036	/	/	[31]
MIL-100(Fe)/TiO$_2$	TC	307	/	85.8	[61]
α-Fe$_2$O$_3$/MIL-101(Cr)	CBZ	949	/	100.0	[59]
MIL-68(In)-NH$_2$/GO	AMX	681	0.50; 1.60	93.0	[75]
UiO-66/CdIn$_2$S$_4$	TCS	311	5.92	92.0	[117]

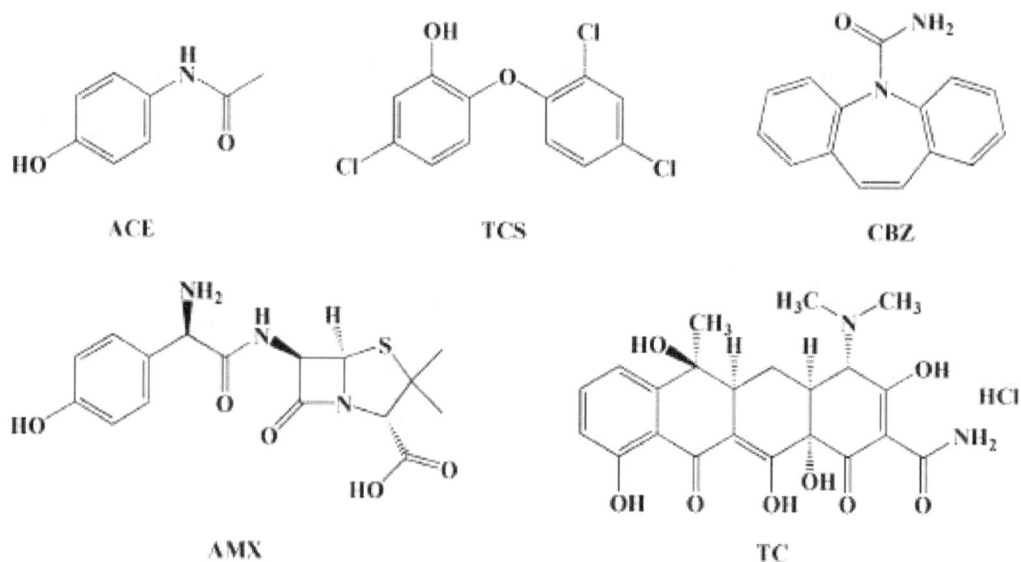

Fig. (5). The chemical structure formulas of common PPCPs in wastewater.

After being applied to animals, most antibiotics are discharged in the form of antibiotic active agents or metabolites through animal excretions [141]. Because it is extremely difficult for microorganisms to degrade antibiotics, it causes antibiotic pollution in surface water, soil, sediment, and groundwater and influences the normal life activities of mankind, mammal, plants, and microorganisms. The long-term existence of PPCPs in a water environment will be likely to cause the generation of antibiotic-resistant genes and antibiotic-resistant bacteria and accelerate the spread of antibiotic resistance. It will not only lead to a vicious circle of antibiotic abuse, interfere with the distribution of bacterial communities, and ultimately threaten the safety of the ecosystem and endanger human life and health [142]. Antibiotic resistance gene is a new type of environmental pollutant, and its risk is more serious and complex than the antibiotic itself. Therefore, how to effectively degrade antibiotics in the environment, especially in water, has gradually gained global attention. Traditional wastewater processing techniques, such as biodegradation, physical adsorption, and chemical precipitation, are the transfer of pollutants and have certain limitations on the removal efficiency of low concentration antibiotics [143]. However, using semiconductor photocatalytic materials to degrade antibiotic wastewater can not only solve the defect of low degradation efficiency of traditional degradation methods but also reduce the degradation cost and shorten the reaction cycle [144]. The modified photocatalytic composites have better photo-responsiveness, reutilization rate, and photocatalytic degradation rate. For instance, the mixed Zr-MIL-125(Ti)-NH$_2$ has been synthesized by the Avilés research group and then was served as a photocatalyst to remove acetaminophen (ACE) under visible light. When the proportion of Zr was low or medium, the crystal structure of the composites was the same as that of NH$_2$-MIL-125, but as Zr^{4+} replaced more Ti^{4+}, the average crystal size of the composites increased and the crystal cell expanded. They showed high S_{BET} values and a basic microporous structure. Crystalline TiZr15 had the highest photocatalytic degradation activity of solar energy for ACE, even higher than non-zirconium MOF, which enabled ACE to complete conversion [31].

In recent years, photocatalytic composite materials prepared with MOF and metal oxides for water purification have also attracted widespread attention. He *et al.* employed surface-coated FeOOH as the precursor to fabricate MIL-100(Fe)/TiO$_2$ composite. The results indicated that MOF grew on TiO$_2$ (Fig. **6**). The obtained MIL-100(Fe)/TiO$_2$ had an enhanced photodegradation capability for tetracycline (TC) over the original TiO$_2$. The degradation efficiency of M/P-10 was improved and reached 85.8%, this was mainly due to enhanced charge separation, which was proved by transient absorption spectroscopy. Part of this enhancement was due to the special chemical structure of the interface, where the Fe–O–Ti bond was produced. Density functional theory calculations attested that this unique structure will produce defect levels near the maximum value of the TiO$_2$ valence band. The UV–vis spectra of TC degraded by M/P-10 were studied. In the process of photocatalysis, the defect level acted as a pool to capture excited carriers and hindered recombination, thus increasing the charge density and improving the photodegradation capacity. At the same time, the close interaction between two components also facilitated the transfer of photoinduced h^+ to Fe–O clusters *via* ligands, thus achieving charge separation [61]. Moreover, Huo *et al*. constructed direct Z-scheme α-Fe_2O_3/MIL-101(Cr) hybrids *via* the hydrothermal method. The obtained samples were applied for degrading CBZ to estimate the photocatalytic activities of the hybrids. The results showed that the removal efficiency of CBZ can reach 100.0% by using the best α-Fe_2O_3(0.3)/MIL-101(Cr). The raised photocatalytic degradation ability may be attributable to the formation of a direct Z-type structure and the good structural properties of the hybrid compounds. In addition, the hybrid material has high efficiency and good reusability in converting CBZ into intermediate products [59].

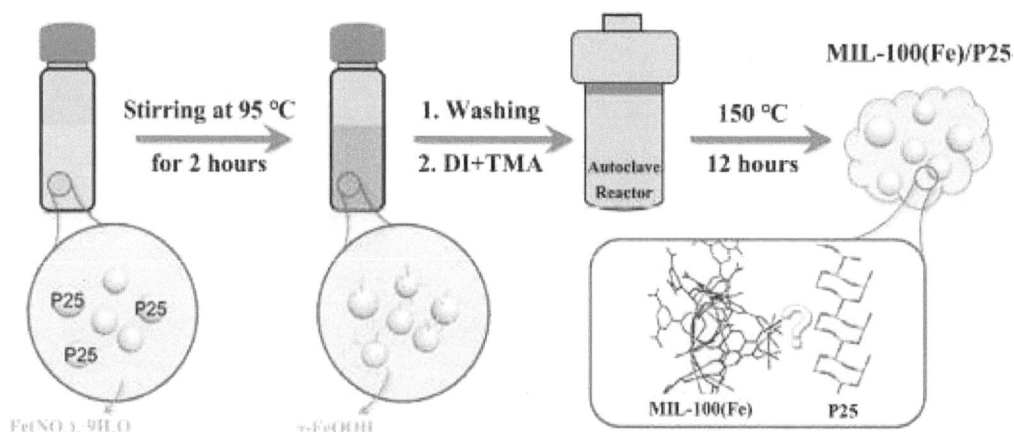

Fig. (6). The preparation process diagram of MIL-100(Fe)/P25. Reproduced with permission from Ref [61]. Copyright 2019, The American Chemical Society.

Recently, Yang *et al*. fabricated MIL-68(In)-NH_2/GO composite and used it as a photocatalyst for the photodegradation of amoxicillin (AMX) under visible light. Compared with pure MOF, the prepared photocatalyst had better photodegradation efficiency. The degradation rate of AMX was 93.0% and the removal rate of TOC was 80.0% after irradiation using 0.6 g L^{-1} prepared photocatalyst at pH 5. This improved photodegradation ability was due to the addition of GO, which can not only be used as an e^- transporter to suppress the recombination of light carriers, but also can be used as a sensitization agent to improve the absorption of visible light. Experimental results confirmed that by altering the pH value of the AMX solution, the photodegradation efficiency of the photocatalyst will also change. The photodegradation of AMX was also affected to some extent. In addition, MIL-68(In)-NH_2/GO also showed good repeatability and stability [75].

The design of MOF-based composites with heterojunction structures can improve their photocatalytic degradation ability. Bariki group has developed UiO-66/CdIn$_2$S$_4$ heterojunction nanocomposite with the help of the hydrothermal method. The evenly dispersed spherical UiO-66 particles (20–40 nm) were anchored CdIn$_2$S$_4$ nanosheets with a high aspect ratio. The characterizations of the prepared nanocomposites manifested the presence of a layered three-dimensional micro-flowered structure with the improved active site of the reaction, better carrier channeling, reduced recombination of photogenerated carriers, and good bandgap alignment between the two components. The photodegradation of triclosan (TCS) by 30UiO-66/CdIn$_2$S$_4$ was best under visible-light irradiation. The photodegradation efficiency of 30UiO-66/CdIn$_2$S$_4$ on TCS was 92.0%, which was 6 times that of CdIn$_2$S$_4$ (15.0%). The growing photodegradation efficiency of photocatalyst prepared can be attributable to its higher S_{BET} and the formed n-n heterojunction, which improved the migration of photo-generated e$^-$. The UV–vis characterization was applied to explore the photocatalytic degradation ability of materials [117].

3.4. Herbicides and Pesticides

Due to the serious impact of organic micro-pollutants such as pesticides and herbicides on public health, the continuous detection of organic micro-pollutants in the water environment has been paid close attention to. Herbicides and pesticides can be found in streams near agricultural activities and the agrochemical industry. In addition, it increases the likelihood of toxicity in the liver and hematopoietic system. Furthermore, researchers have confirmed most herbicides and pesticides are cytotoxic, immune-toxic, and genotoxic [135]. The MOF-based composites have been recently served as photocatalysts to eliminate herbicides and pesticides in aqueous solutions (Table **8**). Furthermore, the chemical formulas for the herbicides and pesticides in the wastewater are provided in Fig. (**7**).

Table 8. Summary of photocatalytic performance of composites based on MOFs for herbicides and pesticides.

Photocatalyst	Reactant	S_{BET} (m^2 g^{-1})	Pore Size (nm)	Photocatalytic Efficiency (%)	Ref.
BiOBr/UiO-66	atrazine	/	/	88.0	[95]
Cu$_2$(OH)PO$_4$-HKUST-1	abamectin	/	/	99.9	[131]
Cu$_2$O/MIL(Fe/Cu)	thiacloprid	1553	/	/	[77]
Fe$_3$O$_4$-COOH@ZIF-8/Ag/Ag$_3$PO$_4$	diazinon	280	1.85	99.7	[92]
UiO-66@WO$_3$/GO	MA	379	8.94	100.0	[108]

(Table 8) cont.....

Photocatalyst	Reactant	S_{BET} $(m^2 g^{-1})$	Pore Size (nm)	Photocatalytic Efficiency (%)	Ref.
UiO-66@ZnO/GO	MA	627	8.94	100.0	[107]
UiO-66@TiO$_2$/GO		257	8.93	87.0	

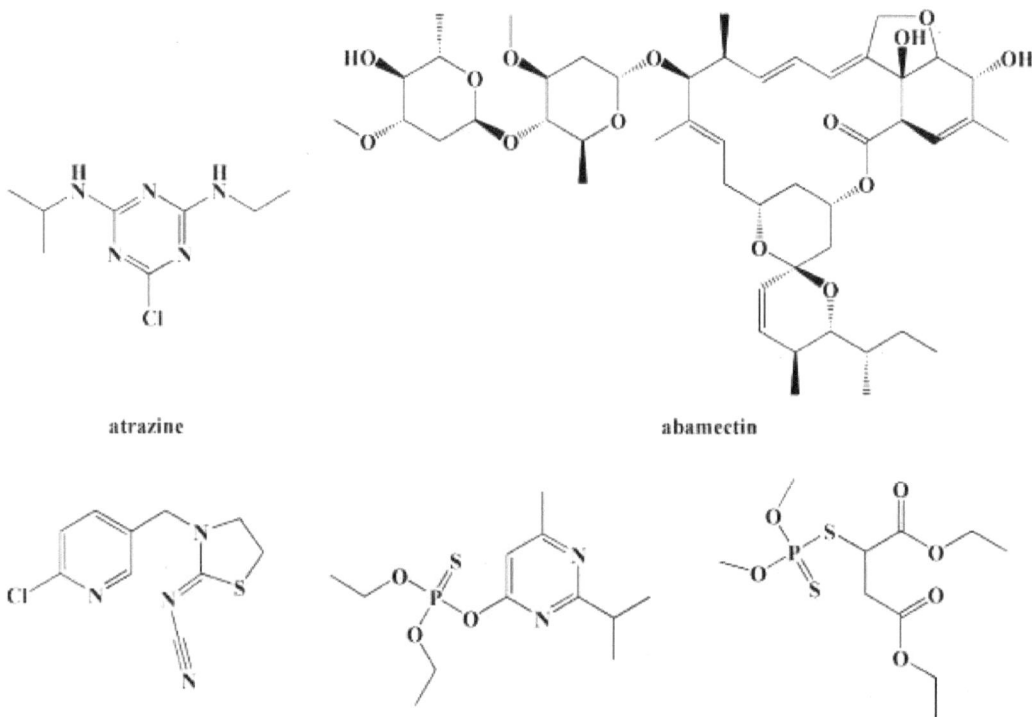

atrazine   abamectin

Fig. (7). The chemical structure formulas of common herbicides and pesticides in wastewater.

The main reason for removing pesticide residues from water is its extremely high toxicity. This can be achieved through mineralization, a process in which organic pesticides release all their inorganic constituents through degradation [145]. Xue *et al.* developed a BiOBr/UiO-66 photocatalyst through an *in-situ* growth method and applied it to the degradation of atrazine. The experimental results confirmed that, compared with pure BiOBr, the prepared photocatalyst exhibited remarkable photocatalytic ability for the removal of atrazine when irradiated with visible light. The optimum ratio was BU-3 (Bi: Zr = 3: 1), displaying a removal effect of 88.0% of atrazine. This may be due to there being a potential interaction between the two single components in the composite, which was conducive to boosting the photodegradation efficiency. In addition, the prepared composites had high photoproduced e$^-$-h$^+$ separation efficiency, which improved the photodegradation and removal ability of atrazine [95].

The removal of abamectin pesticides from water is an important problem in water treatment. Mosleh *et al.* created a visible-driven $Cu_2(OH)PO_4$-HKUST-1 photocatalyst with a visible light response, which was exploited to photodegrade abamectin pesticide with the aid of advanced oxidation processes, mainly containing sonocatalysis and photocatalysis. The characterization results displayed that the direct band gaps of MOF and obtained photocatalysts were calculated as 2.63 and 2.59 eV, respectively, which revealed that they can be activated by using blue light. Under optimized experimental conditions, the degradation efficiency of the abamectin pesticide was the highest (99.9%). This was because of the emergence of synergies between ultrasonic and photocatalysis [131].

Nearly 80% of the pesticides were degraded within a day and the catalyst was shown to be effective by using visible light. Zhong and co-workers constructed Cu_2O/MIL-100(Fe/Cu) composites using the *in-situ* Cu bridging strategy, which could enhance the interfacial synergistic effect of photo-Fenton catalysis, and finally achieved a better degradation efficiency of thiacloprid. The related characterization results exhibited that the highly dispersed photosensitive Cu_2O with enhanced interface could grow well on MIL-100(Fe) with the help of *in situ* doped Cu bridge. The constructed composite showed a conspicuous photodegradation thiacloprid removal rate, which was 7.5 times and 11.0 times faster than pure Cu_2O and MIL-100(Fe), respectively. The TOC removal efficiency reached 82.3%, which was more than 10 times of pure MIL-100(Fe). This tightly bound interface not only took full advantage of their advantages but also promoted the synergistic catalysis of the generation of free radicals by photoinduction of e$^-$ and Fenton. Cu doping into MOF can greatly shorten the bandgap, promote the Fe^{2+}/Fe^{3+} redox reaction, and improve the degree of reversible catalytic redox ability of MIL(Fe). The addition of Cu_2O further increased the light absorption region of MIL(Fe/Cu) and improved the photoelectric conversion efficiency. These advantages promoted the improvement of photocatalytic performance [77].

Diazinon is an organophosphorus pesticide, and it is widely used in agricultural activities to kill various types of insects. But the diazinon can reach humans through the food chain and result in several diseases. More recently, MOF-containing multicomponent photocatalysts have also been used to remove pesticides. For example, Joubani *et al.* fabricated a four-component hetero-structured Fe_3O_4-COOH@ZIF-8/Ag/Ag$_3$PO$_4$ photocatalyst. Firstly, the surface of Fe_3O_4 was modified with carboxylic acid to form Fe_3O_4-COOH. Then, ZIF-8 was induced to grow on its surface using Fe_3O_4-COOH as a template to obtain Fe_3O_4-COOH@ZIF-8 core–shell structures. Then the composites with visible light activity were constructed by using Ag$_3$PO$_4$ and Ag nanoparticles (Ag NPS) as raw materials to combine the prepared Fe_3O_4-COOH@ZIF-8. The prepared quaternary system photocatalyst showed a remarkable effect on the photodegradation of diazinon. Because of the synergy of the SPR of Ag NPs and the continuous energy transfer of the Z-Scheme mechanism, the hetero-structured photocatalyst displayed good activity and stability, which could efficiently prevent the recombination of photogenerated carriers [92]. Besides, Fakhri *et al.* adopted the solvothermal method to construct UiO-66@WO_3/GO (UiO-66@WG) nanocomposite for the first time. The first step was to make WO_3/GO (WG) according to the prepared WO_3 and GO. Afterward, the UiO-66@WG composites were prepared by solvothermal reaction between WG and the precursors of UiO-66. And they selected two representative pollutant models TC and Malathion (MA) to evaluate the capability of photocatalytic degradation. The UiO-66@35WG displayed the best photocatalytic capability for TC (84.0%) and MA (100.0%) by using visible light, respectively. This enhanced photodegradation capability was due to several factors, such as strong response in the visible light region, appropriate matching of CB and VB of MOF and WO_3, resulting in promoted migration of e^--h^+ pairs, and faster photo-produced carrier transfer across the GO layer. Furthermore, this UiO-66@35WG possessed commendable repeatability after four cycles [108]. Later, the research group adopted the same method to form UiO-66@ZnO/GO (UiO-66@ZG) and UiO-66@TiO_2/GO (UiO-66@TG). And these two composites have also been used for eliminating TC and MA under visible light. In the optimum condition, the UiO-66@45 ZG achieved maximum photodegradation efficiency for TC (81.0%) and MA (100.0%), and the UiO-66@45 TG showed a supreme photodegradation efficiency for TC (76.0%) and MA (87.0%). This increase in photocatalytic activity was due to several factors: (1) The Z-scheme was used to achieve the optimal separation of carriers; (2) The excellent electron migration performance of the GO layer; (3) The high surface area. In addition, compared with pure MOF, the band gaps belonging to the heterojunction showed a redshift on the absorption edge, which may lead to the expanded absorption of the solar spectrum [107].

3.5. Other Organics

Except for the photocatalytic removal of the aforementioned common organic pollutants (dyes, phenols, PPCPs, herbicides, and pesticides), other organics have also been explored through the use of MOF-based composites (Table **9**). And the chemical formulas for the other organic contaminants in the wastewater are provided in Fig. (**8**).

Table 9. Summary of photocatalytic performance of MOF-based composites for other organics.

Photocatalyst	Reactant	S_{BET} $(m^2\ g^{-1})$	Pore Size (nm)	Photocatalytic Efficiency (%)	Ref.
CdS-MIL-100(Fe)	benzyl alcohol	327	/	54.3	[53]
NH$_2$-MIL-125(Ti)/TiO$_2$	cyclohexane	571	/	36.0	[21]
GO@MIL-101(Fe)	TCEP	/	/	95.0	[52]
BiOI/BiFeO$_3$/UiO-66(Zr/Ti)	urea	/	/	88.5	[112]

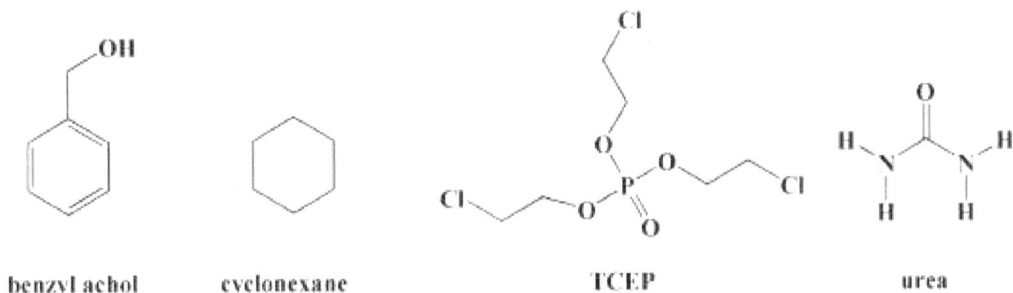

benzyl achol   cyclonexane   TCEP   urea

Fig. (8). The chemical structure formulas of other common organics in wastewater.

For instance, Ke *et al.* created CdS-MIL-100(Fe) photocatalysts by adjusting the amount of MOF added. In the composites, porous MOF was used as a carrier and cadmium acetate was used as a precursor of CdS. The results showed that the prepared photocatalyst can significantly enhance the photocatalytic efficiency for selective oxidation of benzyl alcohol to benzaldehyde at room temperature. In particular, the CdS-MIL-100(Fe)-60 nanocomposite showed the optimal photocatalytic ability and the highest conversion rate (54.3%), and the selectivity to benzaldehyde was about 100.0%. The enhancement of photodecomposition efficiency was mainly attributed to the enhancement of light absorption, which can separate e$^-$-h$^+$ pairs generated by light more effectively, and the addition of MIL-100(Fe) can increase the CdS surface area [53]. In addition, the removal of cyclohexane has also been achieved by a MOF-based composite and was found to be effective. Zhao *et al.* successfully prepared NH$_2$-MIL-125(Ti)/TiO$_2$ composites through the one-pot solvothermal method and utilized them to photocatalytic selective oxidation of cyclohexane. Compared with NH$_2$-MIL-125(Ti), the prepared composite has better catalytic performance. This was due to the energy band coupling between MOF and TiO$_2$ and the tight interfacial contact, which improved the migration of photogenerated carriers and reduced the recombination of e$^-$-h$^+$ pairs. At room temperature and ambient pressure, the activity of the fabricated photocatalyst was 3 times that of MOF through using molecular oxygen as an oxidant [21].

Besides, a light/MOF/H_2O_2 photocatalytic system was presented by the Lin research group based on MIL-101(Fe). Then it was used for photodegradation to remove tris(2-chloroethyl) phosphate (TCEP). The prepared samples revealed narrower bandgap energy (2.17 eV). However, the bandgap of MIL-101(Fe) was 2.41 eV. These results suggested that the visible-light absorption was extended. The 15%GO@MIL-101(Fe) showed the best photocatalytic degradation ability for the removal of TCEP. The TCEP removal rate of the composite was 95.0%, which was higher than that of the pure MOF (50.0%). The prepared composite achieved a faster rapid activation and higher electron mobility because of the high electrical conductivity of GO. The photocatalytic reaction mechanism was that electrons were excited from the HOMO (oxygen orbital in the ligand) to the LUMO (Fe(II)/Fe(III)) in the core node under light excitation [52]. In addition, Bahmani and co-workers fabricated BiOI/BiFeO$_3$/UiO-66(Zr/Ti)-MOF photocatalyst through a facile deposition technique. And the photocatalytic capacity of urea conversion to N_2, H_2O, and CO_2 under blue LED irradiation was studied in a thin film slurry flat plate photocatalytic reactor. And then the central composite design (CCD) and desirability function (DF) were utilized to optimize the reaction parameters. Under the optimum conditions, the photooxidation and removal efficiency of urea reached the highest of 88.5%. Results indicated that adding BiFeO$_3$ and BiOI on the surface of UiO-66 (Zr/Ti) can reduce the recombination of photogenerated carriers, and thus the photodegradation ability was promoted. In addition, suitable energy band matching, photostability, and separation of e$^-$-h$^+$ pair enhanced the decomposition of urea by the obtained samples [112].

3.6. Inorganic Pollutants

Apart from various organic pollutants, there exist hazardous inorganic pollutants in wastewater (Table **10**). The existence of these inorganic pollutants will also cause certain harm to the human body and the water environment.

Table 10. Summary of photocatalytic performance of composites based on MOFs for inorganic pollutants.

Photocatalyst	Reactant	S_{BET} (m^2 g^{-1})	Pore Size (nm)	Photocatalytic Efficiency (%)	Ref.
g-C$_3$N$_4$/MIL-53(Fe)	Cr(VI)	19	1.43; 3.86	/	[36]
Cd$_{0.5}$Zn$_{0.5}$S@ZIF-8	Cr(VI)	174	6.30	100.0	[86]
NH$_2$-UiO-66/BiOBr	Cr(VI)	71	/	88.0	[98]
BUC-21/TNTs	Cr(VI)	/	/	97.6	[125]
BUC-21/N-K$_2$Ti$_4$O$_9$	Cr(VI)	/	/	100.0	[126]
UiO-66-(COOH)$_2$/MoS$_2$/ZnIn$_2$S$_4$	Cr(VI)	140	2.22	98.4	[119]

(Table 10) cont.....

Photocatalyst	Reactant	S_{BET} (m^2 g^{-1})	Pore Size (nm)	Photocatalytic Efficiency (%)	Ref.
ZIF-8/g-C$_3$N$_4$	U(VI)	575	1.16	100.0	[79]

Based on the above research, it is extremely urgent to solve the problem of environmental remediation. Some research groups have established a bunch of novel MOF-based composites. For instance, Huang *et al.* adopted a facile solvothermal method to obtain graphitic carbon nitride (g-C$_3$N$_4$)/MIL-53(Fe) composite and studied their properties of Cr(VI) photoreduction. When the amount of g-C$_3$N$_4$ was 3%, the prepared composites showed the best catalytic efficiency of photoreduction of Cr(VI), which was 2.1 times and 2.0 times higher than that of pure g-C$_3$N$_4$ and MOF, respectively. The related characterization results confirmed that adding specific content of g-C$_3$N$_4$ on MIL-53(Fe) surface can prevent the recombination of photogenerated carriers and increase the migration rate of the photogenerated carrier, thus promoting the improvement of photoreduction efficiency [36].

Recently, researchers have set out to study the Cd-based multiphase catalysts due to their outstanding catalytic ability. Qiu's research group adopted a facile self-assembly strategy to construct Cd$_{0.5}$Zn$_{0.5}$S@ZIF-8 nanocomposite. Compared with pure Cd$_{0.5}$Zn$_{0.5}$S and MOF, the prepared composite showed better photocatalytic activity. In particular, when the amount of ZIF-8 doping was 60 wt%, its photocatalytic capacity was about 1.6 times that of Cd$_{0.5}$Zn$_{0.5}$S. The results of characterization experiments confirmed that the newly formed interfacial Zn–S bond closely connected between two components, and thus significantly improved the migration rate of photoproduced e$^-$ and h$^+$, which will lead to an increase in photocatalytic reduction potential [86].

More recently, Hu *et al.* grew BiOBr nanosheets on NH$_2$-UiO-66 octahedron to design and construct NH$_2$-UiO-66/BiOBr composite heterojunction and evaluated their photo-reductive activity *via* photo-reducing Cr(VI). By regulating and controlling the mass ratio of MOF in composites, the NH$_2$-UiO-66/BiOBr composite with prime photocatalytic performance was obtained. As validated by the experiments, the optimal MOF doping content in the composites was about 15 wt%. After illumination, the NU/BOB-15 exhibited a higher photoreduction performance (88.0%), and the pure BiOBr and MOF reached 56.0% and 34.0%, respectively. The NU/BOB-15 possessed a larger S_{BET}, which enabled the carriers to transfer well between two components. In addition, increased S_{BET} made obtained composites displayed higher adsorption capacity for pollutants. In summary, these results were favorable to the photocatalytic reduction of Cr(VI) [98].

To further address the environmental problems caused by Cr(VI) in the water environment, the Wang research group prepared BUC-21/TNTs through the ball-milling method. And they proved that the obtained samples achieved simultaneous Cr(VI) reduction by employing ultraviolet light and the removal of Cr(III). This was because the introduction of TNTs can capture the Cr(III) formed on the surface of BUC-21, thereby decreasing the local Cr(III) concentration and eliminating the possible precipitation of $Cr(OH)_3$, which thus accelerated the photocatalytic reduction of Cr(VI). The BT-1 composites presented the highest performance (97.6%) compared to that of BUC-21 (92.9%). Through the photoreduction of Cr(VI) and adsorption of Cr(III), the synergy effect between the two components can achieve complete removal of the total Cr [125]. Next, they employed BUC-21 and $N-K_2Ti_4O_9$ to fabricate $BUC-21/N-K_2Ti_4O_9$ composites (B1NX) using a ball-milling process. In addition to studying the effect of reducing Cr(VI) under ultraviolet light, they also studied the effect of reducing Cr(VI) under white light. The obtained B1N0.5 and B1N3 achieved complete removal of Cr(VI). This manifested that the addition of $N-K_2Ti_4O_9$ to MOF not only expanded the absorption range but also increased the mobility of e^- and h^+ [126].

Subsequently, ternary MOF-based composites were also applied to the photoreduction of Cr(VI). Mu *et al.* utilized reflux and hydrothermal methods to fabricate $UiO-66-(COOH)_2/MoS_2/ZnIn_2S_4$ (UMZ) photocatalysts. A large number of characterization results confirmed that $ZnIn_2S_4$ and MoS_2 were in close contact, and the MOF was anchored on $ZnIn_2S_4$ nanosheets. It was found that the $U_{0.30}M_{0.02}Z$ displayed the best photocatalytic Cr(VI), which reached 98.4% upon irradiation. It was probably due to the $U_{0.30}M_{0.02}Z$ possessed a large S_{BET} (139.8 m^2 g^{-1}), thus generating more contact sites to adsorb Cr(VI). The higher S_{BET} was beneficial to accelerating the migration of photoelectric charge, the redox ability between the two components was stronger, and thus the photoreduction capacity can be improved [119].

Additionally, Qiu *et al.* constructed $ZIF-8/g-C_3N_4$ photocatalysts and employed them for the photoreduction of U(VI). The results showed that the prepared composites exhibited a high photocatalytic rate and photocatalytic activity. It still had good activity after 5 times of recycling, which proved its high stability. The quenching experiments and ESR analysis showed that the e^- was responsible for the photoreduction efficiency of U(VI). The efficient removal of U(VI) by prepared photocatalyst was due to the synergistic effect between two components [79].

3.7. Mixed Pollutants

Importantly, MOF-based composites can be utilized not only for the photocatalytic removal of the above-mentioned single pollutant from wastewater but also for the degradation of mixed pollutants (Table **11**). Additionally, the chemical formulas for the involved pollutants in the wastewater are provided in Fig. (**9**).

Table 11. Summary of photocatalytic performance of composites based on MOFs for mixed pollutants.

Photocatalyst	Reactant	S_{BET} $(m^2\,g^{-1})$	Pore Size (nm)	Photocatalytic Efficiency (%)	Ref.
Co/Ni-MOF/BiFeO₃	MO/4-NP	1058	2.20	/	[5]
Ag₃PO₄/Bi₂S₃-HKUST-1	TB/VS	/	/	98.4; 99.4	[132]
TiO₂/mag-MIL-101(Cr)	BPF/AR1	/	/	/	[63]
MIL-125-NH₂@Ag/AgCl	Cr(VI)/RhB/MG	310	/	98.4	[34]

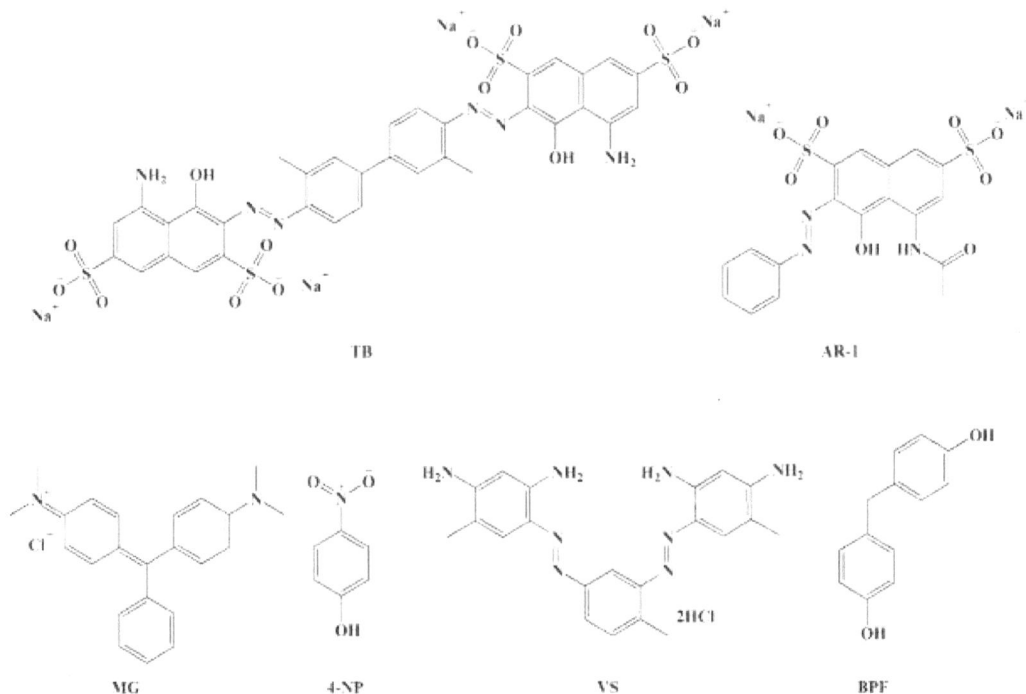

Fig. (9). The chemical structure formula of common mixed pollutants contaminants in wastewater.

Recently, Hamed and co-workers prepared a novel magnetic Co/Ni-MOF/BiFeO₃ composite and applied it to photodegrade MO and 4-nitrophenol (4-NP) from wastewater to elucidate the photocatalytic mechanism. The results of photodegradation experiments manifested that the photodegradation capacity of the prepared samples was improved compared with that of pure MOF and BiFeO$_3$. In the prepared photocatalysts with different molar ratios, the photocatalytic degradation rate of MO and 4-NP reached completion when the molar ratio was 1: 1. The improved photodegradation efficiency can be ascribed to the efficient separation of photogenerated carriers, the emergence of selective pores, high S_{BET}, mass capture of visible light, relatively easy transport of guest molecules, and adsorption capacity [5].

Recently, Mosleh *et al.* reported Ag$_3$PO$_4$/Bi$_2$S$_3$-HKUST-1 composites, as novel visible-light active photocatalysts, and they were devoted to sono-photocatalytic elimination of trypan blue (TB) and vesuvine (VS) in a continuous flow-loop reactor. By optimizing the operating parameters, better photocatalytic removal ability was achieved. Experimental results confirmed that the obtained samples possessed significant synergies in the photodegradation process. Under the optimal conditions, the sono-photodegradation percentage for TB and VS was 98.4% and 99.4%, respectively [132].

In order to further study the adsorption and photocatalytic efficiency of MOF-based composites, Zhang *et al.* synthesized TiO$_2$-supported magnetic MOFs composites (TiO$_2$/mag-MIL-101(Cr)) type adsorptive photocatalyst by means of room temperature reduction-precipitation method. The synthesized TiO$_2$/mag-MIL-101(Cr) was applied to remove bisphenol F (BPF)/acid red 1 (AR1) with the help of adsorption and photocatalysis. About 90.0% of AR1 can be removed by utilizing TiO$_2$/mag-MIL-101(Cr). The BPF removal efficiencies by obtained photocatalyst were both better than AR1. The appearance of these results can be attributed to the high adsorption capacity of MOF and the improved separation efficiency of photo-induced e$^-$ and h$^+$. The bandgap value of the composite material was 1.61 eV because the introduced MOF significantly enhanced the light absorption of TiO$_2$ [63].

Qiu *et al.* exploited a facile deposition-photoreduction method to synthesize MIL-125-NH$_2$@Ag/AgCl photocatalysts firstly and employed them for photodegradation removal of Cr(VI)/RhB/malachite green (MG). When the photocatalyst was irradiated with visible light, the reduction efficiency of Cr(VI) was significantly higher than that of pure MOF as well as Ag/AgCl. When only RhB and MG were used, the reduction capacity of Cr(VI) was significantly improved to 98.4%, which was 3.4 times that of the single system (29.0%), and also 3.4 times that of the binary Cr(VI)/RhB system (69.6%) and Cr(VI)/MG system (67.5%). At the same time, the degradation efficiency of RhB in the ternary pollutant system was also boosted compared with the single system.

According to obtained results, the high efficiency of light absorption and e⁻ transport as well as the synergy effect of redox reaction between Cr(VI) and dye in the composite system prevented the photoinduced recombination of e⁻-h⁺ pairs, leading to its high photodegradation ability [34].

3.8. The Correlation Between Photocatalytic Properties and Physical Properties of MOFs

In order to better design and synthesize MOFs with superior photocatalytic performance, we also discuss the influence of their physical properties on photocatalytic properties. As is known that the electronic properties of MOFs rest with their chemical composition and structure from previous research, that is, the properties of organic ligands and metal ions or metal clusters [146]. Different electronic properties produce different band gap (E_g) values, so the photocatalytic properties of MOFs materials will be diverse. For instance, Kozlova *et al.* studied and analyzed the effect of the nature of metal ions and functional groups in organic ligands on the E_g values and photocatalytic properties of metal-carboxylates, based on experimental DR-UV/vis spectroscopic and catalytic data in their research work. By adjusting the metal ions of the isostructural MIL-100(M) (M = Al, Fe, Cr, V) and the functional groups in organic ligands of isoreticular UiO-66-R (R = H, NH₂, NO₂), the photocatalytic performance was explored. When investigating the influence of metal ions on E_g values, it was found that a larger E_g value ((MIL-100(Al) > MIL-100(Cr) > MIL-100(Fe)≈MIL-100(V)) resulted in worse photocatalytic performance (MIL-100(V)≈MIL-100(Fe) > MIL-100(Cr)). The different band gap values were due to the influence of ion radius and the polarization of the M–O bond. For the effect of functional groups of organic ligands on photocatalytic properties, they employed Hammett analysis to verify the relationship between the E_g values and resonance effects (σ_R) of functional groups. The results showed that there was a linear relationship between the E_g and σ_R, and the specific E_g values order was UiO-66-NH₂ < UiO-66 < UiO-66-NO₂, and the corresponding photocatalytic degradation of MB was UiO-66-NH₂ > UiO-66 > UiO-66-NO₂. The presence of this result was mainly due to the –NH₂ being the electron-donating group, which can activate the aromatic ring by resonance-donating effect increasing the electron density. However, the –NO₂ was an electron-absorbing group, which will reduce the electron density through the resonance-absorbing effect, thus reducing the photocatalytic reaction efficiency [147].

Additionally, Panchenko and co-workers summarized the spectroscopic method (diffusion reflectance ultraviolet-visible spectroscopy) to explore the relationship between electronic properties and photocatalytic performance of MOFs. They noted that the composition of metal clusters had an impact on the magnitude of the E_g value. For MOF-5 and MOF-74, when part of Zn was replaced by Co, the E_g values decreased. Moreover, the adsorption performance for H_2, CO_2, and CH_4 of Co-containing MOF-74 was higher than pure Zn-MOF-74, which was mostly because of the different overlap between the *d*-orbitals of Co and Zn and the *p*-orbitals of O and C atoms. Meanwhile, they also reviewed the influence of functional groups in organic linkers on E_g values. They found that the E_g values of the electron-donating substituents ($-OH$, $-CH_3$, $-Cl$, $-NH_2$) were lower than that of the electron-withdrawing substituents ($-SO_3H$, $-NO_2$). This result was mainly corresponding to Kozlova's study [148].

In summary, it can be found that different metal ions and functional groups will render the E_g values of MOF different, thus affecting its photocatalytic performance.

4. ADSORPTION-PHOTOCATALYSIS SYNERGISTIC WATER PURIFICATION

In addition to preparing MOF-based composites with photocatalytic ability to achieve water purification, some researchers have designed highly efficient composites with both adsorption and photocatalytic degradation capabilities in recent years. This kind of MOF-based composites prepared, on the one hand, can completely remove the organic pollutants in the aqueous solution more efficiently, and on the other hand, it can also save cost (Table **12**).

Table 12. Summary of adsorption-photocatalysis synergistic water purification by composites based on MOFs.

Photocatalyst	Reactant	S_{BET} $(m^2\,g^{-1})$	Pore Size (nm)	Photocatalytic Efficiency (%)	Ref.
TiO_2-MIL-101	MB	531	1.34	/	[62]
TiO_2@NH_2-MIL-88B(Fe)	MB	24	/	100.0	[65]
MIL-53(Al)@SiO_2	BPA	668	0.18	85.0	[48]
TiO_2@HKUST-1	MB	982	2.10	91.0	[129]

Recently, the Chang research group fabricated TiO_2-MIL-101 by using the solvothermal method. MIL-101 was used as a carrier and TiO_2 as a layer to prepare the photocatalyst. The synergistic mechanism of effective adsorption and photodegradation of MB with the prepared nanocomposites was explored. The results showed that the nanocomposite exhibited excellent porosity due to the addition of the MIL-101 matrix, and the TiO_2 crystals were coated on the surface of MOF without agglomeration. Compared with pure large MOF, the prepared nanocomposites had higher adsorption affinity for MB, which was mainly due to the fact that the TiO$_2$ crystal of the coating played a regulating role on the surface charge. According to the experiments, the adsorption of MB reached 59.9% onto nanocomposite, and the adsorption of MB on MOF could reach 42.6%. After reaching the adsorption equilibrium, photocatalytic experiments were performed by irradiation with ultraviolet light. To confirm the capacity of prepared nanocomposites for adsorption and decomposition, they also employed nanocomposites to eliminate RhB and crystal violet (CV). The enhancement of the removal ability was mainly due to the regulating effect of the introduced TiO$_2$ on the surface charge of the nanocomposites [62].

Subsequently, Li and co-workers reported TiO$_2$@NH$_2$-MIL-88B(Fe) heterostructures *via* a one-pot strategy. The obtained samples showed good adsorption affinity, which could improve the photodegradation activity of MB. The existence of synergy effect (adsorption and photocatalytic decomposition) in SU-3 could completely remove MB. In addition, the UV–vis absorption spectra of SU-3 photocatalyzed absorption of MB were studied. The recycling experiment proved that SU-3 was stable and recyclable. The increase in adsorption capacity was mainly due to the electrostatic interaction between dye molecules and prepared samples offered by the negative charge on the photocatalyst surface. Importantly, the existence of a heterojunction between two components provided a powerful platform for rapid interfacial photo-induced carrier migration and improved photodegradation efficiency. Finally, a possible mechanism for the removal of MB by SU-3 was proposed [65].

In order to better reduce the cost of synthesis of MOF-based composite materials, the Chatterjee research group synthesized MIL-53(Al)@SiO$_2$ composites employing coal fly ash and waste aluminum foil (with the best ratio) as raw materials through the solvothermal method, which was effectively used for BPA removal, including adsorption and photocatalytic degradation. The related characterization results demonstrated that the introduction of SiO$_2$ can reduce the possibility of h$^+$ and e$^-$ recombination. The maximum adsorption amount of the prepared sample for BPA could reach 134.68 mg g^{-1}. By means of photocatalytic treatment, 85.0% BPA could be removed [48].

An effective photocatalytic adsorbent TiO$_2$@HKUST-1 was fabricated *via* the single-step hydrothermal method by the Min research group. They altered the amount of TiO$_2$ nanoparticles to prepare composites. The adsorption-photocatalytic degradation ability MB on the obtained samples was measured to assess their removal capacity. The best material for removing dye molecules was prepared 0.02 TiO$_2$@HKUST-1, which was 4.4 times that of pure MOF but 19.3 times that of TiO$_2$. The improved elimination ability was due to the existence of a heterojunction interface between the two components, which enabled the photoinduced h^+-e^- pairs to be effectively separated without recombination. Therefore, the removal of MB by the constructed composite material was characterized by both adsorption and photodegradation under visible light [129].

CONCLUSIONS AND PERSPECTIVES

In conclusion, the chapter has comprehensively summed up the current advances in MOF-based composites used in the synthesis methods and highlighted the applications in photocatalytic water purification in detail. MOFs, a kind of porous crystalline materials, have been widely studied due to their large S_{BET}, adjustable pore size, and functionalized surface. Although MOFs have been used in many fields as reported in the literature, their practical application in the adsorption and photocatalysis of pollutants in the water environment is still limited due to their high cost and poor water stability. Based on the existence of these problems, explosive development has been realized in the preparation of MOF-based composites and their applications in photocatalytic water purification in prior studies, and research on this field is still in progress. The constructed MOF-based composites manifested good photocatalytic degradation performance for pollutants in wastewater. As reviewed in this chapter, several common organic pollutants (dyes, phenols, PPCPs, herbicides, pesticides, *etc*) and other inorganic pollutants in wastewater have been photocatalytic removal by using MOF-based composites. In addition, we also reviewed the synergy of adsorption and photocatalysis of composites prepared for photocatalytic degradation to eliminate pollutants in wastewater.

In spite of the plenteous and significant outcomes that have been realized in this chapter, the research in this field still faces many restrictions and challenges. In order to promote the research area of MOF-based composites more comprehensively, the following questions need to be considered. Firstly, the photocatalytic efficiency of MOF-based composites is still weak compared with inorganic counterparts. Secondly, further investigations are fatal for promoting their utilization as visible light-responsive photocatalysts. Thirdly, the construction of adsorbed photocatalysts is still in the preliminary stage. Hence, it is necessary to put forward a new synthesis method to prepare photocatalysts with better performance, which can not only solve the problems in laboratory research but also be expected to be suitable for large-scale industrial production. Finally, according to the review in this chapter, we can see that a lot of research work mainly focuses on the removal of common pollutants in wastewater by photocatalytic degradation, which should be extended to emerging pollutants in subsequent research. Therefore, we believe that these problems should be seriously considered in future studies to widen the photocatalytic fields of MOF-based composites and improve their photodegradation efficiency. Nowadays, an increasing number of researchers devote themselves to developing new MOF-based composites and exploring their application prospects.

LIST OF ABBREVIATIONS

4-CP	*p*-chlorophenol
4-NP	4-nitrophenol
ACE	acetaminophen
AMX	amoxicillin
AR1	acid red 1
BUC	Beijing University of Civil Engineering and Architecture
BPA	bisphenol A
BPF	bisphenol F
CBZ	carbamazepine
CCD	central composite design
CTC	chlortetracycline
CV	crystal violet
CR	Congo red
DMF	*N,N-dimethylformamide*
DCL	dichlorophenol
DF	desirability function
DCF	diclofenac sodium
EPA	Environmental Protection Agency
EIS	electrochemical impedance spectroscopy
GO	graphene oxide
g-C$_3$N$_4$	graphitic carbon nitride
HKUST	the Hong Kong University of Science & Technology
H$_4$DOBDC	2,5-dihydroxyterephthalic acid
MOFs	metal–organic frameworks
MIL-125	Material Institute Lavoisier framework-125
MIL-100	Material Institute Lavoisier framework-100
MIL-101	Material Institute Lavoisier framework-101
MIL-88	Material Institute Lavoisier framework-88
MIL-68	Material Institute Lavoisier framework-68
MIL-53	Material Institute Lavoisier framework-53
MOF-5	Metal–organic framework-5

MOF-808	Metal–organic framework-808
MB	Methylene blue
MR	Methyl red
MO	Methyl orange
MA	Malathion
MG	Malachite green
NTU	Nanyang Technological University
NPs	Nanoparticles
OTC	Oxytetracycline
org II	Orange II
PPCPs	personal care products
PHIK	poly(heptazine imide)
PNG	*p*-nitrophenol
QDs	quantum dots
RhB	rhodamine B
Rh 6G	rhodamine 6G
SPR	surface plasmon resonance
TNTs	titanate nanotubes
Ti(i-OPr)$_4$	titanium isopropoxide
TCL	trichlorophenol
TC	tetracycline
TCS	triclosan
TCEP	tris(2-chloroethyl) phosphate
TB	trypan blue
UiO-66	University of Olso framework-66
UiO-67	University of Olso framework-67
VS	Vesuvine
ZIF-8	Zeolitic Imidazolate Framework-8
ZIF-67	Zeolitic Imidazolate Framework-67

CONSENT FOR PUBLICATION

Not applicable.

CONFLICT OF INTEREST

The author declares no conflict of interest, financial or otherwise.

ACKNOWLEDGEMENTS

This work was financially supported by the National Key Research and Development Program of China (2019YFC1904304), the Fundamental Research Funds for the Central Universities (2022YQHH09), the Open Fund of State Key Laboratory of Water Resource Protection and Utilization in Coal Mining (WPUKFJJ2019-10), and the Foundation of State Key Laboratory of Structural Chemistry (20190003).

REFERENCES

[1]   Shakoor, M.B.; Nawaz, R.; Hussain, F.; Raza, M.; Ali, S.; Rizwan, M.; Oh, S.E.; Ahmad, S. Human health implications, risk assessment and remediation of As-contaminated water: A critical review. *Sci. Total Environ.,* **2017**, *601-602,* 756-769, 756-769.
[http://dx.doi.org/10.1016/j.scitotenv.2017.05.223] [PMID: 28577410]

[2]   Tang, L.; Lv, Z.; Xue, Y.; Xu, L.; Qiu, W.; Zheng, C.; Chen, W.; Wu, M. MIL-53(Fe) incorporated in the lamellar BiOBr: Promoting the visible-light catalytic capability on the degradation of rhodamine B and carbamazepine. *Chem. Eng. J.,* **2019**, *374,* 975-982.
[http://dx.doi.org/10.1016/j.cej.2019.06.019]

[3]   Elzwayie, A.; Afan, H.A.; Allawi, M.F.; El-Shafie, A. Heavy metal monitoring, analysis and prediction in lakes and rivers: state of the art. *Environ. Sci. Pollut. Res. Int.,* **2017**, *24*(13), 12104-12117.
[http://dx.doi.org/10.1007/s11356-017-8715-0] [PMID: 28353110]

[4]   Chen, D.; Cheng, Y.; Zhou, N.; Chen, P.; Wang, Y.; Li, K.; Huo, S.; Cheng, P.; Peng, P.; Zhang, R.; Wang, L.; Liu, H.; Liu, Y.; Ruan, R. Photocatalytic degradation of organic pollutants using TiO₂-based photocatalysts: A review. *J. Clean. Prod.,* **2020**, *268,* 121725.
[http://dx.doi.org/10.1016/j.jclepro.2020.121725]

[5]   Ramezanalizadeh, H.; Manteghi, F. Immobilization of mixed cobalt/nickel metal-organic framework on a magnetic BiFeO₃ : A highly efficient separable photocatalyst for degradation of water pollutions. *J. Photochem. Photobiol. Chem.,* **2017**, *346,* 89-104.
[http://dx.doi.org/10.1016/j.jphotochem.2017.05.041]

[6]   Tijani, J.O.; Fatoba, O.O.; Babajide, O.O.; Petrik, L.F. Pharmaceuticals, endocrine disruptors, personal care products, nanomaterials and perfluorinated pollutants: a review. *Environ. Chem. Lett.,* **2016**, *14*(1), 27-49.
[http://dx.doi.org/10.1007/s10311-015-0537-z]

[7]   Mon, M.; Bruno, R.; Ferrando-Soria, J.; Armentano, D.; Pardo, E. Metal–organic framework technologies for water remediation: towards a sustainable ecosystem. *J. Mater. Chem. A Mater. Energy Sustain.,* **2018**, *6*(12), 4912-4947.
[http://dx.doi.org/10.1039/C8TA00264A]

[8]   Sha, Z.; Chan, H.S.O.; Wu, J. Ag₂CO₃/UiO-66(Zr) composite with enhanced visible-light promoted photocatalytic activity for dye degradation. *J. Hazard. Mater.,* **2015**, *299,* 132-140.
[http://dx.doi.org/10.1016/j.jhazmat.2015.06.016] [PMID: 26100934]

[9]   Ajmal, A.; Majeed, I.; Malik, R.N.; Idriss, H.; Nadeem, M.A. Principles and mechanisms of photocatalytic dye degradation on TiO₂ based photocatalysts: a comparative overview. *RSC Adv.,* **2014**, *4*(70), 37003-37026.

[http://dx.doi.org/10.1039/C4RA06658H]

[10]   Liu, N.; Jing, C.; Li, Z.; Huang, W.; Gao, B.; You, F.; Zhang, X. Effect of synthesis conditions on the photocatalytic degradation of Rhodamine B of MIL-53(Fe). *Mater. Lett.,* **2019**, *237*, 92-95.
        [http://dx.doi.org/10.1016/j.matlet.2018.11.079]

[11]   Dai, F.; Wang, Y.; Zhou, X.; Zhao, R.; Han, J.; Wang, L. $ZnIn_2S_4$ decorated Co-doped NH_2-MI--53(Fe) nanocomposites for efficient photocatalytic hydrogen production. *Appl. Surf. Sci.,* **2020**, *517*, 146161.
        [http://dx.doi.org/10.1016/j.apsusc.2020.146161]

[12]   López, J.; Chávez, A.M.; Rey, A.; Álvarez, P.M. Insights into the Stability and Activity of MIL-53(Fe) in Solar Photocatalytic Oxidation Processes in Water. *Catalysts,* **2021**, *11*(4), 448.
        [http://dx.doi.org/10.3390/catal11040448]

[13]   Li, P.X.; Yan, X.Y.; Song, X.M.; Li, J.J.; Ren, B.H.; Gao, S.Y.; Cao, R. Zirconium-Based Metal–Organic Framework Particle Films for Visible-Light-Driven Efficient Photoreduction of CO_2. *ACS Sustain. Chem.& Eng.,* **2021**, *9*(5), 2319-2325.
        [http://dx.doi.org/10.1021/acssuschemeng.0c08559]

[14]   Yang, H.; Wang, J.; Ma, J.; Yang, H.; Zhang, J.; Lv, K.; Wen, L.; Peng, T. A novel BODIPY-based MOF photocatalyst for efficient visible-light-driven hydrogen evolution. *J. Mater. Chem. A Mater. Energy Sustain.,* **2019**, *7*(17), 10439-10445.
        [http://dx.doi.org/10.1039/C9TA02357G]

[15]   Rodríguez, N.A.; Savateev, A.; Grela, M.A.; Dontsova, D. Facile Synthesis of Potassium Poly(heptazine imide) (PHIK)/Ti-Based Metal–Organic Framework (MIL-125-NH_2) Composites for Photocatalytic Applications. *ACS Appl. Mater. Interfaces,* **2017**, *9*(27), 22941-22949.
        [http://dx.doi.org/10.1021/acsami.7b04745] [PMID: 28609616]

[16]   Zhu, S.R.; Liu, P.F.; Wu, M.K.; Zhao, W.N.; Li, G.C.; Tao, K.; Yi, F.Y.; Han, L. Enhanced photocatalytic performance of BiOBr/NH_2 -MIL-125(Ti) composite for dye degradation under visible light. *Dalton Trans.,* **2016**, *45*(43), 17521-17529.
        [http://dx.doi.org/10.1039/C6DT02912D] [PMID: 27747336]

[17]   Hu, Q.; Di, J.; Wang, B.; Ji, M.; Chen, Y.; Xia, J.; Li, H.; Zhao, Y. *In-situ* preparation of NH_2-MI--125(Ti)/BiOCl composite with accelerating charge carriers for boosting visible light photocatalytic activity. *Appl. Surf. Sci.,* **2019**, *466*, 525-534.
        [http://dx.doi.org/10.1016/j.apsusc.2018.10.020]

[18]   He, S.; Rong, Q.; Niu, H.; Cai, Y. Platform for molecular-material dual regulation: A direct Z-scheme MOF/COF heterojunction with enhanced visible-light photocatalytic activity. *Appl. Catal. B,* **2019**, *247*, 49-56.
        [http://dx.doi.org/10.1016/j.apcatb.2019.01.078]

[19]   Han, X.; Yang, X.; Liu, G.; Li, Z.; Shao, L. Boosting visible light photocatalytic activity *via* impregnation-induced RhB-sensitized MIL-125(Ti). *Chem. Eng. Res. Des.,* **2019**, *143*, 90-99.
        [http://dx.doi.org/10.1016/j.cherd.2019.01.010]

[20]   Haroon, H.; Majid, K. MnO_2 nanosheets supported metal–organic framework MIL-125(Ti) towards efficient visible light photocatalysis: Kinetic and mechanistic study. *Chem. Phys. Lett.,* **2020**, *745*, 137283.
        [http://dx.doi.org/10.1016/j.cplett.2020.137283]

[21]   Zhao, X.; Zhang, Y.; Wen, P.; Xu, G.; Ma, D.; Qiu, P. NH_2-MIL-125(Ti)/TiO_2 composites as superior visible-light photocatalysts for selective oxidation of cyclohexane. *Mol. Catal.,* **2018**, *452*, 175-183.
        [http://dx.doi.org/10.1016/j.mcat.2018.04.004]

[22]   Wu, Y.; Li, X.; Yang, Q.; Wang, D.; Yao, F.; Cao, J.; Chen, Z.; Huang, X.; Yang, Y.; Li, X. Mxene-modulated dual-heterojunction generation on a metal-organic framework (MOF) *via* surface constitution reconstruction for enhanced photocatalytic activity. *Chem. Eng. J.,* **2020**, *390*, 124519.
        [http://dx.doi.org/10.1016/j.cej.2020.124519]

[23] Miao, S.; Zhang, H.; Cui, S.; Yang, J. Improved photocatalytic degradation of ketoprofen by Pt/MIL-125(Ti)/Ag with synergetic effect of Pt-MOF and MOF-Ag double interfaces: Mechanism and degradation pathway. *Chemosphere,* **2020**, *257*, 127123.
[http://dx.doi.org/10.1016/j.chemosphere.2020.127123] [PMID: 32505037]

[24] Hu, Q.; Yin, S.; Chen, Y.; Wang, B.; Li, M.; Ding, Y.; Di, J.; Xia, J.; Li, H. Construction of MIL-125(Ti)/ZnIn$_2$S$_4$ composites with accelerated interfacial charge transfer for boosting visible light photoreactivity. *Colloids Surf. A Physicochem. Eng. Asp.,* **2020**, *585*, 124078.
[http://dx.doi.org/10.1016/j.colsurfa.2019.124078]

[25] Wang, M.; Yang, L.; Guo, C.; Liu, X.; He, L.; Song, Y.; Zhang, Q.; Qu, X.; Zhang, H.; Zhang, Z.; Fang, S. Bimetallic Fe/Ti-Based Metal-Organic Framework for Persulfate-Assisted Visible Light Photocatalytic Degradation of Orange II. *ChemistrySelect,* **2018**, *3*(13), 3664-3674.
[http://dx.doi.org/10.1002/slct.201703134]

[26] Liu, S.; Zou, Q.; Ma, Y.; Sun, W.; Li, Y.; Zhang, J.; Zhang, C.; He, L.; Sun, Y.; Chen, Q.; Liu, B.; Zhang, H.; Zhang, K. A novel amorphous CoS$_x$/NH$_2$-MIL-125 composite for photocatalytic degradation of rhodamine B under visible light. *J. Mater. Sci.,* **2020**, *55*(34), 16171-16183.
[http://dx.doi.org/10.1007/s10853-020-05210-4]

[27] Zhang, S.; Du, M.; Kuang, J.; Xing, Z.; Li, Z.; Pan, K.; Zhu, Q.; Zhou, W. Surface-defect-rich mesoporous NH$_2$-MIL-125 (Ti)@Bi$_2$MoO$_6$ core-shell heterojunction with improved charge separation and enhanced visible-light-driven photocatalytic performance. *J. Colloid Interface Sci.,* **2019**, *554*, 324-334.
[http://dx.doi.org/10.1016/j.jcis.2019.07.021] [PMID: 31306944]

[28] Abdelhameed, R.M.; Tobaldi, D.M.; Karmaoui, M. Engineering highly effective and stable nanocomposite photocatalyst based on NH$_2$-MIL-125 encirclement with Ag$_3$PO$_4$ nanoparticles. *J. Photochem. Photobiol. Chem.,* **2018**, *351*, 50-58.
[http://dx.doi.org/10.1016/j.jphotochem.2017.10.011]

[29] Emam, H.E.; Ahmed, H.B.; Gomaa, E.; Helal, M.H.; Abdelhameed, R.M. Doping of silver vanadate and silver tungstate nanoparticles for enhancement the photocatalytic activity of MIL-125-NH$_2$ in dye degradation. *J. Photochem. Photobiol. Chem.,* **2019**, *383*, 111986.
[http://dx.doi.org/10.1016/j.jphotochem.2019.111986]

[30] Wang, H.; Cui, P.H.; Shi, J.X.; Tan, J.Y.; Zhang, J.Y.; Zhang, N.; Zhang, C. Controllable self-assembly of CdS@NH$_2$-MIL-125(Ti) heterostructure with enhanced photodegradation efficiency for organic pollutants through synergistic effect. *Mater. Sci. Semicond. Process.,* **2019**, *97*, 91-100.
[http://dx.doi.org/10.1016/j.mssp.2019.03.016]

[31] Gómez-Avilés, A.; Peñas-Garzón, M.; Bedia, J.; Dionysiou, D.D.; Rodríguez, J.J.; Belver, C. Mixed Ti-Zr metal-organic-frameworks for the photodegradation of acetaminophen under solar irradiation. *Appl. Catal. B,* **2019**, *253*, 253-262.
[http://dx.doi.org/10.1016/j.apcatb.2019.04.040]

[32] Yuan, X.; Wang, H.; Wu, Y.; Zeng, G.; Chen, X.; Leng, L.; Wu, Z.; Li, H. One-pot self-assembly and photoreduction synthesis of silver nanoparticle-decorated reduced graphene oxide/MIL-125(Ti) photocatalyst with improved visible light photocatalytic activity. *Appl. Organomet. Chem.,* **2016**, *30*(5), 289-296.
[http://dx.doi.org/10.1002/aoc.3430]

[33] Wu, D.; Han, L. Fabrication of novel Ag/AgBr/NH$_2$ -MIL-125(Ti) nanocomposites with enhanced visible-light photocatalytic activity. *Mater. Res. Express,* **2019**, *6*(12), 125501.
[http://dx.doi.org/10.1088/2053-1591/ab540a]

[34] Qiu, J.; Li, M.; Wang, H.; Yao, J. Integration of plasmonic effect into MIL-125-NH$_2$: An ultra-efficient photocatalyst for simultaneous removal of ternary system pollutants. *Chemosphere,* **2020**, *242*, 125197.
[http://dx.doi.org/10.1016/j.chemosphere.2019.125197] [PMID: 31675592]

[35]   Huang, Z.; Chen, H.; Zhao, L.; He, X.; Fang, W.; Du, Y.; Li, W.; Wang, G.; Zhang, F. CdSe QDs sensitized MIL-125/TiO$_2$@SiO$_2$ biogenic hierarchical composites with enhanced photocatalytic properties *via* two-level heterostructure. *J. Mater. Sci. Mater. Electron.,* **2018**, *29*(14), 12045-12054.
       [http://dx.doi.org/10.1007/s10854-018-9310-y]

[36]   Huang, W.; Liu, N.; Zhang, X.; Wu, M.; Tang, L. Metal organic framework g-C$_3$ N$_4$ /MIL-53(Fe) heterojunctions with enhanced photocatalytic activity for Cr(VI) reduction under visible light. *Appl. Surf. Sci.,* **2017**, *425*, 107-116.
       [http://dx.doi.org/10.1016/j.apsusc.2017.07.050]

[37]   Zhang, C.; Ai, L.; Jiang, J. Solvothermal synthesis of MIL–53(Fe) hybrid magnetic composites for photoelectrochemical water oxidation and organic pollutant photodegradation under visible light. *J. Mater. Chem. A Mater. Energy Sustain.,* **2015**, *3*(6), 3074-3081.
       [http://dx.doi.org/10.1039/C4TA04622F]

[38]   Xiao, H.; Zhang, W.; Yao, Q.; Huang, L.; Chen, L.; Boury, B.; Chen, Z. Zn-free MOFs like MIL-53(Al) and MIL-125(Ti) for the preparation of defect-rich, ultrafine ZnO nanosheets with high photocatalytic performance. *Appl. Catal. B,* **2019**, *244*, 719-731.
       [http://dx.doi.org/10.1016/j.apcatb.2018.11.026]

[39]   Han, Y.; Zhang, L.; Bai, C.; Wu, J.; Meng, H.; Xu, Y.; Liang, Z.; Wang, Z.; Zhang, X. Fabrication of AgI/MIL-53(Fe) Composites with Enhanced Photocatalytic Activity for Rhodamine B Degradation under Visible Light Irradiation. *Appl. Organomet. Chem.,* **2018**, *32*(5), e4325.
       [http://dx.doi.org/10.1002/aoc.4325]

[40]   Xie, L.; Yang, Z.; Xiong, W.; Zhou, Y.; Cao, J.; Peng, Y.; Li, X.; Zhou, C.; Xu, R.; Zhang, Y. Construction of MIL-53(Fe) metal-organic framework modified by silver phosphate nanoparticles as a novel Z-scheme photocatalyst: Visible-light photocatalytic performance and mechanism investigation. *Appl. Surf. Sci.,* **2019**, *465*, 103-115.
       [http://dx.doi.org/10.1016/j.apsusc.2018.09.144]

[41]   Deng, L.; Yin, D.; Khaing, K.K.; Xiao, S.; Li, L.; Guo, X.; Wang, J.; Zhang, Y. The facile boosting sunlight-driven photocatalytic performance of a metal–organic-framework through coupling with Ag$_2$ S nanoparticles. *New J. Chem.,* **2020**, *44*(29), 12568-12578.
       [http://dx.doi.org/10.1039/D0NJ02030C]

[42]   Araya, T.; Jia, M.; Yang, J.; Zhao, P.; Cai, K.; Ma, W.; Huang, Y. Resin modified MIL-53 (Fe) MOF for improvement of photocatalytic performance. *Appl. Catal. B,* **2017**, *203*, 768-777.
       [http://dx.doi.org/10.1016/j.apcatb.2016.10.072]

[43]   Liu, N.; Huang, W.; Tang, M.; Yin, C.; Gao, B.; Li, Z.; Tang, L.; Lei, J.; Cui, L.; Zhang, X. *In-situ* fabrication of needle-shaped MIL-53(Fe) with 1T-MoS$_2$ and study on its enhanced photocatalytic mechanism of ibuprofen. *Chem. Eng. J.,* **2019**, *359*, 254-264.
       [http://dx.doi.org/10.1016/j.cej.2018.11.143]

[44]   Liu, N.; Zheng, Y.; Jing, C.; Gao, B.; Huang, W.; Li, Z.; Lei, J.; Zhang, X.; Cui, L.; Tang, L. Boosting catalytic degradation efficiency by incorporation of MIL-53(Fe) with Ti$_3$C$_2$T$_x$ nanosheeets. *J. Mol. Liq.,* **2020**, *311*, 113201.
       [http://dx.doi.org/10.1016/j.molliq.2020.113201]

[45]   Hu, L.; Deng, G.; Lu, W.; Pang, S.; Hu, X. Deposition of CdS nanoparticles on MIL-53(Fe) metal-organic framework with enhanced photocatalytic degradation of RhB under visible light irradiation. *Appl. Surf. Sci.,* **2017**, *410*, 401-413.
       [http://dx.doi.org/10.1016/j.apsusc.2017.03.140]

[46]   Lin, R.; Li, S.; Wang, J.; Xu, J.; Xu, C.; Wang, J.; Li, C.; Li, Z. Facile generation of carbon quantum dots in MIL-53(Fe) particles as localized electron acceptors for enhancing their photocatalytic Cr(VI) reduction. *Inorg. Chem. Front.,* **2018**, *5*(12), 3170-3177.
       [http://dx.doi.org/10.1039/C8QI01164H]

[47]   Lv, S.W.; Liu, J.M.; Zhao, N.; Li, C.Y.; Wang, Z.H.; Wang, S. Benzothiadiazole functionalized Co-

doped MIL-53-NH$_2$ with electron deficient units for enhanced photocatalytic degradation of bisphenol A and ofloxacin under visible light. *J. Hazard. Mater.,* **2020**, *387*, 122011.
[http://dx.doi.org/10.1016/j.jhazmat.2019.122011] [PMID: 31927354]

[48]   Chatterjee, A.; Jana, A.K.; Basu, J.K. A novel synthesis of MIL-53(Al)@SiO$_2$: an integrated photocatalyst adsorbent to remove bisphenol a from wastewater. *New J. Chem.,* **2020**, *44*(43), 18892-18905.
[http://dx.doi.org/10.1039/D0NJ03714A]

[49]   Hu, L.; Zhang, Y.; Lu, W.; Lu, Y.; Hu, H. Easily recyclable photocatalyst Bi$_2$WO$_6$/MOF/PVDF composite film for efficient degradation of aqueous refractory organic pollutants under visible-light irradiation. *J. Mater. Sci.,* **2019**, *54*(8), 6238-6257.
[http://dx.doi.org/10.1007/s10853-018-03302-w]

[50]   Chen, Y.; Zhai, B.; Liang, Y. Enhanced degradation performance of organic dyes removal by semiconductor/MOF/graphene oxide composites under visible light irradiation. *Diam. Relat. Mater.,* **2019**, *98*, 107508.
[http://dx.doi.org/10.1016/j.diamond.2019.107508]

[51]   He, Y.; Dong, W.; Li, X.; Wang, D.; Yang, Q.; Deng, P.; Huang, J. Modified MIL-100(Fe) for enhanced photocatalytic degradation of tetracycline under visible-light irradiation. *J. Colloid Interface Sci.,* **2020**, *574*, 364-376.
[http://dx.doi.org/10.1016/j.jcis.2020.04.075] [PMID: 32339819]

[52]   Lin, J.; Hu, H.; Gao, N.; Ye, J.; Chen, Y.; Ou, H. Fabrication of GO@MIL-101(Fe) for enhanced visible-light photocatalysis degradation of organophosphorus contaminant. *J. Water Process Eng.,* **2020**, *33*, 101010.
[http://dx.doi.org/10.1016/j.jwpe.2019.101010]

[53]   Ke, F.; Wang, L.; Zhu, J. Facile fabrication of CdS-metal-organic framework nanocomposites with enhanced visible-light photocatalytic activity for organic transformation. *Nano Res.,* **2015**, *8*(6), 1834-1846.
[http://dx.doi.org/10.1007/s12274-014-0690-x]

[54]   Huang, J.; Zhang, X.; Song, H.; Chen, C.; Han, F.; Wen, C. Protonated graphitic carbon nitride coated metal-organic frameworks with enhanced visible-light photocatalytic activity for contaminants degradation. *Appl. Surf. Sci.,* **2018**, *441*, 85-98.
[http://dx.doi.org/10.1016/j.apsusc.2018.02.027]

[55]   Hong, J.; Chen, C.; Bedoya, F.E.; Kelsall, G.H.; O'Hare, D.; Petit, C. Carbon nitride nanosheet/metal–organic framework nanocomposites with synergistic photocatalytic activities. *Catal. Sci. Technol.,* **2016**, *6*(13), 5042-5051.
[http://dx.doi.org/10.1039/C5CY01857A]

[56]   Du, X.; Yi, X.; Wang, P.; Deng, J.; Wang, C. Enhanced photocatalytic Cr(VI) reduction and diclofenac sodium degradation under simulated sunlight irradiation over MIL-100(Fe)/g-C$_3$N$_4$ heterojunctions. *Chin. J. Catal.,* **2019**, *40*(1), 70-79.
[http://dx.doi.org/10.1016/S1872-2067(18)63160-2]

[57]   Ahmad, M.; Chen, S.; Ye, F.; Quan, X.; Afzal, S.; Yu, H.; Zhao, X. Efficient photo-Fenton activity in mesoporous MIL-100(Fe) decorated with ZnO nanosphere for pollutants degradation. *Appl. Catal. B,* **2019**, *245*, 428-438.
[http://dx.doi.org/10.1016/j.apcatb.2018.12.057]

[58]   Zhao, H.; Qian, L.; Lv, H.; Wang, Y.; Zhao, G. Introduction of a Fe$_3$O$_4$ Core Enhances the Photocatalytic Activity of MIL-100(Fe) with Tunable Shell Thickness in the Presence of H$_2$O$_2$. *ChemCatChem,* **2015**, *7*(24), 4148-4155.
[http://dx.doi.org/10.1002/cctc.201500801]

[59]   Huo, Q.; Qi, X.; Li, J.; Liu, G.; Ning, Y.; Zhang, X.; Zhang, B.; Fu, Y.; Liu, S. Preparation of a direct Z-scheme α-Fe$_2$O$_3$/MIL-101(Cr) hybrid for degradation of carbamazepine under visible light

irradiation. *Appl. Catal. B,* **2019**, *255,* 117751.
[http://dx.doi.org/10.1016/j.apcatb.2019.117751]

[60] Wu, W.; Zhu, J.; Deng, Y.H.; Xiang, Y.; Tan, Y.W.; Tang, H.Q.; Zou, H.; Xu, Y.F.; Zhou, Y. TiO_2 nanocrystals with the 001 and 101 facets co-exposed with MIL-100(Fe): an egg-like composite nanomaterial for efficient visible light-driven photocatalysis. *RSC Advances,* **2019**, *9*(54), 31728-31734.
[http://dx.doi.org/10.1039/C9RA06359E] [PMID: 35527976]

[61] He, X.; Fang, H.; Gosztola, D.J.; Jiang, Z.; Jena, P.; Wang, W.N. Mechanistic Insight into Photocatalytic Pathways of MIL-100(Fe)/TiO_2 Composites. *ACS Appl. Mater. Interfaces,* **2019**, *11*(13), 12516-12524.
[http://dx.doi.org/10.1021/acsami.9b00223] [PMID: 30865419]

[62] Chang, N.; Zhang, H.; Shi, M.S.; Li, J.; Yin, C.J.; Wang, H.T.; Wang, L. Regulation of the adsorption affinity of metal-organic framework MIL-101 *via* a TiO_2 coating strategy for high capacity adsorption and efficient photocatalysis. *Microporous Mesoporous Mater.,* **2018**, *266,* 47-55.
[http://dx.doi.org/10.1016/j.micromeso.2018.02.051]

[63] Zhang, C.; Guo, D.; Shen, T.; Hou, X.; Zhu, M.; Liu, S.; Hu, Q. Titanium dioxide/magnetic metal-organic framework preparation for organic pollutants removal from water under visible light. *Colloids Surf. A Physicochem. Eng. Asp.,* **2020**, *589,* 124484.
[http://dx.doi.org/10.1016/j.colsurfa.2020.124484]

[64] Tilgner, D.; Friedrich, M.; Hermannsdörfer, J.; Kempe, R. Titanium Dioxide Reinforced Metal-Organic Framework Pd Catalysts: Activity and Reusability Enhancement in Alcohol Dehydrogenation Reactions and Improved Photocatalytic Performance. *ChemCatChem,* **2015**, *7*(23), 3916-3922.
[http://dx.doi.org/10.1002/cctc.201500747]

[65] Li, Y.; Jiang, J.; Fang, Y.; Cao, Z.; Chen, D.; Li, N.; Xu, Q.; Lu, J. TiO_2 Nanoparticles Anchored onto the Metal–Organic Framework NH_2-MIL-88B(Fe) as an Adsorptive Photocatalyst with Enhanced Fenton-like Degradation of Organic Pollutants under Visible Light Irradiation. *ACS Sustain. Chem.& Eng.,* **2018**, *6*(12), 16186-16197.
[http://dx.doi.org/10.1021/acssuschemeng.8b02968]

[66] Zhao, C.; Wang, J.; Chen, X.; Wang, Z.; Ji, H.; Chen, L.; Liu, W.; Wang, C.C. Bifunctional $Bi_{12}O_{17}Cl_2$/MIL-100(Fe) composites toward photocatalytic Cr(VI) sequestration and activation of persulfate for bisphenol A degradation. *Sci. Total Environ.,* **2021**, *752,* 141901.
[http://dx.doi.org/10.1016/j.scitotenv.2020.141901] [PMID: 33207532]

[67] Zhou, T.; Zhang, G.; Zhang, H.; Yang, H.; Ma, P.; Li, X.; Qiu, X.; Liu, G. Highly efficient visible-light-driven photocatalytic degradation of rhodamine B by a novel Z-scheme Ag_3PO_4/MIL-101/$NiFe_2O_4$ composite. *Catal. Sci. Technol.,* **2018**, *8*(9), 2402-2416.
[http://dx.doi.org/10.1039/C8CY00182K]

[68] Zhao, D.; Cai, C. Adsorption and photocatalytic degradation of pollutants on Ce-doped MIL-101-NH_2/Ag_3PO_4 composites. *Catal. Commun.,* **2020**, *136,* 105910.
[http://dx.doi.org/10.1016/j.catcom.2019.105910]

[69] Lu, W.; Duan, C.; Liu, C.; Zhang, Y.; Meng, X.; Dai, L.; Wang, W.; Yu, H.; Ni, Y. A self-cleaning and photocatalytic cellulose-fiber- supported "Ag@AgCl@MOF-cloth" membrane for complex wastewater remediation. *Carbohydr. Polym.,* **2020**, *247,* 116691.
[http://dx.doi.org/10.1016/j.carbpol.2020.116691] [PMID: 32829819]

[70] Qian, X.; Xu, H.; Zhang, X.; Lei, R.; Gao, J.; Xu, S. Enhanced visible-light-driven photocatalytic activity of Ag_3PO_4/metal–organic framework composite. *Polyhedron,* **2019**, *163,* 1-6.
[http://dx.doi.org/10.1016/j.poly.2019.02.005]

[71] Shao, Z.; Zhang, D.; Li, H.; Su, C.; Pu, X.; Geng, Y. Fabrication of MIL-88A/g-C_3N_4 direct Z-scheme heterojunction with enhanced visible-light photocatalytic activity. *Separ. Purif. Tech.,* **2019**, *220,* 16-24.

[http://dx.doi.org/10.1016/j.seppur.2019.03.040]

[72] Yue, X.; Guo, W.; Li, X.; Gao, X.; Zhang, G. Preparation of Efficient BiOBr/MIL-88B(Fe) Composites with Enhanced Photocatalytic Activities. *Water Environ. Res.,* **2017**, *89*(7), 614-621. [http://dx.doi.org/10.2175/106143017X14839994522821] [PMID: 28105984]

[73] Yuan, R.; Qiu, J.; Yue, C.; Shen, C.; Li, D.; Zhu, C.; Liu, F.; Li, A. Self-assembled hierarchical and bifunctional MIL-88A(Fe)@ZnIn$_2$S$_4$ heterostructure as a reusable sunlight-driven photocatalyst for highly efficient water purification. *Chem. Eng. J.,* **2020**, *401*, 126020. [http://dx.doi.org/10.1016/j.cej.2020.126020]

[74] Xie, A.; Cui, J.; Yang, J.; Chen, Y.; Lang, J.; Li, C.; Yan, Y.; Dai, J. Graphene oxide/Fe(III)-based metal-organic framework membrane for enhanced water purification based on synergistic separation and photo-Fenton processes. *Appl. Catal. B,* **2020**, *264*, 118548. [http://dx.doi.org/10.1016/j.apcatb.2019.118548]

[75] Yang, C.; You, X.; Cheng, J.; Zheng, H.; Chen, Y. A novel visible-light-driven In-based MOF/graphene oxide composite photocatalyst with enhanced photocatalytic activity toward the degradation of amoxicillin. *Appl. Catal. B,* **2017**, *200*, 673-680. [http://dx.doi.org/10.1016/j.apcatb.2016.07.057]

[76] Cao, W.; Yuan, Y.; Yang, C.; Wu, S.; Cheng, J. *in-situ* fabrication of g-C$_3$N$_4$/MIL-68(In)-NH$_2$ heterojunction composites with enhanced visible-light photocatalytic activity for degradation of ibuprofen. *Chem. Eng. J.,* **2020**, *391*, 123608. [http://dx.doi.org/10.1016/j.cej.2019.123608]

[77] Zhong, Z.; Li, M.; Fu, J.; Wang, Y.; Muhammad, Y.; Li, S.; Wang, J.; Zhao, Z.; Zhao, Z. Construction of Cu-bridged Cu$_2$O/MIL(Fe/Cu) catalyst with enhanced interfacial contact for the synergistic photo-Fenton degradation of thiacloprid. *Chem. Eng. J.,* **2020**, *395*, 125184. [http://dx.doi.org/10.1016/j.cej.2020.125184]

[78] Zhang, W.; Shi, S.; Zhu, W.; Huang, L.; Yang, C.; Li, S.; Liu, X.; Wang, R.; Hu, N.; Suo, Y.; Li, Z.; Wang, J. Agar Aerogel Containing Small-Sized Zeolitic Imidazolate Framework Loaded Carbon Nitride: A Solar-Triggered Regenerable Decontaminant for Convenient and Enhanced Water Purification. *ACS Sustain. Chem.& Eng.,* **2017**, *5*(10), 9347-9354. [http://dx.doi.org/10.1021/acssuschemeng.7b02376]

[79] Qiu, M.; Liu, Z.; Wang, S.; Hu, B. The photocatalytic reduction of U(VI) into U(IV) by ZIF-8/g-C$_3$N$_4$ composites at visible light. *Environ. Res.,* **2021**, *196*, 110349. [http://dx.doi.org/10.1016/j.envres.2020.110349] [PMID: 33129860]

[80] Malik, A.; Nath, M. Multicore–shell nanocomposite formed by encapsulation of WO$_3$ in zeolitic imidazolate framework (ZIF-8): As an efficient photocatalyst. *J. Environ. Chem. Eng.,* **2019**, *7*(5), 103401. [http://dx.doi.org/10.1016/j.jece.2019.103401]

[81] Chandra, R.; Singh, V.; Tomar, S.; Nath, M. Multi-core-shell composite SnO$_2$NPs@ZIF-8: potential antiviral agent and effective photocatalyst for waste-water treatment. *Environ. Sci. Pollut. Res. Int.,* **2019**, *26*(23), 23346-23358. [http://dx.doi.org/10.1007/s11356-019-05646-5] [PMID: 31197665]

[82] Mollick, S.; Mandal, T.N.; Jana, A.; Fajal, S.; Desai, A.V.; Ghosh, S.K. Ultrastable Luminescent Hybrid Bromide Perovskite@MOF Nanocomposites for the Degradation of Organic Pollutants in Water. *ACS Appl. Nano Mater.,* **2019**, *2*(3), 1333-1340. [http://dx.doi.org/10.1021/acsanm.8b02214]

[83] Si, Y.; Li, Y.; Xia, Y.; Shang, S.; Xiong, X.; Zeng, X.; Zhou, J. Fabrication of Novel ZIF-8@BiVO$_4$ Composite with Enhanced Photocatalytic Performance. *Crystals (Basel),* **2018**, *8*(11), 432. [http://dx.doi.org/10.3390/cryst8110432]

[84] Xia, Y.; Shang, S.; Zeng, X.; Zhou, J.; Li, Y. A Novel Bi$_2$MoO$_6$/ZIF-8 Composite for Enhanced Visible Light Photocatalytic Activity. *Nanomaterials (Basel),* **2019**, *9*(4), 545.

[http://dx.doi.org/10.3390/nano9040545]

[85]   Chen, W.Q.; Li, L.Y.; Li, L.; Qiu, W.H.; Tang, L.; Xu, L.; Xu, K.J.; Wu, M.H. MoS₂/ZIF-8 Hybrid Materials for Environmental Catalysis: Solar-Driven Antibiotic-Degradation Engineering. *Engineering (Beijing),* **2019**, *5*(4), 755-767.
[http://dx.doi.org/10.1016/j.eng.2019.02.003]

[86]   Qiu, J.; Zhang, X.F.; Zhang, X.; Feng, Y.; Li, Y.; Yang, L.; Lu, H.; Yao, J. Constructing Cd₀.₅Zn₀.₅S@ZIF-8 nanocomposites through self-assembly strategy to enhance Cr(VI) photocatalytic reduction. *J. Hazard. Mater.,* **2018**, *349*, 234-241.
[http://dx.doi.org/10.1016/j.jhazmat.2018.02.009] [PMID: 29428684]

[87]   Liu, A.; Yu, C.; Lin, J.; Sun, G.; Xu, G.; Huang, Y.; Liu, Z.; Tang, C. Construction of CuInS₂@ZIF-8 nanocomposites with enhanced photocatalytic activity and durability. *Mater. Res. Bull.,* **2019**, *112*, 147-153.
[http://dx.doi.org/10.1016/j.materresbull.2018.12.020]

[88]   Si, Y.; Li, X.; Yang, G.; Mie, X.; Ge, L. Fabrication of a novel core–shell CQDs@ZIF-8 composite with enhanced photocatalytic activity. *J. Mater. Sci.,* **2020**, *55*(27), 13049-13061.
[http://dx.doi.org/10.1007/s10853-020-04909-8]

[89]   Abdi, J. Synthesis of Ag-doped ZIF-8 photocatalyst with excellent performance for dye degradation and antibacterial activity. *Colloids Surf. A Physicochem. Eng. Asp.,* **2020**, *604*, 125330.
[http://dx.doi.org/10.1016/j.colsurfa.2020.125330]

[90]   Zhai, B.; Chen, Y.; Liang, Y.; Gao, Y.; Shi, J.; Zhang, H.; Li, Y. Modifying Ag₃VO₄ with metal-organic frameworks for enhanced photocatalytic activity under visible light. *Mater. Chem. Phys.,* **2020**, *239*, 122078.
[http://dx.doi.org/10.1016/j.matchemphys.2019.122078]

[91]   Liu, J.; Li, R.; Wang, Y.; Wang, Y.; Zhang, X.; Fan, C. The active roles of ZIF-8 on the enhanced visible photocatalytic activity of Ag/AgCl: Generation of superoxide radical and adsorption. *J. Alloys Compd.,* **2017**, *693*, 543-549.
[http://dx.doi.org/10.1016/j.jallcom.2016.09.201]

[92]   Naimi Joubani, M.; Zanjanchi, M.A.; Sohrabnezhad, S. The carboxylate magnetic – zinc based metal-organic framework heterojunction: Fe₃O₄-COOH@ZIF-8/Ag/Ag₃PO₄ for plasmon enhanced visible light Z-scheme photocatalysis. *Adv. Powder Technol.,* **2020**, *31*(1), 29-39.
[http://dx.doi.org/10.1016/j.apt.2019.09.034]

[93]   Ma, Y.; Tuniyazi, D.; Ainiwa, M.; Zhu, E. Confined amorphous red phosphorus in metal–organic framework as a superior photocatalyst. *Mater. Lett.,* **2020**, *262*, 127023.
[http://dx.doi.org/10.1016/j.matlet.2019.127023]

[94]   Shao, W.; Chen, Y.R.; Xie, F.; Zhang, H.; Wang, H.T.; Chang, N. Facile construction of a ZIF-67/AgCl/Ag heterojunction *via* chemical etching and surface ion exchange strategy for enhanced visible light driven photocatalysis. *RSC Advances,* **2020**, *10*(63), 38174-38183.
[http://dx.doi.org/10.1039/D0RA06842J] [PMID: 35517549]

[95]   Xue, Y.; Wang, P.; Wang, C.; Ao, Y. Efficient degradation of atrazine by BiOBr/UiO-66 composite photocatalyst under visible light irradiation: Environmental factors, mechanisms and degradation pathways. *Chemosphere,* **2018**, *203*, 497-505.
[http://dx.doi.org/10.1016/j.chemosphere.2018.04.017] [PMID: 29649691]

[96]   Tong, X.; Yang, Z.; Feng, J.; Li, Y.; Zhang, H. BiOCl/UiO-66 composite with enhanced performance for photo-assisted degradation of dye from water. *Appl. Organomet. Chem.,* **2018**, *32*(2), e4049.
[http://dx.doi.org/10.1002/aoc.4049]

[97]   Yang, Z.; Tong, X.; Feng, J.; He, S.; Fu, M.; Niu, X.; Zhang, T.; Liang, H.; Ding, A.; Feng, X. Flower-like BiOBr/UiO-66-NH₂ nanosphere with improved photocatalytic property for norfloxacin removal. *Chemosphere,* **2019**, *220*, 98-106.
[http://dx.doi.org/10.1016/j.chemosphere.2018.12.086] [PMID: 30579953]

[98]   Hu, Q.; Chen, Y.; Li, M.; Zhang, Y.; Wang, B.; Zhao, Y.; Xia, J.; Yin, S.; Li, H. Construction of NH_2-UiO-66/BiOBr composites with boosted photocatalytic activity for the removal of contaminants. *Colloids Surf. A Physicochem. Eng. Asp.,* **2019**, *579*, 123625.
[http://dx.doi.org/10.1016/j.colsurfa.2019.123625]

[99]   Heu, R.; Ateia, M.; Yoshimura, C. Photocatalytic Nanofiltration Membrane Using Zr-MOF/GO Nanocomposite with High-Flux and Anti-Fouling Properties. *Catalysts,* **2020**, *10*(6), 711.
[http://dx.doi.org/10.3390/catal10060711]

[100] Zhang, S.; Wang, Y.; Cao, Z.; Xu, J.; Hu, J.; Huang, Y.; Cui, C.; Liu, H.; Wang, H. Simultaneous enhancements of light-harvesting and charge transfer in UiO-67/CdS/rGO composites toward ofloxacin photo-degradation. *Chem. Eng. J.,* **2020**, *381*, 122771.
[http://dx.doi.org/10.1016/j.cej.2019.122771]

[101] Pan, Y.; Yuan, X.; Jiang, L.; Wang, H.; Yu, H.; Zhang, J. Stable self-assembly AgI/UiO-66(NH_2) heterojunction as efficient visible-light responsive photocatalyst for tetracycline degradation and mechanism insight. *Chem. Eng. J.,* **2020**, *384*, 123310.
[http://dx.doi.org/10.1016/j.cej.2019.123310]

[102] Zhou, Y.C.; Xu, X.Y.; Wang, P.; Fu, H.; Zhao, C.; Wang, C.C. Facile fabrication and enhanced photocatalytic performance of visible light responsive UiO-66-NH_2/Ag_2CO_3 composite. *Chin. J. Catal.,* **2019**, *40*(12), 1912-1923.
[http://dx.doi.org/10.1016/S1872-2067(19)63433-9]

[103] Zeng, H.; Yu, Z.; Shao, L.; Li, X.; Zhu, M.; Liu, Y.; Feng, X.; Zhu, X. Ag_2CO_3@UiO-66-NH_2 embedding graphene oxide sheets photocatalytic membrane for enhancing the removal performance of Cr(VI) and dyes based on filtration. *Desalination,* **2020**, *491*, 114558.
[http://dx.doi.org/10.1016/j.desal.2020.114558]

[104] Zhang, N.; Zhang, X.; Gan, C.; Zhang, J.; Liu, Y.; Zhou, M.; Zhang, C.; Fang, Y. Heterostructural Ag_3PO_4/UiO-66 composite for highly efficient visible-light photocatalysts with long-term stability. *J. Photochem. Photobiol. Chem.,* **2019**, *376*, 305-315.
[http://dx.doi.org/10.1016/j.jphotochem.2019.03.025]

[105] Zhao, W.; Ding, T.; Wang, Y.; Wu, M.; Jin, W.; Tian, Y.; Li, X. Decorating Ag/AgCl on UiO-66-NH_2: Synergy between Ag plasmons and heterostructure for the realization of efficient visible light photocatalysis. *Chin. J. Catal.,* **2019**, *40*(8), 1187-1197.
[http://dx.doi.org/10.1016/S1872-2067(19)63377-2]

[106] Jin, W.; Wang, Y.; Zhao, W.; Du, X.; Tian, Y.; Ding, T.; Li, X. Boosting Visible-Light Photodegradation over Ternary Strategy-Engineered Metal–Organic Frameworks. *Ind. Eng. Chem. Res.,* **2020**, *59*(30), 13491-13501.
[http://dx.doi.org/10.1021/acs.iecr.0c02422]

[107] Fakhri, H.; Bagheri, H. Two novel sets of UiO-66@ metal oxide/graphene oxide Z-scheme heterojunction: Insight into tetracycline and malathion photodegradation. *J. Environ. Sci. (China),* **2020**, *91*, 222-236.
[http://dx.doi.org/10.1016/j.jes.2020.01.013] [PMID: 32172972]

[108] Fakhri, H.; Bagheri, H. Highly efficient Zr-MOF@WO_3/graphene oxide photocatalyst: Synthesis, characterization and photodegradation of tetracycline and malathion. *Mater. Sci. Semicond. Process.,* **2020**, *107*, 104815.
[http://dx.doi.org/10.1016/j.mssp.2019.104815]

[109] Wu, J.; Fang, X.; Zhu, Y.; Ma, N.; Dai, W. Well-Designed TiO_2@UiO-66-NH_2 Nanocomposite with Superior Photocatalytic Activity for Tetracycline under Restricted Space. *Energy Fuels,* **2020**, *34*(10), 12911-12917.
[http://dx.doi.org/10.1021/acs.energyfuels.0c02485]

[110] Du, Q.; Wu, P.; Sun, Y.; Zhang, J.; He, H. Selective photodegradation of tetracycline by molecularly imprinted ZnO@NH_2-UiO-66 composites. *Chem. Eng. J.,* **2020**, *390*, 124614.

[http://dx.doi.org/10.1016/j.cej.2020.124614]

[111] Bargozideh, S.; Tasviri, M.; Shekarabi, S.; Daneshgar, H. Magnetic BiFeO$_3$ decorated UiO-66 as a p–n heterojunction photocatalyst for simultaneous degradation of a binary mixture of anionic and cationic dyes. *New J. Chem.*, **2020**, *44*(30), 13083-13092.
[http://dx.doi.org/10.1039/D0NJ02594A]

[112] Bahmani, M.; Mowla, D.; Esmaeilzadeh, F.; Ghaedi, M. BiFeO$_3$–BiOI impregnation to UiO-66(Zr/Ti) as a promising candidate visible-light-driven photocatalyst for boosting urea photodecomposition in a continuous flow-loop thin-film slurry flat-plate photoreactor. *J. Solid State Chem.*, **2020**, *286*, 121304.
[http://dx.doi.org/10.1016/j.jssc.2020.121304]

[113] Ma, X.; Lou, Y.; Chen, J. New UiO-66/Cu$_x$S heterostructures: surface functionalization synthesis and their application in photocatalytic degradation of RhB. *Bull. Chem. Soc. Jpn.*, **2018**, *91*(4), 515-522.
[http://dx.doi.org/10.1246/bcsj.20170375]

[114] Zhang, X.; Yang, Y.; Huang, W.; Yang, Y.; Wang, Y.; He, C.; Liu, N.; Wu, M.; Tang, L. g-C$_3$N$_4$/UiO-66 nanohybrids with enhanced photocatalytic activities for the oxidation of dye under visible light irradiation. *Mater. Res. Bull.*, **2018**, *99*, 349-358.
[http://dx.doi.org/10.1016/j.materresbull.2017.11.028]

[115] Liang, Q.; Jin, J.; Liu, C.; Xu, S.; Yao, C.; Li, Z. Fabrication of the ternary heterojunction Cd$_{0.5}$Zn$_{0.5}$S@UIO-66@g-C$_3$N$_4$ for enhanced visible-light photocatalytic hydrogen evolution and degradation of organic pollutants. *Inorg. Chem. Front.*, **2018**, *5*(2), 335-343.
[http://dx.doi.org/10.1039/C7QI00638A]

[116] Sha, Z.; Sun, J.; On Chan, H.S.; Jaenicke, S.; Wu, J. Bismuth tungstate incorporated zirconium metal–organic framework composite with enhanced visible-light photocatalytic performance. *RSC Adv.*, **2014**, *4*(110), 64977-64984.
[http://dx.doi.org/10.1039/C4RA13000F]

[117] Bariki, R.; Majhi, D.; Das, K.; Behera, A.; Mishra, B.G. Facile synthesis and photocatalytic efficacy of UiO-66/CdIn$_2$S$_4$ nanocomposites with flowerlike 3D-microspheres towards aqueous phase decontamination of triclosan and H$_2$ evolution. *Appl. Catal. B*, **2020**, *270*, 118882.
[http://dx.doi.org/10.1016/j.apcatb.2020.118882]

[118] Chen, J.; Chao, F.; Mu, X.; Jiang, J.; Zhu, Q.; Ren, J.; Guo, Y.; Lou, Y. ZnIn$_2$S$_4$/UiO-66-(SH)$_2$ composites as efficient visible-light photocatalyst for RhB degradation. *Inorg. Chem. Commun.*, **2019**, *102*, 25-29.
[http://dx.doi.org/10.1016/j.inoche.2019.02.008]

[119] Mu, F.; Cai, Q.; Hu, H.; Wang, J.; Wang, Y.; Zhou, S.; Kong, Y. Construction of 3D hierarchical microarchitectures of Z-scheme UiO-66-(COOH)$_2$/ZnIn$_2$S$_4$ hybrid decorated with non-noble MoS$_2$ cocatalyst: A highly efficient photocatalyst for hydrogen evolution and Cr(VI) reduction. *Chem. Eng. J.*, **2020**, *384*, 123352.
[http://dx.doi.org/10.1016/j.cej.2019.123352]

[120] Subudhi, S.; Paramanik, L.; Sultana, S.; Mansingh, S.; Mohapatra, P.; Parida, K. A type-II interband alignment heterojunction architecture of cobalt titanate integrated UiO-66-NH$_2$: A visible light mediated photocatalytic approach directed towards Norfloxacin degradation and green energy (Hydrogen) evolution. *J. Colloid Interface Sci.*, **2020**, *568*, 89-105.
[http://dx.doi.org/10.1016/j.jcis.2020.02.043] [PMID: 32088455]

[121] Li, Z.; Hu, R.; Ye, S.; Song, J.; Liu, L.; Qu, J. 2-Methylimidazole-modulated UiO-66 as an effective photocatalyst to degrade Rhodamine B under visible light. *J. Mater. Sci.*, **2021**, *56*(2), 1577-1589.
[http://dx.doi.org/10.1007/s10853-020-05267-1]

[122] Wang, L.; Zheng, P.; Zhou, X.; Xu, M.; Liu, X. Facile fabrication of CdS/UiO-66-NH$_2$ heterojunction photocatalysts for efficient and stable photodegradation of pollution. *J. Photochem. Photobiol. Chem.*, **2019**, *376*, 80-87.
[http://dx.doi.org/10.1016/j.jphotochem.2019.03.001]

[123] Wang, J.; Liu, X.; Li, C.; Yuan, M.; Zhang, B.; Zhu, J.; Ma, Y. Fabrication of perylene imide-modified NH$_2$-UiO-66 for enhanced visible-light photocatalytic degradation of tetracycline. *J. Photochem. Photobiol. Chem.,* **2020**, *401*, 112795.
[http://dx.doi.org/10.1016/j.jphotochem.2020.112795]

[124] Lv, S.W.; Liu, J.M.; Li, C.Y.; Zhao, N.; Wang, Z.H.; Wang, S. Two novel MOFs@COFs hybrid-based photocatalytic platforms coupling with sulfate radical-involved advanced oxidation processes for enhanced degradation of bisphenol A. *Chemosphere,* **2020**, *243*, 125378.
[http://dx.doi.org/10.1016/j.chemosphere.2019.125378] [PMID: 31765898]

[125] Wang, X.; Liu, W.; Fu, H.; Yi, X.H.; Wang, P.; Zhao, C.; Wang, C.C.; Zheng, W. Simultaneous Cr(VI) reduction and Cr(III) removal of bifunctional MOF/Titanate nanotube composites. *Environ. Pollut.,* **2019**, *249*, 502-511.
[http://dx.doi.org/10.1016/j.envpol.2019.03.096] [PMID: 30928522]

[126] Wang, X.; Li, Y.X.; Yi, X.H.; Zhao, C.; Wang, P.; Deng, J.; Wang, C.C. Photocatalytic Cr(VI) elimination over BUC-21/N-K$_2$Ti$_4$O$_9$ composites: Big differences in performance resulting from small differences in composition. *Chin. J. Catal.,* **2021**, *42*(2), 259-270.
[http://dx.doi.org/10.1016/S1872-2067(20)63629-4]

[127] Chen, Y.; Li, J.; Zhai, B.; Liang, Y. Enhanced photocatalytic degradation of RhB by two-dimensional composite photocatalyst. *Colloids Surf. A Physicochem. Eng. Asp.,* **2019**, *568*, 429-435.
[http://dx.doi.org/10.1016/j.colsurfa.2019.02.007]

[128] Thi, Q.V.; Tamboli, M.S.; Thanh Hoai Ta, Q.; Kolekar, G.B.; Sohn, D. A nanostructured MOF/reduced graphene oxide hybrid for enhanced photocatalytic efficiency under solar light. *Mater. Sci. Eng. B,* **2020**, *261*, 114678.
[http://dx.doi.org/10.1016/j.mseb.2020.114678]

[129] Min, X.B.; Li, X.Y.; Zhao, J.; Hu, X.X.; Yang, W.C. Heterostructured TiO$_2$@HKUST-1 for the enhanced removal of methylene blue by integrated adsorption and photocatalytic degradation. *Environ. Technol.,* **2021**, *42*(26), 4134-4144.
[PMID: 32188338]

[130] Mosleh, S.; Rahimi, M.R.; Ghaedi, M.; Dashtian, K. HKUST-1-MOF–BiVO$_4$ hybrid as a new sonophotocatalyst for simultaneous degradation of disulfine blue and rose bengal dyes: optimization and statistical modelling. *RSC Adv.,* **2016**, *6*(66), 61516-61527.
[http://dx.doi.org/10.1039/C6RA13837C]

[131] Mosleh, S.; Rahimi, M.R. Intensification of abamectin pesticide degradation using the combination of ultrasonic cavitation and visible-light driven photocatalytic process: Synergistic effect and optimization study. *Ultrason. Sonochem.,* **2017**, *35*(Pt A), 449-457.
[http://dx.doi.org/10.1016/j.ultsonch.2016.10.025] [PMID: 27810164]

[132] Mosleh, S.; Rahimi, M.R.; Ghaedi, M.; Dashtian, K. Sonophotocatalytic degradation of trypan blue and vesuvine dyes in the presence of blue light active photocatalyst of Ag$_3$PO$_4$/Bi$_2$S$_3$-HKUST-1-MOF: Central composite optimization and synergistic effect study. *Ultrason. Sonochem.,* **2016**, *32*, 387-397.
[http://dx.doi.org/10.1016/j.ultsonch.2016.04.007] [PMID: 27150785]

[133] Chen, Y.; Zhai, B.; Liang, Y.; Li, Y. Hybrid photocatalysts using semiconductor/MOF/graphene oxide for superior photodegradation of organic pollutants under visible light. *Mater. Sci. Semicond. Process.,* **2020**, *107*, 104838.
[http://dx.doi.org/10.1016/j.mssp.2019.104838]

[134] Kaur, R.; Rana, A.; Singh, R.K.; Chhabra, V.A.; Kim, K.H.; Deep, A. Efficient photocatalytic and photovoltaic applications with nanocomposites between CdTe QDs and an NTU-9 MOF. *RSC Advances,* **2017**, *7*(46), 29015-29024.
[http://dx.doi.org/10.1039/C7RA04125J]

[135] Samy, M.; Ibrahim, M.G.; Fujii, M.; Diab, K.E.; ElKady, M.; Gar Alalm, M. CNTs/MOF-808 painted plates for extended treatment of pharmaceutical and agrochemical wastewaters in a novel

photocatalytic reactor. *Chem. Eng. J.,* **2021**, *406*, 127152.
[http://dx.doi.org/10.1016/j.cej.2020.127152]

[136] Li, L.; Yu, X.; Xu, L.; Zhao, Y. Fabrication of a novel type visible-light-driven heterojunction photocatalyst: Metal-porphyrinic metal organic framework coupled with PW12/TiO$_2$. *Chem. Eng. J.,* **2020**, *386*, 123955.
[http://dx.doi.org/10.1016/j.cej.2019.123955]

[137] Bai, Y.; Zhang, S.; Feng, S.; Zhu, M.; Ma, S. The first ternary Nd-MOF/GO/Fe$_3$O$_4$ nanocomposite exhibiting an excellent photocatalytic performance for dye degradation. *Dalton Trans.,* **2020**, *49*(31), 10745-10754.
[http://dx.doi.org/10.1039/D0DT01648A] [PMID: 32779671]

[138] Yaseen, D.A.; Scholz, M. Textile dye wastewater characteristics and constituents of synthetic effluents: a critical review. *Int. J. Environ. Sci. Technol.,* **2019**, *16*(2), 1193-1226.
[http://dx.doi.org/10.1007/s13762-018-2130-z]

[139] Khandan, F.M.; Afzali, D.; Sargazi, G.; Gordan, M. Novel uranyl-curcumin-MOF photocatalysts with highly performance photocatalytic activity toward the degradation of phenol red from aqueous solution: effective synthesis route, design and a controllable systematic study. *J. Mater. Sci. Mater. Electron.,* **2018**, *29*(21), 18600-18613.
[http://dx.doi.org/10.1007/s10854-018-9978-z]

[140] Liang, R.; Luo, S.; Jing, F.; Shen, L.; Qin, N.; Wu, L. A simple strategy for fabrication of Pd@MIL-100(Fe) nanocomposite as a visible-light-driven photocatalyst for the treatment of pharmaceuticals and personal care products (PPCPs). *Appl. Catal. B,* **2015**, *176-177*, 240-248.
[http://dx.doi.org/10.1016/j.apcatb.2015.04.009]

[141] Gao, J.; Han, D.; Xu, Y.; Liu, Y.; Shang, J. Persulfate activation by sulfide-modified nanoscale iron supported by biochar (S-nZVI/BC) for degradation of ciprofloxacin. *Separ. Purif. Tech.,* **2020**, *235*, 116202.
[http://dx.doi.org/10.1016/j.seppur.2019.116202]

[142] Yu, Y.; Huang, F.; He, Y.; Wang, F.; Lv, Y.; Xu, Y.; Zhang, Y. Efficient degradation of sulfamethoxazole by catalytic wet peroxide oxidation with sludge-derived carbon as catalysts. *Environ. Technol.,* **2020**, *41*(7), 870-877.
[http://dx.doi.org/10.1080/09593330.2018.1512657] [PMID: 30139300]

[143] Zhang, Q.Q.; Ying, G.G.; Pan, C.G.; Liu, Y.S.; Zhao, J.L. Comprehensive evaluation of antibiotics emission and fate in the river basins of China: source analysis, multimedia modeling, and linkage to bacterial resistance. *Environ. Sci. Technol.,* **2015**, *49*(11), 6772-6782.
[http://dx.doi.org/10.1021/acs.est.5b00729] [PMID: 25961663]

[144] Hou, Y.; Zhao, Y.; Li, Q.; Li, Y. Highly biodegradable fluoroquinolone derivatives designed using the 3D-QSAR model and biodegradation pathways analysis. *Ecotoxicol. Environ. Saf.,* **2020**, *191*, 110186.
[http://dx.doi.org/10.1016/j.ecoenv.2020.110186] [PMID: 31954922]

[145] Gopinath, K.P.; Madhav, N.V.; Krishnan, A.; Malolan, R.; Rangarajan, G. Present applications of titanium dioxide for the photocatalytic removal of pollutants from water: A review. *J. Environ. Manage.,* **2020**, *270*, 110906.
[http://dx.doi.org/10.1016/j.jenvman.2020.110906] [PMID: 32721341]

[146] Horiuchi, Y.; Toyao, T.; Saito, M.; Mochizuki, K.; Iwata, M.; Higashimura, H.; Anpo, M.; Matsuoka, M. Visible-Light-Promoted Photocatalytic Hydrogen Production by Using an Amino-Functionalized Ti(IV) Metal–Organic Framework. *J. Phys. Chem. C,* **2012**, *116*(39), 20848-20853.
[http://dx.doi.org/10.1021/jp3046005]

[147] Kozlova, E.A.; Panchenko, V.N.; Hasan, Z.; Khan, N.A.; Timofeeva, M.N.; Jhung, S.H. Photoreactivity of metal-organic frameworks in the decolorization of methylene blue in aqueous solution. *Catal. Today,* **2016**, *266*, 136-143.
[http://dx.doi.org/10.1016/j.cattod.2015.07.026]

[148] Panchenko, V.N.; Timofeeva, M.N.; Jhung, S.H. Acid-base properties and catalytic activity of metal-organic frameworks: A view from spectroscopic and semiempirical methods. *Catal. Rev., Sci. Eng.,* **2016**, *58*(2), 209-307.
[http://dx.doi.org/10.1080/01614940.2016.1128193]

SUBJECT INDEX